Natural Computing Series

Series Editors: G. Rozenberg (Managing)
Th. Bäck A.E. Eiben J.N. Kok H.P. Spaink
Leiden Center for Natural Computing

Advisory Board: S. Amari G. Brassard M. Conrad
K.A. De Jong C.C.A.M. Gielen T. Head L. Kari
L. Landweber T. Martinetz Z. Michalewicz M.C. Mozer
E. Oja J. Reif H. Rubin A. Salomaa M. Schoenauer
H.-P. Schwefel D. Whitley E. Winfree J.M. Zurada

Springer
Berlin
Heidelberg
New York
Barcelona
Hong Kong
London
Milan
Paris
Singapore
Tokyo

Leila Kallel • Bart Naudts • Alex Rogers (Eds.)

Theoretical Aspects of Evolutionary Computing

With 129 Figures and 15 Tables

Springer

Editors

Dr. Leila Kallel
Ecole Polytechnique
CMAP
91128 Palaiseau Cedex, France
kallel@cmapx.polytechnique.fr

Dr. Bart Naudts
Universiteit Antwerpen (RUCA)
Departement Wiskunde & Informatica
Groenenborgerlaan 171
2020 Antwerpen, Belgium
bnaudts@ruca.ua.ac.be

Alex Rogers
University of Southampton
Department of Electronics and Computer Science
Highfield
SO17 1BJ Southampton, UK
ar97r@ecs.soton.ac.uk

Series Editors

G. Rozenberg (Managing Editor)
Th. Bäck, A.E. Eiben, J.N. Kok,
H.P. Spaink

Leiden Center for Natural Computing
Leiden University
Niels Bohrweg 1
2333 CA Leiden
The Netherlands
rozenber@cs.leidenuniv.nl

Library of Congress Cataloging-in-Publication Data
Theoretical aspects of evolutionary computing/Leila Kallel, Bart Naudts, Alex Rogers (eds.).
   p. cm. - (Natural computing series)
Based on the lectures and workshop contributions of the second EvoNet Summer School on Theoretical Aspects of Evolutionary Computing held at the Middelheim campus of the University of Antwerp, Belgium during the first week of September 1999.
Includes bibliographical references and index.
ISBN 3540673962 (alk. paper)
   1. Evolutionary programming (Computer science) I. Kallel, Leila, 1972- II. Naudts, Bart, 1973- III. Rogers, Alex, 1970- IV. EvoNet Summer School on Theoretical Aspects of Evolutionary Computing (2nd: 1999: University of Antwerp, Belgium) V. Series.
QA76.618.T47 2000
006.3-dc21   00-061910

ACM Computing Classification (1998): F.2.2, I.2.8, I.2.6, G.3, G.1.6, J.3, J.2

ISBN 3-540-67396-2 Springer-Verlag Berlin Heidelberg New York

This work is subject to copyright. All rights are reserved, whether the whole or part of the material is concerned, specifically the rights of translation, reprinting, reuse of illustrations, recitation, broadcasting, reproduction on microfilm or in any other way, and storage in data banks. Duplication of this publication or parts thereof is permitted only under the provisions of the German Copyright Law of September 9, 1965, in its current version, and permission for use must always be obtained from Springer-Verlag. Violations are liable for prosecution under the German Copyright Law.

Springer-Verlag Berlin Heidelberg New York,
a member of BertelsmannSpringer Science+Business Media GmbH
http://www.springer.de

© Springer-Verlag Berlin Heidelberg 2001
Printed in Germany

The use of general descriptive names, trademarks, etc. in this publication does not imply, even in the absence of a specific statement, that such names are exempt from the relevant protective laws and regulations and therefore free for general use.

Cover Design: KünkelLopka, Heidelberg
Typesetting: Camera ready by the editors
Printed on acid-free paper   SPIN 10765204   45/3142SR - 5 4 3 2 1 0

Preface

During the first week of September 1999, the Second EvoNet Summer School on Theoretical Aspects of Evolutionary Computing was held at the Middelheim campus of the University of Antwerp, Belgium. Originally intended as a small get-together of PhD students interested in the theory of evolutionary computing, the summer school grew to become a successful combination of a four-day workshop with over twenty researchers in the field and a two-day lecture series open to a wider audience.

This book is based on the lectures and workshop contributions of this summer school. Its first part consists of tutorial papers which introduce the reader to a number of important directions in the theory of evolutionary computing. *The tutorials are at graduate level and assume only a basic background in mathematics and computer science. No prior knowledge of evolutionary computing or its theory is necessary.* The second part of the book consists of technical papers, selected from the workshop contributions. A number of them build on the material of the tutorials, exploring the theory to research level. Other technical papers may require a visit to the library.

The theory of evolution is at the crossroads of mathematics, computer science, statistics, and the theoretical sides of physics, chemistry, and biology. There is often too little interaction between these disciplines. One of the summer school's goals was to get researchers from many disciplines together. As a result, this book contains papers from researchers with a background in complexity theory, neural networks, probability theory, population genetics, statistical physics, and mathematics. Only a fraction of the authors grew up in the evolutionary computing community itself. This is reflected in the book, which presents a rich variety of approaches to the study of the behavior of evolutionary algorithms.

Overview of the Tutorials

The tutorial part of the book starts with a general introduction to evolutionary computing, from the point of view of an experienced engineering researcher, A. J. Keane, who uses evolutionary algorithms as one of his main tools in practical optimization. This first tutorial has been written in cooperation with A. Rogers.

Before plunging into the theory, we present a second tutorial on evolutionary algorithms, written by A. E. Eiben. He discusses the use of evolutionary algorithms for constrained search problems.

A first course on the microscopic approach to the dynamics of the simple genetic algorithm, as started by M. D. Vose in the early 1990s, is given by J. E. Rowe. Filling in the actual numbers for a concrete example, but also commenting on the latest development, the tutorial is all the reader needs to get started with this exciting mathematical framework.

The statistical physics approach to the dynamics of evolutionary algorithms is a macroscopic one: instead of modeling the exact dynamics, only a few statistics of the dynamics are computed. A. Prügel-Bennett and J. L. Shapiro have written two complementary tutorials on this line of research.

Following the line of macroscopic descriptions is the tutorial on evolution strategies by H.-G. Beyer and D. V. Arnold. Evolution strategies operate on real spaces. Their dynamics on a number of fitness models have now been fully analyzed.

Apart from being successful in practical applications, optimization algorithms based on a model of the fitness distribution can be used to approximate genetic algorithms. H. Mühlenbein and T. Mahnig introduce this approach to studying the dynamics of genetic algorithms, also showing the link with population genetics and learning with Bayesian networks.

Closing the gap between theory and practice is the tutorial by L. Kallel, B. Naudts, and C. R. Reeves on properties of fitness landscapes and their impact on problem difficulty. The tutorial explores issues such as Walsh coefficients, epistasis, correlations in the fitness landscape, basins of attraction, and interaction structure.

Overview of the Technical Papers

In the first technical paper, A. Rogers and A. Prügel-Bennett build on the tutorials by Prügel-Bennett and Shapiro. They analyze the dynamics of a genetic algorithm with ranking selection, uniform crossover, and mutation on the 'basin with a barrier' model, which contains one feature commonly found in hard search problems. A relationship between the optimal mutation rate and population size is found, which is independent of the problem size.

M. Oates and coworkers also examine the relationship between population size and mutation rate. Their exhaustive experimental work on a number of search problems is presented in the second technical paper.

The third and last paper which deals with search in time-static problems is by D. V. Arnold, who presents the state of the art of evolution strategies on noisy fitness models. Beyer and Arnold's tutorial on evolution strategies serves as reference material.

The quasispecies model from theoretical biology is studied by J. E. Rowe, and also by C. Ronnewinkel, C. O. Wilke, and T. Martinetz. Both use the microscopic model from Rowe's tutorial to study the quasispecies adaptability to fitness function changes by computing error thresholds and optimal mutation rates. However, their approach to time-dependence is different. The former identifies cyclic attractors, whereas the latter accumulate the time-dependency of the change cycle.

A. Berny demonstrates the link between statistical learning theory and combinatorial optimization. Rather than searching in the space of bit strings, he demonstrates how to search, using gradient techniques, in the space of distributions over bit strings.

The genetic drift caused by multi-parent scanning crossover is rigorously calculated by C. A. Schippers. He finds that occurrence-based scanning produces a stronger than usual, self-amplifying genetic drift.

Reference material for the next six papers is the tutorial by Kallel, Naudts, and Reeves. Instead of studying the dynamics of evolutionary algorithms, they focus on the representation of search problems and the fitness landscapes induced by the algorithm.

When choosing representation and genetic operators, many implicit assumptions about the epistatic structure of the fitness function to be optimized are made. For example, one-point crossover assumes that adjacent bit positions are highly correlated. F. Burkowski proposes nullifying these assumptions, and learning the epistatic structure on-line.

J. Garnier and L. Kallel develop statistical techniques to estimate the number and sizes of the basins of attraction in a fitness landscape.

Can fitness functions be classified according to the difficulty they present to a genetic algorithm? T. Jansen comments on this issue and details some of the flaws that a number of measures, introduced as predictors of problem difficulty, suffer from.

The subject of symmetry in the representation of the search space has been studied in three independent contributions. First, there is the paper of P. Stagge and C. Igel, which deals with redundancy caused by isomorphic structures in the context of topology optimization of neural networks. In a second contribution, A. Marino introduces specific genetic operators for the graph coloring problem, in order to to search the highly symmetric solution space more efficiently. Finally, C. Van Hoyweghen introduces a technique to detect spin-flip symmetry in arbitrary search spaces.

The last contribution to this book, and also the largest in size, is by P. Del Moral and L. Miclo. They first propose a model for GAs which generalizes Vose's model (introduced in the tutorial by Rowe) to continuous search spaces, inhomogeneous fitness environments, and inhomogeneous mutation operators. In this model a population is seen as a distribution (or measure) on the search space, and the genetic algorithm as a measure-valued dynamical system. Then, they study the convergence properties of the finite population process when the population size tends to infinity, and prove that it converges uniformly with respect to time to the infinite population process. Finally, they show how their model of GAs can be used in the context of nonlinear filtering. They prove the efficiency of finite population GAs to approximate an observed signal in the presence of noise.

Acknowledgments

We would like to thank all the participants of the summer school for attending and sending in valuable contributions afterwards. This book is their work. We were also pleased with the fact that so many senior researchers were willing to write a 20+ page tutorial on their subject.

Finding sponsorship to organize a summer school on a theoretical subject was a lot easier than we thought. We would like to thank the Network of Excellence in Evolutionary Computing (EvoNet), the Doctoral Study Programme Science–Pharmacy of the University of Antwerpen (DOCOP), the Foundation for European Evolutionary Computation Conferences (FEECC), and the Leiden Institute for Advanced Computer Science (LIACS) for their helpful cooperation. The editing of this book was done while B. Naudts was funded as a Postdoctoral Fellow of the Fund for Scientific Research – Flanders, Belgium (F.W.O.).

But above all, we are greatly indebted to our summer school co-organizers: Jennifer Willies of EvoNet, Jano van Hemert of the LIACS, and Ives Landrieu, Luk Schoofs, and Clarissa Van Hoyweghen of the University of Antwerpen. They did a great job!

Paris, Antwerpen, and Southampton, *L. Kallel*
Spring 2001 *B. Naudts*
  A. Rogers

Table of Contents

Part I: Tutorials

Introduction to Evolutionary Computing in Design Search and Optimisation
    A. J. Keane 1

Evolutionary Algorithms and Constraint Satisfaction:
Definitions, Survey, Methodology, and Research Directions
    A. E. Eiben 13

The Dynamical Systems Model of the Simple Genetic Algorithm
    J. E. Rowe 31

Modelling Genetic Algorithm Dynamics
    A. Prügel-Bennett and A. Rogers 59

Statistical Mechanics Theory of Genetic Algorithms
    J. L. Shapiro 87

Theory of Evolution Strategies – A Tutorial
    H.-G. Beyer and D. V. Arnold 109

Evolutionary Algorithms: From Recombination to Search Distributions
    H. Mühlenbein and T. Mahnig 135

Properties of Fitness Functions and Search Landscapes
    L. Kallel, B. Naudts, and C. R. Reeves 175

Part II: Technical Papers

A Solvable Model of a Hard Optimisation Problem
    A. Rogers and A. Prügel-Bennett 207

Bimodal Performance Profile of Evolutionary Search and the Effects
of Crossover
    M. Oates, J. Smedley, D. Corne, and R. Loader 223

Evolution Strategies in Noisy Environments – A Survey of Existing Work
    D. V. Arnold 239

Cyclic Attractors and Quasispecies Adaptability
    J. E. Rowe 251

Genetic Algorithms in Time-Dependent Environments
    C. Ronnewinkel, C.O. Wilke, and T. Martinetz 261

Statistical Machine Learning and Combinatorial Optimization
    A. Berny 287

Multi-Parent Scanning Crossover and Genetic Drift
    C. A. Schippers 307

Harmonic Recombination for Evolutionary Computation
    F. J. Burkowski 331

How to Detect all Maxima of a Function
    J. Garnier and L. Kallel 343

On Classifications of Fitness Functions
    T. Jansen 371

Genetic Search on Highly Symmetric Solution Spaces: Preliminary Results
    A. Marino 387

Structure Optimization and Isomorphisms
    P. Stagge and C. Igel 409

Detecting Spin-Flip Symmetry in Optimization Problems
    C. Van Hoyweghen 423

Asymptotic Results for Genetic Algorithms with Applications to
Nonlinear Estimation
    P. Del Moral and L. Miclo 439

Index 495

An Introduction to Evolutionary Computing in Design Search and Optimisation

A. Keane
Lecture notes produced by A. Rogers

Department of Mechanical Engineering
University of Southampton
Southampton SO17 1BJ, UK
E-mail: *andy.keane@soton.ac.uk*

Abstract. Evolutionary computing (EC) techniques are beginning to find a place in engineering design. In this tutorial we give a brief overview of EC and discuss the techniques and issues which are important in design search and optimisation.

Keywords
Evolutionary computing, design search, optimisation, problem solving environment

1 Introduction

The purpose of this introduction is to present, to a wide audience, some of the ideas which are important in the application of evolutionary computing (EC) algorithms to engineering design. This leads on to an area known as *design search and optimisation* where evolutionary computing techniques are beginning to find real applications.

Perhaps the first question to ask is: why do we do this at all? The general answer is that life is becoming more computationally complex and in the field of engineering we are presented with an increasing number of computationally complex problems to solve. The computational complexity of optimising these systems challenges the capabilities of existing techniques and we seek some other heuristics. When we look around at the real world, we see many examples of complex systems, ourselves included, which have arisen through evolution. This naturally leads to some desire to emulate evolution in our attempts at problem solving.

As an example of a computationally complex problem, we can consider designing aircraft wings – an area from which several of the examples in this tutorial will be drawn. Alternatively, we may be trying to schedule some process in a factory or simply trying to fit curves to data. In some senses, all these things are about design, and by design, we mean the idea that we want to create something. The whole point to evolutionary computing is the creation of designs, particularly in the domain of complex problems and environments.

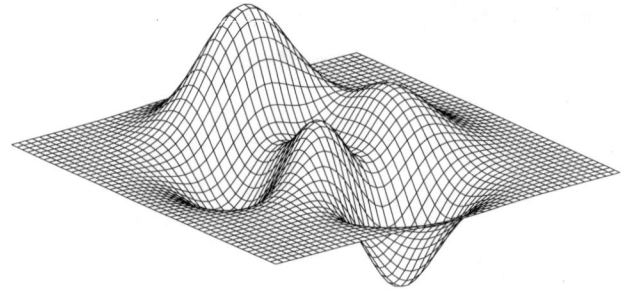

Figure 1. Example of an optimisation landscape

It is useful to briefly consider what design is, how it is done and where evolutionary computing fits in. We are all surrounded every day by extremely complicated technology which we generally take for granted. As an example, consider the cars we drive. Should we have a crash, the airbags will explode and the car will crumple safely around us because someone has thought about and modeled all the processes and components involved in that crash. To do this nowadays, automotive engineers use high-performance computer networks which are highly developed into sophisticated environments.

Designing complex things such as cars requires us to put lots of things together. It requires us to know something about what it is we are trying to design, to have some way of interfacing it to the user and perhaps some type of search and optimisation routine with resource management. The whole thing may be called a *problem solving environment* (PSE) and it is into this environment that evolutionary computing techniques fit.

In summary, designers in the broadest sense – simply people trying to create something – increasingly see the need for a PSE. Ideally, these PSEs contain some sort of search and optimisation and it is here that evolutionary computing techniques find an application. Thus, evolutionary computing is not used in isolation but as a part of a design tool which has to be of some utility.

What follows are some introductory remarks on several topics which will be expanded on later in the course of this tutorial.

Search and Optimisation. Two terms which are often used in evolutionary computing are search and optimisation. We can illustrate what these mean by considering a simple landscape – see figure 1. Optimisation is about trying to walk up a hill and hopefully, through effectively searching the landscape, walking up the highest hill.

At its simplest this is all we are trying to do. We have some landscape which we are trying to walk through. Usually we cannot see the landscape, so the whole business of evolutionary computing is trying to do this blindfold. This introduces questions such as: how do we know we are at the top of the hill and how do we know it is the right hill?

Resources. Search and optimisation always requires some other piece of code to be plugged into it – we apply evolutionary computing to something. It is usually a piece of analysis code and we are trying to capture the intentions of the designer in this code. Given the computational complexity of the problems we are considering, computing power dominates this process. Given overwhelming computing power, most problems can be solved. With finite computing resources we have to be more careful.

Current Techniques. What is the current state of the art in engineering design? Most PSEs now include some sort of hill-climbing optimisation and parallel computation is common. Evolutionary search is known about but it is not generally in everyday use. The use of meta-computing techniques and resource scheduling are beginning to be considered as approaches to the problems of limited computing power.

Representation. Representation is one of the key issues in this work. How do we represent what it is that we are trying to design? Inevitably people represent their designs with numbers, so the method of encoding between the two is important. However, an engineer also has a selection of models which can be used which range in sophistication and cost. It is not simply a question of representation and method but representation, method, and model which seem to be the most important. Designers have aspirations, constraints, and varying and multiple objectives. They often don't know what they want or cannot express what it is exactly they want. But in general, they want robust designs.

Classical search methods have been around for at least fifty years and they are well tried. Crucially, people don't believe they are likely to lead to any more innovation. On the other hand, evolutionary computing methods have some advantages but they have a problem – computational expense. If we come up with an evolutionary computing paradigm which is really very powerful but is too computationally expensive, it will not be used.

2 A Brief Overview of Evolutionary Computing

The history of evolutionary computing goes back to the 1960s with the introduction of ideas and techniques such as genetic algorithms, evolutionary strategies and evolutionary programming [3]. Whilst all differ slightly in their actual implementations, all these evolutionary computing techniques use the same metaphor of mapping problem solving onto a simple model of evolution.

EVOLUTION		PROBLEM SOLVING
Individual	⟷	Candidate Solution
Fitness	⟷	Quality
Environment	⟷	Problem

We have a population which we want to evolve from time step t to time step $t + 1$ by selection, recombination and mutation. There is some contention about which of the two, mutation or recombination, is the more powerful, with there being a whole history in the German academic community of using only mutation in evolutionary strategies. To a certain extent though, the ingredients can be mixed and matched. One is trying to introduce diversity in the designs through mutation and recombination, and then exploit this exploration through selection of better solutions: balancing the growth of ideas alongside the selection of ideas. One interesting aspect is that this balance occurs all the way through but the nature of the algorithms we use implies that there tends to be a large diversity at the beginning and very little towards the end.

Hopefully through the correct choice of operators, we achieve a balance between exploration and exploitation and our solutions steadily get better.

Although a great deal of the evolutionary computing literature concerns comparisons of one algorithm to another, some general points about the advantages of evolutionary computing techniques can be made:

- Widely applicable.
- Low development and application cost.
- Easily incorporated into other methods.
- Solutions are interpretable.
- Can be run interactively and allows incorporation of user-proposed solutions.
- Provide many alternative solutions.

Whilst they have been shown to be useful in many areas, evolutionary computing techniques have several disadvantages when compared to some other techniques:

- No guarantee for optimal solution within finite time.
- Weak theoretical basis.
- May need parameter tuning for good performance.
- Often computationally expensive and thus slow.

Apart from the computational expense, perhaps that which most directly effects the acceptance of evolutionary computing is the weakness of their theoretical basis. With no theory, we are left with a set of rules of thumb for choosing and tuning the algorithms.

3 Evolutionary Computing in Design Search and Optimisation

When actually implementing an evolutionary computing algorithm we are presented with a wide range of choices. Which algorithm should we use? Which operators should we use? How do we set the parameters? There are a number of books available which help with these questions [1–4,7]; however, when using these algorithms in engineering design we have a number of other additional issues which must be considered.

3.1 Representation

Deciding on a good representation is fundamental to the performance of evolutionary computing techniques. The algorithms work on numbers, usually binary, but we are not trying to design a string of numbers. We may be trying to design a wing, a communications network or a schedule, and in some way we have to link the two together. This point turns out to be absolutely critical to the success of the application.

It is self-evidently true that we can trade-off the ability of any representation to be compact, against its ability to represent all possible designs. This is a really fundamental decision. Do we want to describe any possible design or do we want to describe something in a small region around where we currently stand?

If we are using an evolutionary algorithm to design the wing for a jet aircraft, do we want to think about bi-planes? Probably not. Do we want to consider a wing of alternative materials? Maybe. Someone has to make these decisions and it cannot be done without some domain expertise. Experience shows you need a domain expert and an evolutionary computation expert together, as it is this critical stage which often determines the outcome of the project.

If we again consider our wing design problem suggested above, the classical aerodynamics of the wing are based on concepts of wing span, sweep, chord and camber. The interesting thing about these words is that these are ways which aerodynamicists have worked out to describe wings. They are a particular representation of a wing which is relevant when considering the flow of air over it.

An alternative would be to use the x,y and z coordinates for the surfaces. This presents the problem that whilst we can describe a wing, we can also describe virtually anything else as well. Moving down the trade-off in compactness, we are increasing the complexity of the object which we can describe and are thus increasing the domain in which the potential designs can exist. We are making the problem potentially harder and if we do not need the complexity at this level, we run the risk of making the problem almost insoluble.

The opposite danger comes from sticking too rigidly to an existing representation. Through existing design techniques the representation may have become so specific that it traps the possible solutions into a local optimum and does not have the generality to describe wing shapes which may indeed be better.

Choices across this sort of domain make a real difference and it is clear that they cannot be made without some domain knowledge.

3.2 Constraints and Multiple Objectives

Often the objectives of a design are not completely defined. An engineer will often have multiple objectives and in some sort of hierarchy. Unless this appraisal of potential designs is somehow formulated and included in the analysis code, the expectations of the engineer will differ from the designs which our algorithm is actually producing.

Constraints are another issue which is yet to be fully addressed. How do we handle a list of certain criteria which invalidate a particular design? There are a number of techniques, some of which are discussed by Eiben in this volume (page 13), and we must decide which one is most relevant.

3.3 Mutation

Mutation is critical to a lot of these results. It is often said to be there just to ensure that every part of the search space may be reached. However, successful search algorithms are often run just using mutation and selection without crossover. Indeed evolutionary strategies are predominantly used in this fashion.

The idea behind mutation is that we are making local steps in our landscape. Our choice of mutation operator and representation should mean that we are making small changes to our design and not leaping to a radically different solution.

3.4 Recombination

Recombination or crossover allows parents to pass on some of their characteristics to their children. The motivation being that we can use good parents to develop even better children. There are many ways of doing this but all involve the combination of parts of one parent with the complementary parts from another.

There is a trade-off between recombination and mutation which reflects some of the history of the trade-off between the various communities. Recombination tends to be seen as more of an exploitation operator and mutation as more of a exploration operator. Some people would tend to suggest mixing these in, so we do more exploration at the beginning and more exploitation towards the end. These are more examples of parameter and algorithm tuning which are dictated by rules of thumb rather than a solid theoretical basis.

3.5 Niching

When we initially discussed optimisation, we considered trying to find the single global optimum in the problem. In engineering applications this may not be the case and the question has to be asked whether we want the best solution or do we want to know about a number of areas in the problem space where there are good candidate solutions? Depending on our aspirations, we may want to impose some form of niching techniques to allow the population to divide into smaller sub-populations, each focused around a different part of the problem space.

3.6 Repeatability and Elitism

Evolutionary computing techniques are stochastic and thus no two runs will necessarily produce the same results. Whilst in some applications this is acceptable, it can be quite discouraging to produce a particular good solution once and never to be able to find it again.

This leads to many issues concerning the repeatability of evolutionary computing algorithms. Do we just want to get one very good solution or do we want a good solution every time? If we were doing process scheduling we might settle for the former but an automotive engineer designing a car would probably prefer the latter.

Another concern which stems from the stochastic nature of the algorithm, is that a good solution once found within the population may be lost later in the evolution. Whilst it always makes sense to keep a record of the best solutions, elitism strategies ensure that these solutions stay in the population.

3.7 When to Stop?

Having set up the representation and run the evolutionary computing algorithm, the final decision is when to stop it. Without any prior knowledge, it is impossible to tell whether the best solution has been reached. The decision most often comes down to one of time and computing resources.

4 An Example of EC Design

As an example of the techniques and problems discussed we will consider an actual case of an evolutionary computing algorithm being used to solve an engineering problem. Figure 2 shows a photograph from a NASA mission in 1987 as part of a proof of concept programme to show that astronauts could build structures in space.

One of the problems with these structures is that they tend to have very severe vibration problems. They are light regular alloy structures in an environment where there

Figure 2. NASA photograph showing an astronaut constructing a boom in space

is no air to provide damping. We started to look at the vibrational characteristics of structures like these when subject to some vibrational noise at one end [6,5]. The objective was to use evolutionary computing techniques to design the geometry of the beam such that the vibrational noise does not travel through it. This is a straightforward engineering job which is actually critical to both ESA and NASA who plan to launch future missions with booms of this type fifty metres in length.

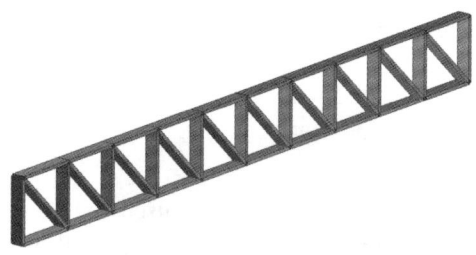

Figure 3. Diagram of simple 2D boom design

Initially a flat two-dimensional beam was considered. As in the cases discussed previously, we are combining computational expensive structural analysis routines with an optimisation technique. Figure 3 shows the initial simple two-dimensional boom design which we are trying to improve upon.

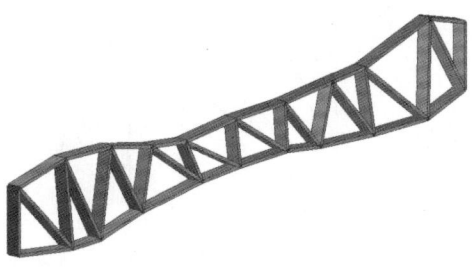

Figure 4. Final GA optimised 2D boom design

A genetic algorithm was used for the optimisation with population size of 300. It was run for fifteen generations which ideally would have been longer but for the high computational cost of evaluating each design. As it was, the complete run took three weeks of computation using eleven parallel workstations. Figure 4 shows the best design found by the GA at the end of the run.

Having obtained this design, it was actually constructed and tested for vibrational performance – a check that the improvements predicted by the analysis code could be realised in actuality. Figure 5 shows a comparison of the initial and final beam vibrational characteristics. The shaded area in the center, from 100–250 Hz, represents the range of frequencies over which the analysis code was run and is thus the range over which the response has been optimised. The interesting thing about this graph is that the vertical scale is in decades of performance and we have something like three decades of improvement in the vibration performance. It is not often that you can improve things by several thousands of percent.

Figure 6 shows the results of applying the same technique to the three-dimensional boom shown in the photograph. Obviously, whilst it has improved vibration characteristics, it is somewhat harder to manufacture. This illustrates the point that unless included in the analysis code, one cannot expect all the engineering criteria to be satisfied.

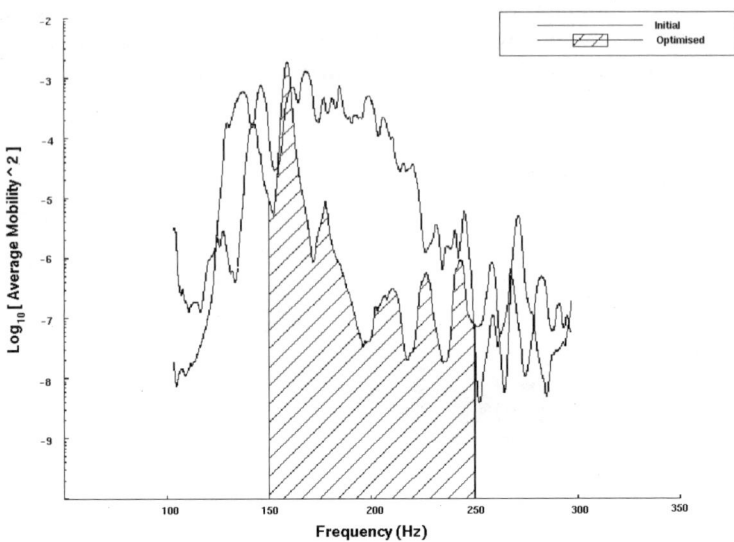

Figure 5. Comparison of vibrational characteristics for the initial and optimised 2D boom design

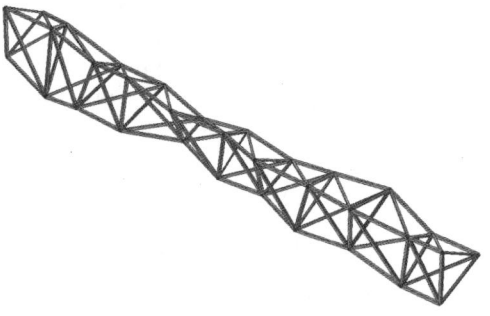

Figure 6. Final GA optimised 3D boom design

5 Conclusions

Evolutionary computing algorithms have been shown to be of great value in design search and optimisation. The concept of a problem solving environment clearly increases the utility of these algorithms but the problems of computational expense must be addressed. This will not come through improvements in hardware as our rising aspirations continually outstrip the improvement in processor performance.

A short-term goal to address this issue, is to look at a layered approach to models in order to enhance speed and better represent what the user wants. A more medium term goal would be the use of meta-computing techniques to fully manage the available corporate computing resources. Long-term goals – possibly twenty years away – are fully agent-based environments which would offer seamless integration of optimisation techniques, computing resources, and access to corporate databases.

However this work progresses, over the next ten years evolutionary search using expensive modelling techniques will doubtless become common-place.

References

1. T. Bäck. *Evolutionary Algorithms in Theory and Practice: Evolution Strategies, Evolutionary Programming, Genetic Algorithms*. Oxford University Press, Oxford, 1996.
2. L. Davis, M. Vose, and K. De Jong. *Evolutionary Algorithms*. Volumes in Mathematics and Its Applications, Vol 111. Springer-Verlag, Berlin Heidelberg New York, 1996.
3. D. Fogel. *Evolutionary Computation: The Fossil Record*. IEEE Press, New York, 1998.
4. D. Goldberg. *Genetic Algorithms in Search, Optimization and Machine Learning*. Addison–Wesley, Reading, MA, 1988.
5. A. J. Keane. Experiences with optimizers in structural design. In I. C. Parmee, editor, *Proceedings of the Conference on Adaptive Computing in Engineering Design and Control 94*, pages 14–27. PEDC, Plymouth, 1994.
6. A. J. Keane and S. M. Brown. The design of a satellite boom with enhanced vibration performance using genetic algorithm techniques. In I. C. Parmee, editor, *Proceedings of the Conference on Adaptive Computing in Engineering Design and Control 96*, pages 107–113. PEDC, Plymouth, 1996.
7. M. Mitchell. *An Introduction to Genetic Algorithms*. MIT Press, Boston, MA, 1996.

Evolutionary Algorithms and Constraint Satisfaction: Definitions, Survey, Methodology, and Research Directions

A. E. Eiben

Free University Amsterdam
The Netherlands
E-mail: *gusz@cs.vu.nl*

Abstract. In this tutorial we consider the issue of constraint handling by evolutionary algorithms (EA). We start this study with a categorization of constrained problems and observe that constraint handling is not straightforward in an EA. Namely, the search operators mutation and recombination are 'blind' to constraints. Hence, there is no guarantee that if the parents satisfy some constraints the offspring will satisfy them as well. This suggests that the presence of constraints in a problem makes EAs intrinsically unsuited to solve this problem. This should especially hold if there are no objectives but only constraints in the original problem specification – the category of constraint satisfaction problems. A survey of related literature, however, discloses that there are quite a few successful attempts at evolutionary constraint satisfaction. Based on this survey we identify a number of common features in these approaches and arrive at the conclusion that the presence of constraints is not harmful, but rather helpful in that it provides extra information that EAs can utilize. The tutorial is concluded by considering a number of key questions on research methodology and some promising future research directions.

Keywords
Constraint handling, evolutionary algorithms

1 Introduction

The interest in constrained problems as the subject of the present study originates from two observations:

1. A great number of practical problems are constrained. This makes such problems practically relevant.
2. A great number of untractable problems (NP-hard, NP-complete, etc.) are constrained. This makes such problems theoretically challenging.

These arguments constitute statements about a problem domain only, i.e., there is nothing said about the technique(s) to solve such problems. So, what motivates our choice of evolutionary algorithms (EAs)? The answer is twofold again:

1. EAs show a good ratio of (implementation) effort to performance.
2. EAs are acknowledged as good solvers for tough problems.

While these arguments are true in general, a quick glance at EAs discloses that no standard EA takes constraints into account. That is, the regular search operators, mutation and recombination, in evolutionary programming, evolution strategies, genetic algorithms, and genetic programming, are 'blind' to constraints. Hence, even if the parents are satisfying some constraints they might very well get offspring violating them. Technically, this means that EAs perform unconstrained search. This observation suggests that the presence of constraints in a problem makes EAs intrinsically unsuited to handle this problem.

In this tutorial we will have a closer look at this phenomenon. We start with formalizing the notion of constrained problems in section 2. Then we investigate some general options for handling constraints within an EA in section 3. In section 4 we give an overview of EAs for constraint satisfaction problems, pointing out the key features that have been added to the standard EA machinery in order to handle constraints. In section 5 we summarize the main lessons learned from the overview and indicate where constraints provide extra information on the problem and how this information can be utilized by an EA. Thereafter, section 6 handles a number of methodological considerations regarding research on constraint satisfaction problem (CSP) solving EAs. The final section concludes this tutorial reiterating that EAs *are* suited to treat constrained problems and touches on a couple of promising research directions.

2 What Is a Constrained Problem?

Consider the traveling salesman problem for n cities and let $C = \{c_1, \ldots, c_n\}$ be the set of city labels. If we define the search space as $S_1 = C^n$, then we need a constraint requiring the uniqueness of each city in an element of S. On the other hand, if we take $S_2 = \{permutations\ of\ c_1, \ldots, c_n\}$ then we need no additional constraint, as each city is unique in each element of this space. Thus, the notion 'constrained problem' seems to depend on what we take as the search space. Note, however, that even in the search space S_2 we cannot apply the standard search operators freely. To establish this property formally we need the following definitions.

Definition 1. We call a mutation operator *mut free*[1] with respect to the search space S if its application does not lead out of S. That is, for the result $\bar{z} = mut(\bar{s})$ of mutating $\bar{s}$ it holds that $\bar{z} \in S$.

Note that if the search space is a Cartesian product of sets $S = D_1 \times \cdots \times D_n$, then a mutation operator *mut* is free w.r.t. this S if for any $\langle x_1, \ldots, x_n \rangle \in S$ it holds

---

[1] Another way of naming the property we define here is to call the search space *closed* under the given operator. The terminology introduced here is to emphasize the operator, rather than the space.

that $mut(\langle x_1, \ldots, x_n \rangle) = \langle x_1', \ldots, x_n' \rangle$, and $x_i' \in D_i$. The usual mutation operators in genetic algorithms and evolution strategies all satisfy this condition. However, the same mutation operators are not free with respect to the above defined space S_2 consisting of all permutations of city labels.

Definition 2. We call a recombination operator *free* with respect to the search space S if its application does not lead out of S. That is, for each offspring $\langle z_1, \ldots, z_n \rangle$ of any $\langle x_1, \ldots, x_n \rangle \in S$ and $\langle y_1, \ldots, y_n \rangle \in S$ (and perhaps more parents in case of multi-parent recombination), we have $z_i \in \{x_i, y_i\}$ in discrete spaces, and $z_i \in [x_i, y_i]$ in continuous spaces.

Note that if the search space is a Cartesian product of sets $S = D_1 \times \cdots \times D_n$, then the usual crossover operators from genetic algorithms (i.e., n-point crossover, uniform crossover) and the usual recombination operators from evolution strategies (discrete recombination, intermediate recombination[2]) are free w.r.t. this S. Returning to the TSP example above, we can now formally express an important difference between the two search spaces: the usual mutation and recombination operators are free w.r.t. S_1, but are not free w.r.t. S_2. With this example in mind now we present a general conceptual framework for constrained problems.

Definition 3. We will call a Cartesian product of sets $S = D_1 \times \cdots \times D_n$ a *free search space*. Here the domains can be discrete or continuous. However, we require that if D_i is continuous, then it is convex.

The most important property of free search spaces is that testing the membership relation of such a space can be done independently on each coordinate and taking the conjunction of the results. Obviously, the usual mutation and recombination operators are free w.r.t. a free search space. Thus, evolutionary search in a free space is free in the sense that no restriction applies to the search operators in order to keep the offspring in the search space.

Definition 4. A *free optimization problem* (FOP) is a pair $\langle S, f \rangle$, where S is a free search space and f is a (real-valued) objective function on S, which has to be optimized (minimized or maximized). A *solution of a FOP* is an $\bar{s} \in S$ with an optimal f-value.

Definition 5. A *constraint satisfaction problem* (CSP) is a pair $\langle S, \phi \rangle$, where S is a free search space and ϕ is a formula (Boolean function on S). A *solution of a CSP* is an $\bar{s} \in S$ with $\phi(\bar{s}) = $ true.

Definition 6. A *constrained optimization problem* (COP) is a triple $\langle S, f, \phi \rangle$, where S is a free search space, f is a (real valued) objective function on S and ϕ is a formula (Boolean function on S). A *solution of a COP* is an $\bar{s} \in S$ with $\phi(\bar{s}) = $ true and an optimal f-value.

---
[2] Note that this latter only holds if the domains are convex.

The above three problem types can be represented in the same scheme as $\langle S, f, \bullet \rangle$, $\langle S, \bullet, \phi \rangle$, $\langle S, f, \phi \rangle$, respectively, where $\bullet$ means the absence of the given component. For CSPs, as well as for COPs we call ϕ the *feasibility condition*, and the set $\{\bar{s} \in S | \phi(\bar{s}) = \text{true}\}$ will be called the *feasible search space*. Usually, a CSP is stated as a problem of finding an instantiation of variables $v_1, \ldots, v_n$ within the finite domains $D_1, \ldots, D_n$ such that constraints (relations) $c_1, \ldots, c_m$ prescribed for (some of) the variables hold. The feasibility condition (the formula ϕ) is then given by the conjunction of the given constraints.

With this terminology, solving a CSP means finding a feasible element, and solving a COP means finding a feasible and optimal element. In general, one may be interested in one, some, or all solutions, or only in the existence of a solution. We restrict our discussion to finding one solution.

3 Constraint Handling in an EA

At a high conceptual level we can distinguish two cases: constraints can be handled indirectly or directly. By *indirect constraint handling* we mean that the problem of satisfying constraints is circumvented by incorporating them in the fitness function f in such a way that the optimality of f implies that the constraints are satisfied. In this case the natural optimization power of the EA can be utilized to find a solution. By *direct constraint handling* we mean that we leave the constraints as they are and modify the EA to enforce them. Later on we return to this question in more detail. Let us note that direct and indirect constraint handling can be applied in combination, i.e., in one application we can handle

(a) all constraints indirectly,
(b) all constraints directly, or
(c) some of the constraints directly and other ones indirectly.

Indirect constraint handling formally means that we transform constraints into optimization objectives, thus relaxing ϕ and modifying f (in case of a COP) or introducing f (in case of a CSP). Hereby we perform a problem transformation. The requirement that the (eliminated) constraints are satisfied if the (new) optimization objectives are at their optimum implies that the given problem is transformed into an equivalent problem, where 'equivalent' means that the two problems have the same solutions. Strictly speaking it is sufficient to require that solutions of the transformed problem are also solutions of the original problem, but this nuance is not relevant for the present discussion. Table 1 shows the problem transformations resulting from constraint relaxation.

Clearly the goal of the problem transformation is to create an equivalent problem that can be solved by an EA. Transforming the original problem is thus only half of the work. Actually, constraint handling in an EA framework has two meanings:

Constraint	Original problem	
handling	CSP	COP
(a) all indirect	$\langle S, \bullet, \phi \rangle \rightarrow \langle S, f, \bullet \rangle$	$\langle S, f, \phi \rangle \rightarrow \langle S, g, \bullet \rangle$
(b) all direct	$\langle S, \bullet, \phi \rangle$	$\langle S, f, \phi \rangle$
(c) mixed	$\langle S, \bullet, \phi \rangle \rightarrow \langle S, f, \psi \rangle$	$\langle S, f, \phi \rangle \rightarrow \langle S, g, \psi \rangle$

Table 1. Problem (type) transformations corresponding to (a) indirect, (b) direct, and (c) mixed constraint handling

1. *How to transform* the constraints in ϕ into f, g, or $\langle f, \psi \rangle$, $\langle g, \psi \rangle$.
2. *How to enforce* the constraints in $\langle S, f, \phi \rangle$, or $\langle S, f, \psi \rangle$, $\langle S, g, \psi \rangle$.

If we treat all constraints indirectly, then constraint handling is only done in the first sense. This case belongs to penalty-based approaches in the common EA terminology. Treating all constraints directly for a COP means constraint handling only in the second sense. Note that for a CSP this option is not suitable, because the resulting problem is a CSP too, having no objective to be optimized by the EA. The mixed direct-indirect approach (case (c) of table 1) means constraint handling in both senses.

For a given constrained problem several equivalent problems can be defined by choosing the subset of the constraints to be eliminated and/or defining the objective function measuring their satisfaction differently. So, there are two important questions to be answered:

1. Which constraints should be handled directly (kept as constraints) and which should be handled indirectly (replaced by optimization objectives)?
2. How shall we define the optimization objectives corresponding to indirectly handled constraints?

3.1 Direct Constraint Handling

Treating constraints directly implies that violating them is not reflected in the fitness function, thus there is no bias towards chromosomes satisfying them. Therefore, the population will not become more and more feasible w.r.t. these constraints[3]. This means that we have to create and maintain feasible chromosomes in the population.

---

[3] At this point we should make a distinction between feasibility in the original problem context (ϕ) and relaxed feasibility in the context of the transformed problem (ψ). For instance, we could introduce the name *allowability* for the conjunction of those constraints that are handled directly [18]. However, to keep the discussion simple we will use the term feasibility in both cases.

The basic problem in this case is that the regular genetic operators are blind to constraints; mutating one or crossing over two feasible chromosomes can result in infeasible offspring. Typical approaches to handle constraints directly are the following:

- Eliminating infeasible candidates.
- Repairing infeasible candidates.
- Preserving feasibility by special operators.
- Decoding, i.e., transforming the search space.

Eliminating infeasible candidates is the most general and the least useful option. Namely, it is very inefficient as the probability of hitting a feasible candidate by chance is practically zero in most applications. Therefore, this way of direct constraint handling is hardly practicable.

Repairing infeasible candidates requires a repair procedure that modifies a given chromosome such that it will not violate constraints. This technique is thus problem dependent, but if a good repair procedure can be developed then it works well in practice, see, for instance, section 4.5 in [31] for a comparative case study on a COP. It is almost impossible to say something in general about repairing mechanisms. Let us only make a remark on the relationship between Lamarckian evolution and repair. In our view, Lamarckian evolution is directed to improve a candidate solution in terms of the given optimization objective(s). Repair procedures aim at improving a quality measure that is *not* reflected in the given fitness function.

The preserving approach amounts to designing and applying problem specific operators that do preserve the feasibility of parent chromosomes. Using such operators the search becomes quasi-free, because the offspring remains in the feasible search space, if the parents were feasible. There is a large class of problems where a candidate solution can be naturally represented by permutations over a certain alphabet, for instance, routing, scheduling, sequencing problems, etc. In all these problems the feasibility of a chromosome means that it contains each allele exactly once. This is called order-based representation and there are many order-based mutations and crossovers that are designed to preserve the property of being a permutation [47,48]. Order-based representation and order-based operators, although limited in use to certain problems, are popular because it is often natural to use them. From the constraint handling point of view the property of being a permutation forms a "nice" feasibility condition in the sense that: (a) it is widely applicable (many problems can be transformed into it); and (b) it is easy to handle (many suitable mutation and crossover operators are at ones disposal). Note that the preserving approach requires the creation of a feasible initial population, which can be NP-hard, e.g., for the traveling salesman problem with time windows.

Finally, decoding can simplify the problem and allow an efficient EA. Formally, decoding can be seen as shifting to a search space that is different from the Cartesian product $D_1 \times \cdots \times D_n$ of the domains of the variables in the original problem formulation. Elements of the new search space S' serve as inputs for a decoding procedure that creates feasible solutions, and it is assumed that a free search (or quasi-free search with preserving operators) can be performed in S' by an EA. For a nice illustration we refer again to section 4.5 in [31].

As the previous paragraphs show, direct constraint handling is not really a single phenomenon, but rather a collection of approaches to treat constraints while solving the problem. A common[4] disadvantage of them is that they are heavily problem dependent. So far there is not enough knowledge collected to give general advice on how to use them. In turn, a widely acknowledged common advantage of them is that they work well. Thus, the extra effort to implement them in a given problem context most probably pays off.

3.2 Indirect Constraint Handling

In the case of indirect constraint handling the optimization objectives replacing the constraints are traditionally viewed as penalties for constraint violation, hence to be minimized. In principle, every penalty function is an (heuristic) estimate of the badness of a given candidate solution. Such heuristics usually try to estimate the distance to feasible solution, or the costs of repairing the infeasibility. Here we discuss two specific types of estimations, thus penalty functions, for they are general and can be used for a broad variety of different problems:

1. Penalty for violated constraints.
2. Penalty for wrongly instantiated variables.

For a formal description let us assume that we have constraints c_i ($i = \{1, \ldots, m\}$), variables v_j ($j = \{1, \ldots, n\}$), and let C^i be the set of constraints involving variable v_i. Then the penalties relative to the two options above described can be expressed as follows:

$$f_1(s) = \sum_{i=1}^{m} w_i \chi(s, c_i) \quad \text{with } \chi(s, c_i) = \begin{cases} 1 & \text{if } s \text{ violates } c_i \\ 0 & \text{otherwise,} \end{cases} \quad (1)$$

$$f_2(s) = \sum_{i=1}^{n} w_i \chi(s, C^i) \quad \text{with } \chi(s, C^i) = \begin{cases} 1 & \text{if } s \text{ violates at least} \\ & \text{one } c \in C^i \\ 0 & \text{otherwise,} \end{cases} \quad (2)$$

respectively. Obviously, for the above functions f_1, f_2 and for each $s \in S$ we have that $\phi(s) = $ true if and only if $f_i(s) = 0$.

Example 1. In the graph 3-coloring problem the nodes of a graph $G = (N, E)$, $E \subseteq N \times N$ have to be colored by three colors in such a way that no neighboring nodes, i.e., nodes connected by an edge, have the same color. This problem can be formalized by means of a CSP with $n = |N|$ variables, each with the same domain $D = \{1, 2, 3\}$. Furthermore, we have $m = |E|$ constraints, one for each edge e,

---
[4] Here we omit the elimination approach because of its very low usefulness.

with $c_e(s)$ = true iff $e = (k, l)$ and $s_k \neq s_l$. Then the corresponding CSP is $\langle S, \phi \rangle$, where $S = D^n$ and $\phi(s) = \bigwedge_{e \in E} c_e$. Using the constraint-oriented penalty function f_1 with $w_i \equiv 1$ we count the incorrect edges that connect two nodes with the same color. The variable-oriented penalty function f_2 with $w_i \equiv 1$ amounts to counting the incorrect nodes that have a neighbor with the same color. □

There are great advantages to indirect constraint handling. First, it is general, or at least there exist problem-independent realizations of the principle, such as the functions f_1 and f_2 discussed above. Second, it reduces the problem to simple optimization where one does not have to bother about constraints any more. Obviously, one still has an (optimization) problem to be solved. This is simple in the sense that EAs have a 'basic instinct' for optimization, but in practice it could be hard to find an optimum, of course. Third, it allows a natural way to express user preferences distinguishing among constraints. To this end, notice that the global quality measure (optimization objective) is composed of elementary units of quality (constraint violation). The composition rule in the two examples f_1 and f_2 is simple, a linear combination. Weights of this linear combination provide a simple and transparent way of emphasizing some constraints' importance.

Certainly, there are also disadvantages to indirect constraint handling. The very compositionality mentioned above has a negative effect, too. It causes a loss of information by packing all detailed knowledge on constraint violations into a single number. This might cause a handicap when trying to solve the problem. Another disadvantage often mentioned is that the penalty-based approach does not work well for sparse problems, where the proportion of feasible solutions is very low. The original source of this claim is most probably section 4.5 in [31] or even the earlier editions of this book. Nevertheless, reports in recent literature seem not to support this claim, so we repeat it here for the sake of completeness only. Third, general as the penalty technique may seem, it remains a problem-dependent issue how to define the optimization objectives representing the constraints and (in case of COPs) how to merge the original objective function with penalties. Some guidelines concerning this issue are discussed in [41].

There are other classification schemes of constraint handling techniques in EC. For instance, the categorization in [33], distinguishes pro-choice and pro-life techniques, where pro-choice encompasses eliminating, decoding, and preserving, while pro-life covers penalty-based and repairing approaches. Overviews and comparisons published on evolutionary computation techniques for constraint handling so far mainly concern continuous domains, [29,30,32,34]. Constraint handling in continuous and discrete domains rely to a certain extent on the same ideas. There are, however, also differences. For instance, in continuous domains constraints can be characterized as linear, non-linear, etc., and in the case of linear constraints special averaging recombination operators can guarantee that offspring of feasible parents are feasible. In discrete domains this is impossible.

4 Evolutionary CSP Solvers

This section contains an overview of a number of EAs for solving CSPs. While giving the necessary pointers to relevant literature, this is not an annotated bibliography. The main objective of this section is to highlight the main ideas on constraint handling (over finite domains) which have been employed in EAs.

All along the EA history there have been attempts to solve specific CSPs, such as the satisfiability problem [22,27]. To our knowledge it is only since 1994 that attempts to treat CSPs in general have been reported in the evolutionary computing literature.

Eiben et al. offered a combination of traditional CSP solving heuristics with genetic algorithms [15]. The underlying motivation is to get the best of two worlds. The greediness of the heuristics (which can lead to dead-ends) and the blindness of the stochastic genetic search (which can lead to wrong areas of the search space) can compensate each other. The heuristics can be incorporated into the genetic operators mutation and crossover. In the mutation operator both the selection of genes to be mutated and the selection of new values for these genes can be heuristic. Tests on the N-queens problem ($100 \leq N \leq 10000$) and on graph coloring with a penalty function counting constraint violations have shown that biased mutations indeed work well. Furthermore, the heuristic GA outperformed Minton's MC algorithm on graph coloring [35].

Paredis proposed a completely different idea: putting the constraints into a second population and co-evolving them with the population of candidate solutions. The driving force of progress is then an arms-race. Each population can only get better at the cost of the other one. On the one hand, a candidate solution has a higher fitness if it satisfies more constraints. On the other hand, a constraint has a higher fitness if it causes more candidate solutions to fail (i.e., if more candidate solutions violate it). A crucial difference between the two populations is that the contents of the constraint population does not change. Fitness information is only used to rank them and the probability that a constraint will be used for evaluating candidate solutions is proportional to its fitness. The population of candidate solutions is, of course, truly evolving. The most interesting aspect of this approach is that it works with a fitness function that is changing over time. Namely, the list of constraints that are used to evaluate candidates changes continuously according to the constraints' changing rank. Paredis compared the co-evolutionary GA with a straightforward GA using a penalty function counting constraint violations on the 50-queens problem and showed that the co-evolutionary approach works better.

Dozier and his co-authors introduced a so-called heuristic-based microgenetic approach in [6], and further refined it in [3,7,8]. In the consecutive publications following the first paper in 1994 several algorithm variants have been reported. We disregard these variations here and summarize the most important specific features that have been introduced along the whole research line:

- Small populations.
- Use of heuristics in the mutation and the crossover operators.

- On-line detection and exploitation of 'nogoods' (i.e., bad values for a couple of variables $\langle v_i, v_j \rangle$).
- Arc-revision: removing a value from a domain if it is proved to be impossible.
- Arc-inconsistency check: terminate the run if a domain becomes empty.

The use of nogoods, also called breakouts, influences the fitness function. Namely, the basic fitness function counting constraint violations is extended with a term that adds an extra penalty if a candidate contains nogoods. Since the nogoods are detected during the run of the GA, this makes the fitness function adaptive. The other remarkable feature is the combination of arc-revision and arc-consistency checks that enables the GA to "realize when to quit". For an evolutionary search process this is a novel feature. In all other CSP solving EAs termination without a solution counts as failure in solving the problem – even though the problem at hand had no solutions at all. The ASGA in [3] is able to report that there is no solution. Therefore, such runs count as success.

A novel approach has been introduced by Riff-Rojas in [40] and further elaborated in [42]. The novelty in this approach lies in using knowledge on the constraint network to assist evolutionary search. This goes further than the use of heuristics in the previously reported papers. One could say that those papers apply 'shallow' heuristics that look one level deep. For instance, a heuristic checks how many possible values remain for variable v_j when instantiating v_i at a given value. In comparison, the heuristics of Riff-Rojas are 'deep', in the sense that they look further than one constraint and take a part of the constraint network into account. It is also noteworthy that she defines heuristic quality estimates of candidate solutions different from counting constraint violations or wrong variable instantiations. Furthermore, different quality estimates are applied in the fitness function, the (heuristic) mutation, and the (heuristic) crossover operators. In [40] she shows the beneficial effects of a new kind of selection that strongly favors chromosomes with a low penalty on random binary CSPs with 30 variables.

In [28] Marchiori introduced a method for solving binary CSPs given in a specific format. A CSP here is defined by equalities and inequalities and solving the problem starts with transforming it. The transformation consists of two steps: elimination of functional constraints and variables, followed by splitting complex constraints into primitive ones. These steps can be seen as problem preprocessing before applying a GA to solve the problem. The elimination step reduces the number of variables (dimension reduction). The splitting step, based on the so-called 'glass-box' approach [46], transforms all constraints into one single syntactical scheme. Information on this scheme is then utilized by the GA solving such a problem. The GA, using a penalty function counting violated constraints, features a specific repair heuristic called dependency propagation. Experiments on the 50-queens problem and the Zebra-puzzle (five houses puzzle) show that this GA outperforms other GAs from the literature.

The *stepwise adaptation of weights* (SAW) as a general mechanism to handle constraints in an EA has been introduced by Eiben and van der Hauw [10] as an improved version of the weight adaptation mechanism of Eiben, Raué and Ruttkay

[16,17]. The basic idea behind the mechanism is that constraints that are not satisfied after a certain number of steps must be hard, thus must be given a high weight (penalty), see the formulas before Example 1. In the earliest attempts, [16,17], the weights were constant during each run, but a series of, say, 100 runs, was performed on the same problem instance. A weight update took place between two runs, raising a weight in fitness function f_n in the n-th run if the best solution in run $n-1$ violated the corresponding constraint. The final version of SAW-ing updates weights during one single run, after a number of fitness evaluations. This technique was very successful in solving 3-SAT problems, graph 3-coloring problems and random binary CSPs [10–14]. Clearly, SAW-ing EAs have an adaptive fitness function. The algorithms in the above papers used a fitness function of type 1 (see equation 1) for the 3-SAT and the random binary CSP application, and a fitness function of type 2 (see equation 2) for graph coloring. A remarkable feature of these algorithms is the very small populations; in some applications population size 1 turned out to be optimal.

5 Lessons Learned

The amount and quality of work in the area of evolutionary CSP solving certainly refutes the initial intuitive hypothesis that EAs are intrinsically unsuited for constrained problems. This raises the question of what makes EAs able to solve CSPs? Looking at the specific features of EAs for CSPs one can distinguish two categories. In the first category we find heuristics that can be incorporated in almost any EA component, the fitness function, the variation operators mutation and recombination, the selection mechanism, or used in a repair procedure. The second category is formed by adaptive features, in particular a fitness function that is being modified during a run. All reported algorithms fall into one of these categories and that of Dozier et al. belongs to both.

A careful look at the above features discloses that they are all based on information related to the constraints themselves. The very fact that the (global) problem to be solved is defined in terms of (local) constraints to be satisfied facilitates the design and usage of 'tricks'. The scope of applicability of these tricks is limited to constrained problems[5], but not necessarily to a particular CSP, like SAT or graph coloring. The first category of tricks is based on the fact that the presence of constraints facilitates measures on sub-individual structures. For instance, one gene (variable) can be evaluated by the number of conflicts its present value is involved in. Such sub-individual measures are not possible for example in a pure function optimization problem, where only a whole individual can be evaluated. These measures are typically used as evaluation heuristics giving hints on how to proceed in constructing an offspring, or in repairing a given individual. The second category is based on the fact that the composite nature of the problem leads to a composite evaluation function, like the ones mentioned in equations 1 and 2. Such a composite function can be

---
[5] Actually, this is not entirely true. For instance, the SAW-ing technique can be easily imported into GP for machine-learning applications, see [9]

tuned during a run by adding new nogoods (Dozier), modifying weights (SAW-ing), or changing the reference set of constraints used to calculate it (coevolution).

Browsing through the literature there are other aspects that (some of) the papers share. Apparently, a CSP ⟶ FOP transformation is common and a CSP ⟶ COP transformation is not popular. Putting it in the terminology we introduced in section 3, this implies that indirect constraint handling is more common practice than direct constraint handling. On the other hand, in almost all applications some heuristics are used even if the transformed problem is an FOP, and these heuristics are meant to increase the chance of satisfying constraints. In other words, constraints are handled directly by these heuristics. This actually suggests that the way we define indirect constraint handling (like using a CSP ⟶ FOP transformation) might not be a 100% match with the practice. With that said, however, we maintain the terminology because a CSP ⟶ FOP transformation does allow the use of the universal EA machinery; applying heuristics is only an extra tool which might or might not be applied.

Another noteworthy property that occurs repeatedly in EAs for CSPs is the small size of the population. Common EA wisdom suggests that big populations are better than small ones for they can keep genetic diversity more easily for longer. From personal communications with authors and our own experience it turns out that using small populations is always justified by experiments. Exactly because small populations contradict one's intuition, such setups are only taken after substantial experimental justification. Such an experimental comparison sometimes leads to surprising outcomes, for instance, that the optimal setup is to use a population of size 1 and only mutation as search operator [2,11]. In this case it is legitimate to ask whether the resulting algorithm is still evolutionary or is it just a hill-climber? Clearly, this is a matter of judgment, but as most people in evolutionary computation accept the (1+1) and the (1,1) evolution strategy as members of the family, it is legitimate to say that one still has an EA in this case.

Summarizing, it seems to be possible to extract some guidelines from existing literature on how to tackle a CSP by EAs. A short list of promising options is:

1. Use, possibly existing, heuristics to estimate the quality of sub-individual entities (like one variable assignment) in the components of the EA: fitness function, mutation and recombination operators, selection, and repair mechanism.
2. Exploit the composite nature of the fitness function and change its composition over time. During the search information is collected (e.g., on which constraints are hard); this information can be very well utilized.
3. Try small populations and mutation-only schemes.

6 Some Methodological Considerations

The foregoing sections have indicated that EAs can solve constrained problems, in particular CSPs. But are these evolutionary CSP solvers competitive with traditional techniques? Some papers draw a comparison between an EA and another technique, for instance, on 3-SAT and graph 3-coloring. In general, however, this question is still open.

Performing an experimental comparison between algorithms, in particular, between evolutionary and other type of problem solvers, implies a number of methodological questions:

1. Which benchmark problems and problem instances should be used?
2. Which competitor algorithms should be used?
3. Which comparative measures should be used?

As for the problems and problem instances one could distinguish two main approaches: the repository and the generator approach. The first one amounts to obtaining prepared problem instances that are freely available from (web-based) repositories, for instance, the Constraints Archive at *http://www.cs.unh.edu/ccc/archive*. The advantage of this approach is that the problem instances are 'interesting' in the sense that other researchers have investigated and evaluated them already. Besides this, an archive often contains performance reports of other techniques, thereby providing a direct feedback on one's own achievements. Using a problem instance generator (which of course could be coming from an archive) means that problem instances are produced on-the-spot. Such a generator usually has some problem specific parameters, for instance the number of clauses and the number of variables for 3-SAT, or the constraint density and constraint tightness for binary CSPs. The advantage of this approach is that the hardness of the problem instances can be tuned by the parameters of the generator. Recent research has shed light on the location of really hard problem instances, the so-called phase transition, for different classes of problems [4,20,21,23,36,38,39,44]. A generator makes it possible to perform a systematic investigation in and around the hardest parameter range. The currently available EA literature mostly follows the repository approach tackling commonly studied problems, like N-queens[6], 3-SAT, graph coloring, or the Zebra puzzle. Dozier et al. use a random problem instance generator for binary CSPs[7] which creates instances for different constraint tightness and density values [6]. Later on this generator has been adopted and reimplemented by Eiben et al. [14].

Advice on the choice of a competitor algorithm boils down to the same suggestion: choose the best one available to represent a real challenge. Implementing this principle is, of course, not always simple. It could be hard to find out which specific algorithm shows the best performance on a given (type of) problem. This is not only

---

[6] This problem has a rather exceptional feature: if its size (the number of queens) is increased, it gets easier [37]. This makes it somewhat uninteresting as the traditional 'scale-up competition' won't work with it.

[7] Binary CSPs (where each constraint concerns exactly two variables) form a nice problem class. While they have a transparent structure it holds that every CSP is equivalent to a binary CSP [45].

due to the difficulties of finding information. Sometimes it is not clear which criteria to use on which to base the choice.

This problem leads us to the third aspect of comparative experimental research: that of the comparative measures. The performance of a problem solving algorithm can be measured in different ways. Speed and solution quality are widely used, and for stochastic algorithms, as EAs are, the probability of finding a solution (of a certain quality) is also a common measure.

Speed is often measured in elapsed computer time, *CPU time* or *user time*. However, this measure depends on the specific hardware, operating system, compiler, network load, etc., and therefore is ill-suited for reproducible research. In other words, repeating the same experiments, possibly elsewhere, may lead to different results. For generate-and-test style algorithms, which EAs are, a common way around this problem is to count the number of points visited in the search space. Since EAs immediately evaluate each newly generated candidate solution, this measure is usually expressed as the number of fitness evaluations. Forced by the stochastic nature of EAs, this is always measured over a number of independent runs and the *average number of evaluations to a solution* (AES) is used. It is important to note that the average is only taken over the successful runs ("to a Solution"), otherwise the maximum number of evaluations actually used would distort the statistics. Fair as this measure seems, there are two possible problems with it. First, it could be misleading if an EA uses 'hidden labor', for instance, some local search heuristics incorporated in the mutation operator. The extra computational efforts within the mutation can increase performance, but are invisible to the AES measure[8]. Second, it can be difficult to apply AES for comparing an EA with search algorithms that do not work in the same search space. An EA is iteratively improving complete candidate solutions, so one elementary search step is the creation of one new candidate solution. However, a constructive search algorithm would work in the space of partial solutions (including the complete ones that an EA is searching through) and one elementary search step is extending the current solution. Counting the number of elementary search steps is misleading if the search steps are different. A common treatment for both of these problems with AES (hidden labor, different search steps) is to compare the scale-up behavior of the algorithms. To this end a problem is needed that is scalable, that is, its size can be changed. The number of variables is a natural scale-up parameter for many problems. Two different types of methods can then be compared by plotting their own speed measure figures against the problem size. Even though the measures used in each curve are different, the steepness information is a fair basis for comparison: the curve that grows at a higher rate indicates an inferior algorithm.

Solution quality of approximative algorithms for optimization is most commonly defined as the distance to an optimum at termination, e.g., $|f_{\text{best}} - f_{\text{opt}}|$, where f is the function to be optimized, f_{best} is the f value of the best candidate solution

---

[8] In the CSP literature the number of constraint checks is used commonly as a speed measure. It seems an interesting option to use this measure in combination with or as an alternative to the AES measure in evolutionary computing.

found in the given run and f_{opt} is the optimal f value. For stochastic algorithms this is averaged over a number of independent runs and in evolutionary computing the *mean best fitness* (MBF) is a commonly used name for this measure. As we have seen in this tutorial, for constraint satisfaction problems it is not straightforward what f to use – there are more sensible options. For comparing the solution quality of algorithms this means that there are more sensible quality measures. The problem is then, that most probably one would use the function f that has been used to find a solution and this can be different for another algorithm. For instance, algorithm A could use the number of unsatisfied constraints as the fitness function and algorithm B could use the number of wrong variable instantiations. It is then not clear what measure to use for comparing the two algorithms. Moreover, in constraint satisfaction it is often not good enough to be close to a solution. A candidate is either good (satisfies all constraints) or bad (violates some constraints). In this case, it makes no sense to look at the distance to a solution as a quality measure, hence the MBF measure is not appropriate.

The third measure which is often used to judge stochastic algorithms, and thus EAs, is the probability of finding a solution (of certain quality). This probability can be estimated by performing a number of independent runs under the same setup on the same problem instance and keep a record on the percentage of runs that did find a solution. This *success rate* (SR) completes the picture obtained by AES and MBF. Note that SR and MBF are related but provide different information, and all different combinations of good/bad SR/MBF are possible. For instance, bad (low) SR and good (high) MBF indicate a good approximization algorithm: it gets close, but misses the last step to hit the solution. Likewise, a good (high) SR and a bad (low) MBF combination is also possible. Such a combination shows that the algorithm mostly performs perfectly, but sometimes it does a very, very bad job. Paraphrasing Murphy: if it goes wrong then it goes very wrong.

7 Concluding Remarks

Let us reiterate the initial observation that a standard EA is 'blind' to constraints. While this is true, the existing literature clearly shows that EAs *are* suited to treat constrained problems, in particular CSPs. The survey of related work disclosed how EAs can be made successful in solving such problems. Roughly classifying the options we encountered, the key features are the utilization of heuristics and/or the adaptation of the fitness function during a run. Both features are based on the structure of the problems in question, so, in a way, the problem of how to treat CSPs carries its own solution.

In particular, constraints facilitate the use of sub-individual measures to evaluate parts of candidate solutions. Such sub-individual measures are not possible, for example, in a pure function optimization problem, where only a whole individual can be evaluated. These measures lead to heuristics that can be incorporated in practically any component of an EA, the fitness function, mutation and recombination operators, selection, or used in a repair mechanism.

Likewise, it is the presence of constraints that leads to a fitness function composed from separate pieces. This composition or the relative importance of the components can be changed over time. During the search information is collected (e.g., on which constraints are hard) and this information can be very well utilized.

The field of evolutionary constraint satisfaction is relatively new. Intensive investigations started approximately in the mid 1990s, while evolutionary computing itself has it roots in the 1960s. Because of the short history coherence is lacking and the findings of individual experimental studies cannot be generalized (yet). There are a number of research directions that should be pursued in the future for further development. These include:

- Study of the problem area. A lot can be learned from the traditional constrained literature about such problems. Existing knowledge should be imported into core EC research.
- Cross-fertilization between the insights concerning EAs for (continuous) COPs and (discrete) CSPs. At present, these two sub-areas are practically unrelated.
- Sound methodology: how to set up fair experimental research, how to obtain good benchmarks, how to compare EAs with other techniques.
- Theory: better analysis of the specific features of constrained problems, and the influence of these features on EA behavior.

References

1. T. Bäck, D. Fogel, and Z. Michalewicz, editors. *Handbook of Evolutionary Computation*. Institute of Physics, Bristol, and Oxford University Press, New York, 1997.
2. T. Bäck, A. E. Eiben, and M. E. Vink. A superior evolutionary algorithm for 3-SAT. In V. William Porto, N. Saravanan, Don Waagen, and A. E. Eiben, editors, *Proc. of the 7th Annual Conference on Evolutionary Programming*, LNCS 1477, pages 125–136. Springer, Berlin Heidelberg New York, 1998.
3. J. Bowen and G. Dozier. Solving constraint satisfaction problems using a genetic/systematic search hybrid that realizes when to quit. In Eshelman [19], pages 122–129.
4. P. Cheeseman, B. Kenefsky, and W. M. Taylor. Where the really hard problems are. In *Proc. of the IJCAI-91*, pages 331–337. Morgan Kaufmann, San Francisco, 1991.
5. A. G. Cohn, editor. *Proc. of the European Conference on Artificial Intelligence*. John Wiley, New York, 1994.
6. G. Dozier, J. Bowen, and D. Bahler. Solving small and large constraint satisfaction problems using a heuristic-based microgenetic algorithm. In IEEE [24], pages 306–311.
7. G. Dozier, J. Bowen, and D. Bahler. Solving randomly generated constraint satisfaction problems using a micro-evolutionary hybrid that evolves a population of hill-climbers. In *Proc. of the 2nd IEEE Conference on Evolutionary Computation*, pages 614–619. IEEE Press, New York, 1995.
8. G. Dozier, J. Bowen, and A. Homaifar. Solving constraint satisfaction problems using hybrid evolutionary search. *IEEE Transactions on Evolutionary Computation*, 2(1):23–33, 1998.

9. J. Eggermont, A. E. Eiben, and J. I. van Hemert. Adapting the fitness function in GP for data mining. In R. Poli, P. Nordin, W. B. Langdon, and T. C. Fogarty, editors, *Genetic Programming, Proc. of EuroGP'99*, LNCS 1598, pages 195–204. Springer-Verlag, Berlin Heidelberg New York, 1999.
10. A. E. Eiben and J. K. van der Hauw. Graph coloring with adaptive evolutionary algorithms. Technical Report TR-96-11, Leiden University, August 1996. Also available at *http://www.wi.leidenuniv.nl/~gusz/graphcol.ps.gz*.
11. A. E. Eiben and J. K. van der Hauw. Solving 3-SAT with adaptive Genetic Algorithms. In IEEE [26], pages 81–86.
12. A. E. Eiben and J. K. van der Hauw. Adaptive penalties for evolutionary graph-coloring. In J.-K. Hao, E. Lutton, E. Ronald, M. Schoenauer, and D. Snyers, editors, *Artificial Evolution'97*, LNCS 1363, pages 95–106. Springer-Verlag, Berlin Heidelberg New York, 1998.
13. A. E. Eiben, J. K. van der Hauw, and J. I. van Hemert. Graph coloring with adaptive evolutionary algorithms. *Journal of Heuristics*, 4(1):25–46, 1998.
14. A. E. Eiben, J. I. van Hemert, E. Marchiori, and A. G. Steenbeek. Solving binary constraint satisfaction problems using evolutionary algorithms with an adaptive fitness function. In A. E. Eiben, T. Bäck, M. Schoenauer, and H.-P. Schwefel, editors, *Proc. of the 5th Conference on Parallel Problem Solving from Nature*, LNCS 1498, pages 196–205. Springer-Verlag, Berlin Heidelberg New York, 1998.
15. A. E. Eiben, P-E. Raué, and Z. Ruttkay. Solving constraint satisfaction problems using genetic algorithms. In IEEE [24], pages 542–547.
16. A. E. Eiben, P.-E. Raué, and Z. Ruttkay. Constrained problems. In L. Chambers, editor, *Practical Handbook of Genetic Algorithms*, pages 307–365. CRC Press, 1995.
17. A. E. Eiben and Z. Ruttkay. Self-adaptivity for constraint satisfaction: learning penalty functions. In IEEE [25], pages 258–261.
18. A. E. Eiben and Z. Ruttkay. Constraint-satisfaction problems. In Bäck et al. [1], pages C5.7:1–C5.7:8.
19. L. J. Eshelman, editor. *Proc. of the 6th International Conference on Genetic Algorithms*. Morgan Kaufmann, San Francisco, 1995.
20. I. Gent, E. MacIntyre, P. Prosser, and T. Walsh. Scaling effects in the CSP phase transition. In *1st International Conference on Principles and Practice of Constraint Programming*, 1995. Also available at *http://www.cs.strath.ac.uk/apes/apepapers.html*.
21. I. Gent and T. Walsh. Phase transitions from real computational problems. In *Proc. of the 8th International Symposium on Artificial Intelligence*, pages 356–364, 1995.
22. J.-K. Hao. A clausal representation and its evolutionary procedures for satisfiability problems. In D. W. Pearson, N. C. Steel, and R. F. Albrecht, editors, *Proc. of the International Conference on Artificial Neural Networks and Genetic Algorithms*, pages 289–292. Springer, Vienna, 1995.
23. T. Hogg and C. Williams. The hardest constraint problems: a double phase transition. *Artificial Intelligence*, 69:359–377, 1994.
24. *Proc. 1st IEEE Conf. on Evolutionary Computation*. IEEE Press, New York, 1994.
25. *Proc. 3rd IEEE Conf. on Evolutionary Computation*. IEEE Press, New York, 1996.
26. *Proc. 4th IEEE Conf. on Evolutionary Computation*. IEEE Press, New York, 1997.
27. K. A. De Jong and W. M. Spears. Using genetic algorithms to solve NP-complete problems. In Schaffer [43], pages 124–132.
28. E. Marchiori. Combining constraint processing and genetic algorithms for constraint satisfaction problems. In T. Bäck, editor, *Proc. of the 7th International Conference on Genetic Algorithms*, pages 330–337. Morgan Kaufmann, San Francisco, 1997.

29. Z. Michalewicz. Genetic algorithms, numerical optimization, and constraints. In Eshelman [19], pages 151–158.
30. Z. Michalewicz. A survey of constraint handling techniques in evolutionary computation methods. In J. R. McDonnell, R. G. Reynolds, and D. B. Fogel, editors, *Proc. of the 4th Annual Conference on Evolutionary Programming*, pages 135–155. MIT Press, Boston, MA, 1995.
31. Z. Michalewicz. *Genetic Algorithms + Data structures = Evolution programs.* Springer-Verlag, Berlin Heidelberg New York, 3rd edition, 1996.
32. Z. Michalewicz and N. Attia. Evolutionary optimization of constrained problems. In A. V. Sebald and L. J. Fogel, editors, *Proc. of the 3rd Annual Conference on Evolutionary Programming*, pages 98–108. World Scientific, Singapore, 1994.
33. Z. Michalewicz and M. Michalewicz. Pro-life versus pro-choice strategies in evolutionary computation techniques. In M. Palaniswami, Y. Attikiouzel, R. J. Marks, D. Fogel, and T. Fukuda, editors, *Computational Intelligence: A Dynamic System Perspective*, pages 137–151. IEEE Press, New York, 1995.
34. Z. Michalewicz and M. Schoenauer. Evolutionary algorithms for constrained parameter optimization problems. *Evolutionary Computation*, 4(1):1–32, 1996.
35. S. Minton, M. D. Johnston, A. Philips, and P. Laird. Minimizing conflicts: a heuristic repair method for constraint satisfaction and scheduling problems. *Artificial Intelligence*, 58:161–205, 1992.
36. D. Mitchell, B. Selman, and H. J. Levesque. Hard and easy distributions of SAT problems. In *Proc. of the AAAI*, pages 459–465. San Jose, CA, 1992.
37. P. Morris. On the density of solutions in equilibrium points for the queens problem. In *Proc. of the 10th National Conference on Artificial Intelligence, AAAI-92*, pages 428–433. AAAI Press/The MIT Press, Cambridge, MA, 1992.
38. P. Prosser. Binary constraint satisfaction problems: some are harder than others. In Cohn [5], pages 95–99.
39. P. Prosser. An empirical study of phase transitions in binary constraint satisfaction problems. *Artificial Intelligence*, 81:81–109, 1996.
40. M. C. Riff-Rojas. Using the knowledge of the constraint network to design an evolutionary algorithm that solves CSP. In IEEE [25], pages 279–284.
41. J. T. Richardson, M. R. Palmer, G. Liepins, and M. Hilliard. Some guidelines for genetic algorithms with penalty functions. In Schaffer [43], pages 191–197.
42. M. C. Riff-Rojas. Evolutionary search guided by the constraint network to solve CSP. In IEEE [26], pages 337–348.
43. J. D. Schaffer, editor. *Proc. of the 3rd International Conference on Genetic Algorithms.* Morgan Kaufmann, San Francisco, 1989.
44. B. M. Smith. Phase transition and the mushy region in constraint satisfaction problems. In Cohn [5], pages 100–104.
45. E. P. K. Tsang. *Foundations of Constraint Satisfaction.* Academic Press, New York, 1993.
46. P. van Hentenryck, V. Saraswat, and Y. Deville. Constraint processing in cc(fd). In A. Podelski, editor, *Constraint programming: basics and trends.* Springer-Verlag, Berlin Heidelberg New York, 1995.
47. D. Whitley. Permutations. In T. Bäck et al. [1], pages C3.2:5 – C3.2:8. Chapter C3.2, Mutation.
48. D. Whitley. Permutations. In T. Bäck et al. [1], pages C3.3:14 – C3.3:20. Chapter C3.3, Recombination.

The Dynamical Systems Model of the Simple Genetic Algorithm

J. E. Rowe

Department of Computer and Information Science
De Montfort University
Milton Keynes MK7 6HP, UK
E-mail: *jrowe@dmu.ac.uk*

Abstract. This tutorial describes the basic theory of the simple genetic algorithm, as developed by Michael Vose. The mathematical framework is established in which the actions of proportionate selection, mutation and crossover can be analysed. The results are illustrated through simple examples. The recently discovered connections between the mathematical form of the genetic operators and the Walsh transform are briefly described. Some current outstanding conjectures are also presented.

Keywords
Population distribution, infinite population model, fixed points, finite population effects

1 Introduction

This tutorial chapter presents some of the basic theorems that are known about the simple genetic algorithm (SGA, simple GA), derived from the theory known as the *dynamical systems model*. This approach is due largely to the work of M. D. Vose, and the most complete description can be found in his book *The Simple Genetic Algorithm: Foundations and Theory* [6]. Many of the results appeared in earlier journal papers and some of these are cited as appropriate. It should be emphasised that this approach is concerned with what can be proved mathematically about the SGA. The reader should therefore expect the results to be of a general nature, but occasionally finding application to real situations. This complements, and provides a foundation for, the approximate statistical dynamics models of Shapiro, Prügel-Bennett and Rogers (described in this volume on pages 59 and 87).

If we imagine that a population is a point in the space of all possible populations, then the effect of one generation of a GA is to move that population to another point. As the generations go by, so a trajectory of populations is mapped out in the space. In this way, we can consider the action of a GA to be a discrete dynamical system, and we wish to study the properties of the trajectories through population space that different genetic operators produce. Of course, in any actual run of a GA, the trajectory is affected by the random nature of the operators. We would therefore like to know about:

1. The probability distribution of the next population.
2. The expected next population.
3. The long-term behaviour of the population.

The dynamical systems model helps us to address these issues for the so-called SGA. That is, it applies to selection by proportional fitness, bitwise mutation and a range of standard crossover operators.

We will first consider the way in which populations are represented mathematically within this model. Then the effects of proportionate selection, mutation and crossover will each be considered in turn.

2 The Space of Possible Populations

Suppose we have a search space Z containing s elements. We can write Z as the set $\{z_0, \ldots, z_{s-1}\}$. Notice that we start counting at element zero. This pays off when we consider the typical search space of fixed-length binary strings of length ℓ. In this case the element z_k corresponds to the binary string whose value when considered as a binary number is k. For example, if $\ell = 2$ then our search space is given by

$$z_0 = 00$$
$$z_1 = 01$$
$$z_2 = 10$$
$$z_3 = 11.$$

The search space size is given by $s = 2^\ell$.

We can represent a population p by considering the number of copies of each element that p contains as a fraction of the total population size. That is, if there are a_k copies of element z_k in a population of size r then we let

$$p_k = \frac{a_k}{r}.$$

This gives us a vector $p = (p_0, \ldots, p_{s-1})$ representing the population. For example, in the case of binary strings with $\ell = 2$, suppose we have a population comprising 2 copies of 00, 1 copy of 01, 5 copies of 10 and 2 copies of 11. The population size is $r = 10$ and the corresponding population vector is

$$p = (0.2, 0.1, 0.5, 0.2).$$

Any population vector satisfies a number of properties. First, it is an element of the vector space $\mathbb{R}^s$. This means that population vectors can be added together and scaled by real numbers to produce other vectors in $\mathbb{R}^s$ (not necessarily representing populations). Second, each entry p_k of a population vector must lie in the range $0 \leq p_k \leq 1$ since it represents a proportion of a population. Third, p has the property

$$\sum_{k=0}^{s-1} p_k = 1.$$

The set of all vectors in $\mathbb{R}^s$ that satisfy these properties is called the *simplex* and is denoted by Λ. That is

$$\Lambda = \left\{ x \in \mathbb{R}^s : \forall k, x_k \geq 0 \text{ and } \sum_{i=0}^{s-1} x_i = 1 \right\}.$$

Clearly, the set of possible populations of a given size r only takes up a finite subset of Λ. Moreover, Λ contains some vectors that could never be real populations as they contain irrational entries. However, as the population size r tends to infinity, the set of possible populations becomes *dense* in Λ. This means that any element of Λ can be made arbitrarily close to some real population by increasing the population size sufficiently.

Let us consider what the simplex Λ looks like in some simple cases. The simplest interesting case is when the search space has two elements, $Z = \{z_0, z_1\}$. Then any population has a certain proportion α of element z_0 where $0 \leq \alpha \leq 1$ and a corresponding proportion $1 - \alpha$ of z_1. These population vectors live in the space $\mathbb{R}^2$. We can draw the simplex within this space as a segment of a straight line, as shown in figure 1. One end of the line segment corresponds to the population $(1, 0)$ which contains only copies of z_0. The other end is the population $(0, 1)$ which contains only z_1. Other positions between these correspond to different mixtures of z_0 and z_1.

When the search space has three elements, $Z = \{z_0, z_1, z_2\}$, the population vectors belong to the space $\mathbb{R}^3$. The simplex Λ is now a triangle with vertices at $(1, 0, 0)$, $(0, 1, 0)$ and $(0, 0, 1)$ each of which corresponds to a population made up exclusively of copies of one element. For example, $(1, 0, 0)$ is the population containing only z_0. A line joining two vertices corresponds to a population containing a mixture of the respective two elements with no copies of the third element. For example, the line joining $(1, 0, 0)$ to $(0, 1, 0)$ contains populations which are a mixture of z_0 and z_1 with no copies of z_2. The interior of the triangle contains populations which have some copies of each of the three elements. Notice that the simplex is a two-dimensional object (a triangle) embedded in a three-dimensional space.

Moving to a search space of four elements, our vectors now belong to $\mathbb{R}^4$ which is hard to imagine. However, the simplex forms a three-dimensional object embedded in this space. In fact it is a tetrahedron with vertices at the points $(1, 0, 0, 0)$, $(0, 1, 0, 0)$, $(0, 0, 1, 0)$ and $(0, 0, 0, 1)$. Again, these vertices correspond to populations containing copies of only one element. Lines joining two vertices correspond

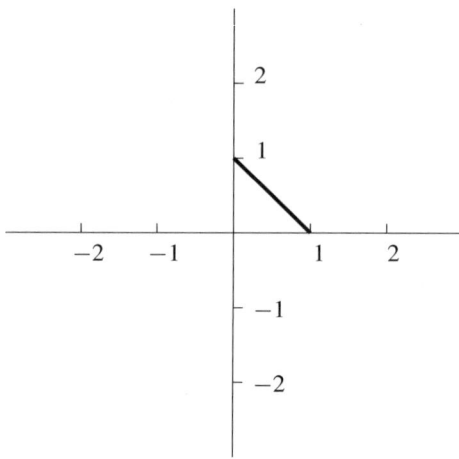

Figure 1. The simplex of possible populations in the case when the search space has only two elements. The axes represent the proportions of each element within the population. The simplex is that set for which these proportions are non-negative and sum to 1.

to mixtures of two elements. The triangular faces correspond to mixtures of three elements, and the interior contains populations with copies of each element. This pattern continues into higher dimensions, where in $\mathbb{R}^s$ the simplex forms a $(s-1)$-dimensional object (a hyper-tetrahedron) whose vertices correspond to populations with copies of only one element. Note that there are therefore s vertices.

When a GA starts, it usually begins with a random population. That is, it starts at a random point in Λ. One generation later, we will have a new population, a new point in Λ. However, because the genetic operators have a random element, we cannot say exactly what the next population will be. What we can say, however, is that there will be a probability distribution over the set of possible populations defined by the genetic operators. We can calculate this, together with the *expected* next population. This will then enable us to study the expected long-term behaviour of the population. It turns out that as the population size r tends to infinity, so the probability that the next population will be the expected one tends to 1. In other words, when the population is infinite, the algorithm is *deterministic* and the trajectory of expected next populations gives the actual behaviour. For this reason, the dynamical systems model is sometimes called the *infinite population* model. However, we can use this model to tell us about the behaviour of finite populations too. The extent to which finite populations approximate the infinite population trajectory is an important issue which we will consider in detail (and see also [1], page 439 in this volume, which addresses the convergence of the finite case to the infinite one).

3 Proportionate Selection

Suppose our current population is $p = (p_0, \ldots, p_{s-1})$, where p_k is the proportion of the population taken up by copies of the element z_k. We wish to calculate the probability that z_k will be selected for the next population. Using fitness proportionate selection, we know this probability is equal to the fitness of z_k divided by the total population fitness, times the number of copies of z_k in the current population. The number of copies of z_k is given by $p_k r$ where r is the population size. Therefore, the probability of selecting z_k is given by

$$\frac{f(z_k) p_k}{\bar{f}(p)}$$

where $f : Z \to \mathbb{R}^+$ is the fitness function and $\bar{f}(p)$ is the *average* fitness of the population. We use this formula as it is independent of the population size r. The average fitness of a population p can be calculated by

$$\bar{f}(p) = \sum_{k=0}^{s-1} f(z_k) p_k.$$

We can therefore create a new vector q containing these probabilities. That is, q_k equals the probability that z_k is selected. We can think of q as being the result of applying an operator $\mathcal{F}$ to p. That is, $q = \mathcal{F} p$. If we define a diagonal matrix S by

$$S_{k,k} = f(z_k)$$

then we can write the following concise formula for q:

$$q = \mathcal{F} p = \frac{1}{\bar{f}(p)} S p. \tag{1}$$

Notice that the probabilities in q define the probability distribution for the next population, if only selection is applied. The next population is found by randomly choosing r elements of the search space Z according to the probabilities in q. This kind of distribution is called a *multinomial* distribution specified by the probabilities $q_0, \ldots, q_{s-1}$. One of the nice properties of a multinomial distribution is that the *expected* next population is simply q itself, interpreted as a population vector. We can therefore apply (1) again to q to get the expected second generation and so on, producing a sequence of populations. It can be proved that, as the population size r tends to infinity, so the actual behaviour of the population gets closer and closer to this sequence [5]. That is, the behaviour of the GA becomes deterministic, and is governed by (1).

Of particular interest to us is the question of what happens to this sequence of populations as selection is applied over and over again. In fact the sequence will converge, and it is straightforward to find the possible limits of the sequence. At the limit of the sequence we have a *fixed-point* of (1). That is, we have some population p such that

$$p = \frac{1}{\bar{f}(p)} Sp$$

and therefore

$$Sp = \bar{f}(p)p.$$

Assume for the moment that all the possible fitness values are different. Since S is a diagonal matrix, it is easy to see that the only solutions to this equation are when p is at a vertex of the simplex. That is, p must comprise copies of just one element. In this case, $\bar{f}(p)$ must be the fitness of that element. There are s such solutions, one for each element of the search space.

It is interesting to ask to which of these fixed-points will the sequence of expected populations converge. Clearly, the answer depends on where the sequence starts. Let us suppose that we start with a population vector that contains some zero entries. That is, there are certain elements of the search space not in the population. The application of (1) to this vector will leave these entries as zero. If there are no copies of an element in one generation, there will also be none in the next. Those entries that are non-zero will be multiplied by $f(z_k)/\bar{f}(p)$. This means that the one with the highest fitness will increase most. Repeated application will continue to magnify this entry until it is the only one left. All the other entries will tend to zero. This means that the population will tend to become full of copies of the element with the highest fitness from the initial population. This corresponds exactly with what we observe experimentally.

To get an idea of what this means in terms of the simplex, let us consider again a search space of four elements. In this case the simplex is a tetrahedron and we have our fixed-points at each vertex. If the initial population is within the simplex, then the sequence of expected populations will converge to the vertex with the highest fitness. If, however, the initial population is on one of the faces, this means that there is one element not present initially. This element will never appear again, and the sequence will take place entirely within that face of the tetrahedron, converging to the vertex of that triangle with the highest fitness. Similarly, if the initial population is on one of the edges, this means that there are only two elements present initially. The sequence will then converge to the vertex of that edge with the greater fitness.

Now let us consider what might happen if some of the fitness values are equal. Then any population which is a mixture of equally-valued elements can be a fixed-point of (1). This means that the sequence of populations will tend to a population containing all the elements that had the highest value in the initial population. In experiments with finite populations, however, sampling errors will lead inevitably to just one of these elements remaining. This is the phenomenon of *genetic drift*.

4 Mutation

We now consider the effects of applying mutation. We want to know the probability that after mutating the individuals that have been selected, we end up with each particular individual. Let's work it out for individual z_i. We can end up with copies of z_i in two ways. First, we select some other individual z_j and then mutate it to produce z_i. The other way is if z_i is selected itself and it is not mutated. We know the probability of selecting any individual z_j is q_j given in (1). Therefore, the probability of ending up with z_i after selection and mutation is

$$\sum_{j=0}^{s-1} U_{i,j} q_j$$

where $U_{i,j}$ is the probability that z_j mutates to form z_i in the case where $i \neq j$. When $i = j$ it is the probability that z_i survives mutation. For example, if we have $\ell = 3$ then $z_0 = 000$ and $z_5 = 101$. In order for 101 to mutate to 000, there needs to be exactly 2 bits mutated (the first and last) and the other bit must not be mutated. If μ is the bitwise mutation rate, then

$$U_{0,5} = \mu^2(1-\mu).$$

We can put all the probabilities $U_{i,j}$ in a matrix, which we will also call U. For example, in the case of $\ell = 2$ we have

$$U = \begin{pmatrix} (1-\mu)^2 & \mu(1-\mu) & \mu(1-\mu) & \mu^2 \\ \mu(1-\mu) & (1-\mu)^2 & \mu^2 & \mu(1-\mu) \\ \mu(1-\mu) & \mu^2 & (1-\mu)^2 & \mu(1-\mu) \\ \mu^2 & \mu(1-\mu) & \mu(1-\mu) & (1-\mu)^2 \end{pmatrix}.$$

Now the probability of ending up with z_i after applying mutation and selection is component i of the vector Uq. Substituting for q from (1) gives us the one time-step equation for a selection plus mutation GA

$$p(t+1) = U \circ \mathcal{F} p(t) = \frac{1}{\bar{f}(p(t))} U S p(t). \tag{2}$$

Once again, we are interested in how this sequence will converge as time t tends to infinity. The population will converge to a fixed-point p satisfying

$$p = \frac{1}{\bar{f}(p)} U S p$$

which, on rearranging, gives

$$U S p = \bar{f}(p) p. \tag{3}$$

This equation has a special form. It says that the matrix US times a vector p produces a multiple of p. In general, whenever one has a matrix A that has the property

$$Ax = \lambda x$$

for some complex number λ and some vector x, then λ is called an *eigenvalue* of A and x is a corresponding *eigenvector*. Notice that any scalar multiple of x is also an eigenvector corresponding to the same eigenvalue λ. Equation 3 states that the fixed-point population p is an eigenvector of the matrix US and that the average fitness of p is a corresponding eigenvalue. We scale p so that its entries sum to 1.

In order that the population p in (3) is a viable one, it must be in the simplex Λ. How do we know which, if any, eigenvalues of US have eigenvectors in Λ? Fortunately, there is a remarkable theorem concerning matrices with positive real entries such as US that helps us answer this question. This is the Perron–Frobenius theorem and it tells us that US will have exactly one eigenvector in Λ, and that this corresponds to the leading eigenvalue (that is, the one with the largest absolute value). The theorem also assures us that this eigenvalue will be real and positive, which is what we require for an average fitness value.

If any of the fitness values are zero, then some entries in US will also be zero. In this case, in addition to the leading eigenvector being in the simplex, there are also the degenerate solutions of populations made up solely of zero-scoring individuals. In these cases the average fitness (and the eigenvalue) is zero. This situation will not arise in practice in a GA, as zero-scoring individuals never get selected under the proportionate selection scheme.

A related term that is sometime encountered is the *spectrum* of a matrix. This is a term that applies to a general class of operators on vector spaces, called *linear operators*. When the vector space concerned has finite dimension (which is the case for us) then linear operators are the same as matrices, and the spectrum of a matrix is the same as the set of eigenvalues. The term *spectral radius* refers to the smallest positive real number r such that $|\lambda| \leq r$ for all elements λ of the spectrum. In our finite-dimensional case, the spectral radius is the same as the leading eigenvalue. The eigenvalues and eigenvectors of a matrix are fundamental to understanding the properties of that matrix.

To summarise what we know about the fixed-point population of the SGA under proportionate selection and bitwise mutation (with rate in the range $0 < \mu < 0.5$), with a positive (non-zero) fitness function:

1. Fixed-points are eigenvectors of US, once they have been scaled so that their components sum to one.
2. Eigenvalues of US give the average fitness of the corresponding fixed-point populations.
3. Exactly one eigenvector of US is in the simplex Λ.
4. This eigenvector corresponds to the leading eigenvalue.

It can also be shown that the sequence $p(0), p(1), p(2), \ldots$ converges to this fixed-point, for any starting population $p(0)$. We say that this GA is *focused*. In fact, this sequence converges to the fixed-point for *any* mutation which, with non-zero probability, can change any member of the search space into any other in a finite number of steps. The mutation matrix is said to be *irreducible* in this case.

Example

Let $\ell = 2$ and consider the following fitness function:

$f(00) = 3$
$f(01) = 2$
$f(10) = 1$
$f(11) = 4.$

The corresponding selection matrix is therefore

$$S = \begin{pmatrix} 3 & 0 & 0 & 0 \\ 0 & 2 & 0 & 0 \\ 0 & 0 & 1 & 0 \\ 0 & 0 & 0 & 4 \end{pmatrix}.$$

Given a mutation rate of 0.1, the mutation matrix U is

$$U = \begin{pmatrix} 0.81 & 0.09 & 0.09 & 0.01 \\ 0.09 & 0.81 & 0.01 & 0.09 \\ 0.09 & 0.01 & 0.81 & 0.09 \\ 0.01 & 0.09 & 0.09 & 0.81 \end{pmatrix}.$$

We are therefore looking for eigenvalues and eigenvectors of the matrix

$$US = \begin{pmatrix} 2.43 & 0.18 & 0.09 & 0.04 \\ 0.27 & 1.62 & 0.01 & 0.36 \\ 0.27 & 0.02 & 0.81 & 0.36 \\ 0.03 & 0.18 & 0.09 & 3.24 \end{pmatrix}.$$

There are a number of computer packages that can calculate the eigensystem of this matrix. After scaling the eigenvectors so that the components add up to 1, we find the following four eigenvectors

(0.0736603, 0.155156, 0.105497, 0.665687)
(0.779133, 0.20572, 0.108063, −0.0929158)
(−0.299921, 1.60114, −0.145272, −0.155948)
(−0.0604414, 0.0239921, 1.0769, −0.0404462)

corresponding to the eigenvectors 3.29954, 2.48524, 1.53345 and 0.781771, respectively. As predicted by the Perron–Frobenius theorem, only one of the eigenvectors (the first) is in the simplex, and it corresponds to the leading eigenvalue. This vector is therefore the predicted fixed-point of the GA, using proportionate selection and a mutation rate of 0.1. The theory predicts that, as the population size tends to infinity, this fixed-point represents the population to which the GA converges. In fact, experiments involving finite populations of modest size will show the GA heading towards this fixed-point.

5 Finite Populations and the Fixed-Point

So far we have predicted that, as the population size tends to infinity, so the GA (with proportionate selection and mutation) will converge to a single fixed-point population. In this section, we will consider some of the effects of having a finite population.

If the population size is r, then each component p_i of a population vector p must be a rational number with r as a denominator. The set of possible finite populations of size r forms a discrete lattice within the simplex Λ. The first consequence of this observation is that the fixed-point population described above might not actually exist as a possible population! How can we be sure, then, that this fixed-point has anything to do with where a finite population GA will end up? Worse still, with non-zero mutation, there is always a chance (even if it is small) that the population will randomly move to any other possible population. In what sense, then, can the algorithm be thought of as converging? We will look at these issues in the case of selection and mutation only. The effects of crossover will be discussed later.

Given a population p, let us define an operator G to be

$$G(p) = U \circ \mathcal{F} p = \frac{1}{\bar{f}(p)} U S p.$$

Then from (2) we have that a population vector at time-step $t + 1$ is

$$p(t + 1) = G(p(t)).$$

This new population vector describes three things:

1. The probability that each individual will be in the next population.
2. The expected next population.
3. The actual next population as $r \to \infty$.

Given an actual (finite) population represented by the vector $p(t)$, we therefore have a probability distribution over all possible next populations, defined by $G(p) = p(t+1)$. It can be shown that the probability that the next population is q given that the current one is p is given by

$$r! \prod_{j=0}^{s-1} \frac{(G(p)_j)^{rq_j}}{(rq_j)!}. \tag{4}$$

Note that $p(t + 1)$ itself might not be an actual population. However, $p(t + 1)$ is the *expected* next population, and we can think of the probability distribution being clustered around that population. Populations that are close to it in the simplex will be more likely to occur as a next population than ones that are far away. As r gets larger, this clustering gets more and more pronounced and, in the limit $r \to \infty$ we have the population $p(t + 1)$ as the certain next actual population. When r is small, though, if the expected next population $p(t + 1)$ is very close to the current population $p(t)$, much closer, in fact, than to any other actual population, then it is probable that the next actual population will be the same as $p(t)$. The GA is likely to stay with this population for some time.

A good way to visualise this is to think of the operator G as defining an arrow at each point in the simplex. The arrow goes from any point p to $G(p)$. What we have just said can now be restated as follows: if a population p has a small arrow coming from it (small, that is, in comparison to the distance to the next possible populations) then it is likely that the GA will stay at p for some time. In particular, we can now use the observation that the operator G is *continuous* when considered as a mapping of the simplex. That is, G will have a very similar effect on points that are close to each other in the simplex. But at the fixed-point of G, there is no arrow at all: it has zero size. Therefore, any actual population that is near the fixed-point will have arrows that are of very small size. This means that if the ends up with a population that is close to the fixed-point calculated as the eigenvector of US, it is likely to stay there for some time. And if it changes, then the next population is likely to also be close to the fixed-point. In other words, the GA is likely to spend much of its time at populations that are in the vicinity of the infinite population fixed-point.

We can formalise the notion of one population being near another, and of arrows being small, by introducing what is called a *norm* on the vector space $\mathbb{R}^s$. A vector $x \in \mathbb{R}^s$ has a norm denoted by $\|x\|$ which is any real-valued function satisfying the following properties:

1. $\|x\| \geq 0$.
2. $\|x\| = 0$ if and only if $x = 0$.
3. $\|\alpha x\| = |\alpha| \|x\|$ for any real number α.
4. $\|x + y\| \leq \|x\| + \|y\|$.

The norm of a vector generalises the idea of 'length'. Notice in particular, the last property which gives a *triangle inequality*. The usual measure of length in $\mathbb{R}^s$ is the formula

$$\|x\| = \left(\sum_{i=0}^{s-1} x_i^2\right)^{\frac{1}{2}}$$

which generalises Pythagoras' theorem. However, we will use a simpler norm:

$$\|x\| = \sum_{i=0}^{s-1} |x_i|$$

which can also be shown to satisfy the four required properties. One of the reasons for preferring this norm is that, for all $p \in \Lambda$, $\|p\| = 1$.

We can use the norm to give us a measure of distance, by defining the distance between two vectors x and y to be $\|x - y\|$. Thus, when we talk about two populations p and q being close to each other, we mean that the value $\|p - q\|$ is sufficiently small. In particular, we are interested in the length of the arrow coming from a population vector p representing the expected next population $G(p)$. This arrow is simply the vector $G(p) - p$ and its length is $\|G(p) - p\|$. This length is also known as the *force* of G at point p. We can now state more formally that the GA is likely to spend some time at populations p such that $\|p - p^*\|$ is small, where p^* is the fixed-point of G (and so the leading eigenvector of US). This is because the force of G will be small for populations that are near to the fixed-point.

Example

Let us consider a simple example involving a 1-bit problem. The fitness function is:

$$f(0) = 1$$
$$f(1) = 2.$$

A mutation rate of 0.1 means that we have

$$US = \begin{pmatrix} 0.9 & 0.1 \\ 0.1 & 0.9 \end{pmatrix} \begin{pmatrix} 1 & 0 \\ 0 & 2 \end{pmatrix} = \begin{pmatrix} 0.9 & 0.2 \\ 0.1 & 1.8 \end{pmatrix}.$$

After scaling (so that their components add to 1), the eigenvectors of US are

$(0.178301, 0.821699)$
$(1.1217, -0.121699)$

corresponding to the eigenvalues 1.8217 and 0.878301, respectively. As previously stated, the only one in the simplex is that corresponding to the leading eigenvalue. This then is the infinite population fixed-point which we will denote p^*. If the population size is 10, then clearly p^* cannot be an actual population. The nearest one to p^* is $(0.2, 0.8)$ which we will denote by p. The distance from p to p^* is

$$\|p - p^*\| = \sum_{j=0}^{1} |p_j - p_j^*|$$
$$= |0.2 - 0.178301| + |0.8 - 0.821699|$$
$$= 0.021699 + 0.021699$$
$$= 0.043398.$$

We can formally justify the observation that this is the closest actual population to p^* by calculating the distances for all possible populations

$\|(0.0, 1.0) - p^*\| = 0.356602$
$\|(0.1, 0.9) - p^*\| = 0.156602$
$\|(0.2, 0.8) - p^*\| = 0.043398$

$\|(0.3, 0.7) - p^*\| = 0.243398$
$\|(0.4, 0.6) - p^*\| = 0.443398$
$\|(0.5, 0.5) - p^*\| = 0.643398$
$\|(0.6, 0.4) - p^*\| = 0.843398$
$\|(0.7, 0.3) - p^*\| = 1.0434$
$\|(0.8, 0.2) - p^*\| = 1.2434$
$\|(0.9, 0.1) - p^*\| = 1.4434$
$\|(1.0, 0.0) - p^*\| = 1.6434.$

According to the earlier discussion, we should expect the distance $\|G(p) - p\|$ to be small, since p is close to p^*. It can easily be calculated that

$$\|G(p) - p\| = \left\| \frac{USp}{\overline{f}(p)} - p \right\|$$
$$= \|(0.188889, 0.811111) - (0.2, 0.8)\|$$
$$= |0.188889 - 0.2| + |0.811111 - 0.8|$$
$$= 0.011111 + 0.011111$$
$$= 0.022222.$$

We would therefore expect the GA to stay in the vicinity of $(0.2, 0.8)$ for some time. In fact, if the current population were $(0.2, 0.8)$ then the probability distribution for the next population can be calculated using formula 4 as follows:

$\Pr[(0.0, 1.0)] = 0.123255$
$\Pr[(0.1, 0.9)] = 0.287032$
$\Pr[(0.2, 0.8)] = 0.300793$
$\Pr[(0.3, 0.7)] = 0.186794$
$\Pr[(0.4, 0.6)] = 0.076125$
$\Pr[(0.5, 0.5)] = 0.021273$
$\Pr[(0.6, 0.4)] = 0.004128$
$\Pr[(0.7, 0.3)] = 0.000549$
$\Pr[(0.8, 0.2)] = 0.000048$
$\Pr[(0.9, 0.1)] = 0.000002$
$\Pr[(1.0, 0.0)] = 0.000000.$

It can be seen that the next population can be expected to be in the vicinity of $(0.2, 0.8)$. And since these neighbouring populations will also have small arrows induced by G then the GA is likely to stay in this area.

6 Metastable States

We have seen that the GA will stay in the vicinity of the infinite population fixed-point, because actual populations have small arrows induced by the operator G. The same argument would apply to any part of the simplex where the force of G is small, even if these areas are nowhere near the fixed-point. Such areas are called *metastable states* and are important in understanding the long-term behaviour of a finite population GA.

The first step in understanding where some of these metastable states may lie, is to think about the operator G in more general terms. Recall that G is defined as

$$G(p) = \frac{1}{\bar{f}(p)} U S p$$

for a given population vector p. We have already used the idea that G can be thought of as acting on the whole of the simplex Λ and not just those points that represent actual finite populations of size r. We now extend G to apply to the whole of $\mathbb{R}^s$. That is, given any vector $x \in \mathbb{R}^s$, we let

$$G(x) = \frac{1}{\bar{f}(x)} U S x$$

where

$$\bar{f}(x) = \sum_{k=0}^{s-1} f(z_k) x_k.$$

There are two observations to make at this point:

1. G is not defined for those vectors x satisfying the equation

    $$\bar{f}(x) = 0$$

    although none of these vectors is in Λ if the fitness function is strictly positive. This set is called the *kernel* of $\bar{f}$, written as

    $$\ker(\bar{f}) = \{x \in \mathbb{R}^s : \bar{f}(x) = 0\}.$$

2. For any $x \in \mathbb{R}^s$ with $x \notin \ker(\bar{f})$

    $$\sum_{k=0}^{s-1} (G(x))_k = 1.$$

    That is, once G has been applied, we always get a vector whose elements sum to 1. This means that, at least after the first application of G to $\mathbb{R}^s$, all the action takes place within the set of vectors whose elements sum to 1. Another way of saying this, is to say that this set of vectors (the image of G) is *invariant* under G. Obviously, Λ lies completely within this set.

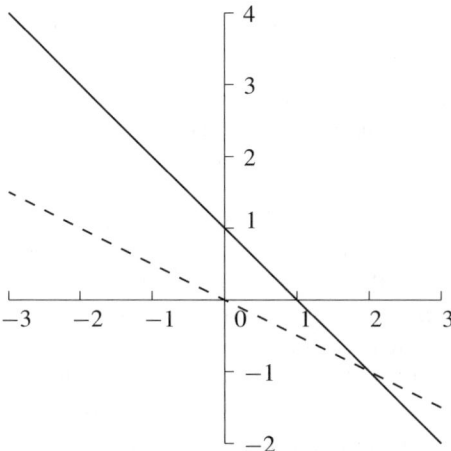

Figure 2. The invariant set of $\mathcal{G}$ (solid line) and the kernel of $\bar{f}$ (dashed line). The axes represent the proportions of each element within the population – actual populations can only exist within the simplex (see figure 1).

This means that we are concerned with the action of G on the set of vectors whose elements sum to 1. However, some of those vectors will have zero fitness, and G is not defined for them. The simplex, Λ, lies within the well-behaved part.

Let us consider what these sets look like for some simple cases. The simplest is when the search space has two elements. We have seen that the simplex is then a line-segment connecting the points $(0, 1)$ and $(1, 0)$. The set of vectors whose elements sum to 1 is just the line which contains that line segment. The set for which G is not defined depends on the fitness function. If we have

$$f(0) = 1$$
$$f(1) = 2$$

then this set is

$$x_0 + 2x_1 = 0$$

which gives us another straight line. These sets are shown in figure 2. Notice that G is not defined for one point of the invariant set, the point $(2, -1)$.

The next simplest case is when the search space contains three elements. Here the simplex is a triangle. The set of vectors whose elements sum to 1 is the plane that contains this triangle. For any fitness function, $\ker(\bar{f})$ defines another plane. These planes will intersect in a straight line. That is, the image set of G contains a straight line on which G is not defined.

In general, when there are s elements in the search space, the simplex Λ forms an $(s-1)$-dimensional tetrahedron. The image set of G, on which G is invariant, is the $(s-1)$-dimensional hyperplane Λ. The set of points for which G is not defined forms another $(s-1)$-dimensional hyperplane. These two hyperplanes intersect in an $(s-2)$-dimensional hyperplane.

Now we can consider an arrow at each point x of the invariant hyperplane (except where G is not defined), pointing to $G(x)$. The length of this arrow is $\|G(x)-x\|$ and is the force of G at x. We are interested in areas for which the force is small. If such areas are near to the simplex then, by the continuity of G, that neighbouring part of the simplex must also have a small force and hence a metastable state. We have already found one such area inside the simplex: it is the vicinity of the fixed-point within the simplex. Recall that the Perron–Frobenius theory predicts exactly one such fixed-point. But now we are considering the action of G on a larger space. So if there are other fixed-points in this larger space, then there will be associated regions of small force. If these fixed-points are close to the simplex then, by continuity, there will be a metastable region in that part of the simplex.

So what are these other fixed-points of G? They are simply the other eigenvectors of US suitably scaled so that their components sum to 1. Suppose that v is any eigenvector of US whose components add to 1. For convenience, we define a function $h : \mathbb{R}^s \to \mathbb{R}$ to be

$$h(x) = \sum_{k=0}^{s-1} x_k.$$

It is simple to check that h is linear. Now we have

$$USv = \lambda v$$

where λ is an eigenvalue of US and $h(v) = 1$. Therefore

$$\frac{USv}{\bar{f}(v)} = \frac{\lambda v}{\bar{f}(v)}$$

and so

$$G(v) = \frac{\lambda v}{\bar{f}(v)}.$$

Applying h to both sides (that is, summing components) gives

$$h(G(v)) = h\left(\frac{\lambda v}{\bar{f}(v)}\right).$$

But we know that the $h(G(x)) = 1$ for any vector x for which G is defined. Therefore

$$h\left(\frac{\lambda v}{\bar{f}(v)}\right) = 1$$

and since h is linear

$$\frac{\lambda}{\bar{f}(v)} h(v) = 1.$$

We have scaled the eigenvector v such that $h(v) = 1$ which must mean that $\lambda = \bar{f}(v)$. Therefore $G(v) = v$ and v is a fixed-point of G.

To find potential metastable states within the simplex then, we simply calculate all the eigenvectors of US and scale them so that their components sum to 1. If any of these are close to the simplex then, by continuity of G, we would expect a metastable region in that part of the simplex nearest the fixed-point.

Example

Suppose our search space is the set of binary strings with $\ell = 3$. The fitness function is based on the hamming distance of a string from the string 000, and is as follows:

$f(000) = 3$
$f(001) = 2$
$f(010) = 2$
$f(011) = 1$
$f(100) = 2$
$f(101) = 1$
$f(110) = 1$
$f(111) = 4.$

The optimum is 111 but there is a 'false optimum' at 000. The mutation rate is set to 0.01. The matrices U and S can be calculated in a straightforward manner, and it will be found that the first two eigenvalues of US are 3.8816 and 2.9127. The corresponding eigenvectors (scaled so that their components sum to 1) are

(0.00003, 0.00033, 0.00033, 0.01294, 0.00033, 0.01294, 0.01294, 0.96019)

and

(0.91485, 0.0277, 0.02767, 0.0007, 0.02767, 0.0007, 0.0007, −0.00004),

respectively. The first of these is the one in the simplex, predicted by the Perron–Frobenius theorem. It consists predominantly of copies of the optimum string (96% of the fixed-point population) with an average fitness of 3.8816. However, the second eigenvector is extremely close to the simplex. According to the argument given above, then, the force of G around this point will be small. This will include a region of the simplex, and we should expect to see a metastable state here. Notice that this eigenvector consists predominantly of copies of the false optimum 000. Since average fitness is a continuous function, we would expect the metastable state to have an average fitness of approximately 2.9127. Checking the other eigenvectors reveals that they are all a long distance from the simplex and so are not the source of local metastable states within the simplex.

The long-term behaviour of the GA can now be described. The population will spend most of the time in a metastable state. At least one such state exists in the vicinity of the infinite population fixed-point. If there are any eigenvectors close to Λ then these too can be sources of other metastable states. Since the GA is a stochastic algorithm, chance mutations will occasionally shift the population from one metastable state to another (see [3] for a particularly clear example). However, it should be realised that there may well be other regions of the simplex which have a small force, which will also produce metastable states. The eigenvectors only account for some of them. For example, see [2] for a whole line segment within the simplex that acts as a metastable region.

7 Crossover and the Mixing-Matrix

So far we have studied the action of proportionate selection and bitwise mutation in the SGA. We have found that it is useful to think of these in terms of operators that act upon the simplex Λ. The operator for proportionate selection is

$$\mathcal{F}p = \frac{1}{\bar{f}(p)} Sp$$

where $\bar{f}(p)$ is the average fitness of the population p and S is a diagonal matrix containing all the possible fitness values on the diagonal. The operator for mutation is a matrix U where $U_{i,j}$ is the probability that individual z_j will mutate to become z_i. The net result of applying these operators is given by the operator $G = U \circ \mathcal{F}$. Given a population $p \in \Lambda$, then $G(p)$ has the following properties:

1. It defines the multinomial probability distribution for the next generation.
2. It gives the *expected* next population.
3. In the infinite population limit, it gives the *actual* next population.

We will now consider the effects of crossover, again in terms of an operator acting upon Λ. This operator we will call $\mathcal{C}$. Given a population $p \in \Lambda$, $(\mathcal{C}p)_k$ will give the probability of individual z_k being in the next generation. The overall action of proportional selection, mutation and crossover is given by the operator $\mathcal{G}$ where

$$\mathcal{G} = \mathcal{C} \circ U \circ \mathcal{F}.$$

$\mathcal{G}(p)$ also has the three properties just given, but now incorporating the effects of crossover. It turns out that for all the usual kinds of crossover that are used in GAs, the order of crossover and mutation doesn't matter. That is, one can apply mutation to two parent individuals and then cross them to give a child, or one can cross two parents and then mutate the child. Either way, the probability of creating a particular individual is the same. It is straightforward then to show that $\mathcal{C} \circ U = U \circ \mathcal{C}$. This combination of crossover and mutation (in either order) gives what is called the *mixing scheme* for the GA, denoted by $\mathcal{M}$. That is

$$\mathcal{M} = \mathcal{C} \circ U = U \circ \mathcal{C}$$

and so $\mathcal{G} = \mathcal{M} \circ \mathcal{F}$.

We will first look at some properties of the crossover operator $\mathcal{C}$ and then extend them to include mutation effects as well. To begin with, let's think about crossover in very general terms. Given a population distribution $p \in \Lambda$ from which the first parent is drawn, and $q \in \Lambda$ for the second parent, let $\omega(p, q) \in \Lambda$ be the population distribution resulting from crossing over random parents from p and q.

Let e_i be the vector with a 1 at position i and zeros elsewhere. Thus e_i represents a population comprised entirely of copies of z_i. Then the kth component of $\omega(e_i, e_j)$ is equal to the probability that crossing z_i and z_j will produce z_k. Therefore

$$\omega(p, q)_k = \sum_{i,j} p_i q_j \omega(e_i, e_j)_k$$

and thus

$$\omega(p, q) = \sum_k \left(\sum_{i,j} p_i q_j \omega(e_i, e_j)_k \right) e_k$$
$$= \sum_{i,j} p_i q_j \sum_k \omega(e_i, e_j)_k e_k$$
$$= \sum_{i,j} p_i q_j \omega(e_i, e_j).$$

$\omega(p, q)$ is thus completely determined by the vectors $\omega(e_i, e_j)$, and this definition can be naturally extended over the whole of $\mathbb{R}^n$. It is an example of a *bilinear* operator. That is, if $x, y, z \in \mathbb{R}^n$ and α, β are scalars, then

$$\omega(\alpha x + \beta y, z) = \alpha \omega(x, z) + \beta \omega(y, z)$$

and

$$\omega(z, \alpha x + \beta y) = \alpha \omega(z, x) + \beta \omega(z, y).$$

An operator must also be bounded to be bilinear, but this condition always holds in finite-dimensions. Note that ω is commutative, since $\omega(x, y) = \omega(y, x)$, and that we usually require $\omega(e_k, e_k) = e_k$ for all k.

Once we have calculated all the vectors $\omega(e_i, e_j)$ for a particular kind of crossover, we can then define the operator $\mathcal{C}$ to be

$$\mathcal{C}(p) = \omega(p, p) = \sum_{i,j} p_i p_j \omega(e_i, e_j).$$

One of the consequence of bilinearity is that for any scalar $\alpha \in \mathbb{R}$

$$\mathcal{C}(\alpha p) = \omega(\alpha p, \alpha p)$$
$$= \alpha \omega(p, \alpha p)$$
$$= \alpha^2 \omega(p, p)$$
$$= \alpha^2 \mathcal{C}(p).$$

In particular, when $\mathcal{C}$ is combined with U and $\mathcal{F}$ we get

$$\begin{aligned}\mathcal{G}(p) &= (\mathcal{M} \circ \mathcal{F})(p) \\ &= (\mathcal{C} \circ U \circ \mathcal{F})(p) \\ &= \mathcal{C}\left(\frac{1}{\bar{f}(p)} U S p\right) \\ &= \frac{1}{(\bar{f}(p))^2} \mathcal{C}(U S p) \\ &= \frac{1}{(\bar{f}(p))^2} \mathcal{M}(S p).\end{aligned}$$

We can put the definition of $\mathcal{C}$ into a more useful form by constructing s matrices, one for each individual z_k. We simply define C_k to be the matrix whose $(i, j)^{\text{th}}$ entry is $\omega_{i,j}$, the probability that crossing z_i with z_j will produce z_k. Then

$$\begin{aligned}\mathcal{C}(p)_k &= \sum_{i,j} p_i p_j \omega(e_i, e_j)_k \\ &= \sum_i p_i \sum_j \omega(e_i, e_j)_k p_j \\ &= p \cdot (C_k p)\end{aligned}$$

where $x \cdot y$ indicates the *dot product* of the vectors x and y defined by

$$x \cdot y = \sum_{k=0}^{s-1} x_k y_k.$$

Example

Let our search space be the set of binary strings with $\ell = 2$

$$\begin{aligned}z_0 &= 00 \\ z_1 &= 01 \\ z_2 &= 10 \\ z_3 &= 11.\end{aligned}$$

Crossover is uniform, and we assume a crossover rate of 1. We therefore need to calculate the entries of the four matrices C_0, C_1, C_2, C_3. The $(i, j)^{\text{th}}$ entry of C_k is $\omega(e_i, e_j)_k$, the probability that z_i and z_j will cross to form z_k. It is easy to calculate

$$C_0 = \begin{pmatrix} 1.0 & 0.5 & 0.5 & 0.25 \\ 0.5 & 0.0 & 0.25 & 0.0 \\ 0.5 & 0.25 & 0.0 & 0.0 \\ 0.25 & 0.0 & 0.0 & 0.0 \end{pmatrix}$$

$$C_1 = \begin{pmatrix} 0.0 & 0.5 & 0.0 & 0.25 \\ 0.5 & 1.0 & 0.25 & 0.5 \\ 0.0 & 0.25 & 0.0 & 0.0 \\ 0.25 & 0.5 & 0.0 & 0.0 \end{pmatrix}$$

$$C_2 = \begin{pmatrix} 0.0 & 0.0 & 0.5 & 0.25 \\ 0.0 & 0.0 & 0.25 & 0.0 \\ 0.5 & 0.25 & 1.0 & 0.5 \\ 0.25 & 0.0 & 0.5 & 0.0 \end{pmatrix}$$

$$C_3 = \begin{pmatrix} 0.0 & 0.0 & 0.0 & 0.25 \\ 0.0 & 0.0 & 0.25 & 0.5 \\ 0.0 & 0.25 & 0.0 & 0.5 \\ 0.25 & 0.5 & 0.5 & 1.0 \end{pmatrix}$$

and so the operator $\mathcal{C}$ is given by

$$\mathcal{C}(p) = (p \cdot C_0 p, \, p \cdot C_1 p, \, p \cdot C_2 p, \, p \cdot C_3 p).$$

Notice the symmetry amongst the matrices in the example. It looks like the matrices are basically the same, except for some swapping of rows and columns. This turns out to be generally true for any of the usual crossovers defined on binary strings, and indeed for higher cardinality strings. The reason for this is that there is an underlying symmetry within the action of crossover itself. Consider the probability that 00 and 11 will cross to produce 10. To achieve this requires taking the first bit of 11 and the second bit of 00, giving a probability of 0.25. The situation is identical to crossing 10 and 01 to produce 00. All that has happened is that the bit values have been *relabelled*. We have performed the following relabelling

$$0* \longmapsto 1*$$
$$1* \longmapsto 0*$$
$$*0 \longmapsto *0$$
$$*1 \longmapsto *1.$$

We can express this relabelling by saying that the string i becomes $i \oplus 10$, where the symbol $\oplus$ means bitwise exclusive-or (or, equivalently, bitwise addition modulo 2). That is

$$00 \longmapsto 00 \oplus 10 = 10$$
$$01 \longmapsto 01 \oplus 10 = 11$$
$$10 \longmapsto 10 \oplus 10 = 00$$
$$11 \longmapsto 11 \oplus 10 = 01.$$

We can now see straight away that 00 and 11 crossing to form 10 is the same question as 10 and 01 crossing to form 00.

Let us take a more complicated example with $\ell = 8$. Suppose we are looking at the probability that 11010011 and 00100001 cross to produce 01110011. If we use this third bitstring as an exclusive-or mask, then we have the following relabelling

$$11010011 \longmapsto 11010011 \oplus 01110011 = 10100000$$
$$00100001 \longmapsto 00100001 \oplus 01110011 = 01010010$$
$$01110011 \longmapsto 01110011 \oplus 01110011 = 00000000.$$

This situation is the same as asking for the probability that 10100000 and 01010010 will cross to produce 00000000.

The same process holds in general. If we are looking for the $(i, j)^{\text{th}}$ entry of C_k (which is the probability that z_i and z_j will cross to produce z_k), then we can use z_k as an exclusive-or mask. The question is thus equivalent to the probability that $z_i \oplus z_k$ and $z_j \oplus z_k$ will cross to produce $z_k \oplus z_k$. But $z_k \oplus z_k = z_0$ for all k. Therefore we have the $(i, j)^{\text{th}}$ entry of C_k equals the $(z_i \oplus z_k, z_j \oplus z_k)^{\text{th}}$ entry of C_0. So all the information about crossover is held within the matrix C_0! All the other matrices are simple permutations of this one. For each $k = 0, \ldots, s - 1$ we define a matrix

$$\sigma_k = [z_i \oplus z_k = z_j]$$

where $[expr]$ evaluates to 1 if $expr$ is true and zero otherwise. Then

$$p \cdot C_k p = (\sigma_k p) \cdot C_0 (\sigma_k p).$$

The matrix σ_k does the required relabelling.

In our example with $\ell = 2$, these matrices take the form

$$\sigma_0 = \begin{pmatrix} 1 & 0 & 0 & 0 \\ 0 & 1 & 0 & 0 \\ 0 & 0 & 1 & 0 \\ 0 & 0 & 0 & 1 \end{pmatrix}$$

$$\sigma_1 = \begin{pmatrix} 0 & 1 & 0 & 0 \\ 1 & 0 & 0 & 0 \\ 0 & 0 & 0 & 1 \\ 0 & 0 & 1 & 0 \end{pmatrix}$$

$$\sigma_2 = \begin{pmatrix} 0 & 0 & 1 & 0 \\ 0 & 0 & 0 & 1 \\ 1 & 0 & 0 & 0 \\ 0 & 1 & 0 & 0 \end{pmatrix}$$

$$\sigma_3 = \begin{pmatrix} 0 & 0 & 0 & 1 \\ 0 & 0 & 1 & 0 \\ 0 & 1 & 0 & 0 \\ 1 & 0 & 0 & 0 \end{pmatrix}.$$

As we would expect, σ_0 is the identity – there is no relabelling required in this case!

We have shown that the crossover operator $\mathcal{C}$ can be written in terms of the matrix C_0 and a set of permutation matrices σ_k

$$\mathcal{C}(p)_k = (\sigma_k p) \cdot C_0(\sigma_k p)$$

In fact, there is nothing special about C_0 here. By associating each k with different exclusive-or masks, we can express $\mathcal{C}$ in terms of any of the matrices C_k and a corresponding set of permutation matrices. Choosing to use C_0 seems natural, though, as then the exclusive-or mask associated with individual z_k is z_k itself.

We now bring mutation back into consideration. This gives us the mixing scheme, $\mathcal{M}$, for the GA

$$\mathcal{M} = \mathcal{C} \circ \mathcal{U}.$$

The kth component of $\mathcal{M}p$ is

$$\mathcal{M}(p)_k = \mathcal{C}(Up)_k = (Up) \cdot (C_k Up)$$

Since U is a real symmetric matrix, it has the property that $(Ux) \cdot y = x \cdot (Uy)$ for any vectors x and y. Therefore we have

$$\mathcal{M}(p)_k = p \cdot (UC_k Up). \tag{5}$$

Now we know that

$$x \cdot (C_k x) = (\sigma_k x) \cdot (C_0 \sigma_k x)$$

for any vector x. The matrices σ_k are *orthogonal* (this means that $\sigma_k^{-1} = \sigma_k^T$) and orthogonal matrices have the property of preserving the dot product. That is, $(\sigma_k x) \cdot (\sigma_k y) = x \cdot y$ for all vectors x and y. Therefore

$$x \cdot (C_k x) = (\sigma_k x) \cdot (C_0 \sigma_k x)$$
$$= x \cdot (\sigma_k^T C_0 \sigma_k x).$$

Since this is true for all x, it follows that

$$C_k = \sigma_k^T C_0 \sigma_k. \tag{6}$$

Substituting in (5) gives

$$\mathcal{M}(p)_k = p \cdot (U \sigma_k^T C_0 \sigma_k U p)$$
$$= p \cdot (\sigma_k^T U C_0 U \sigma_k p)$$
$$= (\sigma_k p) \cdot (U C_0 U \sigma_k p)$$

where we have used the useful property that the matrices σ_k commute with the mutation matrix U.

To help sum up, let us define

$$M_k = U C_k U.$$

The $(i, j)^{\text{th}}$ entry of M_k is the probability that z_i and z_j, after being mutated and crossed, produce z_k. The mixing scheme is given by

$$\mathcal{M}(p)_k = p \cdot (M_k p) = (\sigma_k p) \cdot (M_0 \sigma_k p).$$

All the information about mutating and crossing is held in the matrix M_0. This matrix is called the *mixing matrix* of the mixing scheme $\mathcal{M}$ and is usually denoted by simply M. The action of the GA for one generation is given by the equation

$$\mathcal{G}(p) = \mathcal{M} \circ \mathcal{F}(p).$$

8 Mixing and the Walsh Transform

If we interpret binary strings as being vectors in $\mathbb{R}^\ell$ then we can extend the notion of a dot product. If x and y are binary strings then

$$x \cdot y = \sum_{k=0}^{\ell-1} x_k y_k$$

where x_k is the kth bit of the string x. Using this notation, the *Walsh transform* is defined by the matrix

$$W_{i,j} = \frac{1}{\sqrt{s}} (-1)^{z_i \cdot z_j}$$

for $i, j = 0, \ldots (s-1)$. W is a real symmetric orthogonal matrix. It is therefore its own inverse, that is

$$W^2 = I.$$

We can represent our fitness function $f : Z \to \mathbb{R}^+$ as a vector

$$f = (f(z_0), f(z_1), \ldots, f(z_{s-1}))$$

then $f \in \mathbb{R}^s$. The effect of applying the Walsh transform to f is the vector Wf which contains all the *Walsh coefficients* associated with the fitness function. In addition to producing the Walsh coefficients, however, W has some remarkable properties related to the mixing scheme $\mathcal{M}$.

The first property relates to the permutation matrices σ_k. It can be shown that for each k, the matrix $W\sigma_k W$ is diagonal. Moreover, the entries along the diagonal are the same as the kth row of W, multiplied by $\sqrt{s}$.

The second property is that the transform of the mixing matrix, WMW is sparse. The proportion of non-zero entries in WMW is $(3/4)^\ell$. If there is no mutation, then the mutation matrix $U = I$ and the mixing matrix $M = C_0$. It can be shown that, in this case, $WMW = M$.

These properties give us an efficient way to calculate the effects of the operator $\mathcal{M}$. Recall that

$$\mathcal{M}(p)_k = (\sigma_k p) \cdot (M \sigma_k p)$$

for each k. Now orthogonal matrices (such as W and σk) preserve the dot product, so we can write

$$\mathcal{M}(p)_k = (W \sigma_k p) \cdot (W M \sigma_k p).$$

Now, using the fact that $W^2 = I$ we get

$$\mathcal{M}(p)_k = (W \sigma_k W^2 p) \cdot (W M W^2 \sigma_k W^2 p).$$

It is convenient to denote the Walsh transform of any matrix A by $\hat{A} = WAW$, and the transform of any vector x by $\hat{x} = Wx$. Then we have

$$\mathcal{M}(p)_k = (\hat{\sigma}_k \hat{p}) \cdot (\hat{M} \hat{\sigma}_k \hat{p})$$

where $\hat{M}$ is sparse and $\hat{\sigma k}$ is diagonal. This gives us an efficient method for calculating $\mathcal{M}p$. This means that we can efficiently calculate $\mathcal{G}p$, since

$$\mathcal{G}(p) = \mathcal{M}(\mathcal{F}(p)) = \mathcal{M}\left(\frac{1}{\bar{f}(p)} Sp\right).$$

Therefore we need to take the Walsh transform of $\mathcal{F}(p)$. This is particularly interesting for the first generation of the GA. Let's assume that the population is chosen uniformly at random from the search space. In this case

$$p = (\frac{1}{s}, \frac{1}{s}, \ldots, \frac{1}{s}) = \frac{1}{s}\mathbf{1}$$

where $\mathbf{1}$ is the vector containing all ones. Then the Walsh transform of $\mathcal{F}(p)$ is

$$\begin{aligned}
W \frac{1}{\bar{f}(p)} Sp &= \frac{1}{\bar{f}(p)} W Sp \\
&= \frac{1}{\bar{f}(p)} W S \frac{1}{s} \mathbf{1} \\
&= \frac{1}{s \bar{f}(p)} W S \mathbf{1} \\
&= \frac{1}{s \bar{f}(p)} W f \\
&= \frac{1}{s \bar{f}(p)} \hat{f} \\
&= \frac{\hat{f}}{f \cdot \mathbf{1}}
\end{aligned}$$

which is the Walsh transform of the fitness function, considered as a vector, divided by the sum of all possible fitness values.

When the search space consists of strings whose elements have cardinality $c > 2$, the results concerning the Walsh transform generalise nicely. Full details of this theory can be found in [8].

9 Properties and Conjectures Concerning $\mathcal{G}$

The principle conjecture for the SGA is that $\mathcal{G}$ is *focused* under reasonable assumptions about crossover and mutation. That is, given any population vector $p \in \Lambda$, the sequence $p, \mathcal{G}p, \mathcal{G}^2 p, \ldots$ converges (though remember that this doesn't mean that the *finite* GA converges). This is known to be true if mutation is defined bitwise with a mutation rate < 0.5 and there is no crossover [4]. When there is crossover, it is known to be true provided the fitness function is linear (or near to linear) and the mutation rate is small [7]. Experiments with alternative ways of defining mutation (that is, alternative ways of setting the probabilities in U) can lead to counterexamples [9]. So for the conjecture to be true, there must be something special about using bitwise mutation, with a rate < 0.5 that makes it so.

A second conjecture is that fixed-points of $\mathcal{G}$ will be *hyperbolic* for nearly all fitness functions. A fixed-point x is said to be hyperbolic if the differential of $\mathcal{G}$ at the point x has no eigenvalue with modulus equal to 1. Hyperbolicity is an important concept in determining the stability of fixed-points. Saying that this will be the case for 'nearly all' fitness functions means that the set of fitness functions for which it is true is dense in the set of all possible fitness functions. This means if you happen to construct a fitness function for which it is not true, then there will be another fitness function, arbitrarily close to the first (when they are viewed as vectors) for which fixed-points are hyperbolic. This conjecture is already known to be true for the case of fixed-length binary strings, proportionate selection, any kind of crossover, and mutation defined bitwise with a positive mutation rate (proved by Mary Eberlein in her doctoral dissertation, University of Tennessee, Knoxville, 1996).

Any operator on Λ is called *well-behaved* if it always maps volumes of Λ into other volumes. That is, the image of a volume never has fewer dimensions than that volume. $\mathcal{G}$ is known to be well-behaved if the mutation rate is positive but < 0.5 and if crossover is applied at a rate that is less than 1.

Assuming $\mathcal{G}$ is focused, well-behaved and has hyperbolic fixed-points then the following properties follow (see [4,6] for details):

1. There are only finitely many fixed-points of $\mathcal{G}$.
2. The probability of picking a population p, such that iterates of $\mathcal{G}$ applied to p converge on an *unstable* fixed-point, is zero.
3. The infinite population GA converges to a fixed-point in logarithmic time.

It is clear that the fundamental conjecture is that $\mathcal{G}$ is focused for the SGA, for nearly all fitness functions. Without knowing this, we still do not know that there are only finitely many fixed-points! Of course, should it be proved that there are infinitely many fixed-points, for a set of fitness functions that is not nowhere-dense, then $\mathcal{G}$ cannot be focused.

The possibility of *metastable* states remains as in the case of zero crossover. However, as there are currently no good ways to determine fixed-points of $\mathcal{G}$ beyond repeated iteration using the Walsh transform, this has had very little study in comparison with the mutation-only case.

References

1. P. Del Moral and L. Miclo. Asymptotic results for genetic algorithms with applications to nonlinear estimation. In this volume, page 439.
2. J. E. Rowe. Population fixed-points for functions of unitation. In W. Banzhaf and C. R. Reeves, editors, *Foundations of Genetic Algorithms 5*, pages 69–84. Morgan Kaufmann, San Francisco, 1998.
3. E. van Nimwegen, J. P. Crutchfield, and M. Mitchell. Statistical dynamics of the Royal Road genetic algorithm. Theoretical Computer Science, 229(1-2):41-102, 1999.
4. M. D. Vose. Logarithmic convergence of random heuristic search. *Evolutionary Computation*, 4(4):395–404, 1997.
5. M. D. Vose. Random heuristic search. *Theoretical Computer Science*, 229(1-2):103-142, 1999.
6. M. D. Vose. *The Simple Genetic Algorithm*. MIT Press, Cambridge, MA, 1999.
7. M. D. Vose and A. H. Wright. Simple genetic algorithms with linear fitness. *Evolutionary Computation*, 2(4):347–368, 1995.
8. M. D. Vose and A. H. Wright. The Walsh transform and the theory of the simple genetic algorithm. In S. Pal and P. Wang, editors, *Genetic Algorithms for Pattern Recognition*, pages 25–43. CRC Press, 1996.
9. A. H. Wright and G. Bidwell. A search for counterexamples to two conjectures on the simple genetic algorithm. In R. K. Belew and M. D. Vose, editors, *Foundations of Genetic Algorithms 4*, pages 73–81. Morgan Kaufmann, San Francisco, 1996.

Modelling Genetic Algorithm Dynamics

A. Prügel-Bennett and A. Rogers

Department of Electronics and Computer Science
University of Southampton
Southampton SO17 1BJ, UK
E-mail: *apb@ecs.soton.ac.uk*

Abstract. This tutorial is an introduction to the mathematical modelling of the dynamics of genetic algorithms (GAs). The distinguishing feature of this approach is that we consider *macroscopic* properties of the system. After some brief introductory remarks, we look at a generational GA, with tournament selection, tackling the ones-counting problem. Initially we ignore recombination. We start with a two-parameter model of the evolution. This is sufficient to explain the qualitative features of the dynamics, although it does not give a good quantitative agreement with simulations. We show how the agreement can be improved by using more parameters to describe the population and by introducing finite population corrections. Finally, we come back to recombination and show how this can be modelled.

Keywords
Mathematical modelling, dynamics, ensembles, ones-counting, finite population effects

1 Why Model?

The aim of mathematical modelling is to gain insights into how things work. For genetic algorithms (GAs) we would like to know what problems are amenable to their use, what representation and operators to use, and what are the optimal parameter settings. Modelling will not give precise answers to these questions as they depend on the problem being tackled. Nevertheless, modelling gives us an understanding which we can apply to new problems. To obtain a deeper understanding of how GAs work we must not only model the algorithm but also model different types of problems. This, however, goes beyond the scope of this paper.

This paper is intended as a tutorial in how to accurately model the dynamics of a simple GA. The model is, by necessity, only an approximation of reality. By making an accurate model we can convince ourselves that all the important features have been captured. But accuracy often comes at the price of obfuscating the underlying process. A good model should retain the essential features of the process being modelled but with the minimum of detail necessary. In what follows we start from a minimalist model, adding details as we delve deeper into what is happening.

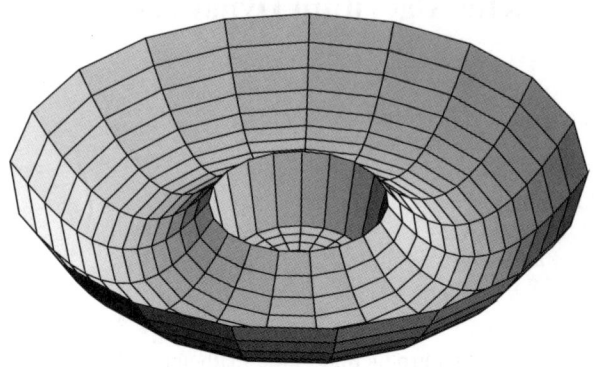

Figure 1. Basin with a barrier in two dimensions

So, what can we reasonably hope to gain from our efforts? Even when modelling simple problems we might expect to compare different strategies such as steady-state versus generational GAs or roulette wheel versus stochastic universal sampling. We might be able to understand how the dynamics is affected by changing a parameter. Although this won't necessarily tell us the optimal parameters for a different problem, it might tell us what we should look for when tuning our parameters. If we can understand how a GA actually searches a complex search space we might even get insights into what are useful representations and which problems can be successfully tackled using GAs.

2 What to Model

Mathematical modelling can be hard work. It makes sense therefore to start simple. Perhaps the simplest problem of all is ones-counting, consisting of a string of ones and zeros, where the fitness is given by the number of ones. In this introduction we will use ones-counting as our example. After all, if we cannot understand ones-counting we have little hope of understanding other models. And we don't understand ones-counting. For example, in a GA with just mutation and selection, how does the equilibrium fitness depend on the string length, the selection rate and the population size? Nobody knows.

Modelling is not, however, confined to ones-counting. The formalism described here has been applied to: a modification of ones-counting where each site contributes differently to the fitness [12]; a model where the fitness depends on the states of the neighbouring loci and has exponentially many local optima [11,12]; the subset-sum problem [13]; learning the weights of a perceptron [14]; and a basin with a barrier [18,17]. This last problem is presented later in this volume, page 207.

3 How to Model

3.1 Naming the Parts

We consider the ones-counting problem where the objective is to maximise the number of ones in a binary string of length L, denoted by $\vec{S} = (S_1, S_2, \ldots, S_L)$. The 'fitness' of a string is given by the number of 1's

$$E = \sum_{i=1}^{L} S_i.$$

To solve this we use a simple generational GA. The size of the population is fixed. We label the members of the population by $\alpha = 1, 2, \ldots, P$. The string of member α we denote by $\vec{S}^\alpha$ and the fitness by E_α. The GA evolves by selection, mutation, and recombination. We use binary tournament selection where two members of the population are drawn at random and the fittest of the two is copied into a mating pool. This is carried out P times. Each member of the mating pool undergoes mutation. In mutation we decide at each site of the string to change its state with a probability u

$$S_i^\alpha \to S_i^{\alpha'} = \begin{cases} S_i^\alpha & \text{with probability } 1 - u \\ 1 - S_i^\alpha & \text{with probability } u. \end{cases}$$

Recombination is performed by pairing the members of the mating pool and generating two children from each pair using uniform crossover. That is, from each pair $\{\vec{S}^\alpha, \vec{S}^\beta\}$ we produce a pair of strings $\{\vec{S}^\mu, \vec{S}^\nu\}$ where

$$S_i^\mu = \begin{cases} S_i^\alpha & \text{with probability } a \\ S_i^\beta & \text{with probability } 1 - a. \end{cases}$$

and $\vec{S}^\nu$ is the complementary child. We usually take $a = 1/2$. If there is no recombination the new population is set equal to the mating pool after selection and mutation. We now have a well-defined system to be modelled. It depends on the parameters L, P, u and whether or not we perform crossover. But how are we going to describe the dynamics? We give an outline of our approach next.

3.2 Modelling Philosophy

Figure 2 shows the distribution of fitnesses for a population evolving under selection and mutation. The overall trend is clear: the population is initially widely distributed around $L/2$ – the mean fitness of a random string. As the population evolves the average fitness increases while the width of the population narrows. Clearly, many of the details, such as the small scale fluctuations, will vary from run to run. The property we are usually most interested in is the maximum fitness. However, the

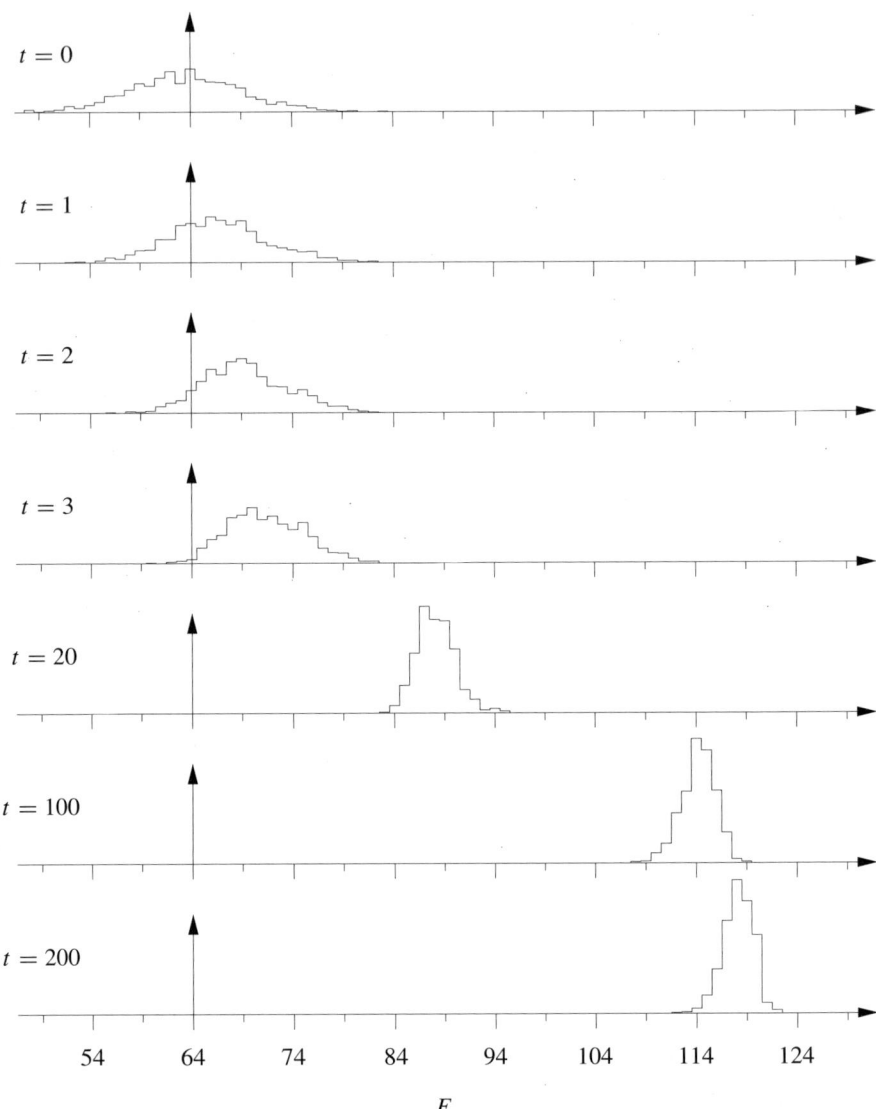

Figure 2. Histograms of the fitnesses of a population after 0, 1, 2, 3, 20, 100, and 200 generations. The figures come from a simulation with $L = 128$, $P = 1000$, and $u = 1/L$. There was no crossover.

evolution of the population will depend more on the pack than the leader. Thus, for modelling, we will focus on coarse statistical properties of the population such as the average fitness and width of the population. If we wish to know the maximum fitness in the population we can estimate this from the statistical properties.

The distinguishing feature of this approach is the use of macroscopic properties of the population to model the dynamics. Because the macroscopic properties don't fully describe the system, we may need to infer other properties. We can do this on the basis of what are the most likely values consistent with the known macroscopic values. One way of making such inferences is using maximum entropy arguments. For the simple model discussed here we can get away without resorting to maximum entropy. The use of maximum entropy is discussed in detail by Shapiro in this volume, page 87. For this brief overview we consider as macroscopic variables the mean fitness, variance in the fitness and later the skewness and correlation. In doing so we are making implicit assumptions about other properties of populations. For example, we are assuming that all the higher order statistics are zero. The results will therefore be approximate, although we can improve the approximation by including more macroscopic variables.

The most important statistical properties of the population are the mean fitness, μ, and the variance, σ^2

$$\mu = \frac{1}{P} \sum_{\alpha=1}^{P} E_\alpha \qquad (1)$$

$$\sigma^2 = \frac{1}{P} \sum_{\alpha=1}^{P} (E_\alpha - \mu)^2 . \qquad (2)$$

The width of the distribution is well characterized by the standard deviation, σ (i.e., the square root of the variance). Figure 3 shows the average fitness and variance in the fitness for five different runs of a GA. The graphs tell us the same story as the histograms of figure 2, although the details are now somewhat clearer. The evolution splits into roughly three regimes. There is an initial transient behaviour which in this case lasts about 10 generations. In this period the variance is rapidly reduced by the relatively strong selection pressure produced by tournament selection. After this transient period the loss in variance due to selection is balanced by the gain in variance due to mutation. Selection however increases the average fitness until the population reaches an equilibrium at around 150 generations where selection and mutation balance exactly. The population then remains at this equilibrium indefinitely.

One obvious feature of figure 3 is that there are substantial fluctuations from one run to another. Which run should we model? We want to know about typical runs. But a typical run has random fluctuations even though these fluctuations differ from run to run. Do we ignore the fluctuations or model a specific run? One way out of this dilemma is to model an ensemble of populations. That is, to model a large (infinite) set of populations evolving in parallel. The statistical properties of the ensemble won't fluctuate, even though the members of the populations will. This is the approach we adopt here. We will consider the ensemble average of μ and σ^2 as a function of number of generations. We denote these by $\bar{\mu}(t)$ and $\overline{\sigma^2}(t)$. As we try to improve the accuracy of our model we will need to include more statistical properties of the ensemble in our model.

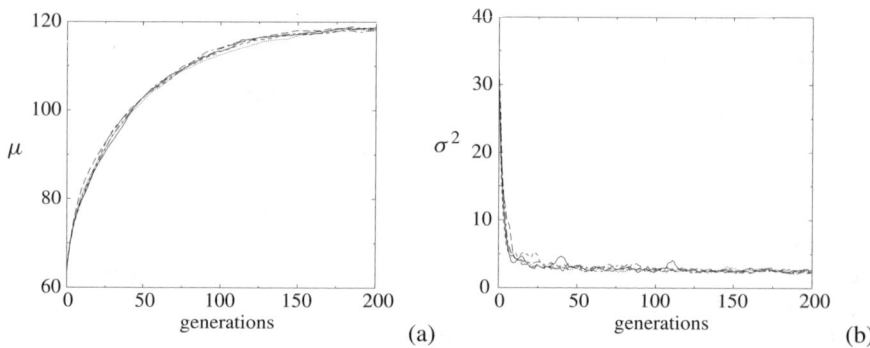

Figure 3. (a) Average fitness and (b) variance versus generations are shown for 5 runs of a GA. The same parameters are used as those in figure 2.

Our approach throughout will be to consider how the ensemble variables are altered by the genetic operators. Let us denote the set of ensemble variables at generation t by $\Xi(t) = \{\bar{\mu}(t), \overline{\sigma^2}(t), \ldots\}$. We then consider applying selection, mutation and recombination. Each genetic operator changes the ensemble variables

$$\Xi(t) \xrightarrow{sel} \Xi_s(t) \xrightarrow{mut} \Xi_{sm}(t) \xrightarrow{cross} \Xi_{smc}(t) = \Xi(t+1).$$

We will calculate how these genetic operators change an arbitrary set of ensemble variables. Then starting from an initial ensemble $\Xi(0)$ we iteratively apply each operator in turn to predict the whole evolution.

3.3 A Zeroth Order Model

We start with the simplest model we can. First, we assume the population is infinite. In this case there are no fluctuations and we can consider the evolution of a single population rather than an ensemble. Second, we assume that the distribution of fitnesses is Gaussian. As a consequence it is sufficient to consider only $\mu(t)$ and $\sigma^2(t)$. We'll also ignore crossover for the time being as that adds some unnecessary complications.

We denote the distribution of fitness as $\rho(E)$, which, by assumption, is Gaussian

$$\rho(E) = \frac{e^{-(E-\mu(t))^2/2\sigma^2(t)}}{\sqrt{2\pi}\,\sigma(t)}.$$

The initial population consists of random strings. As the members of the population are binary strings the fitnesses of the initial population will be binomially distributed with mean $L/2$ and variance $L/4$. If L is large this distribution is very well approximated by a Gaussian. We now consider how a Gaussian distribution would be changed by selection and mutation.

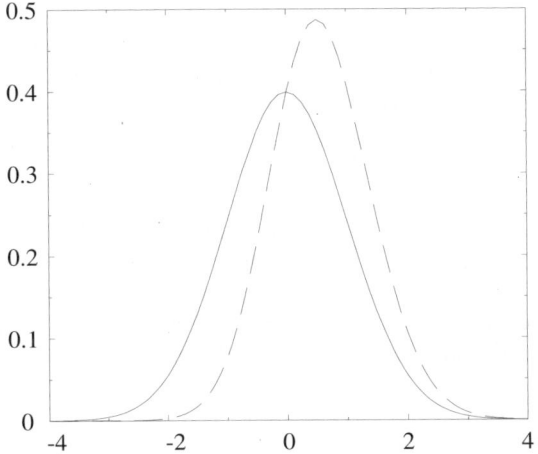

Figure 4. Tournament selection on a zero mean unit Gaussian. The solid curve shows the original Gaussian, while the dashed curve shows the distribution after selection.

Selection. In tournament selection two members are randomly drawn from the population and the fittest member is copied into the mating pool. The probability that a member with fitness E is fitter than a randomly chosen member E' is

$$P(E > E') = \int_{-\infty}^{E} \rho(E') \mathrm{d}E'.$$

This is exactly proportional to the rank of E – it is well known that the two selection mechanism give similar results. The probability distribution of the winner is therefore given by

$$\rho_s(E) = 2\rho(E) \int_{-\infty}^{E} \rho(E') \mathrm{d}E'$$

where the factor 2 ensures $\rho_s(E)$ is properly normalized. The distribution $\rho_s(E)$ is not Gaussian, but it is surprisingly close. Figure 4 shows the effect of selection on a Gaussian.

To calculate the effect of selection on the average and variance we need only perform two integrals

$$\mu_s = \int_{-\infty}^{\infty} E \rho_s(E) = \mu + \frac{\sigma}{\sqrt{\pi}} \tag{3}$$

$$\sigma_s^2 = \int_{-\infty}^{\infty} (E - \mu)^2 \rho_s(E) = \left(1 - \frac{1}{\pi}\right) \sigma^2. \tag{4}$$

These results were first obtain by Blickle and Theile [2]. One feature which is clear from figure 4 and equation (4) is that tournament selection strongly reduces the variance in the population. This will often lead to premature convergence. Tournament selection can be 'softened' by including a probabilistic element. Extension of this type of analysis to soft tournament (or ranking selection) is given in [17].

Mutation. Since we have an infinite population we need only calculate what mutation does on average. So how does mutation change a site variable S_i? (For the moment we drop the member label, α.) Mutation leaves S_i unchanged with probability $(1-u)$ and changes it to $1-S_i$ with probability u. Denoting the site variable after mutation by S_i^m, and the average over all possible mutation by $\langle \ldots \rangle_m$, then

$$\langle S_i^m \rangle_m = (1-u)S_i + u(1-S_i) = u + (1-2u)S_i.$$

In fact, this is the only average we need since $S_i^m \in \{0, 1\}$ so for any power of S_i^m we have $(S_i^m)^n = S_i^m$. The average fitness of a string after mutation is given by

$$\langle E_m \rangle_m = \left\langle \sum_i S_i^m \right\rangle_m = \sum_i \langle S_i^m \rangle_m$$

$$= u \sum_i 1 + (1-2u) \sum_i S_i = uL + (1-2u)E.$$

Averaging over all members of the population gives

$$\mu_m = uL + (1-2u)\mu. \tag{5}$$

Similarly, we can calculate how mutation changes E^2

$$\langle E_m^2 \rangle_m = \left\langle \left(\sum_i S_i^m \right)^2 \right\rangle_m = \sum_i \langle (S_i^m)^2 \rangle_m + \sum_{i \neq j} \langle S_i^m S_j^m \rangle_m$$

but $(S_i^m)^2 = S_i^m$ and $\langle S_i^m S_j^m \rangle = \langle S_i^m \rangle_m \langle S_j^m \rangle_m$, since mutation acts independently on each spin, therefore

$$\langle E_m^2 \rangle_m = \sum_i \langle S_i^m \rangle_m + \sum_{i \neq j} \langle S_i^m \rangle_m \langle S_j^m \rangle_m$$

$$= \sum_i \left(\langle S_i^m \rangle_m - \langle S_i^m \rangle_m^2 \right) + \left(\sum_i \langle S_i^m \rangle_m \right)^2.$$

To find the variance we average $\langle E_m^2 \rangle_m - \langle E_m \rangle_m^2$ over all members of the population. After a little algebra we find

$$\sigma_m^2 = u(1-u)L + (1-2u)^2 \sigma^2. \tag{6}$$

The algebra can be simplified by redefining the string variables S_i^α to be ± 1 – when extending the analysis it pays to make this redefinition.

The equations for mutations are exact. We can use them to compute the initial distribution. If we mutate the states of each locus with probability $1/2$ we immediately reach a random distribution. Putting $u = 1/2$ in (5) and (6) gives $\mu = L/2$ and $\sigma^2 = L/4$.

Putting it Together. From (3–6) we can compute the whole evolution. The mean and variance from one generation to the next is related by

$$\mu(t+1) = uL + (1-2u)\mu(t) + \frac{(1-2u)}{\sqrt{\pi}} \sigma(t) \tag{7}$$

$$\sigma^2(t+1) = u(1-u)L + (1-2u)^2 \left(1 - \frac{1}{\pi}\right) \sigma^2(t). \tag{8}$$

The equation for the variance is a linear difference equation of the form $x(t+1) = a + bx(t)$, which has the general solution

$$x(t) = x(\infty) + \left(x(0) - x(\infty)\right) e^{-t/\tau}$$

where $x(\infty) = a/(1-b)$ and $\tau = -\log(b)$. This describes a trajectory starting at $x(0)$ and converging at an exponential rate to $x(\infty)$. For the variance $a = u(1-u)L$, $b = (1-2u)^2(1-1/\pi)$, the initial variance is $\sigma^2(0) = L/4$, while the final variance is given by $\sigma^2(\infty) = \sigma_{eq}^2 = u(1-u)L/(1-b)$ and the characteristic time decay time is $\tau_{\sigma^2} = -\log(b)$. For small mutation rates ($u \ll 1$) $\sigma_{eq}^2 \approx \pi u L$ and $\tau_{\sigma^2} \approx \log(1 - 1/\pi) \approx 0.38$. This describes the loss of variance during the initial transient period.

If we ignore the transient period then $\sigma(t)$ remains constant and the equation for μ is a linear difference equation with $\mu(\infty) = \mu_{eq} = L/2 + (1-2u)\sigma_{eq}/(2u\sqrt{\pi})$ and with a characteristic time $\tau_\mu = -\log(1-2u)$. For small mutation rates $\mu_{eq} \approx (L + \sqrt{L/u})/2$ and $\tau_\mu = 2u$. Note, that if $u < 1/L$ this predicts $\mu_{eq} > L$, which is clearly impossible – our approximations are too crude. Nevertheless, this simple model does explain the overall behaviour. There is an initial transient behaviour as the variance decays rapidly to its equilibrium value. Then there is a long period as the mean approaches its equilibrium, after which the population reaches a fixed distribution. Furthermore, we have obtained estimates for the characteristic times associated with each phase and estimates for the equilibrium values of μ and σ^2.

Figure 5 shows a comparison between the theory and simulations. The qualitative features of the evolution are well captured although the details are clearly wrong. In particular, the equilibrium average fitness is over-estimated, although we know from our discussion above that the approximations must break down at some point. Still, not bad for such a simple model.

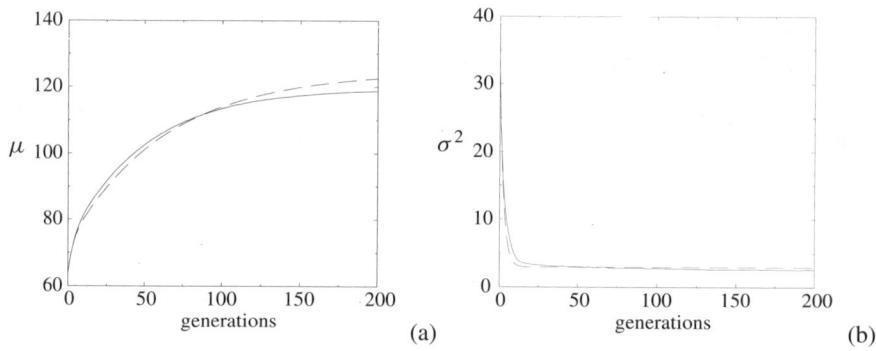

Figure 5. Simulations (solid line) and theory (dashed line) for (a) the average fitness and (b) the variance. The curves are for the same system as that used in figure 2. The simulations are average over 1000 runs. The theory is computed by iterating (7) and (8).

3.4 Increasing the Accuracy

We made two unjustified assumptions: the population was approximated by a Gaussian and the population was infinite. In this section and section 3.5 we indicate how to relax these restrictions.

A distribution can be described by a cumulant expansion (see appendix A). The first two cumulants are the average and variance. The third cumulant which measures the 'skewness' is defined by

$$\kappa_3 = \frac{1}{P} \sum_{\alpha=1}^{P} (E_\alpha - \mu)^3.$$

The actual skewness is defined to be κ_3/σ^3. Higher cumulants are slightly more complicated. For a Gaussian all the high-order cumulants (κ_n for $n > 2$) are zero. Figure 6 shows three distribution with a negative, zero and positive third cumulant.

We can include higher cumulants in our model by using the Gram–Charlier expansion (see appendix A)

$$\rho(E) = \frac{e^{-(E-\mu)^2/2\sigma^2}}{\sqrt{2\pi}\,\sigma} \left(1 + \frac{\kappa_3}{\sigma^3} h_3\left(\frac{E-\mu}{\sigma}\right)\right)$$

where $h_3(x) = (x^3 - 3x)/3!$. This is easily extensible to higher cumulants. The Gram–Charlier expansion has the correct first three cumulants but it does not guarantee that $\rho(E) > 0$. We can compute the effect of selection in an analogous way to

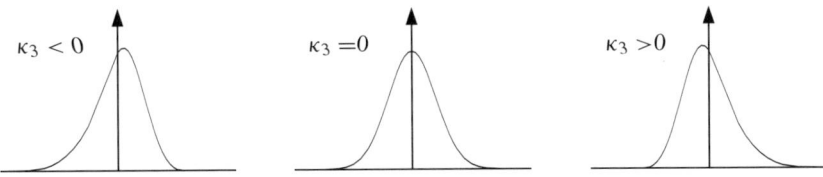

Figure 6. Three distribution with skewness, $\kappa_3 < 0$, $\kappa_3 = 0$, and $\kappa_3 > 0$

what we did above, although the integrals are hairier. We find

$$\mu_s = \mu + \frac{\sigma}{\sqrt{\pi}}\left(1 - \frac{s^2}{96}\right) \tag{9}$$

$$\sigma_s^2 = \sigma^2\left(1 - \frac{1}{\pi} + \frac{s}{2\sqrt{\pi}} + \frac{s^2}{48\pi} - \frac{s^4}{9216\pi}\right) \tag{10}$$

$$\kappa_3^s = \sigma^3 \left(\frac{1}{2\sqrt{\pi}}\left(\frac{4}{\pi} - 1\right) + s\left(1 - \frac{3}{2\pi}\right) - \frac{s^2}{192\sqrt{\pi}}\left(\frac{12}{\pi} + 1\right) \right.$$
$$\left. + \frac{s^3}{64\pi} + \frac{s^4}{1536\pi^{3/2}} - \frac{s^6}{442368\pi^{3/2}} \right) \tag{11}$$

where $s = \kappa_3/\sigma^3$.

The mutation results for the first two cumulants, (5) and (6), are exact (i.e., we did not assume a Gaussian). We can compute the result of mutation on the third cumulant in an analogous way (again the algebra becomes tedious)

$$\kappa_3^m = (1 - 2u)^3 \kappa_3 - u(1 - u)(1 - 2u)(2\mu - L). \tag{12}$$

Higher-order results can be found in [10] (in fact, the effect of mutation to all orders can be obtained relatively painlessly from a generating function).

The initial population will be symmetrically distributed around a fitness $L/2$. Consequently, $\kappa_3(0) = 0$. The evolution equations are less easy to understand by inspection, but to calculate the full evolution we need only iterate selection (9–11) and the mutation (5), (6) and (12). Figure 7 compares the first two cumulants computed using the three cumulant approximation with simulation results. Introducing the third cumulant improves the accuracy although this good agreement is rather fortunate.

The evolution of the third cumulant is shown in figure 8. The third cumulant initially grows. We might expect this as selection positively skews the distribution of a Gaussian – see (11) (this is true for ranking and tournament selection, but not in general). Eventually, however, the population becomes negatively skewed – we can just about detect this in the histograms of figure 2. Mutation skews the distribution negatively when μ is greater than $L/2$, as can be seen from (12). Note that the third

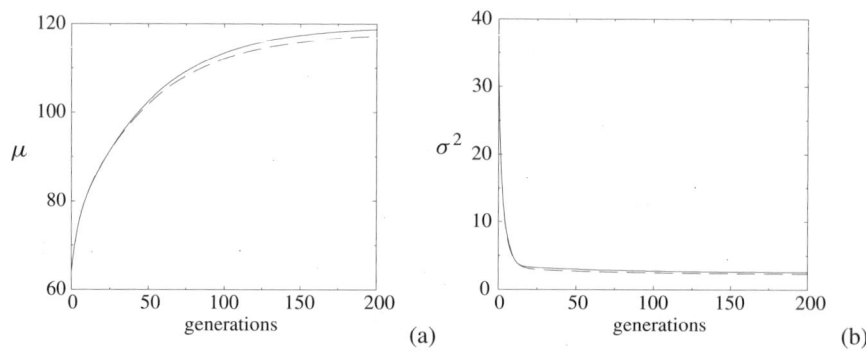

Figure 7. Simulations (solid line) and theory (dashed line) for (a) the average fitness and (b) the variance. The curves are for the same system as that used in figures 2 and 5. The theory curve is now computed using three cumulants.

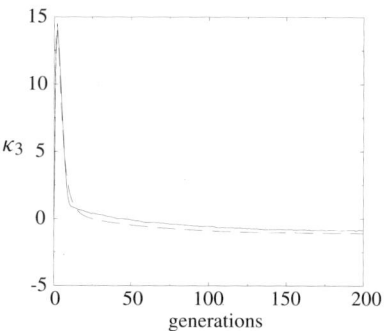

Figure 8. The evolution of the third cumulant is shown for simulations (solid line) and theory (dashed line). The results are shown for the same system as that use in all the previous graphs.

cumulant tells us little about the skewness on its own. The quantity κ_3/σ^3 (which is the technical definition of skewness) gives a 'dimensionless' measure of skewness. Thus, even though initially κ_3 has a large positive value the actual skewness is small because the width of the distribution, σ, is large. Later on when κ_3 is negative the skewness is more marked because the distribution is then relatively narrow.

The theoretical prediction for κ_3, although qualitatively similar to the simulations, is not particularly accurate. We could improve accuracy by including the fourth cumulant, but we would then find κ_4 was inaccurate. We can continue *ad infinitum*. For some infinite population models we can do better and solve the model exactly. But for finite population models we have the unsatisfying feature that we must truncate the expansion somewhere and hope for the best. Still this often gives a very respectable approximation.

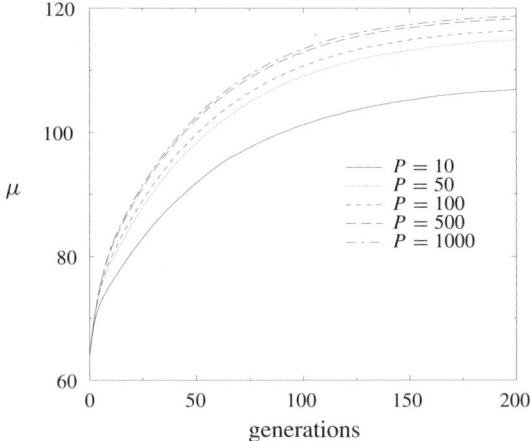

Figure 9. Example of finite population effects. The evolution of the average fitness is shown for a population of size $P = 10, 50, 100, 500$, and 1000. The other parameters are $L = 128$ and $U = 1/N$. The results are from simulations averaged over 1000 runs.

3.5 Finite Population Effects

Finite population corrections can be significant. Figure 9 shows the evolution of the average fitness for different sizes of population. There is a dramatic deterioration in the quality of the solutions as the population size becomes very small.

Modelling a finite population introduces a number of new complications. Still, the size of the population can substantially alter the dynamics, so we have to understand how to model them. Finite populations alter selection in two ways. First, selection now involves random sampling – that is, the number of offspring an individual has is proportional to its fitness only on average. Second, the evolution becomes stochastic, so we must consider the evolution of an ensemble. (At first this just involves adding a bar to the variables, but for highly accurate modelling the fluctuations give rise to many correction terms.)

Sampling. We can model selection as a two-stage process: reproduction and sampling. In reproduction, an infinite population, the mating pool, is generated by reproducing each individual exactly in proportion to their fitness. We then choose a random sample of P members from this infinite population to create a new population. We can think of this in biology terms. Each individual produces a large number of seeds proportional to their fitness, but only a random sample of P seeds actually survive. Furthermore, rather than considering mutation on the finite population we can mutate the infinite population before sampling. In terms of biology, the mutations occur in the seeds, but do not affect which ones survive. This means that we don't have to calculate finite population corrections for mutations.

There are two sources of sampling corrections. One comes because we draw a finite sample from an infinite distribution. The second comes from approximating a discrete sum by an integral (we should probably give this correction a different name). We compute first the corrections coming from drawing a finite sample from a probability distribution.

We assume that we have an infinite population with a distribution $\rho(E)$, with mean μ, variance σ^2, etc. We now draw a sample of P individuals. We denote the number of individuals with fitness E by $n(E)$. The average fitness after sampling is given by

$$\mu_{\text{smpl}} = \sum_E n(E) E.$$

Taking the average over all ways of drawing the population we find

$$\bar{\mu}_{\text{smpl}} = \frac{1}{P} \sum_E \langle n(E) \rangle_{\text{smpl}} E$$

where $\langle \ldots \rangle_{\text{smpl}}$ signifies an average over all ways of doing sampling. But

$$\langle n(E) \rangle_{\text{smpl}} = P \rho(E)$$

so

$$\bar{\mu}_{\text{smpl}} = \sum_E \rho(E) E = \mu.$$

Sampling does not change the average fitness, but the story is different with other cumulants. The variance is given by

$$\sigma^2_{\text{smpl}} = \frac{1}{P} \sum_E n(E) E^2 - \frac{1}{P^2} \sum_{E,E'} n(E) n(E') \, E \, E'.$$

We average over all ways of drawing samples. For roulette wheel selection the probability of a particular draw is given by a multinomial distribution. It is straightforward to show for a multinomial distribution (see appendix B)

$$\langle n(E) n(E') \rangle_{\text{smpl}} = (P^2 - P) \rho(E) \rho(E') - P \rho(E) \delta_{E,E'}.$$

Substituting this into the average of σ^2_{smpl} we find

$$\overline{\sigma^2}_{\text{smpl}} = \left(1 - \frac{1}{P}\right) \left(\sum_E \rho(E) E^2 - \sum_{E,E'} \rho(E) \rho(E') \, E \, E' \right)$$

$$= \left(1 - \frac{1}{P}\right) \sigma^2.$$

This is the familiar formula for the unbiased estimate of the variance. A similar calculation for the third cumulant after sampling gives

$$\bar{\kappa}_3^{smpl} = \left(1 - \frac{1}{P}\right)\left(1 - \frac{2}{P}\right)\kappa_3.$$

Fisher called these sampled cumulants k-statistics [4]. They can be found efficiently using a generating function.

We are still not done. There is a further finite population effect that arises when turning a sum into an integral. One way to obtain an approximation for this correction is to consider the correction for neutral selection. In this case the ranks are randomly assigned to the individuals. We need to calculate

$$\kappa_2' = \sum_\alpha (p_\alpha - p_\alpha^2) E_\alpha^2 - \sum_{\alpha \neq \beta} p_\alpha p_\beta E_\alpha E_\beta$$

where we are assuming neutral selection so we can average over p_α and E_α separately. This gives

$$\kappa_2' = (1 - \langle p_\alpha^2 \rangle_\alpha) \left(\frac{1}{P} \sum_\alpha E_\alpha^2 - \frac{1}{P(P-1)} \sum_{\alpha \neq \beta} E_\alpha E_\beta \right)$$

where we have use $\sum_\alpha p_\alpha = 1$ and $1 = (\sum_\alpha p_\alpha)^2 = \sum_\alpha p_\alpha^2 + \sum_{\alpha \neq \beta} p_\alpha p_\beta$. Now for tournament selection each member of the population is drawn according to its rank. In neutral selection the ranks are arbitrarily assigned. Each rank will occur once, so (see appendix C)

$$\langle p_\alpha^2 \rangle = \frac{\sum_{r=0}^{P-1} r^2}{\left(\sum_{r=0}^{P-1} r\right)^2} = \frac{4P - 2}{3P(P-1)}.$$

Similarly, the correction for the third cumulant is given by

$$\langle p_\alpha - 3p_\alpha^2 + 2p_\alpha^3 \rangle_\alpha = 1 - \frac{4P - 6}{P(P-1)}.$$

Putting together these corrections with the correction due to the randomness in sampling we find for sampling

$$\bar{\mu}_{smpl} = \mu_s \tag{13}$$

$$\overline{\sigma^2}_{smpl} = \left(1 - \frac{1}{P}\right)\left(1 - \frac{4P - 2}{3P(P-1)}\right)\sigma_s^2 \tag{14}$$

$$\bar{\kappa}_3^{smpl} = \left(1 - \frac{1}{P}\right)\left(1 - \frac{2}{P}\right)\left(1 - \frac{12P - 10}{P(P-1)}\right)\kappa_3^s \tag{15}$$

where μ_s, σ_s^2 and κ_3^s are our calculation from the statistics after sampling.

The main consequence of sampling is that it reduces the variance in the population. This limits the search and leads to a poorer performance. The size of the sampling corrections depend on how we perform the selection. Ranking selection together with stochastic universal sampling [1] would reduce the size of the sampling corrections giving improved performance. The effect of stochastic universal sampling has been studied by the authors [17] and is presented later in this volume, page 207. Other types of GAs, such as steady state GAs, give different sampling corrections. A comparison of these schemes has been previously published by the authors [15].

Fluctuations. So far we have ignored fluctuations. However, to deal with fluctuations we must consider an ensemble of populations. Each population in the ensemble will have its own set of statistics μ, σ^2, κ_3, etc. The ensemble can be thought of as a probability distribution, $P(\mu, \sigma^2, \kappa^3)$, over the population statistics. So far, we have calculated the means of this distribution. However, we should average all quantities over the whole ensemble. For formulas that are linear in the statistical variables, such as mutation and sampling, this average makes no difference. However, selection is non-linear in the statistics. For example, we found that selection depends on the skewness and higher powers of the skewness. But

$$\overline{\left(\frac{\kappa_3}{\sigma^3}\right)} \neq \frac{\bar{\kappa}_3}{\bar{\sigma}^3}$$

where the bar denotes the ensemble average. In fact, even in the two cumulant system $\bar{\sigma}^2 \neq \overline{\sigma^2}$. Thus, there will be systematic corrections to the average fitness coming from fluctuations in σ^2. To calculate the corrections we need to know how big the fluctuations are. To do so requires keeping information such as the covariance between the statistical variables

$$v_{n,m} = \overline{\kappa_n \kappa_m} - \bar{\kappa}_n \bar{\kappa}_m$$

where $\kappa_1 = \mu$ and $\kappa_2 = \sigma^2$. These variables may be of interest in themselves. For example, $v_{1,1}$ tells us about the fluctuations in the average fitness between runs.

Fortunately, the corrections from these fluctuations are often small and they can usually be neglected. A full treatment of fluctuations was given in an earlier publication [10]. There it was shown that by using a sufficient number of ensemble parameters very good agreement with simulations could be achieved.

Putting in Sampling Corrections. The sampling corrections can be put into our model by applying the sampling equations (13–15) after selection and mutation. If we consider the two variable system (i.e., we ignore the skewness), then the sampling term reduces the equilibrium variance by approximately $1 - 7/(3P)$. As a consequence, the equilibrium fitness is reduced by approximately $7\sqrt{L/u}/(3P)$ – if $u = 1/L$ then μ_{eq} is reduced by approximately $7L/3P$. Decreasing P reduces

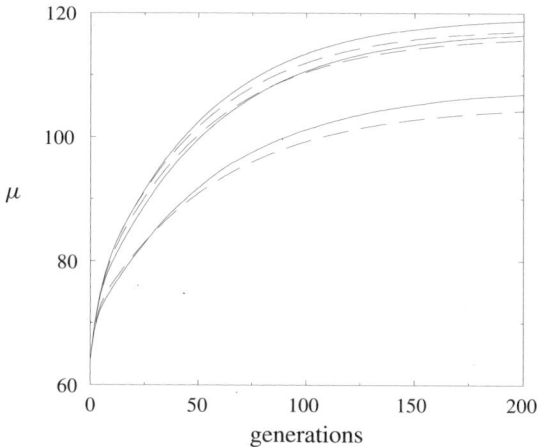

Figure 10. Comparison of simulations (solid curves) and theory (dashed curves) for a GA with population size $P = 10$, 100, and 1000. In both cases the higher curves represent large populations. The other parameters are $L = 128$ and $u = 1/L$. There is no crossover.

the average fixed point fitness, but this only becomes significant when P is small. Figure 10 shows a comparison of simulations with theory for populations of size 10, 100, and 1000. The theory captures the dominant behaviour, although the theory tends to underestimate the equilibrium average fitness. Partly this is due to truncating the cumulant expansion (the theory over-estimates the average fitness when using an even number of cumulants and under-estimates it when using an odd number).

3.6 Recombination

So far we have not used recombination. In the remainder of this paper we describe how to treat this important operator.

Recombination is achieved by pairing each string in the mating pool and from every pair of parents $\{\vec{S}^\alpha, \vec{S}^\beta\}$ producing a pair of children $\{\vec{S}^\mu, \vec{S}^\nu\}$ where

$$S_i^\mu = \begin{cases} S_i^\alpha & \text{with probability } a \\ S_i^\beta & \text{with probability } 1 - a. \end{cases}$$

and $\vec{S}^\nu$ is the complement of $\vec{S}^\mu$. In this way we conserve all the variables in the mating pool and prevent further convergence due to sampling in crossover. We could also have generated a mating pool of $2P$ individuals and produced a single child from each pair of parents. This would have reduced the convergence due to sampling in selection but at the price of introducing sampling into recombination. The sampling effects would be similar in the two strategies.

Recombination is a mixed blessing: it complicates the theory, but it substantially reduces the higher cumulants, allowing much better agreement between theory and simulations. Up to now we did not need to know the state of the strings. The only thing that mattered was their fitness. The additional complications arise when using recombination because the effect of crossover depends on the states of the strings. To see this consider two strings having the same fitness. These strings may be identical or they may be very different. With no crossover this made no difference, but when we perform crossover, the results will be very different. Combining two identical strings will give a child with exactly the same configuration, and hence the same fitness as the parents. Crossing over non-identical strings can give rise to a child with a very different fitness to the parents.

Recombination therefore depends on the pairwise correlations

$$q_{\alpha,\beta} = \frac{1}{L}\sum_{i=1}^{L}(2S_i^\alpha - 1)(2S_i^\beta - 1).$$

With this definition, $q_{\alpha,\beta} = 1$ if the strings α and β are identical at each site, and $q_{\alpha,\beta} = -1$ if the two strings differ at each site. Let $\vec{S}^\mu$ be the string produced by combining $\vec{S}^\alpha$ and $\vec{S}^\beta$. We can calculate the effect of recombination in a similar manner that we calculated mutation. We suppose that with a probability a we choose an allele from parent α and with probability $1 - a$ we choose the allele from parent β. Thus, on average

$$\langle S_i^\mu \rangle_c = a S_i^\alpha + (1 - a) S_i^\beta$$

where $\langle \ldots \rangle_c$ signifies the average over all ways of performing crossover. Just as for mutation, this is the only average we need to know to calculate the effect of recombination, since $(S_i^\mu)^n = S_i^\mu$. We now proceed as we did for mutation. The fitness on average after crossover is given by

$$\langle F_\mu^c \rangle_c = \sum_{i=1}^{L} \langle S_i^\mu \rangle_c = a \sum_{i=1}^{L} S_i^\alpha + (1-a) \sum_{i=1}^{L} S_i^\beta$$

$$= a F_\alpha + (1-a) F_\beta.$$

Averaging over all children and all ways of pairing the mating pool we find

$$\mu_c = \mu. \tag{16}$$

That is, crossover does not change the average fitness. We can compute $\langle (F_\mu^c)^2 \rangle_c$, just as we did for mutation in (6), and use it to calculate σ_c^2. The algebra is straightforward if tedious (it would simplify if we had used $S_i^\alpha \in \{-1, 1\}$). We find

$$\sigma_c^2 = \left(a^2 + (1-a)^2\right)\sigma^2 + 2a(1-a)\frac{L(1-q)}{4} \tag{17}$$

where

$$q = \frac{1}{P(P-1)} \sum_{\alpha \neq \beta} q_{\alpha\beta}.$$

Equation (17) tells us that recombination drives the variance towards $L(1-q)/4$ at a rate $\tau_c = -\log(a^2 + (1-a)^2)$. This rate reaches a maximum of $\log(2)$ when $a = 1/2$, that is, when there is maximal mixing of the parents. A similar calculation gives for the third cumulant

$$\kappa_3^c = \left(a^3 + (1-a)^3\right)\kappa_3 + \frac{L}{2P(P-1)} \sum_{\alpha \neq \beta}(E_\alpha - \mu)q_{\alpha\beta}. \tag{18}$$

A reasonable assumption is that the correlation and fitnesses are independent, in which case

$$\kappa_3^c = \left(a^3 + (1-a)^3\right)\kappa_3. \tag{19}$$

When $a = 1/2$ then $\kappa_3^c = \kappa_3/4$.

We see explicitly the complication added by recombination. We now need to know about the mean correlation q and how the pair correlation and fitnesses are related. Thus to model the dynamics accurately we must model q. We show how we can model q below. However, if we are just interested in the fixed point fitness then we can make the approximation $\sigma^2 = L(1-q)/4$. This is the value which the variance is being driven to by recombination. From the definition of the variance

$$\sigma^2 = \frac{L(1-q)}{4} + \sum_{i \neq j}\left(\langle S_i^\alpha S_j^\alpha \rangle_\alpha - \langle S_i^\alpha \rangle_\alpha \langle S_j^\alpha \rangle_\alpha\right)$$

where $\langle \ldots \rangle_\alpha$ denotes the average over all members of the population. Our fixed point assumption is that $\langle S_i^\alpha S_j^\alpha \rangle_\alpha - \langle S_i^\alpha \rangle_\alpha \langle S_j^\alpha \rangle_\alpha$ vanishes on average. This is known in the biological literature as *linkage equilibria*. Selection pushes the population out of linkage equilibrium – as we are using a selection mechanism that has a strong selection pressure the system will quickly move out of linkage equilibrium. However, later in the evolution, when selection is balanced by mutation, recombination has a chance of pushing the population back towards linkage equilibria. If we assume linkage equilibria, then recombination leaves the variance unchanged. Recombination decreases the higher cumulants towards their linkage equilibrium values. In general the cumulants, κ_n, are driven towards their linkage equilibria values at a rate $(n-1)\log(2)$ (assuming $a = 1/2$). As a consequence, the fitness distribution will be much closer to a Gaussian for a sexual population than for an asexual population.

To understand the full dynamics we need to model the correlation, q. This approach was first used by Rattray [13] and is used later in this volume by Rogers. To model the correlation we introduce the idea of common ancestry. If two variables, S_i^α and S_i^β, from the same site but in different individuals, are inherited from the same ancestor we write $S_i^\alpha \sim S_i^\beta$. We define the degree of common ancestry to be

$$C_{\alpha\beta} = \frac{1}{L}\sum_{i=1}^{L}[S_i^\alpha \sim S_i^\beta]$$

where

$$[predicate] = \begin{cases} 1 & \text{if } predicate \text{ is true} \\ 0 & \text{otherwise.} \end{cases}$$

$C_{\alpha\beta}$ is the proportion of the two strings that come from the same ancestors. The correlation between two strings can be broken down into a piece where the states come from a common ancestor and a piece where the states come from different ancestors. If the states come from the same ancestor they are obviously identical. We assume that the part of the strings that come from different ancestors are randomly correlated. The proportion of ones in string α is E_α/L. Thus we would expect if $S_i^\alpha \not\sim S_i^\beta$ that

$$\langle S_i^\alpha S_i^\beta \rangle = \frac{E_\alpha E_\beta}{L^2}.$$

Putting this together we would expect the average correlation between two strings to be

$$q_{\alpha\beta} = \frac{1}{L}\sum_{i=1}^{L}\left([S_i^\alpha \sim S_i^\beta] + [S_i^\alpha \not\sim S_i^\beta](2S_i^\alpha - 1)(2S_i^\beta - 1)\right)$$

$$= C_{\alpha\beta} + (1 - C_{\alpha\beta})\left(1 - \frac{2(E_\alpha + E_\beta)}{L} + \frac{4E_\alpha E_\beta}{L^2}\right).$$

Assuming that the degree of common ancestry is independent of the fitness of string we find the average correlation is given by

$$q = C + (1 - C)\frac{(L - 2\mu)^2}{L^2}$$

where

$$C = \frac{1}{P(P-1)}\sum_{\alpha \neq \beta} C_{\alpha\beta}.$$

We must now compute how the correlation is changed by selection, mutation, and crossover. The easiest operator is crossover as this does not change the correlation – it just shuffles the states variables S_i^α between strings. It is easy to calculate how selection causes a change in the degree of common ancestry. If we assume string α is selected n_α times then

$$C_s = \frac{1}{P(P-1)}\left(\sum_{\alpha=1}^{P} n_\alpha(n_\alpha - 1)C_{\alpha\alpha} + \sum_{\alpha \neq \beta} n_\alpha n_\beta C_{\alpha\beta}\right)$$

but $C_{\alpha\alpha} = 1$. We assume that the number of times a string is selected, n_α, is independent of $C_{\alpha\beta}$. We can average over all ways of doing selection. The average decouples from the sum. The expression can then be simplified by noting

$$P^2 = \left(\sum_\alpha n_\alpha\right)^2 = \sum_\alpha n_\alpha^2 + \sum_{\alpha \neq \beta} n_\alpha n_\beta.$$

After a little algebra we find

$$C_s = 1 - \frac{P - \langle n_\alpha^2 \rangle_\alpha}{P - 1}(1 - C).$$

We still have to compute $\langle n_\alpha^2 \rangle_\alpha$. The probability of choosing n_α copies of member α is given by a multinomial distribution (see appendix B). Thus on average

$$\langle n_\alpha^2 \rangle_s = P(P-1)p_\alpha^2 + P p_\alpha.$$

Averaging over all members of the population

$$\langle n_\alpha^2 \rangle_\alpha = \frac{1}{P} \sum_{\alpha=1}^{P} \left(P(P-1)p_\alpha^2 + P p_\alpha\right).$$

Now the probability of choosing an individual is proportional to its rank – $p_\alpha = 2r_\alpha/P(P+1)$ where $r_\alpha \in \{0, 1, 2, \ldots, P-1\}$. Since we are averaging over all members of the population, all ranks are included so

$$\langle n_\alpha^2 \rangle_\alpha = \frac{1}{P} \sum_{r=0}^{P-1} P(P-1)\frac{4r^2}{P^2(P-1)^2} + 1 = \frac{7}{3} - \frac{2}{3P}$$

and

$$C_s = 1 - \frac{(3P-1)(P-2)}{3P(P-1)}(1 - C). \tag{20}$$

Finally, we must calculate how mutation changes the correlation. Remember the correlation is defined by

$$q_{\alpha,\beta} = \frac{1}{L} \sum_{i=1}^{L} (2S_i^\alpha - 1)(2S_i^\beta - 1).$$

Mutation causes a change of state $(2S_i^\alpha - 1) \to -(2S_i^\alpha - 1)$ with probability u. Thus on average

$$\langle (2S_i^{\alpha'} - 1) \rangle_m = (1 - 2u)(2S_i^\alpha - 1)$$

and consequently

$$q_m = (1 - 2u)^2 q. \tag{21}$$

It is straightforward to turn this into an expression relating the ancestry C before and after mutation.

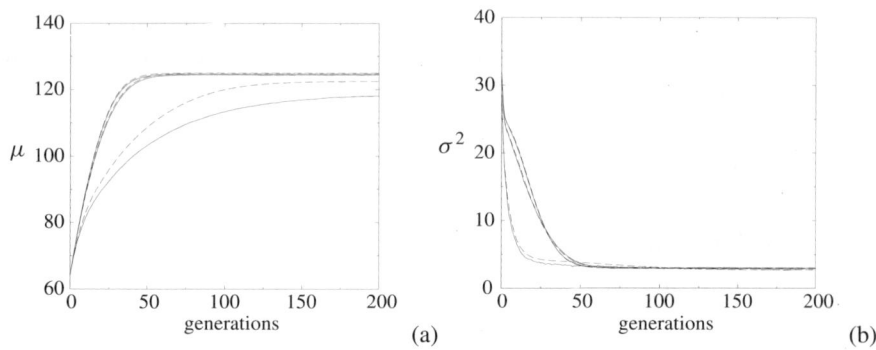

Figure 11. (a) the average fitness and (b) the variance versus generation measured from simulations (solid lines) and theory results (dashed lines) for populations of size $P = 10$, 100, and 1000. The curves with higher average fitness correspond to larger populations. The results are for $L = 128$ and $u = 1/L$. The simulations are averaged over 1000 runs.

Putting it Together. Figure 11 shows the simulations and theory for a GA with recombination. Introducing recombination dramatically increases the speed of the search because a large variance is maintain for much longer. Only for very small populations (e.g., $P = 10$) does the variance fall rapidly towards zero. The agreement with simulations becomes worse for small populations because of the growth in the higher cumulants. (For $P = 10$ we have $\kappa_4 > \sigma^4$, while for $P = 100$, $\kappa_4 < \sigma^4$.)

4 Historical Note

Modelling of biological populations began in the 1920s and 1930s with the seminal works of Ronald Aylmer Fisher [5] and Sewall Wright [24]. However, the main interest of population geneticists was the frequency of particular traits. These, after all, are what could be measured experimentally. Still, Fisher, Wright and their colleagues have developed a huge body of theory. Fisher and Wright introduced the diffusion approximation for describing models close to linkage equilibrium. These were further developed by Motoo Kimura [7]. The use of cumulants to describe evolving populations dates back at least to Bulmer [3] and used more recently by Turelli and Barton [19].

Heinz Mühlenbein has modelled the dynamics of GAs for many years [8] (see also this volume, page 135). Investigations of the effect of selection were carried out by Blickle and Theile [2]. A similar approach to that discussed here has been used on the 'royal road function' by Erik van Nimwegen, James P. Crutchfield, and Melanie Mitchell [20]. We have not touched on modelling using Markov chain analysis, which has a long history both in biology [24] and also for modelling GAs [6,9,23,21,22].

5 Discussion

We have gone through a lot of calculations very rapidly. Let us recap on what we have done. We started looking at a two-parameter (mean and variance) model. This gave a reasonable qualitative picture of the evolution, although quantitatively it didn't give very good agreement. We might have made much more use of the model, for instance, by examining annealing schedules for the mutation rate. We showed that we could improve the accuracy of the model by introducing higher cumulants, but we do so at the price of obtaining more complicated equations. We then computed the finite population corrections. These corrections can become quite substantial, especially for small populations. Finite population effects are particularly difficult to calculate for tournament selection. Finally, we looked at recombination. This has the effect of maintaining a broad population, while reducing the higher cumulants. One result of this is that we don't need to calculate so many cumulants. However, to understand the initial dynamics we need to know the average correlation in the population. This again complicates the analysis.

We have learned a lot about the dynamics for this type of GA and this particular problem. Other types of GA have been studied, such as the steady-state GA [16]. By comparing algorithms even on relatively simple problems we can understand how they differ. For example, in the steady-state GA there are effectively two selections (choosing the fit member and the unfit member) and two samplings going on at each step. Consequently, to a first approximation, the steady-state GA appears to go at twice the rate of a generation GA, although, it also correlates at twice the rate, so it may not explore the solution space so well. This is apparent, even in a simple problem such as ones-counting, from carefully modelling the dynamics. This illustrates that it is not always necessary to consider realistic problems to obtain deep insights into how GAs work.

Ones-counting is, however, a pathologically easy problem to learn. If we were really interested in efficiently solving such a problem, we certainly wouldn't use a GA. Ones-counting tells us relatively little about how we should select parameters in more realistic problems. To understand that, we have to look at different problems. As we described in section 2 many models have been examined within this formalism, although there is still much more work to do in this direction. The formalism discussed in the paper is relatively new and many of the techniques are still being developed. The results obtained are approximations, but generally very accurate. More importantly, by modelling GAs we can begin to understand how they work.

Acknowledgments

Thanks to Mathematica for doing the hairy integrals.

Appendix A Cumulant Expansion

The characteristic function of a distribution is defined to be its Fourier transform

$$\phi(\omega) = \sum_E \rho(E)\, e^{i\omega E}.$$

The Taylor expansion of the characteristic function is given by

$$\phi(\omega) = \sum_{n=0} \frac{\mu_n (i\omega)^n}{n!}$$

where μ_n are the moments of the distribution define by

$$\mu_n = \frac{\partial^n \phi(\omega)}{\partial (i\omega)} = \sum_E \rho(E)\, E^n.$$

Since $\rho(E)$ is normalized $\mu_0 = 1$. The first moment, μ_1, is the average of the distribution.

The cumulant generating function is defined to be the logarithm of the characteristic function

$$G(\omega) = \log(\phi(\omega)) = \sum_{n=1} \frac{\kappa_n (i\omega)}{n!}.$$

The cumulants are related to the moments

$$\kappa_1 = \sum_E \rho(E)\, E = \mu_1$$

$$\kappa_2 = \sum_E \rho(E)\, (E - \kappa_1)^2 = \mu_2 - \mu_1^2$$

$$\kappa_3 = \sum_E \rho(E)\, (E - \kappa_1)^3 = \mu_3 - 3\mu_2\mu_1 + 2\mu_2^3$$

$$\kappa_4 = \sum_E \rho(E)\, (E - \kappa_1)^4 - 3\kappa_2^2$$

$$= \mu_4 - 4\mu_2\mu_1 - 3\mu_2^2 + 12\mu_2\mu_1^2 - 6\mu_1^4.$$

Although the moments and cumulants are trivially related for a single distribution, their relation is very non-trivial when we have to average over an ensemble of distributions. Cumulants, although apparently complicated, all scale linearly with L, whereas the nth moment will be of order L^n. This makes it easier to model with the cumulants than the moments.

Gram–Charlier expansion. The Gram–Charlier expansion is given by

$$\rho(E) = \frac{e^{-(E-\mu)^2/2\sigma^2}}{\sqrt{2\pi}\,\sigma}\left(1 + \sum_{n=3}^{\infty} \frac{\kappa_n}{\sigma^n} h_n\left(\frac{E-\mu}{\sigma}\right)\right)$$

where $h_3(x) = (x^3 - 3x)/3!$, $h_4(x) = (x^4 - 6x^2 + 3)/4!$, etc., are related to the Hermite polynomials. They can be generated from the recurrence relation

$$h_n(x) = \frac{1}{n}(xh_{n-1}(x) - h_{n-2}(x))$$

with $h_0(x) = 1$ and $h_1(x) = x$.

Appendix B Multinomial Distribution

A multinomial distribution arises when we consider making M random draws (with replacement) from N different objects. We label these objects $\alpha = 1, \ldots, N$ and denote the probability of drawing object α as p_α. If we make M draws the probability of drawing n_α objects of type α is given by

$$p(\vec{n}) = M!\prod_{\alpha=1}^{N} \frac{p_\alpha^{n_\alpha}}{n_\alpha!}\,\delta\!\left(\sum_\alpha n_\alpha - M\right).$$

To calculate averages of multinomial distributions we introduce a generating function

$$F(\vec{l}) = \left\langle e^{\sum_\alpha n_\alpha l_\alpha}\right\rangle$$

where $\langle \ldots \rangle$ denotes the average over the multinomial distribution. To calculate averages, such as $\langle n_\alpha n_\beta \rangle$, we take the appropriate derivatives of the generating function

$$\langle n_\alpha n_\beta \rangle = \left.\frac{\partial^2 F(\vec{l})}{\partial l_\alpha \partial l_\beta}\right|_{\vec{l}=0}.$$

The multinomial coefficients occur when expanding the product of a sum

$$\left(\sum_{\alpha=1}^{N} a_\alpha\right)^M = \sum_\alpha M!\prod_{\alpha=1}^{N}\frac{a_\alpha^{n_\alpha}}{n_\alpha!}\,\delta\!\left(\sum_\alpha n_\alpha - M\right)$$

but this is just what we need to compute the generating function. Setting $a_\alpha = p_\alpha e^{l_\alpha}$ we find

$$F(\vec{l}) = \left(\sum_\alpha p_\alpha e^{l_\alpha}\right)^M.$$

It is now quite straightforward to compute all the averages we need

$$\langle n_\alpha \rangle = P p_\alpha$$

$$\langle n_\alpha^2 \rangle = P(P-1) p_\alpha^2 + P p_\alpha$$

$$\langle n_\alpha n_\beta \rangle = P(P-1) p_\alpha p_\beta .$$

Appendix C Sums

We make use of the following sums

$$\sum_{r=1}^{P-1} r = \frac{P(P-1)}{2}$$

$$\sum_{r=1}^{P-1} r^2 = \frac{P(P-1)(2P-1)}{6}$$

$$\sum_{r=1}^{P-1} r^3 = \frac{P^2(P-1)^2}{4}.$$

References

1. J. E. Baker. Reducing bias and inefficiency in the selection algorithm. In *Proceedings of the Second International Conference on Genetic Algorithms*, pages 14–21. Lawrence Erlbaum, Hillsdale, NJ, 1987.
2. T. Blickle and L. Theile. A mathematical analysis of tournament selection. In L. J. Eshelman, editor, *Proceedings of the Sixth International Conference on Genetic Algorithms*, pages 9–16. Morgan Kaufmann, San Francisco, 1995.
3. M. G. Bulmer. *The Mathematical Theory of Quantitative Genetics*. Clarendon Press, Oxford, 1980.
4. R. A. Fisher. Moments and product moments of sampling distributions. *Proceedings of the London Mathematical Society*, series 2, 30:199–238, 1929.
5. R. A. Fisher. *The Genetical Theory of Natural Selection*. Oxford University Press, Oxford, 1930.
6. D. E. Goldberg and P. Segrest. Finite Markov chain analysis of genetic algorithms. In J. J. Grefenstette, editor, *Proceedings of the Second International Conference on Genetic Algorithms*. Lawrence Erlbaum, Hillsdale, NJ, 1987.
7. M. Kimura and N. Takahata. *Population Genetics, Molecular Evolution, and the Neutral Theory: Selected Papers*. University of Chicago Press, 1994.
8. H. Mühlenbein and D. Schlierkamp-Voosen. Predictive models for the breeder genetic algorithm. *Evolutionary Computation*, 1(1):25–49, 1993.

9. A. Nix and M. D. Vose. Modeling genetic algorithms with Markov chains. *Annals of Mathematics and Artificial Intelligence*, 5:79–88, 1991.
10. A. Prügel-Bennett. Modelling evolving populations. *J. Theor. Biol.*, 185:81–95, 1997.
11. A. Prügel-Bennett and J. L. Shapiro. An analysis of genetic algorithms using statistical mechanics. *Phys. Rev. Lett.*, 72(9):1305–1309, 1994.
12. A. Prügel-Bennett and J. L. Shapiro. The dynamics of a genetic algorithm for simple random Ising systems. *Physica D*, 104:75–114, 1997.
13. M. Rattray. The dynamics of a genetic algorithm under stabilizing selection. *Complex Systems*, 9(3):213–234, 1995.
14. M. Rattray and J. Shapiro. The dynamics of genetic algorithms for a simple learning problem problem. *J. Phys. A*, 29:7451–7473, 1996.
15. A. Rogers and A. Prügel-Bennett. Genetic drift in genetic algorithm selection schemes. *IEEE Transactions on Evolutionary Computation*, 3(4), 1999.
16. A. Rogers and A. Prügel-Bennett. Modelling the dynamics of steady-state genetic algorithms. In W. Banzhaf and C. Reeves, editors, *Foundations of Genetic Algorithms 5*, pages 57–68. Morgan Kaufmann, San Francisco, 1999.
17. A. Rogers and A. Prügel-Bennett. The dynamics of a genetic algorithm on a model hard optimization problem. *Complex Systems*. To appear.
18. J. L. Shapiro and A. Prügel-Bennett. Genetic algorithms dynamics in two-well potentials with basins and barriers. In R. K. Belew and M. D. Vose, editors, *Foundations of Genetic Algorithms 4*, pages 101–139. Morgan Kaufmann, San Francisco, 1997.
19. M. Turelli and N. H. Barton. Genetic and statistical analysis of strong selection on polygenic traits: what, me normal? *Genetics*, 138:913, 1994.
20. E. van Nimwegen, J. P. Crutchfield, and M. Mitchell. Finite populations induce metastability in evolutionary search. *Physics Letters A*, 229:144–150, 1997.
21. M. D. Vose. Modelling simple genetic algorithms. In D. Whitley, editor, *Foundations of Genetic Algorithms 2*, pages 63–74. Morgan Kaufmann, San Francisco, 1993.
22. M. D. Vose. *The Simple Genetic Algorithm: Foundations and Theory*. MIT Press, Cambridge, MA, 1999.
23. M. D. Vose and G. E. Liepins. Punctuated equilibria in genetic search. *Complex Systems*, 5:31–44, 1991.
24. S. Wright. *Evolution and the Genetics of Populations*. University of Chicago Press, 1968.

Statistical Mechanics Theory of Genetic Algorithms

J. L. Shapiro

Department of Computer Science
Manchester University
Manchester, UK
E-mail: *jls@cs.man.ac.uk*

Abstract. This tutorial gives an introduction to the statistical mechanics method of analysing genetic algorithm (GA) dynamics. The goals are to study GAs acting on specific problems which include realistic features such as: finite population effects, crossover, large search spaces, and realistic cost functions. Statistical mechanics allows one to derive deterministic equations of motion which describe average quantities of the population after selection, mutation, and crossover in terms of those before. The general ideas of this approach are described here, and some details given via consideration of a specific problem. Finally, a description of the literature is given.

Keywords
Macroscopic dynamics, maximum entropy, statistical mechanics, crossover

1 Overview

One of the difficulties in understanding how genetic algorithms (GAs) work is that it is difficult to solve specific examples. General notions, such as building-blocks, schema processing, big jumps in state space, and so forth, cannot be used to make quantitative predictions for particular problems. Alternatively, exact formulations of the dynamics, such as the Markov chain formulation, are too complicated to solve in specific cases containing many realistic aspects. This is in contrast with other optimisation methods. For example, if one wonders what effect the step-size has in gradient descent algorithms, one can solve the quadratic case exactly to study this. Such simple models do not exist for genetic algorithms. One road to understanding GAs is to find a way to study and solve specific examples. These solvable examples should contain some of the features found in realistic problems: large search spaces, relatively small population sizes, strong selection, realistic crossover models, fitness functions containing some complexity found in real problems, and so forth.

The statistical mechanics formulation of GAs allows one to study the dynamics of GAs searching for solutions to a number of simple combinatorial problems. This approach addresses some of the challenges raised above. The method can describe problems of realistic size and can include finite population effects, which prove to be essential to understanding how the GA searches. Results are possible for all values of selection pressure, although there are simplifications for weak selection. Both transient and equilibrium behaviour can be modelled under the formalism and fluctuations about the expected trajectory can also be calculated. Comparisons with simulations show accurate agreement with high significance for most of the problems considered. The examples on which the GA has been analysed using the approach contain some aspects of realistic problems, including multiple optima and interactions between sites. However, they are nonetheless toy problems. In particular, the problems considered so far lack the types of spatial structure which reveal substantial differences between the different forms of crossover.

As an example of what the statistical mechanics formalism can achieve, consider the following problem: there are L files which are to be backed up to a disk of size D. The size of the i-th file is f_i, and the total size of all of the files exceeds the size of the disk. Which of the files should be stored to come closest to filling the disk? More formally, which subset of the f_i's comes closest to summing to D? This is an example of the *subset sum* problem, a simple version of a bin packing problem. It is strictly NP-hard, although for many sets of file sizes it can be solved in polynomial time using standard methods [7].

The subset sum problem was analysed using a statistical mechanics approach by Rattray [14,15]. Figure 1 shows the prediction of this analysis. What is shown in the left-hand figure are the population statistics of total amount stored on the disk – the mean amount stored (κ_1) and the variance (κ_2). A measure of the similarity between the strings in the population (φ) is also shown. The right-hand figure shows the squared difference between the amount stored by the best member of the population and the disk size D which is denoted E_{best}. Also shown for comparison is the average of 2000 simulations of a GA. The predictions of the analysis are accurate, but there is some deviation between theory and simulation. Also observe that the best member of the population increases initially, but decreases towards the end of the run. This is due to a decrease in the effective population size due to loss of diversity in the population. This is an important finite population effect which is captured by this approach.

This tutorial presents an introductory overview of the statistical mechanics approach to analysing GAs. First, the general ideas of the approach are described. Then, these ideas are illustrated in more detail by consideration of a simple problem. Finally, some of the applications from the literature are briefly described.

2 What is a Statistical Mechanics Theory of GAs?

The difficulties in studying the dynamics of evolutionary search come from two sources. First, the dynamics is stochastic – selection involves sampling, and mutation and crossover involve probabilistic changes to the strings. Second, the dynamics acts on the space of possible populations, a space of astronomical dimensionality. We need a way to derive a *deterministic* set of equations describing the evolution which are sufficiently simple to allow for some type of study.

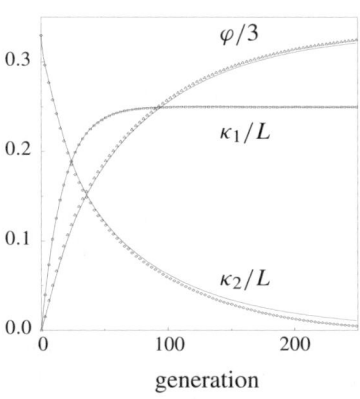 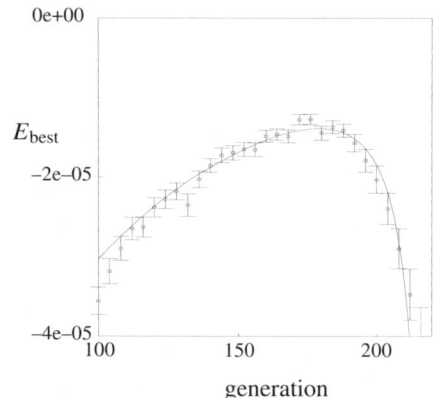

Figure 1. Comparison of prediction of a statistical mechanics analysis (solid lines) with an average over 2000 runs of a genetic algorithm (symbols) for the subset sum problem. The left figure shows the total amount stored on the disk averaged over the population (κ_1), the population variance in this quantity (κ_2), and a measure of the similarity of the strings in the population (φ). The first two are scaled by the string length L. The right figure shows the squared difference between the amount stored by the best member of the population and D. The disk size was $D = L/4$; other parameters were population size of 80, string length $L = 150$, selection strength $s = 0.03$, and uniform crossover was used. Figures are from Rattray [15].

A similar problem is faced by a physicist trying to use molecular properties to understand characteristics of matter, for example the heat capacity of a vessel of water. The state space of the system is of the order of the number of molecules which is horrendously large. And, although the dynamics may be deterministic (in classical physics), the initial conditions are unknown, this lack of knowledge introduces randomness.

Both situations can be treated using statistical mechanics. To get a simple description, the system is described in terms of a few *macroscopic* quantities, which are large-scale properties of the system. To get deterministic equations, these macroscopics will be averages over all possible realisation of the randomness.

There is an additional ingredient required to treat systems in this way. In order to do the averaging, one requires knowledge of the probability distribution of the states of the system. However, the formulation does not give access to this distribution. All that is known are the macroscopics which are statistical averages over this distribution. A mechanism is required to get from the known macroscopics other statistical quantities without knowledge of the underlying distribution of states. The statistical mechanics approach is to use an assumption about the distribution of states – that it is a maximum entropy distribution. Justification of this assumption is explained with respect to evolutionary algorithms in [23]; the general justification is the work of E. T. Jaynes [8]. It will discussed briefly in the next two sections.

2.1 Macroscopic Versus Microscopic Descriptions

The first issue is to justify the use of a macroscopic description. Since the genetic operators take a population at time t and produce a new population at time $t + 1$, an obvious level of description would consider precisely which strings are in the population, and would attempt to predict how the constitution of the population changes in time. A description which takes into account precisely which individuals are in the population at each time step is *microscopic*, as it considers all of the small-scaled details of the population.

The advantages of a microscopic approach are clear. It acts directly on the atomic variables of the system, and since all of the information required to predict the changes are available at every time-step, it could be rigorous. However, there are drawbacks. There are an astronomical number of possible populations, many differing in insignificant ways, and a model at this level captures tiny changes as well as significant ones. Typically the types of questions one wishes to ask about GA behaviour involves much higher-level properties – the evolution of the best, the diversity, and so forth – properties which are considerably removed from a full description of all strings. In addition, high-dimensional, non-linear systems are notoriously difficult to analyse, and the dimension of a GA described microscopically is horrendously large.

An alternative approach is to describe the GA in terms of statistical properties of the population. This is an example of a *macroscopic* level of description, and the particular statistical properties which are used to characterise the population, are called *macroscopics*. The advantages of a macroscopic formulation are evident: dynamics in a huge number of degrees of freedom is reduced to dynamics in a few quantities. A nonlinear system of a few degrees of freedom can be readily solved or numerically iterated. The number of macroscopic quantities need not scale with the size of the problem or the population, so perhaps systems of realistic sizes could be tackled. As the macroscopics could be chosen to be closely related to the properties which one is interested in studying, it is perhaps easier to learn about how the system works at this level of description.

There are obvious drawbacks. Since a small number of global quantities contains far less information about the population than a complete description, no matter how accurate the approach may turn out to be, it will not be derived from first principles, it will be approximate. Since a macroscopic description throws information away, it is essential that the appropriate macroscopics are chosen. Thus, it is not a mechanical, procedural method. Some insight about what is important and what is inessential is required. Finally, since the macroscopic quantities may not provide all of the information required to determine the effects of the genetic operators an inference mechanism is required. The principle of statistical mechanics is that any quantity which cannot be determined trivially from known macroscopics is computed using a maximum entropy distribution with the known macroscopics as constraints.

2.2 Statistical Mechanics

These ideas are formalised as follows. Imagine a system which can be in states $s_1, s_2, \ldots, s_{N_\Omega} \in \Omega$, where Ω is the state space and N_Ω is the number of states in the state space, which is typically huge. For a GA, a state could correspond to a particular population with all strings specified. Assume that the system is stochastic, either in the initial conditions, or in the dynamics, or in both.

A natural way to deal with the stochasticity is to express the system in terms of the probability $P(s_i)$ of being in state s_i. For a GA, this probability would correspond to the frequency of runs which result in that population (at the generation in question). This probability expresses as much information about the system as is possible. If one could predict these N_Ω numbers, the $P(s_i)$'s, for all times, one would have a complete characterisation of the system. This is the goal of microscopic formulations.

In a macroscopic description, the $P(s_i)$'s will not be known. What will be known is some average (macroscopic) quantity Q.

$$Q = \sum_{s_i \in \Omega} \tilde{Q}(s_i) P(s_i). \tag{1}$$

Here $\tilde{Q}$ expresses the functional relationship between the state s_i and the value of the statistic associated with that state. Suppose, in addition, knowledge is required about some other quantity, G

$$G = \sum_{s_i \in \Omega} \tilde{G}(s_i) P(s_i)$$

since the P's are not known, how can G be found from Q?

The answer is, of course, it cannot without an assumption being introduced. In statistical mechanics, maximum entropy inference is used. Choose the assignment of the $\{P(s_i)\}$ which is consistent with Q and otherwise *has the most 'uncertainty'* (or *assumes the least information*). The uncertainty of a distribution is given by the entropy

$$H = -\sum_{s_i \in \Omega} P(s_i) \log P(s_i). \tag{2}$$

This determines the amount of information required to remove the uncertainty.

The maximum entropy distribution is that which optimises the entropy (2) subject to the constraint (1), i.e., we optimise

$$H = -\sum_{s_i \in \Omega} P(s_i) \log P(s_i) + \lambda \sum_{s_i \in \Omega} \tilde{Q}(s_i) P(s_i).$$

The solution of this is the Gibbs distribution

$$P(s_i) = \frac{e^{-\lambda \tilde{Q}(s_i)}}{\sum_{s_i' \in \Omega} e^{-\lambda \tilde{Q}(s_i')}}$$

where the Lagrange multiplier, λ is chosen to satisfy the constraint

$$\frac{\sum_{s_i \in \Omega} e^{-\lambda \tilde{Q}(s_i)} \tilde{Q}(s_i)}{\sum_{s'_i \in \Omega} e^{-\lambda \tilde{Q}(s'_i)}} = Q.$$

The maximum entropy distribution is a reasonable one to assume, because it is the assumption which introduces the least assumptions beyond that which is known, the known macroscopic Q. It also occupies almost all of the probability space as $N_\Omega \to \infty$ [6]. But is it an accurate assumption? It will work well if the macroscopic Q explains most of the non-random aspects of the system, and if Q itself does not fluctuate too widely. In physics, the average energy is often chosen as the macroscopic. Since energy is conserved, in many situations all of the randomness is in other combinations of degrees of freedom. If the macroscopic Q does not explain most of the non-random aspects, or if it is fluctuating greatly, additional macroscopic quantities may be required. This occurs, for example, in the vicinity of phase transitions, where order parameters may be required to determine the phase of the system.

2.3 Application to Evolutionary Dynamics

How this is applied to evolutionary dynamics is the topic of most of this tutorial. The principles can be illustrated by a simple example. A simple set of macroscopics could be the mean and variance of the fitness in the population averaged over all possible runs. If selection is sufficiently weak, and if every run of the GA is statistically similar (see section 3.3), one can easily derive the effects of selection on these two quantities. This is not so for crossover, however. The effectiveness of crossover will depend on how similar the strings are to each other, a quantity called the correlation. If the strings are very similar crossover will make very little change to the population; if the strings are very dissimilar crossover can have a big effect. One way to predict the effect of crossover is to infer the correlation from the mean and variance. How can this be done, given the fact that it is possible for strings to have similar fitness without being correlated? It can be done using a maximum entropy assumption. Consider all possible populations of the same population size and with the same mean fitness and fitness variance as the GA population. For each possible population, one can compute the average correlation in the population. For each correlation value one can assign an entropy, that is the minus the log of the number of populations which have that value. The correlation value which maximises the entropy is the one which is produced by the most populations. The correlation will be that of a population which is as random as possible given the values of the mean and variance. Crossover effects on such a population can be found.

The situation is slightly more complicated than described above. The above would work in an infinite population. In a finite population, higher cumulants than the mean and variance must be considered to characterise selection. Not all of the correlation can be deduced from the maximum entropy distribution. The increase in correlation due to founder effects must be treated separately. The basic idea is otherwise as described above. We will illustrate this in the next section by working through one of the simplest examples where maximum entropy inference is an issue – an inhomogeneous version of *one-max*.

3 Illustration of the Details

3.1 The Problem

The GA will be applied to a class of problems consisting of L binary variables, $x_i \in \{0, 1\}$. An instance of the problem is defined by L 'weights' $J_i \in \mathbb{R}$, which determine the importance of each site. The J's are fixed (for a given problem); the task is to optimise a cost function E over the x's, where

$$E = \sum_{i=1}^{L} J_i x_i.$$

Obviously, this is a trivial problem. The solution is to set x_i to 1 on the sites where J_i is positive, and to 0 at the other sites. It is a generalisation of the *one-max* problem in which each site can contribute a different amount to the cost of a solution (which has important consequences). The methods described in this section can also be applied to cost functions which are non-linear functions of $\sum_{i=1}^{L} J_i x_i$, which includes hard problems. This is described later.

The GA will consist of a population of size N of binary strings of length L; the i-th character of a strings is just x_i. Selection will be a form of roulette-wheel selection. At each generation, N parents will be selected from the current population, where the probability of choosing string α is

$$p^\alpha = \frac{W(E^\alpha)}{\sum_\gamma W(E^\gamma)}. \tag{3}$$

Here $W(E^\mu)$ is the 'selection weight' for a string with cost function E^μ. The selection weight depends on the selection operator, for example:

Fitness proportional selection: $W(E) = E$.
Boltzmann selection: $W(E) = \exp(sE)$, where s is a parameter controlling the selection strength.
Ranking selection: $W(E) = \int_{-\infty}^{E} \rho(E')dE'$.

Much of the analysis has used Boltzmann selection, but tournament selection, ranking selection, and Baker selection have also been looked at [15,19].

For reproduction, two parents are selected at random to produce an offspring, and this is done N times. Uniform crossover will be used in which at each site independently the bit is taken from the first parent with probability a. ($a = 1/2$ corresponds to the usual definition of uniform crossover.) The usual mutation operator will be used. Each site of each of the selected parents will be mutated with a probability γ.

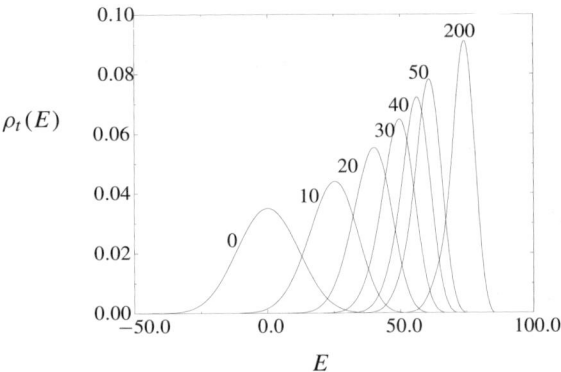

Figure 2. The distribution of costs in a population, averaged over 10,000 runs of the GA. The string length $L = 127$, population size $N = 50$, selection strength $s = 0.1$, mutation rate $\gamma = 1/(2L)$. Crossover probability is 1. Cumulants of this distribution are what we will predict. This figure is from [13].

3.2 Choice of Macroscopics

In order to apply the statistical mechanics approach, we need to decide on a few macroscopic quantities with which to characterise the population. We are obviously interested in the the values of the cost function in the population. Figure 2 shows the distribution of cost function values in the population at different generations, averaged over many runs of the GA. We would like to evolve *statistics* of this distribution. The average distribution of costs in the population at time t will be denoted $\rho_t(E)$.

The distribution of cost function values in the population will be characterised by its *cumulants* [1]. The first two cumulants are the mean and the variance respectively. The third cumulant is related to the 'skewness' of the population; the fourth cumulant is related to the 'kurtosis' or 'excess'. See figure 3 for an illustration of these cumulants. Cumulants are chosen because they average well, and because the complete set of cumulants contains all of the information of the distribution itself. If we take the first few cumulants, we will lose information about the distribution, particularly about the tails. We denote the n-th cumulant by K_n.

The value of the cumulants will vary in each run of the genetic algorithm. In order to make a deterministic system, we will predict the *average* cumulants, denoted κ_n for the n-th one. The averaging is over all runs of the GA (or all realizations of the stochasticity). These average cumulants will serve as our first set of macroscopic quantities.

We will also require a measure of the similarity of the strings in the population. The effect of crossover will depend on the similarity of the strings, as well the expected best member of the population. The obvious measure of similarity is a correlation

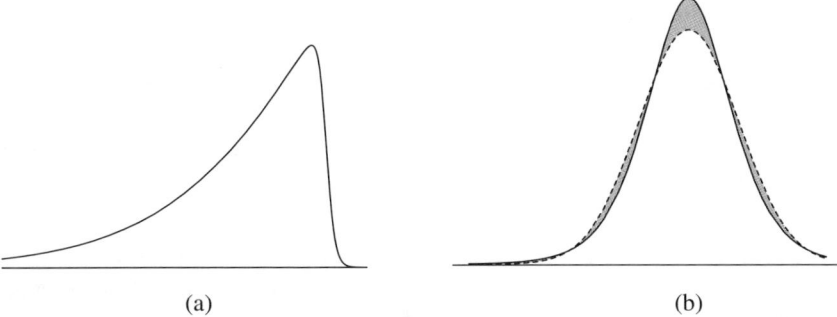

Figure 3. (a) A distribution with negative skewness ($\kappa_3 < 0$). (b) The solid line is a distribution with positive kurtosis; the dashed line is a Gaussian. It is only barely visible, but the Gaussian falls off more quickly for large values.

measure. The one we use is

$$q = \frac{1}{LN^2} \sum_i \sum_\mu \sum_\nu J_i^2 (2x_i^\mu - 1)(2x_i^\nu - 1).$$

This is the average over all pairs of strings of a measure which counts $+J_i^2$ where the strings are the same, and $-J_i^2$ where they differ. This slightly strange form is chosen because it couples most directly to crossover. A completely correlated population will have $q = \sum_i J_i^2/L$; a completely uncorrelated population will have q reduced by a factor $1/N$. Once again, this will need to be averaged over ensembles; the averaged value of q will be denoted φ.

The first goal of the formulation is to produce a set of equations of motion, expressing the macroscopics after selection, mutation, and crossover in terms of those before the operations. At generation t, we have a set of macroscopics:

$$\varphi(t), \kappa_1(t), \kappa_2(t), \ldots, \kappa_n(t),$$

where n is the truncation order (typically 4). What is needed are equations taking us to time $t + 1$:

$$\begin{pmatrix} \varphi(t) \\ \kappa_1(t) \\ \vdots \\ \kappa_n(t) \end{pmatrix} \xrightarrow{\text{selection}} \begin{pmatrix} \varphi^{(s)}(t) \\ \kappa_1^{(s)}(t) \\ \vdots \\ \kappa_n^{(s)}(t) \end{pmatrix} \xrightarrow{\text{mutation}} \begin{pmatrix} \varphi^{(m)}(t) \\ \kappa_1^{(m)}(t) \\ \vdots \\ \kappa_n^{(m)}(t) \end{pmatrix}$$

$$\xrightarrow{\text{crossover}} \begin{pmatrix} \varphi^{(c)}(t) \\ \kappa_1^{(c)}(t) \\ \vdots \\ \kappa_n^{(c)}(t) \end{pmatrix} = \begin{pmatrix} \varphi(t+1) \\ \kappa_1(t+1) \\ \vdots \\ \kappa_n(t+1) \end{pmatrix}.$$

Deriving these equations is the main topic of this tutorial. These must then be iterated or solved.

3.3 Selection Dynamics

We need to compute the effect of selection on the macroscopics. Remember that the macroscopics are ensemble averages, averages over all realizations of the stochasticity in the dynamics. Thus, there are two averages which must be done. First, we must average over all ways of averaging over selecting from a given population. Second, we must average over possible populations.

These two stages could be expressed as follows. Let $\{E^\mu\}$ denote the set of costs in a particular population. Let n^μ be the number of times string μ is selected in a particular population. As an example calculation, consider the mean in the population after selection, which equals

$$K_1^{(s)} = \sum_\mu n^\mu E^\mu.$$

The first stage is to average over the n^μ's. The probability of these, $P(\{n^\mu\}|\{E^\mu\})$, is just a multinomial distribution, with probabilities given in (3). Thus, after the first step

$$K_1^{(s)} = \sum_\mu \frac{W^\mu}{\sum_\nu W^\nu} E^\mu. \qquad (4)$$

The second stage is to average over all populations, i.e., to average (4) over $P(\{E^\mu\})$. This is a problem, however, because this distribution is unknown. All that is known is the average distribution of costs, averaged over all populations. $\rho_t(E)$. We therefore make the following assumption:

$$P(\{E^\mu\}) = \prod_\mu^N \rho_t(E^\mu). \qquad (5)$$

The assumption is that the costs found in a specific population is statistically equivalent to that of the ensemble of populations, i.e., that runs of the GA are statistically similar. This would be false, for example, if for half the runs, the population converged about one value of the cost function, and for the other runs the population converged about another. Covariation across populations is ignored in this approximation, which is in the spirit of the maximum entropy assumption mentioned earlier. A good example where this approximation breaks down is the *Royal Road* problem [24].

Making this assumption allows expressions for the average cumulants after selection. For example, the average mean cost function is

$$\kappa_1^{(s)} = \left\langle \frac{\sum_\mu E^\mu W^\mu}{\sum_\mu W^\mu} \right\rangle \qquad (6)$$

where

$$\langle \cdot \rangle = \prod_\mu^N \int dE^\mu \rho_t(E^\mu).$$

This is an N-dimensional integral, and several ways of computing these integrals have been described elsewhere [13]. I will describe here a simple approach, which is to expand around the infinite population result. In an infinite population, the sum in (6) would be equivalent to an average over all populations. Thus, in the infinite population limit

$$\kappa_1^{(s)} = \frac{\overline{EW}}{\overline{W}}$$

where $\overline{\cdot}$ denotes averaging over the distribution $\rho_t(E)$. To expand around this, write the expression as

$$\kappa_1^{(s)} = \left\langle \frac{\sum_\mu E^\mu W^\mu}{N\overline{W} + \sum_\mu (W^\mu - \overline{W})} \right\rangle$$

and expand in small $\sum_\mu (W^\mu - \overline{W})$

$$\kappa_1^{(s)} = \left\langle \frac{\sum_\mu E^\mu W^\mu}{N\overline{W}} \left[1 - \frac{\sum_\mu (W^\mu - \overline{W})}{N\overline{W}} + \frac{\sum_\mu (W^\mu - \overline{W})^2}{(N\overline{W})^2} - \cdots \right] \right\rangle.$$

This expansion gives the infinite population result as the leading term, and the corrections are powers of the inverse population size. This is an asymptotic expansion, and although it is not obvious from these equations, it is valid if selection is weak. If selection is strong, the population after selection will not be close to that found in an infinite population limit, for example, the population might consist of N identical strings. In this limit, the corrections are not powers of N but logarithmic, see [12,13].

To give an explicit example, let $W(E) = 1 + sE$ with selection strength s. If s is very small, selection is almost random; if s is very large, selection is proportional to the cost function value. (This is the weak selection expansion of Boltzmann selection to first order in the selection strength s.) Then, the first two cumulants are

$$\kappa_1^{(s)} = \kappa_1 + s\left(1 - \frac{1}{N}\right)\kappa_2$$

$$\kappa_2^{(s)} = \left(1 - \frac{1}{N}\right)\kappa_2 + s\left(1 - \frac{3}{N} + \frac{2}{N^2}\right)\kappa_3.$$

The increase in the mean is proportional to the variance in the population, and the variance is reduced through finite population sampling, but can be increased through

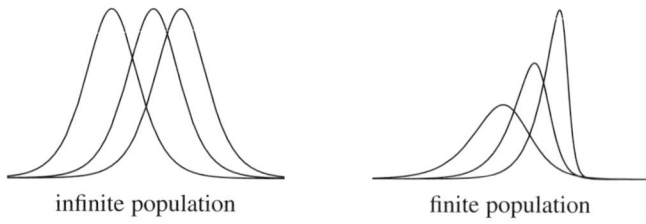

Figure 4. In an infinite population, selection takes an approximately Gaussian distribution of cost functions and produces an approximately Gaussian distribution with a shifted mean. In a finite population, however, selection introduces higher cumulants because selection can only amplify the frequency of the better strings.

coupling with the third cumulant. Likewise for the next two cumulants

$$\kappa_3^{(s)} = \left(1 - \frac{3}{N} + \frac{2}{N^2}\right)\kappa_3 + s\left(1 - \frac{7}{N} + \frac{12}{N^2} - \frac{6}{N^3}\right)\kappa_4$$
$$- s\frac{6}{N}\left(1 - \frac{1}{N}\right)\kappa_2^2$$
$$\kappa_4^{(s)} = \left(1 - \frac{7}{N} + \frac{12}{N^2} - \frac{6}{N^3}\right)\kappa_4 - \frac{6}{N}\left(1 - \frac{1}{N}\right)\kappa_2^2$$
$$+ s\left(1 - \frac{15}{N} + \frac{50}{N^2} - \frac{60}{N^3}\right)\kappa_5 - \frac{60s}{N}\left(1 - \frac{3}{N} + \frac{6}{N^2}\right)\kappa_2\kappa_3.$$

These are reduced through finite population sampling, and can be increased through coupling to the next higher cumulant. However, there is also a non-linear coupling to lower cumulants, for example, the third cumulant couples to the variance squared. This is a finite population effect. It corresponds to the fact that the population becomes more and more skewed (and non-Gaussian) through selection alone. This is because selection can only amplify the strings which are in the population. In an infinite population, all strings are in the population, so the distribution can remain approximately Gaussian while increasing the mean. In a finite population, the population becomes increasingly non-Gaussian under selection. Selection introduces higher cumulants in a finite population; the stronger the selection pressure, the more important these effects are. This difference is illustrated schematically in figure 4.

We still have to consider the effect of selection on the average correlation measure, φ. This is difficult because selection does not act directly on the configuration of strings; it acts on the cost functions. As discussed in section 1, there is no functional relationship between the cost functions and the configurations. Many different populations of strings could give rise to the same set of cost function values, and these different populations could be correlated to differing degrees.

In general, we can use maximum entropy inference to infer the distribution of strings from the distribution of cost function values [13] as described in section 2.2. If $\{x^\mu\}$ is the set of strings in a population, and $\tilde{\kappa}_i(\{x^\mu\})$ denotes the functional relationship between the cumulant and the strings, the maximum entropy distribution is

$$P_{\text{ME}}(\{x\}|\kappa_1,\ldots) = Z^{-1} e^{-\lambda_1 \tilde{\kappa}_1(\{x\}) - \lambda_2 \tilde{\kappa}_2(\{x\}) - \ldots}$$

where Z is the normalisation and the λ's are Lagrange multipliers whose values are chosen to give the known cumulants. This distribution can be used to compute q; averaging over all ensembles would yield the desired macroscopic φ. For calculational reasons, we use

$$P_{\text{ME}}(x|\kappa_1, \varphi) = Z^{-1} e^{-\lambda_1 \tilde{\kappa}_1(x) - \lambda_2 \tilde{\varphi}(x) - \ldots} \tag{7}$$

and use it to calculate κ_2.

To apply this to the current task, the prediction of the effects of selection on the correlation in the population, we could compute the effect of selection on the cumulants, as described above, and use those to compute the correlation. However, there is a problem with this, as pointed out by Rattray [14]. This approach would underestimate the correlations in a finite population. In a finite population, selection can only increase fitness by duplicating strings, and this correlates the population more than a maximum entropy distribution would predict. As a simple example, consider a cost function with two peaks, and the population in a valley between them. As the cost function increases, the population will climb one or both of the peaks. The maximum entropy distribution would predict that both peaks will be populated, because that gives the maximum randomness. However, in a finite population, only one peak may be populated due to 'founder effects'. A string part way up one of the peaks is generated stochastically, and selection duplicates that string.

In order to treat founder effects, Rattray divided the correlation after selection into two parts

$$q^{(s)} \to \underbrace{\frac{1}{L} \sum_i J_i^2 \sum_\mu n^\mu n^\mu}_{\text{founder effect due to duplication}} + \underbrace{\sum_\mu \sum_\nu q^{\mu\nu} n^\mu n^\nu}_{\text{natural increase treated via MaxEnt}}.$$

The duplication term is treated as the cumulants, by averaging over all selections and populations, and expanding around the infinite population,

$$\left\langle \sum_\mu \left(\frac{W^\mu}{\sum_\nu W^\nu} \right)^2 \right\rangle \sim \frac{\overline{W^2}}{N \overline{W}^2} + \ldots$$

$$\sim \frac{1}{N}\left(1 + \kappa_2 s^2 + \ldots\right).$$

The 'natural increase' term would occur in an infinite population due to the decrease in state space for higher fitness states. This is calculated from the maximum entropy distribution, using the cumulants after selection to predict the correlation after selection, as in (7).

To summarise the effects of selection, an expansion has been produced which predicts the macroscopics after selection in terms of those before selection. The results depend on the selection model; we have written down the equations for $W = 1+sE$. In an infinite population, each cumulant couples to the next higher – the mean increases through the variance, the variance is changed via the skewness, etc. The correlation is decreased through the decrease in state space associated with a population of increased value. Important finite population effects have been computed. Higher cumulants are increased through a coupling with lower ones. So selection produces skewness and kurtosis in the population, which makes the better members of the population worse than would be predicted in a Gaussian approximation. The population is more correlated than predicted by an infinite population theory due to founder effects.

3.4 Crossover Dynamics

Next we calculate the effects of crossover on the macroscopics. First, introduce some new crossover variables

$$C_i^{\mu\nu} = \begin{cases} 1 & \text{if site } i \text{ in } \mu \text{ is replaced by site } i \text{ in } \nu \\ 0 & \text{otherwise.} \end{cases} \qquad (8)$$

The strings crossover can be expressed as

$$x^\mu \to x^\mu + (x^\nu - x^\mu) \cdot C^{\mu\nu}.$$

This must be averaged over all realizations of the C's and all populations. The averaging is straightforward, for example

$$K_1^{(c)} = \frac{1}{N^2} \sum_{\mu\nu} \sum_i J_i \left[x_i^\mu + C_i^{\mu\nu} \left(x_i^\nu - x_i^\mu \right) \right]$$

which must be averaged over the C's, the x_i's in the population, and over populations. The statistics of C's are determined by the crossover model. For simplicity, consider uniform crossover with probability a of crossing at each site.

It is clear that crossover does not affect the average correlation. For this problem, crossover will not effect the average value of the cost function either. Thus

$$\kappa_1^{(c)} = \kappa_1$$
$$\varphi^{(c)} = \varphi.$$

The higher cumulants are found by averaging; the variance obeys

$$\kappa_2^{(c)} = \kappa_2 - 2a(1-a)\left[\kappa_2 - \frac{1}{4}\left(\sum_i J_i^2 - L\varphi\right)\right].$$

The variance can be increased due to crossover, but that increase is reduced if the population is correlated. This shows the importance of using the correlation as one of the macroscopics, and of getting the finite population effects. The correlating duplication effects of selection mediate the ability of crossover to restore variation into the population, and this interplay is predicted quantitatively in this formulation.

The higher cumulants are also driven towards fixed point values by crossover. It is useful to think about the nature of the crossover fixed point. Obviously, if crossover is applied repeatedly, the population evolves to a state in which there is no memory of which string each bit came from. In other words, any correlation between different sites in individual strings is removed. The probability that any bit of any string is 1 is just determined by the frequency of 1's in the population before crossover was repeatedly applied. Such a population is said to be at *linkage equilibrium*. (The linkage equilibrium fixed point is what results from one application of *bit-simulated crossover* [22].)

The effects of crossover on the higher cumulants is expressed as follows:[1]

$$\kappa_3^{(c)} = \kappa_3 - 3a(1-a)\left(\kappa_3 - \kappa_3^{\text{LE}}\right) \tag{9}$$

$$\kappa_4^{(c)} = \kappa_4 - 2a(1-a)(2-a(1-a))\left(\kappa_4 - \kappa_4^{\text{LE}}\right) \tag{10}$$

where the superscript LE denotes the cumulants of the fixed-point of crossover, a distribution where every site is drawn independently from a binomial

$$P_{\text{LE}}(x_i^\mu = 1) = \frac{1}{N}\sum_\nu x_i^\nu \equiv \tau_i.$$

This does not completely characterise crossover, however, because we do not know the frequency of 1's in the current population. In particular, we do not know how the frequency of 1's depends upon the sites. Each site contributes differently to the performance of the string through J_i. We do not know how τ_i depends upon (J_i), and neither the average correlation nor the cost function statistics determine this unambiguously.

Again the principle of maximum entropy must be applied. The maximum entropy distribution decouples by site, so it is a linkage equilibrium distribution

$$P_{\text{ME}}(\{x_i^\mu\}) = \prod_i P_{\text{ME}}(\tau_i|J_i)$$

$$P_{\text{ME}}(\tau_i|J_i) = Z_i^{-1} e^{-\lambda_1 J_i \tau_i - \lambda_2 J_i^2 \tau_i^2}.$$

The λ's are chosen to give the values of κ_1 and φ, and then the third and fourth cumulant of this distribution are the predicted fixed point values to put into (9) and (10). Details of how to solve the maximum entropy model are in [13].

To summarise the effects of crossover, it has no effect on the average correlation in the population, and for this simple problem, no effect on the mean of the cost

---
[1] These equations are not exact, but involve approximations I do not wish to explain here.

function, either. The diversity in the population (the variance) is increased typically, but that is related to how similar the strings in the population are. Higher cumulants are returned towards the linkage equilibrium values, and these values must be computed for a set of macroscopics from maximum entropy, since they are functions of the frequency of 1's at each site, which is not known from the chosen macroscopics.

3.5 Mutation Dynamics

Mutation is treated in a similar way to crossover. Mutation variables are introduced, as in (8). These are averaged over, and the ensemble average is taken to produce, for mutation rate γ,

$$q^{(m)} = (1 - 2\gamma)^2 q$$
$$\kappa_1^{(m)} = \kappa_1 + 2\gamma(\kappa_1^{\text{random}} - \kappa_1)$$
$$\kappa_2^{(m)} = \kappa_2 + 4\gamma(1 - \gamma)(\kappa_2^{\text{random}} - \kappa_2).$$

The superscript 'random' denotes cumulants of a random population.

The higher cumulants, like those of crossover, relate to the frequency of 1's as a function of J_i. This is computed from the maximum entropy distribution precisely as in the crossover case. The details are in [13]. The reason that mutation depends on the frequency of 1's as a function of the J_i's is not hard to understand. Consider a string of particular cost E. The effect of mutation will depend on whether the strings obtain this cost by getting a few of the important sites correct (those with J_i's of large magnitude), or by getting many of the less important sites correct. On average it does not matter, but the higher cumulants are effected.

3.6 Putting it Together

This process can be iterated to find the dynamics of the GA. At each stage, the non-linear difference equations of selection, mutation, and crossover must be iterated, along with the maximum entropy model which couples the cumulants to the correlation, and effects of crossover and mutation. Figure 5 shows the result of this with four cumulants and φ for small and intermediate values of the mutation rate. The predictions for the cumulants are very accurate, although the predictions for the average correlation is less so. These figures come from Rattray [15].

For this problem, a picture emerges about the way in which the GA works. Selection increases the mean, reduces the variance, and introduces skewness and higher cumulants. These higher cumulants mean that the high fitness tail of the cost distribution is not very populated, while the low fitness tail is more populated. A schematic picture of this is shown in figure 6(a). Crossover restores higher cumulants such as skewness to the linkage equilibrium value. This repopulates the high fitness tail, as shown schematically in figure 6(b). The ability for crossover to have this effect is mediated by the degree of correlation in the population.

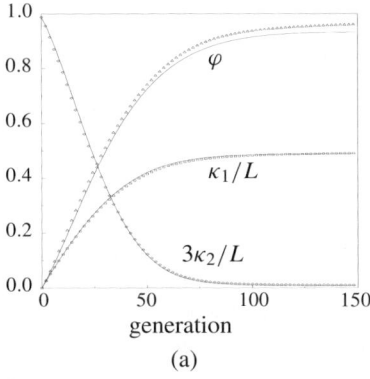 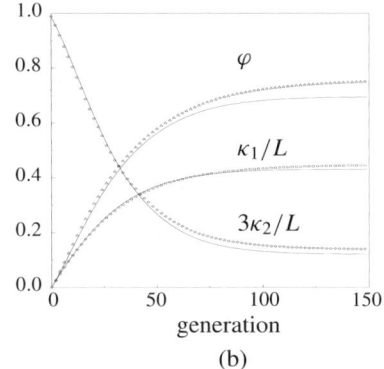

Figure 5. (a) The theory is compared to averaged results for the paramagnet with a low mutation rate, averaged over 5000 runs, with solid lines showing the theory. The weights were chosen from a uniform distribution in the range [0, 1] so that the optimum was $N/2$ on average. The other parameters were $P = 80$, $N = 120$, $\gamma = 0.001$, $s = 0.25$ and uniform crossover was used with $a = 0.5$. (b) The theory is compared to averaged results for the paramagnet with a moderate mutation rate, averaged over 5000 runs, with solid lines showing the theory. All details are as in (a), except that the mutation rate is $\gamma = 0.005$. Figures are from [15].

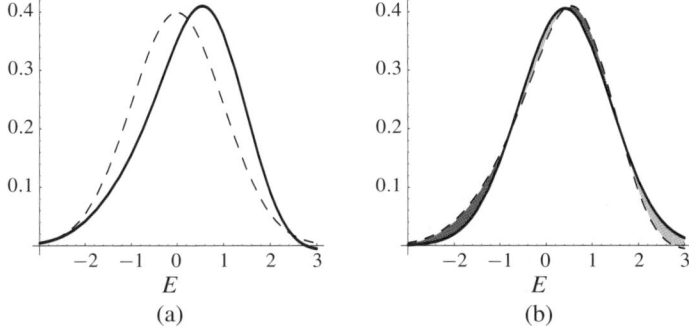

Figure 6. (a) The dashed line represents the distribution of costs before selection; the solid line is that after selection. Although the mean is increased, there is little increase in the high fitness tail. (b) The dashed line represents the distribution of costs before crossover but after selection; the solid line is that after crossover. The gray area shows the increase in the better members of the population.

As the statistical mechanics approach is approximate, it is worth pointing out the sources of error and methods for reducing these where possible. The first source is truncation error. Only a small number of the cumulants are used. If selection is strong or crossover is weak it may be important to use higher cumulants. This is because selection introduces higher cumulants, but crossover reduces the importance of these. This was the case for GAs without crossover in the limit of long strings, in which eight cumulants were used [10].

Selection introduces non-linearities, e.g., κ_2^2. The average of κ_2^2 is not the same as the square of the average of κ_2, which is an assumption we have made. An approach to treat this more rigorously was developed by Prügel-Bennett [10]. He introduced another set of macroscopic quantities, which are the fluctuations around the cumulants.

An important assumption used in describing the effects of selection is encapsulated in (5). The assumption described in this equation is that there is no significant variation across populations, every run of the GA carries out similar dynamics. This is not the case in many problems. For example, if there is a path to the optimum which is hard to find and the optimum has a very high value, those populations which have found the optimum will be very different from those which have not. This is related to the problem discussed in the previous paragraph, because it is caused by the fluctuations of populations across different runs of the genetic algorithm. This type of problem has been well-studied in the Ph.D. thesis of Erik van Nimwegen [23,24]. He used macroscopics which were not averages over GA runs, and maximum inference applied to populations of a single run. In these types of problems, less of the dynamics can be predicted, but this is because fewer aspects of it are the same for different runs.

Finally, it is not always the case that the distribution will satisfy maximum entropy. In principle, it may always be possible to use a maximum entropy assumption of the appropriate macroscopics found, but it may not be clear what these are. From this point of view, the maximum entropy assumption is not a controlled approximation.

4 Application to Harder Problems: A Review of the Literature

The statistical mechanics formalism has been applied to a number of problems. Some examples are in these proceedings. In this section, a brief description of some of this work will be given. The goal is to give the reader an introduction to the literature. In addition to the literature described here, there is a literature on the application of statistical mechanics to biological population genetics. See, for example, the review by Baake [2] and references therein.

First, let us return to the problem presented at the beginning of the paper, the subset sum problem. This problem can be analysed as the simple problem studied in the previous section, but now the cost function is *a function* of the weighted sum of the bits, e.g.,

$$E = \left(\sum_i f_i x_i - D\right)^2.$$

The same macroscopics can be used, that is, cumulants of $\sum_i f_i x_i$ and φ. The only difference is that the selection weight is now a more complicated function. If we define $h = \sum_i f_i x_i$, then the cumulants are statistics of the distribution of h, and the selection weight is a function of $(h - D)^2$.

Details are worked out in [14]. Selection is 'stabilising', which means that after an initial transient the mean stops changing. Improvement occurs through the reduction in variance. However, as the population correlates, the effective population size decreases. This explains the decrease in the expected best string seen in figure 1(b).

Other problems can also be treated as a functions of a weighted-sum of bits. Baum [3] showed that learning in a perceptron with binary (± 1) weights is equivalent to a noisy version on *one-max*. The noise is due to the the fact that the training examples constitute a finite sample of the possible examples. Baum analysed this and a simpler problem, and argued that a GA with strong culling selection performed better than a hill-climbing algorithm, but not as well as the theoretical optimum. Rattray and Shapiro [16] analysed this same problem. They showed that noise increases the finite population effects which have a deleterious effect on the search algorithm. They also showed that the number of training examples could be optimised to minimise the computation costs. The training set needs to be large to decrease the noise-induced finite-population effects, but the cost of the fitness evaluations increase with the size of the training set. The relationship between these quantities obeys a scaling relationship which allows a simplification of the analysis of this trade-off.

Another problem which can be cast as a function of the weighted sum of bits in the string, involves transitions between barriers in a multi-modal landscape. This was originally considered by Shapiro and Prügel-Bennett [21], and has been analysed more thoroughly by Rogers and Prügel-Bennett [18] in this volume, page 207. Using methods related to those described above, one can analyse a cost function of the form $V(\sum_i x_i)$ where V had two peaks. Since transitions between wells is well understood for thermally-induced transitions, this is a case in which a comparison between GAs and simulated annealing can be carried out analytically. For certain forms of V, crossover-induced transitions can occur at a substantially higher rate than thermally-induced ones; a situation in which a GA will outperform simulated annealing.

One topic which is an important test of theory is the question of finding the optimal parameters of GA search. Insufficient work has been done on this problem. Since we have derived equations of motion, we can do *greedy* search on the parameters, i.e., find the parameters which make the biggest improvement at each time-step.

In the weak selection limit shown here, optimal selection strength appears to be ∞. However, a more careful look [13] shows that up to a certain selection strength s^*, the mean increases with little loss diversity ($\sim 1/N$). Beyond that point diversity is lost with almost no increase in the mean

$$s^* = \alpha\sqrt{\kappa_2/(2\log N)}$$
$$\alpha \sim 0.3.$$

This gives an appropriate rescaling of the selection pressure to compensate for the loss of diversity in the population.

In these simple problems, crossover causes very little disruption. Thus, very strong crossover is optimal. To study how to optimise crossover, very different types of problems will need to be considered.

There have been some considerations of the effects of time-varying mutation rates. We studied a the one-max problem in a special limit [17]. Crossover occurs to linkage equilibrium at each time-step (massive crossover). The diffusion limit is taken, $N \to \infty$, $sN \to$ constant, $\gamma N \to$ constant, and discrete dynamics goes to a continuum limit. In this limit, selection + mutation dynamics is linear *even with finite population effects* and the only errors are truncation errors. For this problem, one can choose the mutation rate γ to get the largest increase in κ_1 after mutation + selection. One finds that the mutation rate decays monotonically to zero as the search brings the population towards the optimum.

Bornholdt [4] proposed a phenomenological method for determining the mutation rate at each generation. He defines

$$\tilde{m} = \frac{\langle E E^{(m)} \rangle - \langle E \rangle \langle E^{(m)} \rangle}{\langle E^2 \rangle - \langle E \rangle^2}$$

where $\langle \cdot \rangle$ denotes population and mutation averaging, and $E^{(m)}$ is the objective function after some mutation. (This is like a parent-child correlation function of Manderick.) He also assumes the cumulants after mutation are linear combinations of cumulants before (which is approximately true for one-max):

$$\kappa_1^{(m)} = \tilde{m}\kappa_1 + (1 - \tilde{m})\kappa_1^{\text{random}}$$
$$\kappa_2^{(m)} = \tilde{m}^2\kappa_2 + (1 - \tilde{m}^2)\kappa_2^{\text{random}}$$
$$\kappa_3^{(m)} = \tilde{m}^3\kappa_3$$
$$\kappa_4^{(m)} = \tilde{m}^4\kappa_4.$$

The optimisation criterion is to maximise the best in the population at each generation. Assume

$$E_{\text{best}} \approx \kappa_1^{(m)} + \sqrt{2\kappa_2^{(m)} \log \tilde{N}}$$

where

$$\tilde{N} \sim \begin{cases} N & \text{stochastic hill-climbing} \\ 1 & \text{steady-state GA}. \end{cases}$$

Then one can solve $\tilde{m}$ as a function of the mutation rate for specific problems and optimise it. Bornholdt applied these ideas to the one-max problem, and to Kauffman's NK-landscape which is a rugged fitness function.

4.1 Future Work

The statistical mechanics formulation of GA dynamics produces accurate predictions on complex problems. But many of the questions which motivate theoretical research into GAs remain unanswered. One important question involves the relationship between the search parameters and how to optimise them. We need to study weaker forms of crossover on problems in which the trade-offs between disruption and search are more important. We also need to develop a better understanding of possible annealing schedules, for mutation, selection, and especially for crossover.

GAs are often used in learning, particularly in reinforcement learning. We have only studied the simplest learning system; these methods could be applied to harder learning problems. Of particular interest would be evolutionary reinforcement learning, and how it performs relative to other reinforcement learning algorithms (a question which has not been adequately studied experimentally, either). It would also be interesting to study co-evolving systems or interacting systems [20]. This is a very difficult problem, but is applicable to a wide range of situations.

References

1. M. Abramowitz and I. A. Stegun, editors. *Handbook of Mathematical Functions.* Dover Publications, New York, 1967.
2. E. Baake and W. Gabriel. Biological evolution through mutation, selection, and drift: an introductory review. *Ann. Rev. Comp. Phys. VII,* pages 203–264, 2000.
3. E. Baum, D. Boneh, and C. Grant. *Where Genetic Algorithms Excel.* NEC Research Institute Publication, 1995.
4. S. Bornholdt. Annealing schedule from population dynamics. *Phys. Rev. E,* 59:3942, 1999.
5. M. G. Bulmer. *Mathematical Theory of Quantitative Genetics.* Clarendon Press, Cambridge, 1980.
6. T. Cover and J. Thomas. *Elements of Information Theory.* John Wiley, New York, 1991.
7. M. R. Garey and D. S. Johnson. *Computers and Intractability – A Guide to the Theory of NP-Completeness.* W. H. Freeman, San Francisco, 1979.
8. E. T. Jaynes. *E. T. Jaynes: Papers on Probability, Statistics, and Statistical Physics.* Kluwer, London, 1989.
9. H. Mühlenbein and D. Schlierkamp-Voosen. Analysis of selection, mutation and recombination in genetic algorithms. In W. Banzhaf and F. H. Eckman, editors, *Evolution and Biocomputation,* volume 899 of *LNCS,* pages 142–168. Springer-Verlag, Berlin Heidelberg New York, 1995.
10. A. Prügel-Bennett. Modelling evolving populations. *J. Theoretical Biology,* 185:81–96, 1997.
11. A. Prügel-Bennett. On the long string limit. In W. Banzhaf and C. R. Reeves, editors, *Foundations of Genetic Algorithms 5,* pages 45–56. Morgan Kaufmann, San Francisco, 1999.
12. A. Prügel-Bennett and J. L. Shapiro. Analysis of genetic algorithms using statistical mechanics. *Physical Review Letters,* 72(9):1305–1309, 1994.
13. A. Prügel-Bennett and J. L. Shapiro. The dynamics of a genetic algorithm for the ising spin-glass chain. *Physica D,* 104:75–114, 1997.
14. L. M. Rattray. Dynamics of a genetic algorithm under stabilizing selection. *Complex Systems,* 9:213–234, 1995.

15. L. M. Rattray. *Modelling the Dynamics of Genetic Algorithms Using Statistic Mechanics*. PhD thesis, Computer Science Dept., University of Manchester, Oxford Road, Manchester M13 9PL, UK, 1996.
16. L. M. Rattray and J. L. Shapiro. The dynamics of a genetic algorithm for a simple learning problem. *Journal of Physics A*, 29:7451–7473, 1996.
17. L. M. Rattray and J. L. Shapiro. Cumulant dynamics in a finite population: linkage equilibrium theory. Preprint, available from LANL preprint server as adap-org/9907009, *http://xxx.lanl.gov/abs/adap-org/9907009*, 1999.
18. A. Rogers and A. Prügel-Bennett. A solvable model of a hard optimisation problem. This volume, pages 207–221.
19. A. Rogers and A. Prügel-Bennett. Modelling the dynamics of a steady-state genetic algorithm. In W. Banzhaf and C. R. Reeves, editors, *Foundations of Genetic Algorithms 5*, pages 57–68. Morgan Kaufmann, San Francisco, 1999.
20. J. L. Shapiro. Does data-model co-evolution improve generalization of evolving learners? In A. E. Eiben, Th. Bäck, M. Schoenauer, and H.-P. Schwefel, editors, *Proceedings of the 5th Conference on Parallel Problem Solving from Nature*, volume 1498 of *LNCS*, pages 140–149. Springer-Verlag, Berlin Heidelberg New York, 1998.
21. J. L. Shapiro and A. Prügel-Bennett. Genetic algorithm dynamics in a two-well potential. In R. K. Belew and M. D. Vose, editors, *Foundations of Genetic Algorithms 4*, pages 101–116. Morgan Kaufmann, San Francisco, 1997.
22. G. Syswerda. Simulated crossover in genetic algorithms. In L. D. Whitley, editor, *Foundations of Genetic Algorithms 2*, pages 239–256. Morgan Kaufmann, San Francisco, 1993.
23. E. van Nimwegen. *The Statistical Dynamics of Epochal Evolution*. PhD thesis, Universiteit Utrecht, The Netherlands, 1999.
24. E. van Nimwegen, J. Chrutchfield, and M. Mitchell. Finite populations induce metastability in evolutionary search. *Physics Letters A*, 229:144–150, 1997.

Theory of Evolution Strategies – A Tutorial

H.-G. Beyer and D. V. Arnold

Department of Computing Science XI
University of Dortmund
44221 Dortmund, Germany
E-mail: *{beyer,arnold}@ls11.cs.uni-dortmund.de*

Abstract. Evolution strategies form a class of evolutionary optimization procedures the behavior of which is comparatively well understood theoretically, at least for simple cases. An approach commonly adopted is to view the optimization as a dynamical process. Local performance measures are introduced so as to arrive at quantitative results regarding the performance of the algorithms. By employing simple fitness models and approximations exact in certain limits, analytical results can be obtained for multi-parent strategies including recombination and even for some simple strategies employing self-adaptation of strategy parameters. This tutorial outlines the basic algorithms and the methods applied to their analysis. Local performance measures and the most commonly used fitness models are introduced and results from performance analyses are presented. Basic working principles of evolution strategies – namely the evolutionary progress principle, the genetic repair principle, and the phenomenon of speciation – are identified and discussed. Finally, open problems and areas of future research are outlined.

Keywords
Evolution strategy, dynamics, progress rate

1 Introduction

Like genetic algorithms (GA), evolutionary programming (EP), and genetic programming (GP), evolution strategies (ES) form a class of evolutionary algorithms (EA) whose attractiveness lies in their robustness and relative ease of implementation. Unlike the current situation for some of the other variants, there exists a comparatively large body of theoretical, quantitative knowledge regarding the optimization behavior of ES at least in simple situations. This tutorial presents an overview of that knowledge and outlines the methods employed to obtain it. It focuses on ES research in a line propelled by, among others, Rechenberg [14,15], Schwefel [19,21], Scheel [18], Rudolph [16], and Beyer [2–7], and employs an approach that has the evaluation of local performance at its base. It is important to note that other approaches do exist. The problem of ascertaining global convergence has been addressed in Rudolph [16,17]. Recently, Droste et al. [10–12] have approached the field of evolutionary computation from a computational complexity perspective.

This tutorial is organized as follows. Section 2 introduces the $(\mu/\rho \stackrel{+}{,} \lambda)$-algorithm as a canonical ES, discusses a number of fitness models that have been the subject of investigations, and presents the ideas applied to the performance analysis of variants of the $(\mu/\rho \stackrel{+}{,} \lambda)$-algorithm on the fitness models. Section 3 outlines the most important results from investigations of the performance of ES on real-valued spaces and discusses the insight gained. In particular, the evolutionary progress principle, the genetic repair principle, and the phenomenon of speciation are identified as basic working principles of ES. Section 4 concludes with a brief summary and suggestions for future work.

2 Algorithms, Fitness Models, and Methods of Analysis

This section first outlines the kind of optimization problem to be dealt with in the following and introduces notational conventions. Then it presents the $(\mu/\rho \stackrel{+}{,} \lambda)$-algorithm with self-adaptation of strategy parameters as a canonical ES. Self-adaptation refers to the idea of making the strategy parameters of the ES part of the genetic information of the individuals, and to thereby implicitly subject them to the process of evolutionary optimization along with the object variables of the problem. Then, fitness models that have been used for the analysis of ES are outlined. Finally, a view of ES as dynamical systems is presented and corresponding concepts and methods employed for analyzing ES are described. In particular, local performance measures are introduced and their importance for the understanding of ES is discussed.

2.1 Preliminaries

ES were originally conceived as an algorithm for solving complex technical optimization problems. In all of the following, we assume real-valued parameter and fitness spaces. That is, the general problem considered is of the form

$$f(\hat{\mathbf{y}}) = \min\{f(\mathbf{y}) | \mathbf{y} \in M \subseteq \mathbb{R}^n\}$$

where f is the fitness function mapping vectors $\mathbf{y} = (y_1, \ldots, y_n)^{\mathrm{T}}$ with real-valued components $y_i \in \mathbb{R}$ to values $f(\mathbf{y}) \in \mathbb{R}$. The support of f may be restricted to a proper subset of $\mathbb{R}^n$ by means of various constraints. It is emphasized that as the task is minimization, high fitness corresponds to low values of the fitness function and vice versa. Note also that the restriction to real-valued problems is not a necessary one. While most of the results of the approach outlined in this tutorial have been obtained for real-valued optimization problems, studies of the convergence behavior of ES in Boolean, integer, and mixed parameter spaces do exist. Instead of presenting their results we simply refer to Rudolph [16] and Droste et al. [10–12].

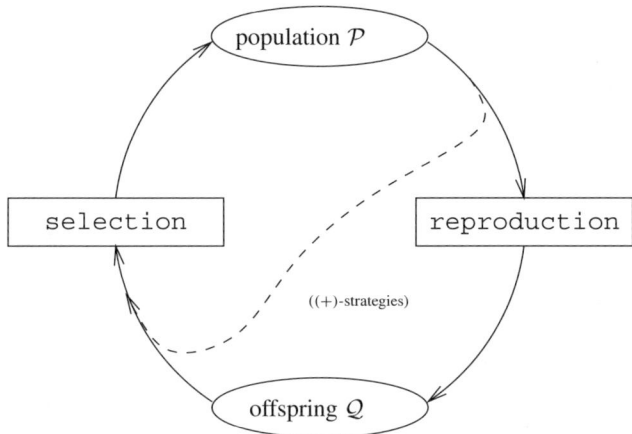

Figure 1. The general ES algorithm. For (,)-strategies selection is from the pool of offspring $\mathcal{Q}$, for (+)-strategies it is from the union $\mathcal{P} \cup \mathcal{Q}$ as indicated by the dashed line.

2.2 The General ES-Algorithm

A $(\mu/\rho \overset{+}{,} \lambda)$-ES operates on a population $\mathcal{P}$ of μ individuals. Each individual is a pair $(\mathbf{y}, \mathbf{s})$, comprised of an object variable vector $\mathbf{y}$ in search space $\mathbb{R}^n$ and a strategy parameter set $\mathbf{s}$. The purpose of the strategy parameter set will be outlined in connection with the discussion of mutations below.

In ES, as in other EA as well, time proceeds in discrete steps often referred to as generations. In every time step t, a pool $\mathcal{Q}^{(t)}$ of λ offspring individuals is created from $\mathcal{P}^{(t)}$ by means of reproduction. Reproduction involves, for each individual to be generated, the random selection of ρ parents, the recombination of their genetic information, and the subsequent mutation of that information. The individuals to form the population $\mathcal{P}^{(t+1)}$ of time step $t+1$ are selected – depending on the selection type – either from $\mathcal{P}^{(t)} \cup \mathcal{Q}^{(t)}$ or from $\mathcal{Q}^{(t)}$ on the basis of individual fitness. Figure 1 illustrates the algorithm. The following paragraphs introduce the genetic operators of selection, recombination, and mutation. As their design is a critical factor for the performance of ES, an understanding of their behavior is desirable.

Selection. Selection in ES is deterministic, with the $(\overset{+}{,})$ symbolism denoting the two mutually exclusive selection types. Using (+)-selection, the μ best of the $\mu + \lambda$ individuals in $\mathcal{P}^{(t)} \cup \mathcal{Q}^{(t)}$ are selected to form $\mathcal{P}^{(t+1)}$. Using (,)-selection, the life-time of an individual is restricted to a single time step and it is the μ best of the λ individuals in $\mathcal{Q}^{(t)}$ that form $\mathcal{P}^{(t+1)}$. Obviously, (,)-strategies require $\lambda \geq \mu$. According to Schwefel [20], empirical investigations have shown that for real-valued optimization problems often (,)-selection seems superior.

Recombination. Recombination is a process in which both the object components and the strategy components of ρ randomly selected parent individuals are combined to form the corresponding components of an offspring individual. For a particular offspring individual to be generated, let $i_k \in \{1, \ldots, \mu\}$, $k = 1, \ldots, \rho$, denote the indices of the parent individuals in the population. Note that this does not exclude the possibility of one individual being drawn several times as a parent for a single offspring individual. However, for the $(\mu/\mu, \lambda)$-ES considered below it is assumed that all μ individuals in the parent population participate in the generation of any single offspring individual, effectively ruling out the possibility of an individual being drawn more than once. As currently no theoretical results regarding self-adaptation in multi-parent ES exist, we refrain from discussing the recombination of strategy parameters here.

Recombination can be either intermediate or dominant, with the dominant variant sometimes being referred to as discrete recombination. Using intermediate recombination, the object component y of an offspring individual is generated by averaging the parental components to yield

$$y = \frac{1}{\rho} \sum_{k=1}^{\rho} y_{i_k}.$$

Using dominant recombination, for each component of the object component y of the offspring individual to be generated, one of the ρ parent individuals is randomly selected to exclusively transfer its respective component to the offspring individual, yielding

$$y = \sum_{j=1}^{n} \left(e_j^T y_{i_{k_j}}\right) e_j$$

where the k_j are numbers randomly drawn from $\{1, \ldots, \rho\}$. Note that in the special case of $\rho = 1$ both recombination variants are identical and effectively there is no (nontrivial) recombination. The corresponding strategy is usually referred to as $(\mu \stackrel{+}{,} \lambda)$-ES.

Mutation. In the process of mutation both the object and the strategy components of a newly created individual are subjected to small random disturbances as a source of innovation and genetic variation. In that process, the strategy parameters are used for determining the size and direction of the disturbances to be applied to the object variables. The central idea of self-adaptation of the strategy parameters is to mutate the strategy component of an individual first, and then to use the newly generated set of strategy parameters in the process of applying disturbances to the object component. The underlying reasoning is that a well-adapted set of strategy parameters is more likely to result in a fitness improving mutation of the object component than an ill-adjusted set, and that it will therefore be more likely to survive the process of selection. The discussion of mutation operators for strategy parameters is deferred until section 3.5.

Mutations of the object component of an individual are usually accomplished by adding a normally distributed random vector $z = (z_1, \ldots, z_n)^T \in \mathbb{R}^n$ with density

$$p(z) = \frac{1}{(\sqrt{2\pi})^n \sqrt{\det[C]}} \exp\left(-\frac{1}{2} z^T C^{-1} z\right)$$

to y[1]. Note that this way mutations are unbiased as z has zero mean, and are scalable in the sense that adapting the covariance matrix C provides control of both mean step size and direction. For optimal performance of the algorithm a control scheme for adapting the covariance matrix C to the local structure of the fitness function is required. One possibility is to make C part of the strategy parameter component of the individuals, thereby employing self-adaptation for the control of mean mutation step size and direction.

To simplify things, both in many applications as well as in theoretical analyses, it is assumed that mutations are isotropic, i.e., that $C = \sigma^2 \mathbf{1}_{n \times n}$, where $\mathbf{1}_{n \times n}$ denotes the $n \times n$ unity matrix. In that case, only the single parameter σ, in the following referred to as the mutation strength, needs to be controlled.

2.3 Fitness Models

The performance of ES is analyzed on selected fitness functions. Although certainly not the kind of problems ES are the most appropriate optimization methods for, these functions provide valuable insights into the working principles of ES as they are simple enough to allow for an analytical evaluation of ES performance. Moreover, they can be used to locally approximate aspects of much more complex fitness functions. For example, it is often the case that the vicinity of a local optimum of a multimodal fitness landscape can be approximated by a simple model such as the hypersphere.

The hypersphere model is the most commonly used and at the same time one of the simplest fitness models. In this model the fitness of an individual depends on its distance to the optimum alone, making

$$f(y) = g(\|y - \hat{y}\|) \qquad (1)$$

where g is a function of a single parameter. Usually, as in what follows, g is additionally required to be strictly increasing. The theoretical results reported in section 3 have been derived for this model.

Provided that the ES considered is invariant with respect to coordinate rotations, the results from the hypersphere theory can immediately be extended to the hyperplane

$$f(y) = c^T y$$

by formally letting the distance from the optimum tend to infinity while considering finite mutation strengths.

---

[1] For a discussion of Cauchy distributed mutations see [17].

Situations with an open success region in which progress towards the optimum is along a single direction while deviations from that direction are penalized, have been investigated in the frameworks of the corridor and the discus models, both being inclined hyperplanes with different kinds of constraints, and the ridge function model

$$f(\mathbf{y}) = y_1 + d \left(\sum_{i=2}^{n} y_i^2 \right)^{\frac{\alpha}{2}}, \quad \alpha, d > 0$$

analyzed by Oyman et al. [13].

An approach to locally approximating any differentiable fitness function f is to consider its Taylor expansion

$$f(\mathbf{y} + \mathbf{z}) = f(\mathbf{y}) + \sum_{i=1}^{n} \frac{\partial f}{\partial y_i} z_i + \frac{1}{2} \sum_{i=1}^{n} \sum_{j=1}^{n} \frac{\partial^2 f}{\partial y_i \partial y_j} z_i z_j + \ldots$$

up to and including the quadratic term. This naturally leads to the definition of the local quality function

$$q_{\mathbf{y}}(\mathbf{z}) = f(\mathbf{y} + \mathbf{z}) - f(\mathbf{y}) = \mathbf{a}^T \mathbf{z} + \mathbf{z}^T \mathbf{Q} \mathbf{z}$$

where $\mathbf{a} = \nabla f(\mathbf{y})$ and $\mathbf{Q} = \nabla^2 f(\mathbf{y})$ are the gradient and the Hessian matrix of f at location $\mathbf{y}$, respectively, as a general quadratic fitness model. The usefulness of this model for the quality gain analysis of general quadratic fitness functions has been demonstrated in [3]. Alternatively, results from the hypersphere theory can be applied to more general fitness functions by employing a differential geometric approach developed in [5] that uses the mean radius

$$r_{(x)} = \frac{n-1}{2} \frac{\|\mathbf{a}\|}{\left| \text{Tr}[\mathbf{Q}] - \mathbf{a}^T \mathbf{Q} \mathbf{a} / \|\mathbf{a}\|^2 \right|}$$

of the fitness landscape at location $\mathbf{y}$ as a surrogate radius in the hypersphere theory. This approach has been found useful at least for quadratic fitness functions with a not too degenerate eigenvalue spectrum of the Hessian.

Finally, the effects of noise can be included in theoretical investigations by adding a normally distributed term with mean zero to the fitness of an individual to model the effects of influences ranging from measurement errors and the use of stochastic simulation procedures to user interaction. A survey of noise related results from ES theory can be found in [1].

2.4 Methods of Analysis

An ES together with a fitness function forms a dynamical system the behavior of which can be studied in the time domain. The goal of much of the research in the field of ES is to obtain quantitative results regarding the dynamics of different strategies when applied to selected fitness functions, thereby providing an understanding of the working principles of the genetic operators and making it possible to predict their performance. In a picture used in [7], the goal of ES research is to formulate and solve the "equations of motion" governing the dynamics of ES and to identify in the process the microscopic "forces" at work that drive the ES towards better and better solutions.

Computing the full dynamics of ES turns out to be intractable even in the simplest cases. The individuals of a population $\mathcal{P}$ are points in state space $\mathcal{A} = \mathbb{R}^n \times \mathbb{R}_+$[2]. Evolution takes place in the product space

$$\mathcal{A}^\mu = \underbrace{\mathcal{A} \times \mathcal{A} \times \cdots \times \mathcal{A}}_{\mu \text{ times}}.$$

The evolutionary operators of recombination, mutation, and selection define a Markovian process, stochastically mapping $\mathcal{A}^\mu$ into itself, which is readily described by the Chapman–Kolmogorov equation

$$p^{(t+1)}(\mathcal{P}) = \int p^{(t)}(\mathcal{P}')p(\mathcal{P}|\mathcal{P}')d\mathcal{P}'. \qquad (2)$$

Due to the mathematical difficulties involved this equation cannot be used to determine the population probability distribution. Already the task of formally expressing the transition probabilities $p(\mathcal{P}|\mathcal{P}')$ in closed form is impossible except in the simplest cases.

Fortunately, however, much of the dynamics of ES can be recovered by observing statistical parameters and their expected rates of change from one generation to another. Two important such parameters are the average distance of the population from the optimum in object variable space and in the space of fitness values. Their respective rates of change, the progress rate

$$\varphi = \mathrm{E}\left[r^{(t)} - r^{(t+1)}\right] \qquad (3)$$

where $r^{(t)} = \|\hat{\mathbf{y}} - \langle \mathbf{y}^{(t)} \rangle_\mu\|$ denotes the distance of the center of mass of the population from the optimum at time t, and the expected quality gain

$$\overline{Q} = \mathrm{E}\left[\langle f^{(t)} \rangle_\mu - \langle f^{(t+1)} \rangle_\mu\right] \qquad (4)$$

measure the expected improvement within a single generation. Note that in general progress rate and quality gain are by no means equivalent, and that it is possible to have negative progress rate and positive quality gain at a time and vice versa. Figure 2 displays two situations in which high progress rate and large quality gain are conflicting goals. Also note that the quality gain is an observable quantity that can be

---
[2] It is assumed that there is only a single endogenous strategy parameter, the mutation strength σ.

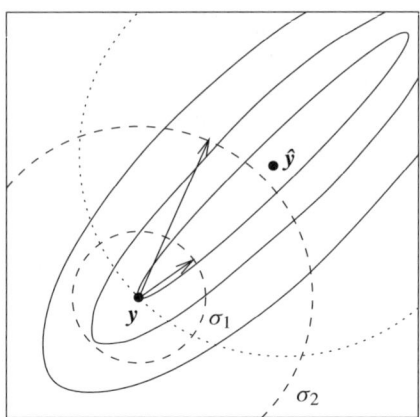 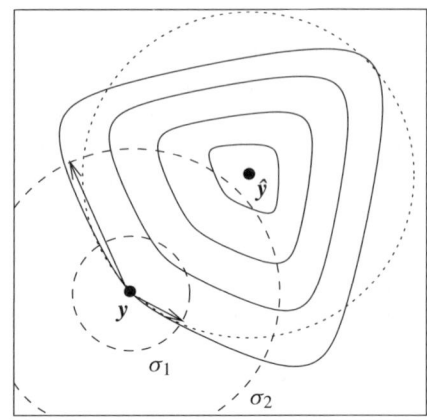

Figure 2. Progress rate versus quality gain. Solid lines indicate surfaces of constant fitness. For large n mutations are on hypersphere surfaces with center $\mathbf{y}$. The left part of the figure displays a situation in which the maximal progress rate is obtained for a mutation strength much higher than that for which the quality gain is optimized. Using mutation strength σ_1 the quality gain is higher than for mutation strength σ_2, but the progress rate is lower. The right part of the figure shows an example for the opposite case. Here, there are mutations resulting in a positive quality gain while increasing the distance from the optimum.

used for mutation strength adaptation algorithms while the progress rate generally is not accessible as the optimum $\hat{\mathbf{y}}$ is, of course, unknown. Throughout the following, only progress rates will be considered.

In section 3, the results from progress rate analyses of different strategies operating on the hypersphere model will be reported. Using normalized mutation strength $\sigma^* = \sigma n/r$ it will be shown that within the limits of the approximation the normalized progress rate $\varphi^* = \varphi n/r$ does not depend on either n or r. Therefore, using continuous time in an approximation to the discrete character of the optimization process, from the definition of the progress rate it follows that

$$\frac{dr}{dt} = -\frac{\varphi^*(\sigma^*(t))}{n} r(t). \qquad (5)$$

Assuming some mechanism that assures that σ^* is constant over time, the normalized progress rate $\varphi^*(\sigma^*)$ is constant and it can easily be verified that the differential equation (5) is solved by

$$r(t) = r(0) \exp\left(-\frac{\varphi^*(\sigma^*)}{n} t\right)$$

revealing that an appropriate adjustment of mutation strength results in linear convergence order. As the number of time steps required to achieve a given accuracy r

equals

$$t = \frac{n}{\varphi^*(\sigma^*)} \log\left(\frac{r(0)}{r}\right)$$

the time required to achieve this accuracy is linear in n.

Mathematical tools used to derive the results cited below include some basic results from order statistics. In particular, use is made of the fact that the probability density function of the ith smallest of λ independent and identically distributed continuous random variables with distribution function P and corresponding probability density p is

$$p_{i:\lambda}(x) = i\binom{\lambda}{i} p(x) [P(x)]^{i-1} [1 - P(x)]^{\lambda-i}. \tag{6}$$

Another mathematical tool employed for the derivation of the following results are Gram-Charlier expansions of probability distributions. Gram-Charlier expansions make it possible to approximate probability distributions with known moments which are approximately normal by successively adding correction terms to a normal distribution function. In the derivation of progress rates and quality gain of ES these expansions make it possible to obtain approximations of increasing accuracy for which the occurring integrals can be analytically evaluated.

3 Results

This section presents results and insight from theoretical analyses of the behavior of ES on the hypersphere model. Although with difficulty, approximations for progress rate and quality gain in dependence on parameters such as n, σ, μ, and λ can be obtained. Because emphasis is not so much on the accuracy of an approximation but rather on the basic working principles of ES, we restrict ourselves here to results obtained in the limit $n \to \infty$. Better approximations that show good accordance with experiments for moderate parameter space dimensions which generalize most of the results cited here can be found in [7].

Section 3.1 discusses the evolutionary progress principle introduced in [8] instances of which will reappear throughout this section. Sections 3.2 and 3.3 present results from investigations of single- and multi-parent ES without recombination and self-adaptation. In particular, (1 + 1)-ES and (1, λ)-ES as the most commonly used single-parent strategies and (μ, λ)-ES as a multi-parent strategy are discussed. Section 3.4 deals with the effects of recombination. It introduces the genetic repair principle and discusses the phenomenon of speciation. Finally, section 3.5 includes self-adaptation of the mutation strength σ in (1, λ)-ES into the analysis.

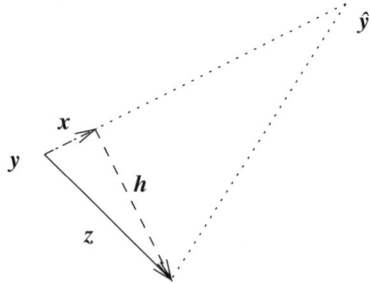

Figure 3. Decomposition of a mutation *z* into a component *x* in direction of the optimum and a component *h* perpendicular to that

3.1 The Evolutionary Progress Principle

The *evolutionary progress principle* as formulated in [8] states that any mutation *z* can be decomposed into a component *x* towards the optimum and a component *h* perpendicular to that. Figure 3 illustrates the situation. The component towards the optimum provides a progress gain as it decreases the distance to the optimum, while the component perpendicular to it increases that distance. Increasing the mutation strength increases both the gain and the loss components, but as the loss component usually grows at a faster rate there is an optimal mutation strength for which the difference between gain and loss attains a maximum. Performance-improving measures can alternatively aim at increasing the gain or at decreasing the loss of mutations. The following paragraphs discuss the evolutionary progress principle in connection with several instances of ES.

3.2 Single-Parent Strategies

Single-parent strategies are the easiest to investigate analytically as there is no need to keep track of a population distribution. The two most commonly used and investigated single-parent strategies are the $(1+1)$-ES and the $(1, \lambda)$-ES.

The (1+1)-ES. In $(1+1)$-ES the single individual forming the population generates one offspring individual at a time which replaces the parent individual if and only if it has a higher fitness. The progress rate law

$$\varphi^*_{1+1}(\sigma^*) = \frac{\sigma^*}{\sqrt{2\pi}} e^{-\frac{1}{8}\sigma^{*2}} - \frac{\sigma^{*2}}{2}\left[1 - \Phi\left(\frac{\sigma^*}{2}\right)\right] \quad (7)$$

was first derived by Rechenberg [14]. The term in brackets is the probability of an offspring having higher fitness than its parent and is referred to in what follows as success probability. The first term in (7) is the gain term identified in the evolutionary progress principle, the second term is the loss term. For small σ^* the gain term increases linearly with σ^*, but is quickly muted by the exponential term as σ^* grows. The loss term is quadratic in σ^* for small σ^*, but is moderated by the decreasing success probability for larger σ^*. For $\sigma^* \to \infty$ both terms approach zero. Figure 4 depicts the progress rate φ^*_{1+1} as a function of mutation strength σ^* for several values of the parameter space dimension n, demonstrating that the deviation from (7) which is valid exactly only in the limit $n \to \infty$ is moderate for $n \geq 30$. For $n \to \infty$ the maximal progress rate $\hat{\varphi}^*_{1+1} = \max_{\sigma^*}[\varphi^*_{1+1}(\sigma^*)]$ that can be achieved is

$$\hat{\varphi}^*_{1+1} \approx 0.202 \quad \text{for} \quad \hat{\sigma}^*_{1+1} \approx 1.224. \tag{8}$$

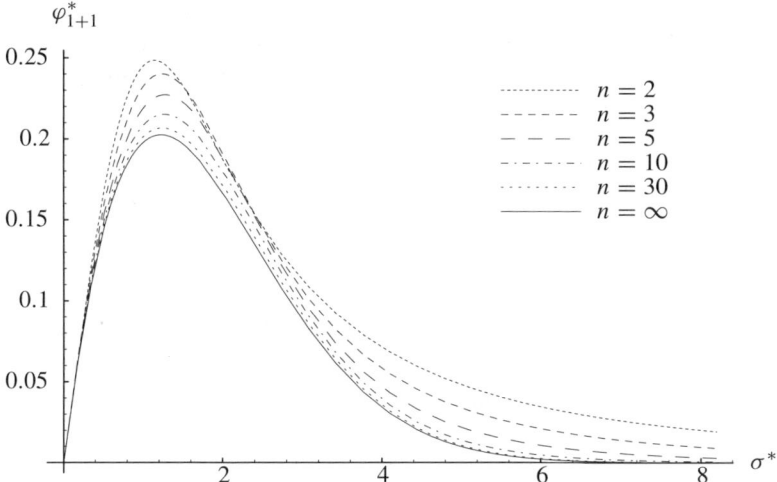

Figure 4. Progress rate φ^*_{1+1} as a function of mutation strength σ^* for the $(1+1)$-ES on hyperspheres of several parameter space dimensions n. The results for $n < \infty$ have been obtained by numerical integration of the exact progress rate integrals. See [7] for details.

An approach due to Rechenberg [14], the 1/5th success rule, uses the observation that the success probability is a strictly decreasing function of σ^* for a mutation strength control scheme. The success probability in case of maximal progress is about 0.270. For the corridor model analyzed in [14] that probability is approximately 0.184. Assuming that for other fitness models the optimal success probability is within that range as well, it seems reasonable to measure the success probability by averaging over a number of generations, and to increase the mutation strength if the measured success probability is too high and decrease it if it is too low.

The $(1, \lambda)$-ES. $(1, \lambda)$-ES differ from $(1 + 1)$-ES in that λ offspring individuals are generated at a time – naturally allowing for parallel processing – the best of which replaces the parent even if it is inferior. Obviously, a $(1, 1)$-ES merely performs a random walk. The progress rate law

$$\varphi^*_{1,\lambda}(\sigma^*) = c_{1,\lambda}\sigma^* - \frac{\sigma^{*2}}{2} \tag{9}$$

for the $(1, \lambda)$-ES on the infinite dimensional hypersphere has been obtained by a number of authors, including Scheel [18]. The factor $c_{1,\lambda}$ in (9) is the $(1, \lambda)$-progress coefficient defined by

$$c_{1,\lambda} = \frac{\lambda}{\sqrt{2\pi}} \int_{-\infty}^{\infty} x e^{-\frac{1}{2}x^2} [\Phi(x)]^{\lambda-1} \, dx.$$

It captures the influence of selection and can be obtained by numerical integration. The first few $(1, \lambda)$-progress coefficients are listed in table 1. A comparison with (6) reveals that $c_{1,\lambda}$ is the expectation of the maximum of λ independent standard normally distributed random variables[3]. In [2] it has been proven that

$$c_{1,\lambda} = \mathcal{O}(\sqrt{\log \lambda}) \tag{10}$$

showing that the progress coefficients grow only slowly with λ.

λ	1	2	3	4	5	6	7	8
$c_{1,\lambda}$	0.000	0.564	0.846	1.029	1.163	1.267	1.352	1.424

Table 1. The first few $(1, \lambda)$-progress coefficients

The first term in (9) is the gain term identified in the evolutionary progress principle, the second term is the loss term. Obviously, the gain is linear in σ^* while the loss is quadratic. The maximal progress rate that can be achieved is

$$\hat{\varphi}^*_{1,\lambda} = \frac{c^2_{1,\lambda}}{2} \quad \text{for} \quad \hat{\sigma}^*_{1,\lambda} = c_{1,\lambda}. \tag{11}$$

For $\sigma^* > 2c_{1,\lambda}$ the progress rate is negative and the quality of the solution deteriorates over time. Increasing the number of offspring λ increases the gain term by increasing the progress coefficient $c_{1,\lambda}$. However, from (10) and (11) it follows that

$$\hat{\varphi}^*_{1,\lambda} = \mathcal{O}(\log \lambda). \tag{12}$$

---

[3] Rudolph [16], p.181, notes that the approximation using normal random variables employed in the derivation of (9) leads to asymptotically incorrect behavior for large λ and suggests that it is necessary to employ order statistics of Beta-distributed random variables for determining the asymptotic behavior of $(1, \lambda)$-ES for $\lambda \to \infty$. This is correct, however, only if λ tends to ∞ much faster than n does. If $\log \lambda / n \to 0$ then $c_{1,\lambda}$ as defined above correctly describes the asymptotic behavior.

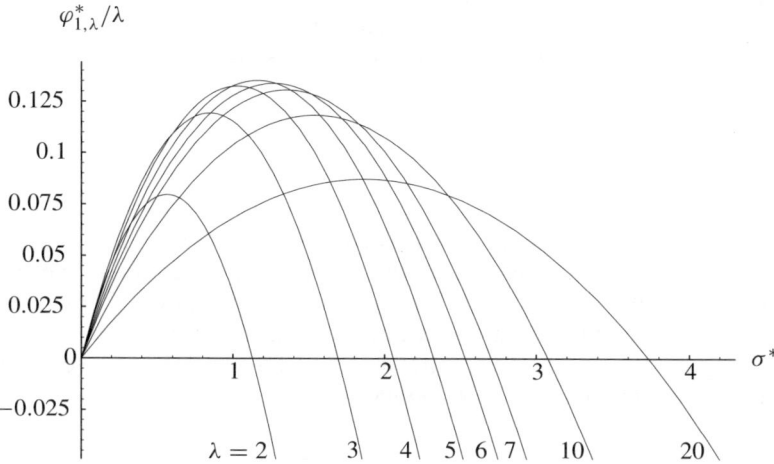

Figure 5. Progress rate per offspring $\varphi^*_{1,\lambda}/\lambda$ as a function of mutation strength σ^* for different values of the number of offspring λ

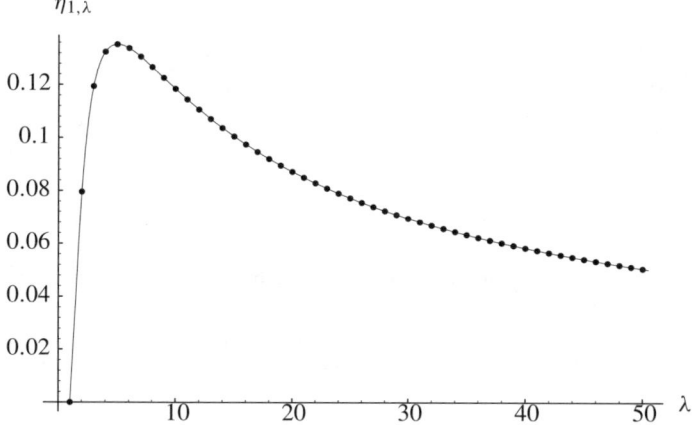

Figure 6. Fitness efficiency $\eta_{1,\lambda}$ of the $(1, \lambda)$-ES as a function of the number of offspring λ

Thus, the speedup that can be achieved by increasing the number of offspring is only logarithmic in λ while the computational effort is linear, making it uneconomical to increase the number of offspring beyond a certain value. Figure 5 shows the progress rate per offspring $\varphi^*_{1,\lambda}/\lambda$ as a function of σ^* for different values of λ. Figure 6 shows the fitness efficiency

$$\eta_{1,\lambda} = \frac{\hat{\varphi}^*_{1,\lambda}}{\lambda} = \frac{c^2_{1,\lambda}}{2\lambda}$$

which measures the maximal progress per offspring that can be obtained using a $(1, \lambda)$-ES as a function of λ. It can be seen that $\eta_{1,\lambda}$ is maximal for $\lambda = 5$, making five the optimal number of offspring for a $(1, \lambda)$-ES on the hypersphere.

3.3 Multi-Parent Strategies

In contrast to the single-parent strategies considered so far, (μ, λ)-ES with $\mu > 1$ operate on a population of individuals that is formed in the course of the evolutionary process and the distribution of which is not known a priori. Consequently, the mathematical difficulties encountered during their analysis are much higher than for the single-parent strategies. The central problem of the analysis is the determination of a self-consistent population distribution.

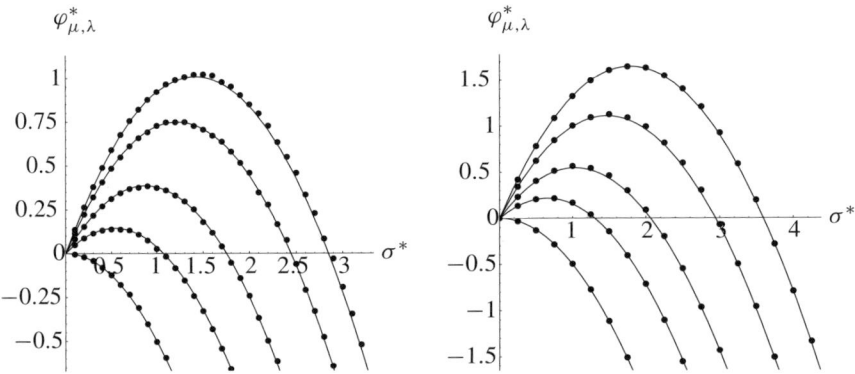

Figure 7. Progress rate $\varphi^*_{\mu,\lambda}$ of the (μ, λ)-ES on the hypersphere as a function of mutation strength σ^*. The left graph compares results of computer experiments with theoretically obtained predictions for $\lambda = 8$ and, from left to right, $\mu = 8, 6, 4, 2, 1$, the right graph shows the corresponding results for $\lambda = 80$ and $\mu = 80, 60, 40, 20, 10$. For details see [4].

Rechenberg [15] gives the progress rate law

$$\varphi^*_{\mu,\lambda}(\sigma^*) = c_{\mu,\lambda}\sigma^* - \frac{\sigma^{*2}}{2} \tag{13}$$

which occurs also as the limiting case for $n \to \infty$ of the analysis presented by Beyer [4]. Figure 7 shows comparisons of the theoretical predictions with computer experiments, demonstrating good accordance. The (μ, λ)-progress coefficients $c_{\mu,\lambda}$ can be computed numerically using the results in [4] or be determined in computer experiments. Table 2 lists some of them. Note that the $(1, \lambda)$-progress coefficients occur as special instances of the (μ, λ)-coefficients for $\mu = 1$.

$\lambda\backslash\mu$	1	2	3	4	5	6	7	8	9	10
5	1.16	0.91	0.67	0.40	0.00					
10	1.54	1.35	1.19	1.04	0.90	0.77	0.63	0.47	0.30	0.00
20	1.87	1.71	1.58	1.47	1.37	1.29	1.20	1.13	1.05	0.98
50	2.25	2.12	2.01	1.93	1.85	1.79	1.73	1.68	1.62	1.57
100	2.51	2.39	2.30	2.22	2.16	2.10	2.05	2.00	1.96	1.92

Table 2. Some of the $c_{\mu,\lambda}$-progress coefficients

As in the case of single-parent strategies, the gain and loss terms in (13) are easily identified. The maximal attainable progress rate is

$$\hat{\varphi}^*_{\mu,\lambda} = \frac{c^2_{\mu,\lambda}}{2} \quad \text{for} \quad \hat{\sigma}^*_{\mu,\lambda} = c_{\mu,\lambda}. \tag{14}$$

As for fixed λ, the $c_{\mu,\lambda}$ monotonically decrease with increasing μ, for the hypersphere in the absence of fitness noise a simple $(1, \lambda)$-ES is more efficient than any (μ, λ)-ES with $\mu > 1$. Accordingly, no speedup greater than of order $\log \lambda$ can be achieved.

3.4 Recombination

While the previous section has shown that simple multi-parent strategies do not have advantages in comparison with single-parent strategies, at least in the case of the basic hypersphere model, the situation changes when recombination is employed to merge the genetic information of parent individuals during the generation of an offspring individual. Theoretical results exist for $(\mu/\mu, \lambda)$-ES with either intermediate or dominant recombination. The two strategies are denoted as $(\mu/\mu_I, \lambda)$-ES and $(\mu/\mu_D, \lambda)$-ES, respectively, and are studied separately in the following paragraphs.

Intermediate Recombination. In $(\mu/\mu_I, \lambda)$-ES at every generation the center of mass

$$\langle y \rangle = \sum_{i=1}^{\mu} y_{i;\lambda} \tag{15}$$

of the μ best of the λ offspring individuals generated in the previous generation is computed and used as the starting point for the offspring of the subsequent generation. As a result, the analysis of the $(\mu/\mu_I, \lambda)$-ES resembles that of the $(1, \lambda)$-ES and does not require the determination of a self-consistent population distribution. In [5] an n-dependent progress law for $(\mu/\mu_I, \lambda)$-ES has been derived that, in the limit $n \to \infty$, agrees with the result in [15] and reads

$$\varphi^*_{\mu/\mu_I,\lambda}(\sigma^*) = c_{\mu/\mu,\lambda}\sigma^* - \frac{\sigma^{*2}}{2\mu}. \tag{16}$$

The $(\mu/\mu, \lambda)$-progress coefficients $c_{\mu/\mu,\lambda}$ defined by

$$c_{\mu/\mu,\lambda} = \frac{\lambda-\mu}{2\pi}\binom{\lambda}{\mu}\int_{-\infty}^{\infty} e^{-x^2}[\Phi(x)]^{\lambda-\mu-1}[1-\Phi(x)]^{\mu-1}\,dx$$

can be obtained by numerical integration. Some of these are listed in table 3. It can be shown that the $(1, \lambda)$-progress coefficients $c_{1,\lambda}$ occur as special instances of the $(\mu/\mu, \lambda)$-progress coefficients for $\mu = 1$. In [5] the law

$$c_{\mu/\mu,\lambda} = \mathcal{O}\left(\sqrt{\log\frac{\lambda}{\mu}}\right) \tag{17}$$

governing the growth of the $(\mu/\mu, \lambda)$ progress coefficients for large μ and λ has been derived.

$\lambda\backslash\mu$	1	2	3	4	5	6	7	8	9	10
5	1.16	0.83	0.55	0.29	0.00					
10	1.54	1.27	1.07	0.89	0.74	0.60	0.46	0.32	0.17	0.00
20	1.87	1.64	1.47	1.33	1.21	1.11	1.02	0.93	0.85	0.77
50	2.25	2.05	1.91	1.80	1.71	1.62	1.55	1.49	1.43	1.37
100	2.51	2.33	2.20	2.10	2.02	1.95	1.88	1.83	1.78	1.73

Table 3. Some of the $(\mu/\mu, \lambda)$-progress coefficients

Equation (16) shows a close resemblance to the progress rate law for (μ, λ)-ES in (13). As in the case of the strategy without recombination the gain and loss parts of the evolutionary progress principle are linear and quadratic in σ^*, respectively. As a comparison of tables 2 and 3 reveals, the progress coefficients of the (μ, λ)-ES are always moderately greater than or – if $\mu = 1$, in which case the two strategies are identical – equal to those of the $(\mu/\mu_I, \lambda)$-ES. Thus, for equal mutation strength σ^*, the strategy with recombination does not have a larger gain component than that without recombination. However, the loss component of the strategy with recombination is lower by a factor of $1/\mu$, making it possible to use a higher mutation strength. Indeed it follows from (16) that the $(\mu/\mu_I, \lambda)$-ES exhibits maximal progress

$$\hat{\varphi}^*_{\mu/\mu_I,\lambda} = \mu\frac{c^2_{\mu/\mu,\lambda}}{2} \quad \text{for} \quad \hat{\sigma}^*_{\mu/\mu_I,\lambda} = \mu c_{\mu/\mu,\lambda} \tag{18}$$

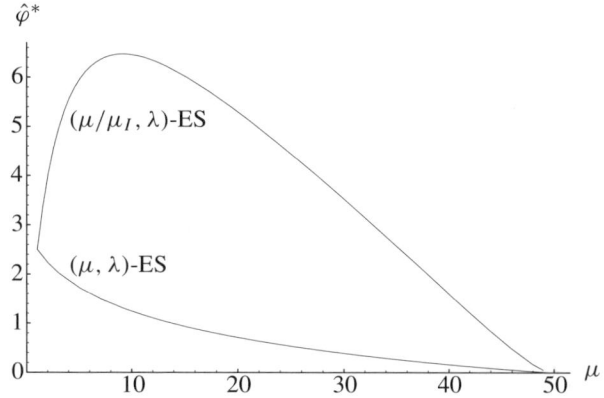

Figure 8. Maximally achievable progress rate $\hat{\varphi}^*$ of the $(\mu/\mu_I, \lambda)$-ES and of the (μ, λ)-ES as a function of the population size μ for $\lambda = 50$ and $n = 100$

and thus at a mutation strength which is almost μ times as large as that of the optimally tuned (μ, λ)-ES.

Figure 8 compares the maximal achievable performance of (μ, λ)-ES and $(\mu/\mu_I, \lambda)$-ES on the hypersphere, revealing that the strategy employing recombination exhibits better or equal performance than that without recombination for any population size μ. Moreover, it shows the existence of a population size $\hat{\mu}$ for which the performance of the $(\mu/\mu_I, \lambda)$-ES is optimal and far superior to that of the $(1, \lambda)$-ES. From (17) and (18) it follows that

$$\hat{\varphi}^*_{\mu/\mu_I, \lambda} = \mathcal{O}\left(\mu \log \frac{\lambda}{\mu}\right). \tag{19}$$

For large λ, and in the limit $n \to \infty$, it has been shown numerically in [5] that the fitness efficiency $\eta_{\mu/\mu_I, \lambda} = \hat{\varphi}^*_{\mu/\mu_I, \lambda}/\lambda$ attains a maximum of 0.202 for a population size $\hat{\mu} = 0.270\lambda$. These numbers look familiar. According to (8) the fitness efficiency is exactly that of an optimally tuned $(1 + 1)$-ES, and the ratio of $\hat{\mu}/\lambda$ for the $(\mu/\mu_I, \lambda)$-ES is equal to the optimal success probability of the $(1 + 1)$-ES. Consequently, if the evaluation of the offspring is performed on λ processors in parallel and if it can be assumed that the overhead due to recombination and selection is negligible compared with the effort of fitness evaluation, the $(\mu/\mu_I, \lambda)$-ES achieves a λ-fold speedup over the $(1 + 1)$-ES, making optimal use of the parallel resources. Note however that the speedup is less than linear for finite n.

An interesting question is where does the greatly improved speedup as compared with the (μ, λ)-ES result from? Beyer [5] traces the root of the factor $1/\mu$ in the loss term in (16), which is ultimately responsible for the speedup, back to the averaging of the h components when computing the center of mass of the selected offspring in (15). Recall from section 3.1 that mutations z can be decomposed into a component x delivering a progress gain and a component h resulting in a loss. While the x

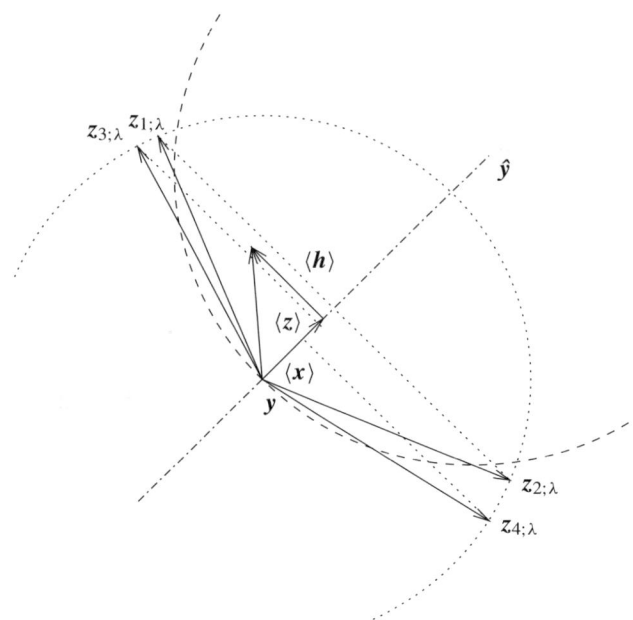

Figure 9. Illustration of the genetic repair effect for a $(\mu/\mu_I, \lambda)$-ES with $\mu = 4$ in two dimensions. Each of the selected z vectors $z_{i;\lambda}$, $i = 1, \ldots, 4$, can be decomposed into an x component and an h component. The result of averaging these vectors is a vector z with an h component which is smaller by about a factor of $1/\sqrt{4}$ than the expected size of the h components of the $z_{i;\lambda}$. Note that in the depicted case, even though none of the selected offspring individuals has a higher fitness than the parent, their center of mass does.

components of the selected individuals are correlated due to the selective pressure the h components are statistically independent if the mutations are independent. Therefore, the averaging performed in (15) makes the length of the h component of the resulting mutation vector smaller by a factor of about $1/\sqrt{\mu}$ than that of the h components of the individual mutations. The length of the x component of the resulting mutation vector, on the other hand, is of the same order of magnitude as that of the individual mutations. Figure 9 illustrates the effect in two dimensions.

This observation has led to formulation of the *genetic repair principle* in [5]. The beneficial influence of recombination for ES on the hypersphere does not stem from the combination of the good traits of the parent individuals as claimed for GA in the building block hypothesis but instead results from statistical error correction. The "good traits" of the mutations are their x components, but obviously the x component of the result of application of (15) cannot be greater than that of the best individual participating in the recombination. Instead, it is the reduction of the harmful component h of the mutations which allows for the use of a higher mutation strength and therefore for improved progress. In that sense, intermediate recombination has the effect of an extraction of similarities of the participating individuals.

Two conclusions outlined in [5] can be drawn from this perspective. First, the statistical independence of mutations is a necessary prerequisite for the genetic repair principle to hold. This may serve as an explanation for the incest taboo observed in nature. Second, while a mathematical analysis of the effects of having $\rho < \mu$ individuals participating in the process of generating an offspring individual does not yet exist, the genetic repair principle makes it plausible to assume that for given μ and λ the choice $\rho = \mu$ is optimal, as any other choice would lead to a lesser reduction of the harmful part of the mutations.

Dominant Recombination. While algorithmically appealing, intermediate recombination bears little resemblance to biological recombination mechanisms. Dominant recombination more closely models biological reality. Using dominant recombination in $(\mu/\mu, \lambda)$-ES, the ith component of an offspring individual is created from the ith component of a randomly selected individual of the parental population. Consequently, in contrast to the analysis of intermediate recombination, the exact treatment of dominant recombination requires the determination of a self-consistent population distribution. Such an analysis exceeds in difficulty even that of the (μ, λ)-ES and has not yet been carried out. Instead, Beyer [5] suggests a simplified scheme that relies on the use of surrogate mutations. Surrogate mutations include the effects of both recombination and mutation into a single operator that is applied to the center of mass of the population as in the case of intermediary recombination. In the derivation of a self-consistent distribution for the surrogate mutations, selection can be neglected in the limit of $n \to \infty$, thereby making the surrogate mutations isotropic.

As in the case of the intermediate recombination the theoretical approach to dominant recombination in [5] agrees for $n \to \infty$ with the relationship presented in [15]. The progress rate law for $(\mu/\mu_D, \lambda)$-ES reads

$$\varphi^*_{\mu/\mu_D,\lambda}(\sigma^*) = \sqrt{\mu} c_{\mu/\mu,\lambda} \sigma^* - \frac{\sigma^{*2}}{2}. \tag{20}$$

Again, gain and loss terms are easily identified. Compared with the strategy employing intermediate recombination the gain term is higher by a factor of $\sqrt{\mu}$, but the factor of $1/\mu$ in the loss term is missing. The $(\mu/\mu_D, \lambda)$-ES exhibits maximal performance

$$\hat{\varphi}^*_{\mu/\mu_D,\lambda} = \mu \frac{c^2_{\mu/\mu,\lambda}}{2} \quad \text{for} \quad \hat{\sigma}^*_{\mu/\mu_D,\lambda} = \sqrt{\mu} c_{\mu/\mu,\lambda}. \tag{21}$$

Comparison with (18) shows that while the maximal performance is the same as for intermediate recombination the mutation strength required to achieve it is smaller by a factor of $1/\sqrt{\mu}$.

A byproduct of the analysis is the observation that the combination of dominant recombination and mutation, summarized in the surrogate mutations, results in a stable offspring distribution of finite standard deviation reminiscent of the biological concept of a species. Even in the absence of selection, the individuals of the population crowd around an imaginary wild-type corresponding to the center of mass of the population. This is in striking contrast to the diffusion-like random walk the individuals perform in the presence of mutation alone, and to the contraction of the population to a single point resulting from the presence of dominant mutation alone[4]. It has also been shown that the distribution of the individuals in the population is such that the effective mutation strength of the surrogate mutations is $\sqrt{\mu}\sigma$, leading to perfect agreement with the optimal mutation strength in the case of intermediary recombination.

Beyer concludes his analysis by demonstrating that genetic repair is present in dominant recombination as it is in intermediate recombination. While in intermediate recombination genetic repair stems from the explicit averaging of the parental components in the computation of the center of mass, genetic repair in dominant recombination is a more implicit effect. Dominant recombination can be viewed as a statistical sampling process in which an approximation of the first moment of the parental distribution is implicitly obtained. The sampling error resulting from the statistical approximation in a finite population explains the observably somewhat better performance of intermediate recombination as compared with dominant recombination.

3.5 Self-Adaptation

To achieve linear convergence of an ES with isotropic mutations on the hypersphere model, the mutation strength needs to be dynamically adjusted. Optimally, an adaptation algorithm would adjust σ such that at any point in time $\sigma^* = \sigma n/r$ is in the vicinity of $\hat{\sigma}^*$. Using self-adaptation, σ is made part of the genetic information of the individuals and subjected to recombination and mutation along with the object variables, giving each individual its own personal strategy parameter. The object parameter values of an offspring individual are generated using this personal mutation strength, and selection based on local quality gain determines whether the offspring individual and, along with it, its mutation strength survive. The underlying reasoning is that a good mutation strength is likely to generate a successful object component and will therefore probably prevail in the selection process. Some theoretical results regarding the performance of self-adaptation in the case of $(1, \lambda)$-ES which are to be summarized here have been reported in [6].

As the expected change in distance to the optimum is φ, to achieve constant σ^* it has to be ensured on average that

$$\sigma^{(t+1)} = \sigma^{(t)} \left(1 - \frac{\varphi^*(\sigma^{(t)})}{n}\right). \tag{22}$$

Therefore, mutation of the mutation strength should be multiplicative rather than additive. In practice, it is achieved by multiplication of the parental mutation strength

---
[4] A more thorough discussion of the *mutation-induced speciation by recombination* effect can be found in [8,9].

with a random variable ξ. Two frequently employed mutation rules are log-normal mutation, where $\xi = \exp(\tau \mathcal{N})$ and $\mathcal{N}$ is a standard normally distributed random variable, and two-point mutation, where ξ takes on values $1 + \beta$ and $1/(1 + \beta)$ with equal probability. The exogenous strategy parameters τ and β, respectively, need to be set appropriately, but can remain constant throughout the evolution[5]. As they govern the rate with which the mutation strength is adapted they are often referred to as learning rates. In [6] it has been shown that within reasonable limits both mutation rules generate the same dynamics if

$$\tau^2 = \beta^2 (1 - \beta). \tag{23}$$

Therefore, all results regarding log-normal mutation of the mutation strength can easily be converted to the case of two-point mutations and vice versa.

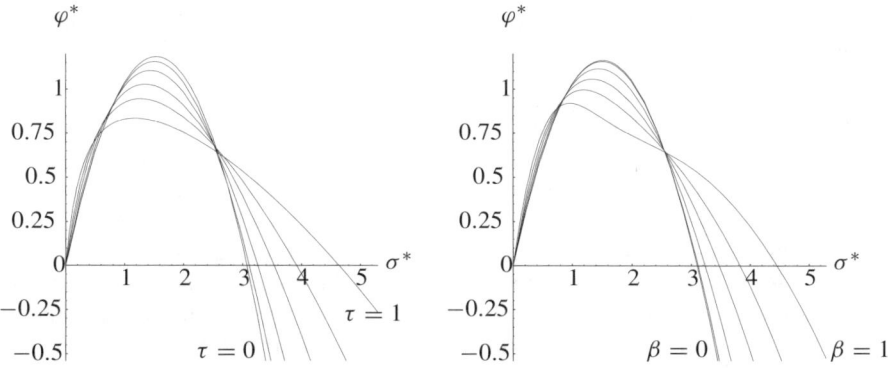

Figure 10. Progress rate φ^* of the $(1, 10)$-ES with self-adaptation as a function of mutation strength σ^* for $n = 30$. The left graph shows results obtained by numerical integration for the log-normal operator with $\tau = 0.0, 0.1, 0.3, 0.5, 0.7$, and 1.0, the right graph for the two-point operator with $\beta = 0.0, 0.1, 0.3, 0.5, 0.7$, and 1.0. For details see [6].

Figure 10 shows the progress rate as a function of the mutation strength for a number of learning parameter values for both log-normal and two-point mutation of σ, revealing that non-zero learning parameters lead to a decrease in the maximal progress rate that can be achieved. However, theoretical investigations in [6] confirm the long-standing recommendation made in [19] that for optimal performance τ should be approximately $c_{1,\lambda}/\sqrt{n}$, with the exact choice being rather uncritical. Therefore, for large n the learning parameter should be very close to zero, making the loss in performance indicated in figure 10 negligible.

---

[5] On a practical note, in [6] it has been shown that the time it takes to drive an initially ill-chosen mutation strength into the range of positive progress rates grows linearly with the problem dimension n if the learning parameter is optimally chosen. To avoid long adaptation times for large n it is possible to start with a high values of τ or β which are reduced to their recommended values after a number of generations.

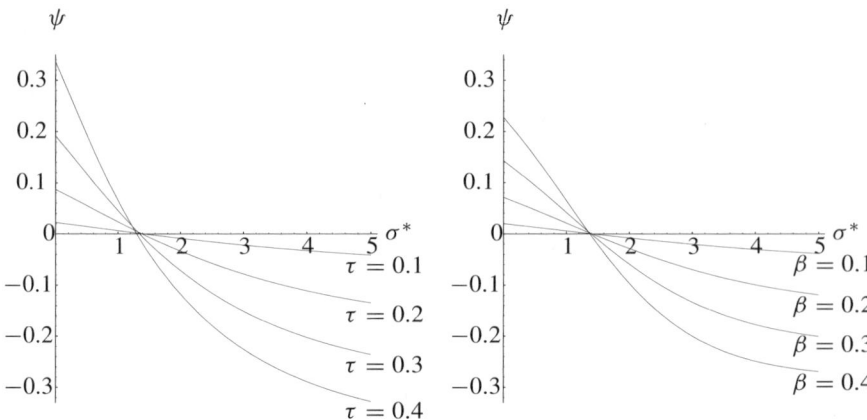

Figure 11. Self-adaptation response ψ as a function of mutation strength σ^* for the (1, 10)-ES for $n = 30$, obtained by numerical integration. The left graph shows results for the log-normal operator, the right one for the two-point operator. For details see [6].

The state of a $(1, \lambda)$-ES without self-adaptation is described by the distance to the optimum r because it is assumed in the analysis that σ is explicitly given. The change in r from one generation to the next is governed by a stochastic difference equation, and the dynamics of the evolution are adequately described by considering mean values such as the progress rate φ^*. This is no longer so if self-adaptation is used for the control of the mutation strength. In that case, the mutation strength σ becomes a second state variable, and the state change from one generation to the next is described by two coupled stochastic difference equations, one for the change in r and one for the change in σ. Moreover, mean values no longer adequately describe the dynamics of the evolutionary process, making it necessary to take fluctuations around the mean values into account. While the fluctuations of r around the mean value average out over several generations, the fluctuations of σ^* cannot be neglected as values both to the left and to the right of the mean lead to a decrease in the expected progress.

Just as φ describes the expected change in r from one generation to the next, the self-adaptation response

$$\psi = E\left[\frac{\sigma^{(t+1)} - \sigma^{(t)}}{\sigma^{(t)}} \Big| \sigma^{(t)}, r^{(t)}\right] \qquad (24)$$

describes the expected change in mutation strength. Figure 11 shows the self-adaptation response as a function of mutation strength. Obviously, there is a mutation strength σ_0^* for which the self-adaptation response is zero. For mutation strengths higher than σ_0^*, the self-adaptation response is negative, leading to a reduction of σ, and for mutation strengths lower than σ_0^* the self-adaptation response is positive, leading to an increase in σ. Consequently, the dynamics lead to a mutation strength fluctuating around σ_0^*. Comparison with figure 10 shows that the mutation strength is in the vicinity of the optimal mutation strength $\hat{\sigma}^*$ as desired.

According to [6], a good approximation for the progress rate of the $(1, \lambda)$-ES with self-adaptation on the hypersphere is

$$\varphi^*(\sigma^*) = c_{1,\lambda} \sigma^* - \frac{\sigma^{*2}}{2} - \frac{1}{2} D^2(\sigma^*) \tag{25}$$

where $D^2(\sigma^*)$ denotes the variance of σ^*. Thus, the fluctuations in mutation strength are responsible for a significant performance decrease that should be addressed. Surprisingly, the analysis in [6] shows that decreasing the learning rate is not an effective counter-measure against the fluctuations as the influence of τ on their magnitude is rather small. The fluctuations are not a result of a high learning parameter but of the stochastic dynamics of the system and they need to be addressed as such. It is likely that an improvement can be expected from the implementation of a fading memory for the mutation strength. Whether the principle of genetic repair holds for strategy parameters if a population of size $\mu > 1$ is used is currently unknown.

4 Summary and Outlook

The present tutorial has offered an introduction to the theory of ES based on a view of ES in combination with fitness functions as dynamical systems. Local performance measures quantifying the expected change in the state of the system from one generation to the next have been introduced as vehicles for the characterization for the overall dynamics. Results from the computation of the progress rate as one such local performance measure have been reported for a variety of strategy types on a spherical fitness model. Such computations make it possible to compare different strategy variants as to their effectiveness for the model at hand and offer insight valuable to the understanding of the genetic operators and the basic working principles of ES. The distinction between the harmful and the beneficial components of mutations which underlies the evolutionary progress principle has been discussed, and the value of the principle for understanding ES has been verified for a number of strategies. It has been demonstrated that for the simple case of the sphere model in the absence of noise a single-parent strategy outperforms multi-parent strategies if there is no recombination, but that the introduction of recombination results in a qualitative speedup. The genetic repair principle, which states that the benefit of recombination lies not in the combination of advantageous traits but in the statistical correction of errors resulting from mutation, has been identified as the origin of the speedup. Finally, a theoretical framework for the analysis of self-adaptation of the mutation strength has been outlined and applied to the simple case of a single-parent strategy.

Theoretical studies of ES offer a vast field of opportunity for future research. Even within the existing framework much remains to be done. The performance of ES on fitness models other than the sphere needs to be analyzed. Investigations of the corridor and ridge functions in [14] and [13], respectively, are but a start. The effects of noise on the performance of ES are worth further investigation as there is a hope that noisy environments are a field in which EA outperform other optimization methods. Strategies different from those analyzed so far, such as $(\mu/\rho \stackrel{+}{,} \lambda)$-ES with $1 < \rho < \mu$ and strategies employing other operators for mutation or for selection may contribute to a greater understanding. Finally, self-adaptation of strategy parameters in multi-parent strategies as an area of extraordinary practical interest remains entirely uncharted.

Acknowledgments

The first author is a Heisenberg fellow of the Deutsche Forschungsgemeinschaft (DFG) under grant Be 1578/4-1. The second author is supported by the DFG under grant Be 1578/6-1. Both authors would like to thank Bart Naudts for carefully revising the manuscript.

References

1. D. V. Arnold. Evolution Strategies in Noisy Environments – A Survey of Existing Work. This volume.
2. H.-G. Beyer. Toward a theory of evolution strategies: some asymptotical results from the $(1 \stackrel{+}{,} \lambda)$-theory. *Evolutionary Computation*, 1(2), pages 165–188, 1993.
3. H.-G. Beyer. Towards a theory of 'evolution strategies': progress rates and quality gain for $(1 \stackrel{+}{,} \lambda)$-Strategies on (nearly) arbitrary fitness functions. In Y. Davidor, R. Männer, and H.-P. Schwefel, editors, *Parallel Problem Solving from Nature*, 3, pages 58–67. Springer-Verlag, Berlin Heidelberg New York, 1994.
4. H.-G. Beyer. Toward a theory of evolution strategies: the (μ, λ)-theory. *Evolutionary Computation*, 2(4), pages 381–407, 1995.
5. H.-G. Beyer. Toward a theory of evolution strategies: on the benefit of sex – the $(\mu/\mu, \lambda)$-theory. *Evolutionary Computation*, 3(1), pages 81–111, 1995.
6. H.-G. Beyer. Toward a theory of evolution strategies: self-adaptation. *Evolutionary Computation*, 3(3), pages 311–347, 1996.
7. H.-G. Beyer. *Zur Analyse der Evolutionsstrategien*. Habilitationsschrift, University of Dortmund, 1996.
8. H.-G. Beyer. An alternative explanation for the manner in which genetic algorithms operate. *BioSystems*, 41, pages 1–15, 1997.
9. H.-G. Beyer. On the dynamics of GAs without selection – the MISR principle. In W. Banzhaf and C. Reeves, editors, *Foundations of Genetic Algorithms*, 5, pages 20–41. Morgan Kaufmann, San Francisco, 1998.
10. S. Droste, T. Jansen, and I. Wegener. A rigorous complexity analysis of the $(1+1)$ evolutionary algorithm for separable functions with boolean inputs. *Evolutionary Computation*, 6(2), pages 185–196, 1998.
11. S. Droste, T. Jansen, and I. Wegener. On the optimization of unimodal functions with the $(1+1)$ evolutionary algorithm. In A. E. Eiben, T. Bäck, M. Schoenauer, and H.-P. Schwefel, editors, *Parallel Problem Solving from Nature*, 5, pages 13–22. Springer-Verlag, Berlin Heidelberg New York, 1998.

12. S. Droste, T. Jansen, and I. Wegener. A rigorous complexity analysis of the (1 + 1) evolutionary algorithm for linear functions with boolean inputs. In *Proceedings of the IEEE Congress on Evolutionary Computation*, pages 499–504. IEEE Press, 1998.
13. A. I. Oyman, H.-G. Beyer, and H.-P. Schwefel. Where Elitists Start Limping: Evolution Strategies at Ridge Functions. In A. E. Eiben, T. Bäck, M. Schoenauer, and H.-P. Schwefel, editors, *Parallel Problem Solving from Nature 5*, LNCS 1498, pages 34–43. Springer, Heidelberg, 1998.
14. I. Rechenberg. *Evolutionsstrategie: Optimierung Technischer Systeme nach Prinzipien der biologischen Evolution*. Frommann-Holzboog, Stuttgart, 1973.
15. I. Rechenberg. *Evolutionsstrategie '94*. Frommann-Holzboog, Stuttgart, 1994.
16. G. Rudolph. *Convergence Properties of Evolutionary Algorithms*. Verlag Dr. Kovač, Hamburg, 1997.
17. G. Rudolph. Local convergence rates of simple evolutionary algorithms with Cauchy mutations. *IEEE Transactions on Evolutionary Computation*, 1(4), pages 249–258, 1997.
18. A. Scheel. *Beitrag zur Theorie der Evolutionsstrategie*. Dissertation, TU Berlin, 1985.
19. H.-P. Schwefel. *Numerische Optimierung von Computer-Modellen mittels der Evolutionsstrategie*. Birkhäuser, Basel, 1977.
20. H.-P. Schwefel. Collective phenomena in evolutionary systems. In P. Checkland and I. Kiss, editors, *Problems of Constancy and Change – the Complementarity of Systems Approaches to Complexity*, pages 1025–1033, 1987.
21. H.-P. Schwefel. *Evolution and Optimum Seeking*. John Wiley, New York, 1995.

Evolutionary Algorithms: From Recombination to Search Distributions

H. Mühlenbein and T. Mahnig

RWCP* Theoretical Foundations GMD[†] Laboratory
53754 Sankt Augustin, Germany
E-mail: *muehlenbein@gmd.de*

Abstract. First we show that all genetic algorithms can be approximated by an algorithm which keeps the population in linkage equilibrium. Here the genetic population is given as a product of univariate marginal distributions. We describe a simple algorithm which keeps the population in linkage equilibrium. It is called the univariate marginal distribution algorithm (UMDA). Our main result is that UMDA transforms the discrete optimization problem into a continuous one defined by the average fitness $\tilde{W}(p_1, \ldots, p_n)$ as a function of the univariate marginal distributions p_i. For proportionate selection UMDA performs gradient ascent in the landscape defined by $W(\boldsymbol{p})$. We derive a difference equation for p_i which has already been proposed by Wright in population genetics. We show that UMDA solves difficult multimodal optimization problems. For functions with highly correlated variables UMDA has to be extended. The factorized distribution algorithm (FDA) uses a factorization into marginal and conditional distributions. For decomposable functions the optimal factorization can be explicitly computed. In general it has to be computed from the data. This is done by LFDA. It uses a Bayesian network to represent the distribution. Computing the network structure from the data is called learning in Bayesian network theory. The problem of finding a minimal structure which explains the data is discussed in detail. It is shown that the Bayesian information criterion is a good score for this problem.

Keywords
Linkage equilibrium, selection, factorization of distributions, replicator equation, Bayesian networks, Boltzmann distribution

1 Introduction

Simulating evolution as seen in nature has been identified as one of the key computing paradigms for the next decade. Today evolutionary algorithms have been successfully used in a number of applications. These include discrete and continuous optimization problems, synthesis of neural networks, synthesis of computer programs from examples (also called genetic programming) and even evolvable hardware. But in all application areas problems have been encountered where evolutionary algorithms performed badly. Therefore a mathematical theory of evolutionary

---

* Real World Computing Partnership
[†] GMD – Forschungszentrum Informationstechnik

algorithms is urgently needed. Theoretical research has evolved from two opposite ends: from the theoretical approach there are theories emerging that are getting closer to practice; from the applied side ad hoc theories have arisen that often lack theoretical justification.

In this tutorial we concentrate on evolutionary algorithms for optimization. Here results from classical population genetics and statistics can be used. The outline of the paper is as follows. In section 2 we summarize research which indicates that all simple genetic algorithms (GAs) can be approximated by an algorithm keeping the population in linkage equilibrium. The simplest algorithm keeping the population in linkage equilibrium is the *univariate marginal distribution algorithm* UMDA. UMDA uses search distributions instead of recombination. Its mathematical analysis is done in section 3.

In section 4 we briefly survey algorithms using univariate marginal distributions. The mathematical behavior of UMDA with proportionate selection is described by a differential equation called the *replicator equation*. This equation is discussed in section 5. Then numerical results for UMDA are presented. They confirm that UMDA is able to solve difficult multimodal optimization problems. Certain optimization problems need marginal distributions of higher order. The *factorized distribution algorithm* FDA uses a general factorization of the distribution. It is introduced in section 7.

The mathematical theory of UMDA and FDA is based on infinite populations. The problem of finite samples is discussed in section 8. In the final section LFDA, the *learning* FDA, is introduced. This algorithm computes a plausible factorization of the search distribution from a finite sample of data.

2 The Simple GA

In this section we discuss the standard GA, also called simple GA (SGA). The algorithm is described by Holland [12] and Goldberg [9]. It consists of

- fitness proportionate selection
- recombination/crossover
- mutation.

In this section we will analyze selection and recombination only. Mutation is considered to be a background operator. It can be analyzed by known techniques from stochastics [21,18]. We will investigate two widely used recombination/crossover schemes.

Definition 1. Let two strings x and y be given. In *one-point crossover* the string z is created by randomly choosing a crossover point $0 < l < n$ and setting $z_i = x_i$ for $i \leq l$ and $z_i = y_i$ for $i > l$. In *uniform crossover* z_i is randomly chosen with equal probability from $\{x_i, y_i\}$.

Let $x = (x_1, \ldots, x_n)$ denote a binary vector. For notational simplicity we restrict the discussion to binary variables $x_i \in \{0, 1\}$. We use the following conventions. Capital letters X_i denote variables, small letters x_i assignments. Let a function $f : X \to R^{\geq 0}$ be given. We consider the optimization problem $x_{\text{opt}} = \text{argmax } f(x)$.

Definition 2. *Let $p(x, t)$ denote the probability of x in the population at generation t. Then $p(x_i, t) = \sum_{x, X_i = x_i} p(x, t)$ defines a univariate marginal distributions.*

We write $p(x_i)$ if just one generation is discussed. In this notation the average fitness of the population and the variance is given by

$$\bar{f}(t) = \sum_x p(x, t) f(x)$$

$$V(t) = \sum_x p(x, t) \left(f(x) - \bar{f}(t)\right)^2.$$

The *response to selection $R(t)$* is defined by

$$R(t) = \bar{f}(t+1) - \bar{f}(t). \tag{1}$$

2.1 Proportionate Selection

Proportionate selection changes the probabilities according to

$$p(x, t+1) = p(x, t) \frac{f(x)}{\bar{f}(t)}. \tag{2}$$

Lemma 1. *For proportionate selection the response is given by*

$$R(t) = \frac{V(t)}{\bar{f}(t)}. \tag{3}$$

Proof. We have

$$R(t) = \sum_x p(x, t) \frac{f^2(x)}{\bar{f}(t)} - \bar{f}(t)$$

$$= \sum_x p(x, t) \frac{f^2(x) - \bar{f}^2(t)}{\bar{f}(t)}$$

$$= \frac{V(t)}{\bar{f}(t)}.$$

With proportionate selection the average fitness never decreases. This is true for every selection scheme.

2.2 Recombination

For the analysis we introduce a special distribution, called Robbins' proportions.

Definition 3. Robbins' proportions are given by the distribution π

$$\pi(x, t) := \prod_{i=1}^{n} p(x_i, t). \tag{4}$$

A population in Robbins' proportions is said to be in *linkage equilibrium* in population genetics.

Geiringer [8] has shown that all reasonable recombination schemes lead to the same limit distribution.

Theorem 1 (Geiringer). *Recombination does not change the univariate marginal frequencies, i.e., $p(x_i, t+1) = p(x_i, t)$. The limit distribution of any complete recombination scheme is Robbins' proportions $\pi(x)$.*

Complete recombination means that for each subset S of $\{1, \ldots, n\}$, the probability of an exchange of genes by recombination is greater than zero. Convergence to the limit distribution is very fast. We will prove this for $n = 2$ loci.

Theorem 2. *Let $D(t) = p(0, 0, t)p(1, 1, t) - p(0, 1, t)p(1, 0, t)$. If there is no selection then we have for two loci and uniform crossover*

$$D(t) = (-1)^{|x|^2} \bigl(p(x, t) - p_1(x_1, 0)p_2(x_2, 0) \bigr). \tag{5}$$

Furthermore the factor $D(t)$ is halved each generation

$$D(t+1) = \frac{1}{2} D(t). \tag{6}$$

Proof. Without selection the univariate marginal frequencies are independent of t, because in an infinite population a recombination operator does not change them. Then from

$$p(1, 1, t) - p_1(1, 0)p_2(1, 0)$$
$$= p(1, 1, t) - \bigl(p(1, 0, t) + p(1, 1, t)\bigr)\bigl(p(0, 1, t) + p(1, 1, t)\bigr)$$
$$= p(1, 1, t) - p(0, 1, t)p(1, 0, t) - p(1, 1, t)(1 - p(0, 0, t))$$

we obtain

$$D(t) = p(1, 1, t) - p_1(1, 0)p_2(1, 0).$$

This gives (5) for $x = (1, 1)$. The other cases are proven in the same way.

The gene frequencies after recombination are obtained as follows. We only consider $p(1, 1, t)$. The probability of $p(1, 1, t + 1)$ can be computed from the probability that recombination generates string $(1, 1)$. The probability is given by

$$p(1, 1, t + 1) = p(1, 1, t) \cdot \left(\tfrac{1}{2}p(0, 0, t) + p(0, 1, t) + p(1, 0, t) + p(1, 1, t)\right)$$
$$+ \frac{1}{2}p(0, 1, t)p(1, 0, t)$$
$$= p(1, 1, t) - \tfrac{1}{2}\left(p(1, 1, t)p(0, 0, t) - p(0, 1, t)p(1, 0, t)\right)$$
$$= p(1, 1, t) + (-1)^{|x|^2+1}\frac{1}{2}D(t).$$

By computing $D(t + 1)$, (6) is obtained.

We will use as a measure for the deviation from Robbins' proportions the mean square error $DSQ(t)$

$$DSQ(t) = \sum_{x}\left(p(x, t) - p_1(x_1)p_2(x_2)\right)^2. \tag{7}$$

From the above theorem we obtain:

Corollary 1. *For two loci the mean square error is reduced each step by one fourth*

$$DSQ(t + 1) = \frac{1}{4}DSQ(t).$$

For more than 2 loci the equations for uniform crossover and one-point crossover get more complicated. Uniform crossover converges faster to linkage equilibrium, because it mixes the genes much more than one-point crossover.

Table 1 gives numerical results for $n = 8$ loci. For the initial probabilities $q_0 = q_7 = 0.5$ linkage equilibrium is given by $q_i = 1/8$. One-point crossover converges slowly to linkage equilibrium; uniform crossover converges very fast.

We have to mention an important fact. In a finite population linkage equilibrium cannot be exactly achieved. We take the uniform distribution as example. Here linkage equilibrium is given by $p(x) = 2^{-n}$. This value can only be obtained if the size of the population N is substantially larger than 2^n. The finite size effect is demonstrated in table 2. There the numerical value for $DSQ(t)$ is shown for different population sizes. In addition $c = DSQ(t + 1)/DSQ(t)$ is displayed. From theorem 2 a factor of $c = 0.25$ is expected.

For a population of $N = 1000$ the minimum deviation DSQ_{min} from Robbins' proportions is already achieved after four generations, then DSQ slowly increases due to stochastic fluctuations by *genetic drift*. Ultimately the population will consist of one genotype only. Genetic drift has been analyzed by Asoh and Mühlenbein [1]. It will not be considered here.

t	q_0	q_1	q_2	$q_0 - 1/2^8$	$q_1 - 1/2^8$	$q_2 - 1/2^8$
0	0.5	0.0	0.0	0.4961	-3.906E-3	-3.906E-3
1	0.3774	0.0177	0.0	0.3735	1.369E-2	-3.906E-3
2	0.2879	0.0262	0.0012	0.2840	2.229E-2	-2.706E-3
3	0.2225	0.0303	0.0028	0.2186	2.639E-2	-1.106E-3
4	0.1768	0.0314	0.0042	0.1729	2.749E-2	+0.294E-3
5	0.1421	0.0298	0.0050	0.1382	2.589E-2	+1.094E-3
0	0.5	0.0	0.0	0.4961	-3.906E-3	-3.906E-3
1	0.2533	0.0020	0.0023	0.2494	-1.927E-3	-1.646E-3
2	0.0895	0.0097	0.0101	0.0856	+5.834E-3	+6.174E-3
3	0.0323	0.0093	0.0102	0.0283	+5.434E-3	+6.244E-3
4	0.0148	0.0074	0.0072	0.0108	+3.574E-3	+3.254E-3
5	0.0090	0.0057	0.0056	0.0051	-1.794E-3	+1.794E-3

Table 1. Comparison of convergence to Robbins' proportions for $n = 8$ loci, one-point (upper half) and uniform crossover (lower half), $q_0 = p(0, \ldots, 0)$, $q_1 = p(0, \ldots, 1)$, $q_2 = p(0, \ldots, 1, 0)$, population size $N = 1000$, averages over 100 runs, $q_0 = q_7 = 0.5$

	$N = 1000$		$N = 10000$		$N = 20000$	
t	DSQ	c	DSQ	c	DSQ	c
0	9.00E-2		9.00E-2		9.00E-2	
1	2.28E-2	0.25	2.25E-2	0.25	2.25E-3	0.25
2	6.37E-3	0.28	5.77E-3	0.25	5.59E-3	0.25
3	2.34E-3	0.37	1.55E-3	0.27	1.45E-3	0.26
4	1.62E-3	0.70	4.96E-4	0.32	4.24E-4	0.29
5	1.91E-3	1.03	2.43E-4	0.49	1.67E-4	0.39
8	2.62E-3	1.13	2.25E-4	1.10	1.41E-4	1.10

Table 2. Convergence to linkage equilibrium for $n = 2$ loci

2.3 Selection and Recombination

We have shown that the average $\bar{f}(t)$ never decreases after selection and that any complete recombination scheme moves the genetic population to Robbins' proportions. Now the question arises: what happens if recombination is applied *after* selection? The answer is very difficult. The problem still puzzles researchers on population genetics [23].

Formally the difference equations can be written compactly. Let a recombination distribution R be given. $R_{x,yz}$ denotes the probability that y and z produce x after recombination. Then

$$p(x, t+1) = \sum_{y,z} R_{x,yz} p^s(y) p^s(z). \tag{8}$$

$p^s(x)$ denotes the probability of string x after selection. For n loci the recombination distribution R consists of $2^n \times 2^n$ parameters. Recently Christiansen and Feldman [4] have written a survey about the mathematics of selection and recombination from the viewpoint of population genetics. A new technique to obtain the equations has been developed by Vose [31]. In both frameworks one needs a computer program to compute the equations for a given fitness function.

We discuss the problem for a special case only, uniform crossover for $n = 3$ loci. For notational convenience we use the integer representation i_x for x. Furthermore we set $q_{i_x} := p(x)$. In the next theorem we only give the equations for q_7 and q_3. From these expressions the reader can extrapolate the remaining five equations.

Theorem 3. *For proportionate selection and uniform crossover the probabilities are given by*

$$q_7(t+1) = q_7(t)\frac{f_7}{\bar{f}(t)} + \frac{1}{\bar{f}(t)^2}\Big(-\frac{1}{2}f_7q_7(f_1q_1 + f_2q_2 + f_4q_4) - \frac{3}{4}f_7q_7 f_0q_0$$
$$+ \frac{1}{2}(f_3q_3 f_5q_5 + f_3q_3 f_6q_6 + f_5q_5 f_6q_6)$$
$$+ \frac{1}{4}(f_1q_1 f_6q_6 + f_2q_2 f_5q_5 + f_3q_3 f_4q_4)\Big)$$

and

$$q_3(t+1) = q_3(t)\frac{f_3}{\bar{f}(t)} + \frac{1}{\bar{f}(t)^2}\Big(-\frac{1}{4}f_3q_3(f_0q_0 + f_5q_5 + f_6q_6) - \frac{3}{4}f_3q_3 f_4q_4$$
$$+ \frac{1}{2}(f_1q_1 f_7q_7 + f_2q_2 f_7q_7 + f_1q_1 f_2q_2)$$
$$+ \frac{1}{4}(f_0q_0 f_7q_7 + f_1q_1 f_6q_6 + f_2q_2 f_5q_5)\Big).$$

Proof. We outline the proof. After proportionate selection and recombination we obtain

$$q_7(t+1) = q_7(t)\frac{f_7}{\bar{f}(t)^2}\left(\sum_{j=1}^{7} 2R_{7,7j}q_j(t)f_j - q_7 f_7\right) + \text{rest}$$

$$\text{rest} = \sum_{i \neq 7, j \neq 7} R_{7,ij} q_i(t)\frac{f_i}{\bar{f}(t)} q_j(t)\frac{f_j}{\bar{f}(t)}.$$

For uniform crossover we compute $R_{7,70} = 1/8$, $R_{7,71} = R_{7,72} = R_{7,74} = 1/4$, $R_{7,73} = R_{7,75} = R_{7,76} = 1/2$. Inserting these numbers and rearranging the terms we obtain

$$q_7(t+1) = q_7(t)\frac{f_7}{\bar{f}(t)} + \frac{1}{\bar{f}(t)^2}\Big(-\frac{1}{2}f_7q_7(f_1q_1 + f_2q_2 + f_4q_4) - \frac{3}{4}f_7q_7 f_0q_0\Big)$$
$$+ \text{rest}.$$

The term *rest* is computed in the same way.

We have split the difference equation into the selection term and a recombination term. The recombination term consists of two terms: the probabilities that x will be reduced by recombination with other genotypes and the probabilities that recombination of two strings different from x will produce string x. A mathematical analysis of the mathematical properties of 3 loci systems is difficult. But we easily obtain an interesting result for a special case.

Definition 4. A fitness function is called *multiplicative* if

$$f(x) = C \cdot \prod_i e^{a_i x_i}. \tag{9}$$

We easily obtain from theorem 3 the following corollary. The proof is left to the reader.

Corollary 2. *If the fitness function is multiplicative, if the population is in linkage equilibrium, and if the assumptions of theorem 3 are valid, then the population remains in linkage equilibrium and the probabilities are given by (2).*

We *conjecture that the corollary is true for arbitrary n.* It seems to be known in population genetics, but we have not found a proof. Unfortunately for all other fitness functions, selection and recombination leads to *linkage disequilibrium*.

The question is whether linkage disequilibrium is important for evolutionary optimization. The answer is no. We provide evidence for this statement by citing a theorem from [18]. It describes the difference equations for the univariate marginal frequencies.

Theorem 4. *For any complete recombination/crossover scheme used after proportionate selection, the univariate marginal frequencies are determined by*

$$p(x_i, t+1) = \sum_{x|X_i=x_i} \frac{p(x,t)f(x)}{\bar{f}(t)}. \tag{10}$$

Proof. After selection the univariate marginal frequencies are given by

$$p^s(x_i, t) = \sum_{x|X_i=x_i} p^s(x, t) = \sum_{x|X_i=x_i} \frac{p(x,t)f(x)}{\bar{f}(t)}.$$

Now the selected individuals are randomly paired. Since complete recombination does not change the allele frequencies, these operators do not change the univariate marginal frequencies. Therefore

$$p_i(x_i, t+1) = p^s(x_i, t).$$

This theorem can be formulated in the terms of Holland's schema theory[12]. Let $H(x_i) = (*, \ldots, *, x_i, *, \ldots, *)$ be a first-order schema at locus i. This schema includes all strings where the gene at locus i is fixed to x_i. The univariate marginal frequency $p(x_i, t)$ is obviously identical to the frequency of schema $H(x_i)$. The fitness of the schema at generation t is given by

$$f(H(x_i), t) = \frac{1}{p(x_i, t)} \sum_{x | X_i = x_i} p(x, t) f(x). \tag{11}$$

From theorem 4 we obtain:

Corollary 3 (First-order schema theorem). *For a GA with proportionate selection using any complete recombination scheme, the frequency of first-order schemata changes according to*

$$p_i(x_i, t+1) = p_i(x_i, t) \frac{f(H(x_i), t)}{\bar{f}(t)}. \tag{12}$$

Note that the above corollary is valid for an infinite population with proportionate selection and recombination. Holland's famous schema theorem implies for first-order schemata only an inequality [12]:

$$p_i(x_i, t+1) \geq p_i(x_i, t) \frac{f(H(x_i), t)}{\bar{f}(t)}.$$

We summarize the results. All complete recombination schemes lead to the same univariate marginal distributions after one step of selection and recombination. If recombination is used for a number of times without selection, then the genotype frequencies converge to linkage equilibrium. This means that *all GAs are identical if after after one selection step recombination is done without selection a sufficient number of times*. This fundamental algorithm keeps the population in linkage equilibrium. In the next section we will characterize the fundamental algorithm.

3 UMDA – The Univariate Marginal Distribution Algorithm

Instead of performing recombination a number of times in order to converge to linkage equilibrium, one can achieve this in one step by *gene pool recombination* [22]. In gene pool recombination a new string is computed by randomly taking for each loci a gene from the distribution of the selected parents. This means that gene x_i occurs with probability $p^s(x_i)$ in the next population, where $p^s(x_i)$ is the distribution of x_i in the selected parents. Thus new strings x are generated according to the distribution

$$p(x, t+1) = \prod_{i=1}^{n} p^s(x_i, t). \tag{13}$$

One can simplify the algorithm still more by directly computing the univariate marginal frequencies from the data. Then (13) can be used to generate new strings. This method is used by the UMDA.

UMDA

STEP 0: Set $t \Leftarrow 1$. Generate $N \gg 0$ points randomly.

STEP 1: Select $M \leq N$ points according to a selection method. Compute the marginal frequencies $p^s(x_i, t)$ of the selected set.

STEP 2: Generate N new points according to the distribution $p(\boldsymbol{x}, t+1) = \prod_{i=1}^{n} p^s(x_i, t)$. Set $t \Leftarrow t+1$.

STEP 3: If termination criteria are not met, go to STEP 1.

UMDA needs $2n$ parameters, the marginal distributions $p(x_i)$. $\bar{f}(t)$ can also be seen as a function which depends on $p(x_i)$. To emphasize this dependency we write

$$W(p(x_1 = 0), p(x_1 = 1), \ldots, p(x_n = 1)) := \bar{f}(t). \tag{14}$$

W formally depends on $2n$ parameters; $p(x_i = 1)$ and $p(x_i = 0)$ are considered as two independent parameters despite the constraint $p(x_i = 0) = 1 - p(x_i = 1)$. We abbreviate $p_i := p(x_i = 1)$. If we insert $1 - p_i$ for $p(x_i = 0)$ into W, we obtain $\tilde{W}$. $\tilde{W}$ depends on n parameters. Now we can formulate the main theorem.

Theorem 5. *For infinite populations and proportionate selection, the difference equations for the gene frequencies used by UMDA are given by*

$$p(x_i, t+1) = p(x_i, t) \frac{\bar{f}_i(x_i, t)}{W(t)} \tag{15}$$

where $\bar{f}_i(x_i, t) = \sum_{\boldsymbol{x}, X_i = x_i} f(\boldsymbol{x}) \prod_{j \neq i}^{n} p(x_j, t)$. *The equations can also be written as*

$$p(x_i, t+1) = p(x_i, t) + p(x_i, t) \frac{\frac{\partial W}{\partial p(x_i)} - W(t)}{W(t)} \tag{16}$$

$$p_i(t+1) = p_i(t) + p_i(t)(1 - p_i(t)) \frac{\frac{\partial \tilde{W}}{\partial p_i}}{\tilde{W}(t)}. \tag{17}$$

Proof. Equation (15) has been proven in [18]. Equation (16) directly follows. We only have to prove (17). Note that

$$p_i(t+1) - p_i(t) = p_i(t) \frac{\bar{f}_i(x_i = 1, t) - \tilde{W}(t)}{\tilde{W}(t)}.$$

Obviously we have

$$\frac{\partial \tilde{W}}{\partial p_i} = \bar{f}(x_i = 1, t) - \bar{f}(x_i = 0, t).$$

From

$$p_i(t)\bar{f}_i(x_i = 1, t) + (1 - p_i(t))\bar{f}_i(x_i = 0, t) = \tilde{W}(t)$$

we obtain

$$\bar{f}(x_i = 1, t) - \tilde{W}(t) - (1 - p_i(t))\bar{f}(x_i = 1, t) + (1 - p_i(t))\bar{f}(x_i = 0, t) = 0.$$

This gives

$$\bar{f}_i(x_i = 1, t) - \tilde{W}(t) = (1 - p_i(t))\frac{\partial \tilde{W}}{\partial p_i}.$$

Inserting this equation into the difference equation gives (17).

The above equations completely describe the dynamics of UMDA with proportionate selection. Mathematically, UMDA performs gradient ascent in the landscape defined by W or $\tilde{W}$.

Corollary 4. *UMDA solves the continuous optimization problem* argmax $\tilde{W}(p)$ *on the unit cube* $[0, 1]^n$. *The continuous problem is an extension of the discrete optimization problem* argmax $f(x)$.

Equation (17) is especially suited for theoretical analysis. It has first been proposed by Wright [32]. Wright's remarks are still valid today: "The appearance of this formula is deceptively simple. Its use in conjunction with other components is not such a gross oversimplification in principle as has sometimes been alleged ... Obviously calculations can be made only from rather simple models, involving only a few loci or simple patterns of interaction among many similarly behaving loci ... Apart from application to simple systems, the greatest significance of the general formula is that its form brings out properties of systems that would not be apparent otherwise."

The restricted application lies in the following fact. In general the difference equations need the evaluation of 2^n terms. The computational complexity can be drastically reduced if the fitness function has a special form. This is discussed next.

3.1 Computing the Average Fitness

In mathematical terms the discrete optimization problem with variables x is transformed into a continuous optimization problem with variables p. We will take a closer look at this transformation. For notational convenience, we introduce a multi-index $\alpha = (\alpha_1, \ldots, \alpha_n)$, and define

$$x^\alpha := \prod_i x_i^{\alpha_i}.$$

Definition 5. The representation of a binary discrete function using the ordering according to function values is given by

$$f(x) = f(0, \ldots, 0)(1 - x_1) \cdots (1 - x_n) + \cdots + f(1, \ldots, 1) x_1 \cdots x_n. \quad (18)$$

The representation using the ordering according to variables is

$$f(x) = \sum_\alpha a_\alpha x^\alpha. \quad (19)$$

$\max\{|\alpha|_1 : a_\alpha \neq 0\}$ is called the order of the function.

In both representations the function is linear in each variable x_i. The following two lemmas are obvious.

Lemma 2. *The two representations are unique. There exist a unique matrix A of dimension $2^n \times 2^n$ such that*

$$a_\alpha = (Af)_\alpha.$$

Lemma 3. $\tilde{W}(p) := \bar{f}(t)$ *is an extension of $f(x)$ to the unit cube $[0,1]^n$. There exist two representations for $\tilde{W}(p)$, given by*

$$\tilde{W}(p) = f(0, \ldots, 0)(1 - p_1) \cdots (1 - p_n) + \cdots + f(1, \ldots, 1) p_1 \cdots p_n \quad (20)$$
$$\tilde{W}(p) = \sum_\alpha a_\alpha p^\alpha. \quad (21)$$

Equation (21) can also be used to compute the derivative. It is given by

$$\frac{\partial \tilde{W}(p)}{\partial p_i} = \sum_{\alpha | \alpha_i = 1} a_\alpha p^{\alpha'} \quad (22)$$

with $\alpha'_i = 0, \alpha'_j = \alpha_j$. If the function is of a low order, the partial derivatives can be easily evaluated. This allows us to compute the difference equations for p_i. In special cases the difference equation can even be solved analytically. We will discuss examples later. First we have to show that proportionate selection has a serious drawback, both for breeding of livestock as well for evolutionary algorithms. It selects too weakly for optimization purposes.

3.2 The Selection Problem

Fitness proportionate selection is the undisputed selection method in population genetics. It is considered to be a model for *natural selection*. But the selection strongly depends on the fitness values. When the population approaches an optimum, selection gets weaker and weaker, because the fitness values become similar. Therefore breeders of livestock use other selection methods. For large populations they mainly apply *truncation selection*. It works as follows. A truncation threshold τ is fixed. Then the τN best individuals are selected as parents for the next generation. These parents are then randomly mated.

We use mainly truncation selection in our algorithms. Another popular scheme is *tournament selection of size k*. Here k individuals are randomly chosen. The best individual is taken as parent. Unfortunately the mathematics for both selection methods is more difficult. Analytical results for tournament selection have been obtained by Mühlenbein [18].

3.3 Tournament Selection

We model binary tournament selection as a game. Two individuals with genotype x and y "play" against each other. The one with the larger fitness gets a payoff of 2. If the fitness values are equal, both will win half of the games. This gives a payoff of 1. The game is defined by a *payoff matrix* with coefficients

$$a_{xy} = \begin{cases} 2 & \text{when } f(x) > f(y) \\ 1 & \text{when } f(x) = f(y) \\ 0 & \text{when } f(x) < f(y). \end{cases}$$

With some effort one can show that

$$\sum_x \sum_y p(x,t) a_{xy} p(y,t) = 1. \tag{23}$$

After a round of tournaments the genotype frequencies are given by

$$p^s(x, t+1) = p(x,t) \sum_y a_{xy} p(y,t). \tag{24}$$

If we set

$$b(x,t) = \sum_y a_{xy} p(y,t)$$

then the above equation is similar to proportionate selection using the function $b(x,t)$. But b depends on the genotype frequencies. Furthermore, the average $\bar{b}(t) = \sum p(x,t) b(x,t)$ remains constant, $\bar{b}(t) \equiv 1$.

The difference equations for the univariate marginal frequencies can be derived in the same manner as for proportionate selection. They are given by

$$p(x_i, t+1) = p(x_i, t) \cdot \bar{B}_i(t) \tag{25}$$

$$\bar{B}_i(t) = \sum_{\boldsymbol{x}, X_i = x_i} b(\boldsymbol{x}, t) \prod_{\substack{j=1 \\ j \neq i}}^{n} p(x_j, t). \tag{26}$$

The difference equation for binary tournament selection is more difficult than for proportionate selection: $\bar{B}_i$ is quadratic in $p(x_j)$. Tournament selection uses only the order relation of the fitness values. The fitness values themselves do not change the outcome of a tournament. Therefore the evolution of the univariate marginal frequencies depends on the order relation only.

The analysis is still more difficult for truncation selection. Therefore breeders have developed a macroscopic theory using average fitness and variance of the population.

3.4 The Science of Breeding

For a single trait the theory can be easily summarized. Starting with the fitness distribution, the *selection differential* $S(t)$ is introduced. It is the difference between the average of the selected parents and the average of the population

$$S(t) = W(\boldsymbol{p}^s(t+1)) - W(\boldsymbol{p}(t)). \tag{27}$$

Similarly the response $R(t)$ is defined:

$$R(t) = W(\boldsymbol{p}(t+1)) - W(\boldsymbol{p}(t)). \tag{28}$$

Next a linear regression is done

$$R(t) = b(t)S(t) \tag{29}$$

where $b(t)$ is called *realized heritability*. The selection differential can often be approximated by

$$S(t) \approx I_\tau V^{\frac{1}{2}}(t) \tag{30}$$

where I_τ is called the *selection intensity*. V is the variance of the fitness distribution. Combining the two equations we obtain the *famous equation for the response to selection*:

$$R(t) = b(t) I_\tau V^{\frac{1}{2}}(t). \tag{31}$$

The most difficult problem is to estimate $b(t)$. Breeders use the estimate

$$b(t) \approx \frac{V_A(t)}{V(t)} \tag{32}$$

where $V_A(t)$ denotes the additive genetic variance. For UMDA it is defined as

$$V_A(t) = \sum_{i=1}^{n} \sum_{x_i} p(x_i, t) \left(\frac{\partial W}{\partial p(x_i)} - W(t) \right)^2. \tag{33}$$

These equations are discussed in depth in [18]. For the special case that all univariate marginal distribution are equal, i.e., $p_i := p$, the response to selection equation gives a difference equation for p. Thus it might be possible to obtain an analytical solution for $p(t)$.

We cite from [18] the analytical solutions for the linear function $OneMax(n) = \sum_i x_i$. The difference equation for binary tournament selection has also been computed [18].

Theorem 6. *If in the initial population all univariate marginal frequencies are identical to $p_0 > 0$, then we obtain for UMDA and OneMax*

$$R(t) = 1 - p(t) \tag{34}$$

$$p(t) = 1 - (1 - p_0)(1 - \frac{1}{n})^t \tag{35}$$

for proportionate selection, and

$$R(t) \approx I_\tau \sqrt{np(t)(1 - p(t))} \tag{36}$$

$$p(t) \approx 0.5 \left(1 + \sin\left(\frac{I_\tau}{\sqrt{n}} t + \arcsin(2p_0 - 1) \right) \right) \tag{37}$$

for truncation selection.

These solutions perfectly match the figures obtained from actual UMDA runs. In figure 1, the analytical results for proportionate selection and truncation selection with $\tau = 0.3$ are almost identical to the simulation results.

Using proportionate selection, it takes the population a long time to approach the optimum. In contrast, truncation selection and tournament selection lead to much faster convergence: p increases almost linearly until near the optimum. Truncation selection with $\tau = 0.6$ behaves very similarly to tournament selection. This is known from [18].

The theory of breeding uses macroscopic variables, the average and the variance of the population. But there exists only one equation, the response to selection equation. We need a second equation connecting the average fitness and the variance in order to be able to compute the time evolution of the average fitness and the variance. There have been many attempts in population genetics to find a second equation. But all equations assume that the variance of the population continuously decreases. This is not the case for arbitrary fitness functions. Recently, Prügel-Bennett

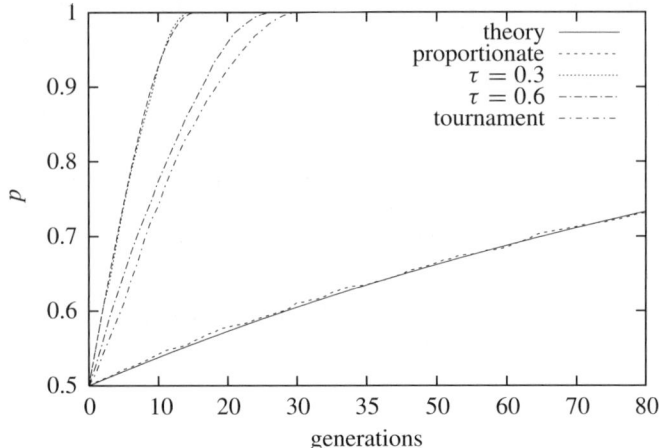

Figure 1. Comparison of selection methods for *OneMax*(128)

and Shapiro [26] have independently proposed using moments for describing GAs. They apply methods of statistical physics to derive equations for higher moments for special fitness functions.

Before we numerically show the optimization power of UMDA, we briefly discuss other algorithms which also use univariate marginal distributions.

4 Optimization Methods Using Univariate Distributions

The importance of using univariate marginal distributions has been independently discovered by several researchers. We just discuss PBIL of Baluja and Caruana [2] and *ant colony optimization* by Dorigo and Di Caro [5]. PBIL does not use strict Darwinian selection in populations, but the the probabilities are updated according to

$$p(x_i, t+1) = p(x_i, t) + \lambda\big(p^s(x_i, t) - p(x_i, t)\big). \tag{38}$$

The string x is generated as before:

$$p(x, t+1) = \prod_{i=1}^{n} p(x_i, t+1). \tag{39}$$

The convergence speed of this algorithm critically depends on λ. For $\lambda = 0$ we have a random search, for $\lambda = 1$ we obtain UMDA. The smaller λ, the slower the convergence speed. This problem is discussed in [18]. Our numerical experiments indicate that the UMDA method is to be preferred, because it is very difficult to choose λ for a given problem.

In principle, univariate marginal distributions are also used in ant colony optimization (ACO, [5]). For each ant k, a probability p_{ij} is computed to move from state i to state j. The equation is given by

$$p_{ij}^k := \begin{cases} \frac{\tau_{ij}(t)}{\sum_{j \in N(i)} \tau_{ij}(t)} & \text{for } j \in N(i) \\ 0 & \text{for } j \notin N(i). \end{cases} \quad (40)$$

The variable τ_{ij} is updated according to

$$\tau_{ij}(t+1) = \tau_{ij}(t) + \Delta \tau_{ij}. \quad (41)$$

Here $p_{ij} = p_i(x_j)$ plays the role of our univariate marginal distributions. If all states j are contained in the neighbourhood $N(i)$ then we have an UMDA algorithm with integer variables. A solution x is generated with probability

$$p(x) = \prod_i p_i(x_j).$$

Ant colony optimization is mainly applied to constrained combinatorial optimization problems like the travelling salesman problem (TSP) or the quadratic assignment problem. Here the neighbourhoods have to be dynamically changed. We take TSP as example. If city l is chosen to be at a certain place of the tour, it is not allowed to be chosen again. ACO solves this problem by setting all p_{il}^k with $i > l$ to 0. The other p_{ij}^k values are renormalized, so that the sum of the probabilities of all feasible moves is 1. Thus ACO constructs feasible solutions. But the UMDA theory does not apply because of the renormalization of the probabilities.

5 The Replicator Equation and Combinatorial Optimization

For mathematical analysis the equations for discrete generations are often approximated by equations with continuous time. That is, the *difference equations* are approximated by *differential equations*. The reason is that the mathematical analysis of differential equations is easier. In general, this approximation is a complicated issue. We consider only the proportionate selection equation (2).

Let $n_x(t)$ denote the number of occurrences of string x at generation t. Let $N(t)$ denote the size of the population. Then we have $p(x, t) = n_x(t)/N(t)$.

Lemma 4. *If the occurrences grow according to their fitness*

$$n_x(t+1) = f(x) n_x(t)$$

then

$$p(x, t+1) = p(x, t) \frac{f(x)}{\bar{f}(t)}.$$

This is just the equation describing proportionate selection. It is obtained from

$$N(t+1) = \sum_x f(x)\frac{n_x(t)}{N(t)} N(t) = \bar{f}(t) N(t).$$

The corresponding lemma for continuous t is as follows:

Lemma 5. *If the occurrences grow according to their fitness*

$$\frac{dn_x(t)}{dt} = f(x) n_x(t)$$

then

$$\frac{dp(x,t)}{dt} = p(x,t)\big(f(x) - \bar{f}(t)\big). \tag{42}$$

Proof. We have

$$\frac{dp(x,t)}{dt} = \frac{\frac{dn_x}{dt} N - \frac{dN}{dt} n_x}{N^2}$$

$$= \frac{n_x(t)}{N(t)} \big(f(x) - \bar{f}(t)\big).$$

Equation (42) is a special case of a differential equation defined by

$$\frac{dp_\alpha}{dt} = p_\alpha \left(f_\alpha(p) - \sum_\alpha p_\alpha f_\alpha(p) \right) \tag{43}$$

where p_α now denotes $p(x)$. This equation is called the *replicator equation*. It is of great importance in many fields connected to biology. For a general investigation of these equations the reader is referred to [11]. Interesting discussions can also be found in [6] and [25].

We introduce an extension of the replicator equation, called the *diversified replicator equation* [30].

Definition 6. Let $p_{\alpha k} \geq 0$ be defined for $1 \leq k \leq m$ with $\sum_k^m p_{\alpha k} = 1$. Then a diversified replicator equation is defined for discrete time by

$$p_{\alpha k}(t+1) - p_{\alpha k}(t) = p_{\alpha k}(t) \frac{f_{\alpha k}(p(t)) - \sum_{k=1}^m p_{\alpha k}(t) f_{\alpha k}(p(t))}{\sum_{k=1}^m p_{\alpha k} f_{\alpha k}(p(t))}. \tag{44}$$

The corresponding differential equation is given by

$$\frac{dp_{\alpha k}}{dt} = p_{\alpha k} \left(f_{\alpha k}(p) - \sum_{k=1}^m p_{\alpha k} f_{\alpha k}(p) \right). \tag{45}$$

The replicator and the diversified replicator equation differ in the constraints. We have $\sum_\alpha p_\alpha = 1$ for the replicator equation and $\sum_k p_{\alpha k} = 1$ for the diversified replicator equation. Our central UMDA equation (16) is a special case of a diversified replicator equation. This can be seen by setting $k \in \{0, 1\}$, $\alpha \in \{1, \ldots, n\}$, $p_{\alpha k} := p(x_\alpha = k)$ and

$$f_{\alpha k}(\boldsymbol{p}) = \frac{\partial W}{\partial p_{\alpha k}}. \tag{46}$$

Thus (16) defines a *gradient system* with the potential $W(\boldsymbol{p})$. Gradient systems have nice properties. We just give one example.

Theorem 7. *If the diversified replicator equation is a gradient system, the potential W never decreases, i.e.,*

$$\frac{dW}{dt} \geq 0. \tag{47}$$

Proof. We compute

$$\begin{aligned}
\frac{dW}{dt} &= \sum_\alpha \sum_k \frac{\partial W}{\partial p_{\alpha k}} \frac{dp_{\alpha k}}{dt} \\
&= \sum_\alpha \sum_k \frac{\partial W}{\partial p_{\alpha k}} p_{\alpha k} \left(f_{\alpha k}(\boldsymbol{p}) - \sum_{k=1}^m p_{\alpha k} f_{\alpha k}(\boldsymbol{p}) \right) \\
&= \sum_\alpha \left(\sum_k p_{\alpha k} f_{\alpha k}^2(\boldsymbol{p}) - (\sum_k p_{\alpha k} f_{\alpha k}(\boldsymbol{p}))^2 \right) \\
&\geq 0.
\end{aligned}$$

The diversified replicator equation has been used by Voigt [30] and Mühlenbein [19] to solve combinatorial problems. In their method the difference equations are iterated until all probabilities $p_{\alpha k}$ have converged. This method poses a major difficulty. It stops at local maxima in the interior of the unit cube. But these points are not feasible. Therefore Voigt [30] has developed adaptive techniques which drive the probabilities to the corner of the simplices.

Using the UMDA algorithm is a much simpler solution to the problem. We interpret $p_{\alpha k}$ as the probability that x_α is set to k. We generate a population of solutions, select the better ones and compute the probabilities $p^s(x_\alpha)$ of the selected strings. The new population is generated by the method used by UMDA

$$p(\boldsymbol{x}, t+1) = \prod_\alpha p^s(x_\alpha). \tag{48}$$

The diversified replicator equation has been used in [30,19] to solve combinatorial problems like the *graph partitioning problem* (GPP) and the TSP. It is worthwhile to test UMDA on this problem.

We will numerically investigate UMDA in the next section. The application domain is the optimization of discrete functions.

6 Numerical Results for UMDA

This section solves the problem put forward by Mitchell et al. [14], i.e., to understand the class of problems for which GAs are most suited, and in particular, for which they will outperform other search algorithm. We start with the *Royal Road* function, which was erroneously believed to lay out a royal road for the GA to follow to the optimal string.

6.1 Royal Road Function

We discuss the Royal Road function R_1, which was used by Mitchell et al. [14]. It is defined as follows:

$$R_1(l, x) = \sum_{i=0}^{l-1} \prod_{j=1}^{8} x_{8i+j}. \tag{49}$$

The function is of order 8. The building block hypothesis (BBH, [12]) states that "the GA works well when instances of low-order, short schemas that confer high fitness can be recombined to form instances of larger schemas that confer even higher fitness." In our terminology a schema defines a marginal distribution. Thus a first-order schema defines a univariate marginal distribution. Our analysis has shown that only the first half of the BBH is correct: first-order schemata of high fitness are recombined. Larger schemata play no role.

1+1	SGA	UMDA p	UMDA $\tau = 0.3$	UMDA $\tau = 0.05$	FDA $\tau = 0.3$
6,334	61,334	55,586	28,000	14,264	7,634

Table 3. Mean function evaluations for Royal Road(8)

Table 3 confirms and extends the results of Mitchell et al. [14]. The really bad performance of SGA is mainly a result of proportionate selection. UMDA with proportionate selection (UMDA p) needs slightly less evaluations. With very strong selection, UMDA needs only about twice as many function evaluations as the $(1 + 1)$-algorithm. This algorithm performs a random bit flip and accepts a new configuration if its fitness is equal or better. The good performance of this algorithm has already been shown in [17]. But it performs well only if the fitness function never decreases with increasing number of bits. Almost identical performance to the $(1+1)$-algorithm can be obtained by FDA. It uses marginal distributions of size 8 instead of univariate marginal distributions. This will be explained in section 7.

Figure 2 shows once more the importance of selection. Proportionate selection performs very well in the beginning, because the fitness values of all strings containing

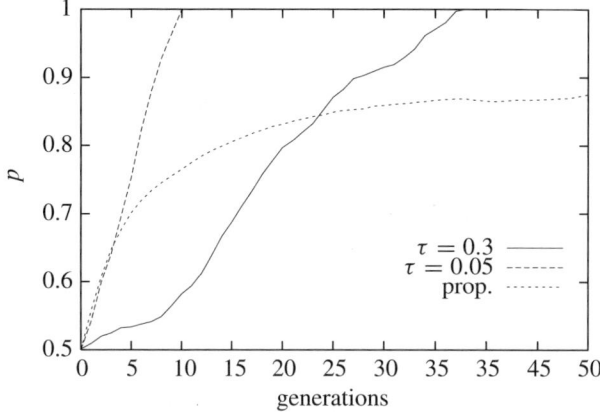

Figure 2. Convergence of Royal Road

no building block are zero. These strings are not reproduced. But after 5 generations proportionate selection gets weaker. Truncation selection with $\tau = 0.3$ overtakes it after 23 generations. We just mention that the numerical results would be much worse for proportionate selection, if we added 1 to the Royal Road function. In this case proportionate selection also selects many strings without a building block.

We will now explain the results by using our theory to analytically solve the equations. We have

$$\tilde{W}(p) = \sum_{i=0}^{l-1} \prod_{j=1}^{8} p_{8i+j}$$

$$\frac{\partial \tilde{W}}{\partial p_k} = \prod_{\substack{j=1 \\ 8i+j \neq k}}^{7} p_{8i+j}, \qquad \text{for } 8i \leq k < 8i + 8.$$

For truncation selection we will apply the response to selection equation. Therefore we have to compute the variance $V_l(t)$. We simplify the computation by observing that the blocks of 8 variables are independent and therefore

$$V_l(t) = l \cdot V_1(t).$$

We recall that all function values are 0 except $f(1, \ldots, 1)$. Therefore

$$V_1(t) = \sum_x p(x, t) f^2(x) - W^2$$
$$= \prod p_i - (\prod p_i)^2.$$

If we assume that $p_i = p$ for all i, we obtain

$$V_8(t) = 8p(t)^8(1 - p(t)^8). \tag{50}$$

We can now formulate the theorem.

Theorem 8. *If all univariate marginal distributions are identical to $p(t)$, and if $p(0) = p_0$, then we obtain for proportionate selection*

$$p(t+1) - p(t) = \frac{1 - p(t)}{8} \tag{51}$$

$$p(t) = 1 - (1 - p_0)(\tfrac{7}{8})^t. \tag{52}$$

For truncation selection with threshold τ we approximately get

$$R(t) \approx b(t) I_\tau \sqrt{8 p(t)^8 (1 - p(t)^8)} \tag{53}$$

$$p(t)^8 \approx 0.5 \left(1 + \sin\left(\frac{b(t) I_\tau}{\sqrt{8}} t + \arcsin(2 p_0^8 - 1) \right) \right). \tag{54}$$

Proof. The conjectures for proportionate selection directly follow from (17). From the response to selection equation we obtain

$$8 p(t+1)^8 - 8 p(t)^8 \approx b(t) I_\tau \sqrt{8 p(t)^8 (1 - p(t)^8)}.$$

If we set $q(t) = p(t)^8$, the above equation is identical to the one for *OneMax*(8). The approximate solution is given by (37).

In figure 3, a comparison between the theoretical solution and a simulation run is made. The simulation run gives slightly larger values of W. The reason is that in finite populations there are random fluctuations. UMDA is not able to keep $p_i = p_j := p$. We use the estimate $p(t) = 1/n \sum_i p_i(t)$ to compute W for the figure. But the similarity between the theory and the simulations is still impressive.

In order to apply (54) we need an estimate for the realized heritability $b(t)$. Experiments show that $b(t)$ increases approximately linearly from about 0 to 1. Thus we set $b(t) \propto t$. Figure 4 shows a comparison between (54) and a simulation with truncation threshold 0.05. The coincidence between theory and simulation is very good.

This example shows that the response to selection equation can in special cases be used to compute an analytical solution for $p(t)$. The difficulty is to determine the heritability $b(t)$.

6.2 Multimodal Functions Suited for UMDA Optimization

Equation (16) shows that UMDA performs a gradient ascent in the landscape given by W. This helps our search for functions best suited for UMDA. We take as the first example the function *BigJump*. It is defined as follows, with $|x|_1 = \sum x_i$ equal to the number of 1-bits:

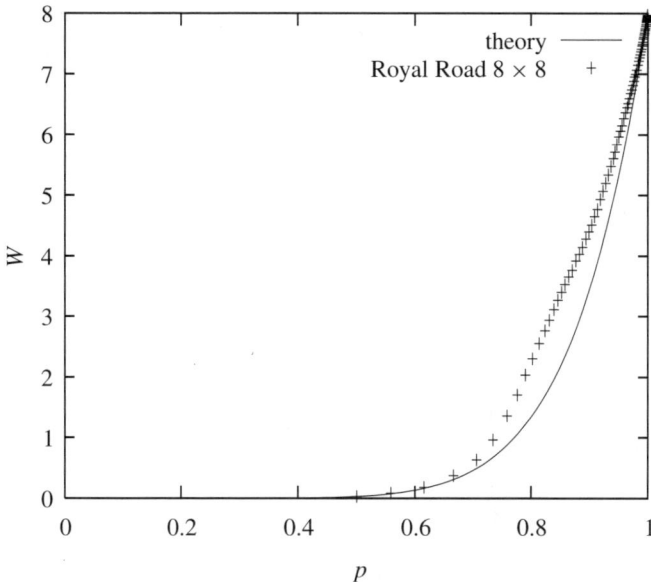

Figure 3. Proportionate selection for Royal Road: theory and simulation

$$BigJump(n, m, k, x) := \begin{cases} |x|_1 & \text{for } 0 \leq |x|_1 \leq n - m \\ 0 & \text{for } n - m < |x|_1 < n \\ k \cdot n & \text{for } |x|_1 = n. \end{cases} \quad (55)$$

The bigger m, the wider the valley. The parameter k can be increased to give bigger weight to the maximum. For $m = 1$ we obtain the popular *OneMax* function defined by $OneMax(n) = |x|_1$.

BigJump depends only on the number of bits. We assume that all $p(x_i = 1)$ are identical to a single value denoted as $p(t)$. Then W depends only on one parameter, p. $W(p)$ is shown for $m = 30$ and $k = 20$ in figure 5. In contrast to the discrete function the average fitness $W(p)$ looks fairly smooth. The open circles are the values of $p(t)$ determined by an UMDA run, setting $p(t) := 1/n \sum_i p_i(t)$. Note how closely the simulation follows the theoretical curve. Because we use discrete generations in UMDA the population is able to pass the local minimum at about $p = 0.83$.

This simple example confirms the results of our theory in a nutshell. *Evolutionary algorithms transform the original fitness landscape given by $f(x)$ into a fitness landscape defined by $\tilde{W}(p)$. This transformation smooths the rugged fitness landscape $f(x)$. In these landscapes simple evolutionary algorithms will find the global optimum.*

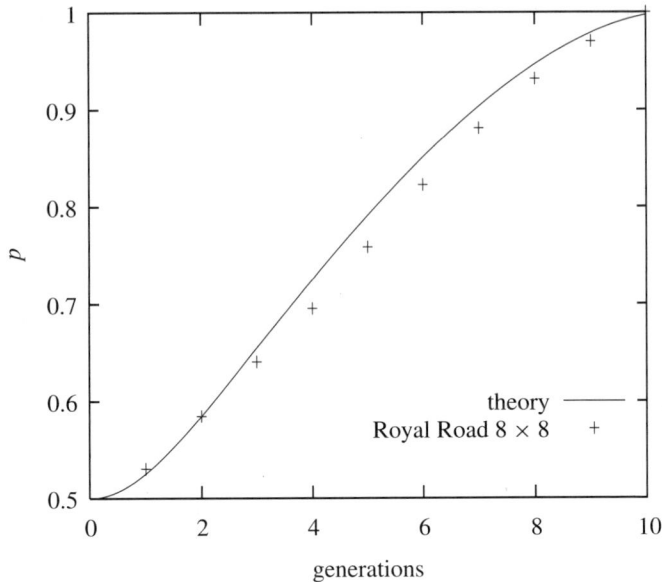

Figure 4. Truncation selection for Royal Road: theory and simulation

A still more spectacular example is the *Saw* landscape. The definition of the function can be extrapolated from figure 6. In $Saw(n, m, k)$, n denotes the number of bits and $2m$ the distance from one peak to the next. The highest peak is multiplied by k (with $k \leq 1$), the second highest by k^2, then k^3 and so on. The landscape is very rugged. In order to get from one local optimum to another one, one has to cross a deep valley.

But again the transformed landscape $W(p)$ is fairly smooth. An example is shown in figure 7. Whereas $f(x)$ has 5 isolated peaks, $W(p)$ has three plateaus, a local peak and the global peak. Therefore we expect that UMDA should be able to cross the plateaus and terminate at the local peak. This behavior can indeed be observed in figure 7. Furthermore, as predicted by (16), the progress of UMDA slows down on the plateaus.

Next we will investigate UMDA with truncation selection. We have not been able to derive precise analytical expressions. In figure 8 the results are displayed.

In the simulation two truncation thresholds, $\tau = 0.05$ and $\tau = 0.01$, have been used. For $\tau = 0.05$ the probability p stops at the local maximum for $\tilde{W}(p)$. It is approximately $p = 0.78$. For $\tau = 0.01$ UMDA is able to converge to the optimum $p = 1$. It does so by even going downhill!

These two examples show that UMDA can solve difficult multimodal optimization problems. It is obvious that any search method using a single search point like the $(1 + 1)$-algorithm needs an almost exponential number of function evaluations.

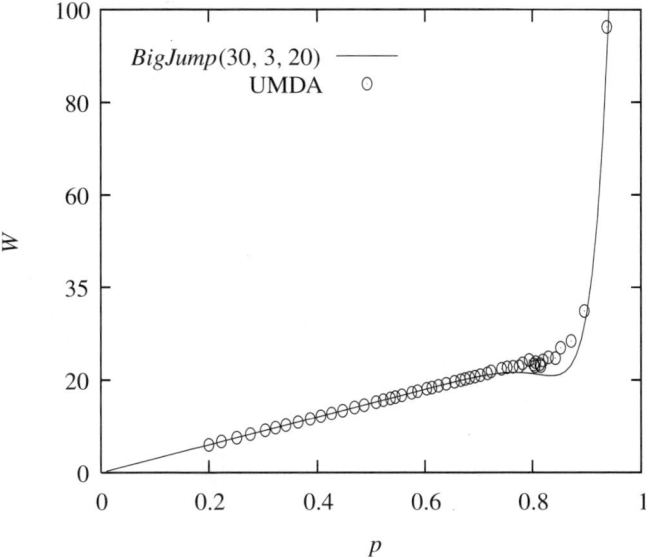

Figure 5. *BigJump*(30,3,20), UMDA, p versus average fitness, population size 2000

6.3 Deceptive Functions

There are many optimization problems, however, where UMDA is misled. UMDA will converge to local optima, because it does not use correlations between the variables. We demonstrate this problem by a deceptive function. We use the definition

$$Deceptive(\boldsymbol{x}, k) := \begin{cases} k - 1 - |\boldsymbol{x}|_1 & \text{for } 0 \leq |\boldsymbol{x}|_1 < k \\ k & \text{for } |\boldsymbol{x}|_1 = k. \end{cases} \quad (56)$$

The global maximum is isolated at $x = (1, \ldots, 1)$. The function can be analyzed as before by using the representation

$$Deceptive(\boldsymbol{x}, k) = k - 1 - \sum_i x_i + (k+1) \prod_i x_i.$$

$\tilde{W}(\boldsymbol{p})$ is obtained by exchanging x_i with p_i. The local minimum of $\tilde{W}(\boldsymbol{p})$ is at $p = (0.2)^{1/3} \approx 0.58$.

We simplify the optimization problem by adding l distinct *Deceptive(k)*-functions to give a fitness function of size $n = l \times k$

$$Deceptive(n, k) = \sum_{i=1, k+1, \ldots}^{n} Deceptive\big((x_i, x_{i+1}, \ldots, x_{i+k-1}), k\big). \quad (57)$$

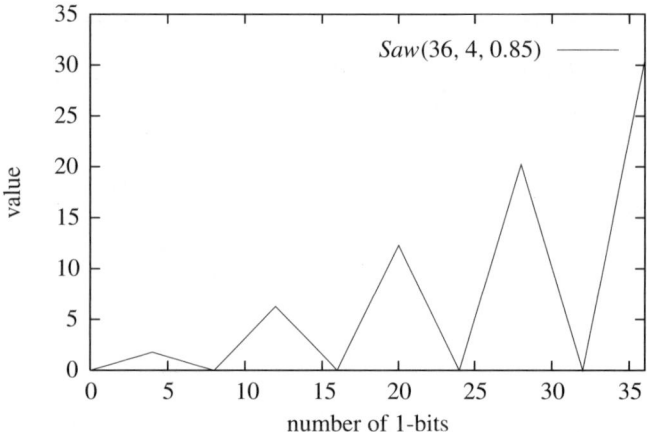

Figure 6. Definition of *Saw*(36,4,0.85)

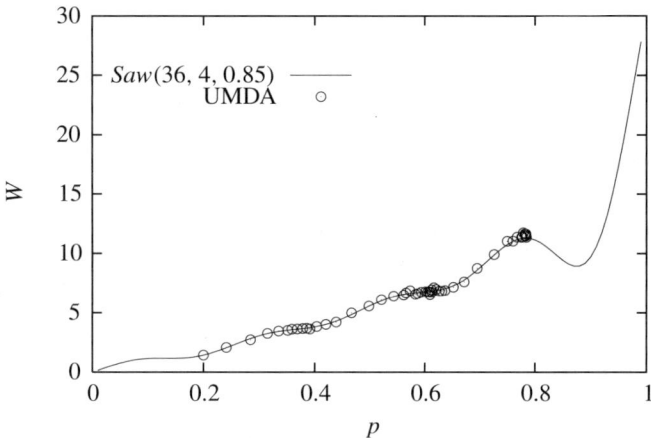

Figure 7. *Saw*(36,4,0.85), UMDA, p versus average fitness, population size 2000

This function is also deceptive. The local optimum $x = (0, \ldots, 0)$ is surrounded by good fitness values, whereas the global optimum is isolated.

In figure 9, we show the average fitness $W(p)$ and an actual UMDA run. Starting at $p(0) = 0.5$, UMDA converges to the local optimum $x = (0, \ldots, 0)$. UMDA will converge to the global optimum if it starts to the right from the local optimum, e.g., $p(0) \geq 0.59$. Also shown is a curve derived from FDA. FDA uses fourth-order marginal distributions. It converges to the global optimum, even if the initial population is generated randomly. But one can see in the figure that $p(t)$ also decreases first. FDA is discussed in section 7.

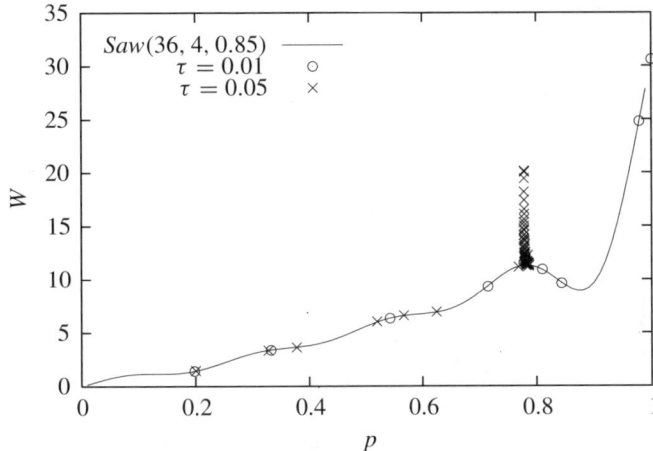

Figure 8. Results with normal and strong selection

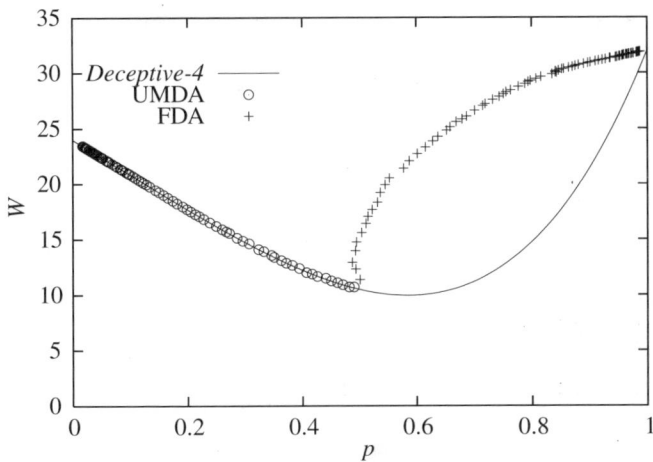

Figure 9. Average fitness $W(p)$ for UMDA and FDA for *Deceptive*(36,4)

6.4 Numerical Investigations of the Science of Breeding

In this section we show that the science of breeding can be very usefully applied to evolutionary optimization. Linear functions are the ideal case for the theory. The heritability $b(t)$ is 1 and the additive genetic variance is identical to the variance. We skip this trivial case and start with a multiplicative fitness function $f(x) = \prod_i (1-s)^{1-x_i}$.

Figure 10 confirms the theoretical results from section 2 (V_A and V are multiplied by 10 in this figure). Additive genetic variance is identical to the variance and the heritability is 1. The function is highly nonlinear of order n, but nevertheless it is

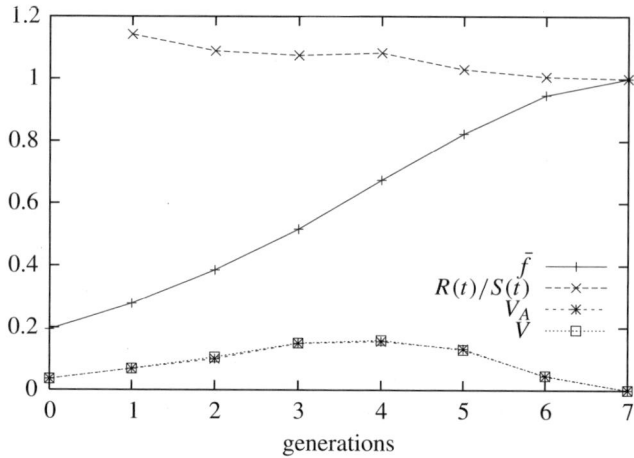

Figure 10. Heritability and variance for a multiplicative function

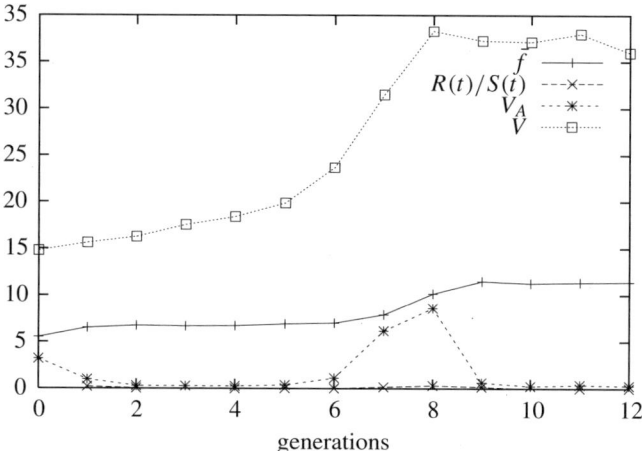

Figure 11. Heritability and variance for function *Saw*

easy to optimize. The function has also been investigated by Rattray and Shapiro [27]. They have not observed that the population remains in linkage equilibrium, making their calculations very difficult.

The function *Saw* is difficult to optimize. We see in figure 11 that for a long time there is no progress. An increase of the average fitness occurs at generations 6 till 9. During this time the additive genetic variance V_A is higher. But the heritability is almost zero almost anywhere.

An interesting case is the function *Deceptive-4*. In figure 12, the function is optimized for 32 bits. As predicted by the theory, UMDA converges to the local optimum $x = (0, \ldots, 0)$. Heritability is almost zero at the beginning, indicating that the

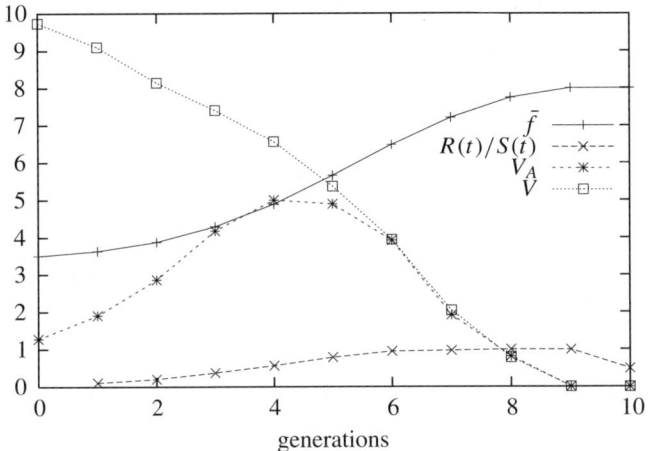

Figure 12. Heritability and variance for *Deceptive-4*: $\tau = 0.3$

optimization problem is difficult. In the beginning the competition between setting the genes to 0 or to 1 is undecided. UMDA decides to go in the direction of 0. If there is a high percentage of zeros in the population, then heritability increases to almost 1. In this area the fitness function is almost linear.

The examples demonstrate that it is worthwhile to compute the quantities used for a scientific breeding programme. They clearly indicate how difficult the optimization problem is. In breeding of livestock heritability is normally greater than than 0.2. If we optimize arbitrary fitness functions the heritability can be almost 0. But because we can easily compute 1000 generations on a computer in a few minutes, UMDA can be used for problems with very low heritability.

We have shown that UMDA can optimize difficult multimodal functions, thus explaining the success of GAs in optimization. We have also shown that UMDA can easily be deceived by simple functions called deceptive functions. These functions need marginal distributions of higher order.

7 FDA – The Factorized Distribution Algorithm

For the mathematical analysis we will use Boltzmann selection. Boltzmann selection can be seen as proportionate selection applied to the transformed function $F(x) = \exp(\beta f(x))$.

Definition 7. For Boltzmann selection the distribution after selection is given by

$$p^s(x, t) = p(x, t) \frac{e^{\beta f(x)}}{W_\beta} \tag{58}$$

where $\beta > 0$ is a parameter, also called the inverse temperature, and

$$W_\beta = \sum p(x,t) e^{\beta f(x)}$$

is the weighted average of the population.

For Boltzmann distributions we have proven a factorization theorem for the distribution $p(x,t)$ and convergence for an algorithm using this factorization [20]. The proof is simple, because if $p(x,t)$ is a Boltzmann distribution with factor β_1 and Boltzmann selection is done with factor β_2, then $p(x,t+1) = p(x,t)$ is a Boltzmann distribution with factor $\beta = \beta_1 + \beta_2$.

Theorem 9. *Let $p(x, 0)$ be randomly distributed. Let $\beta_1, \ldots, \beta_{t-1}$ be the schedule of the inverse temperature for Boltzmann selection. Then the distribution is given by*

$$p(x,t) = \frac{e^{\beta f(x)}}{Z_\beta} \tag{59}$$

where $\beta = \sum_{i=1}^{t-1} \beta_i$. Z_β is the partition function $Z_\beta = \sum_x e^{\beta f(x)}$.

Equation (59) is a complete analytical solution of the dynamics. But it cannot be used for an algorithm. $p(x,t)$ consists of $2^n - 1$ variables. Therefore the amount of computation is exponential. But there are many cases where the distribution can be factored into conditional marginal distributions each depending only on a small number of parameters. We recall the definition of conditional probability.

Definition 8. *The conditional probability $p(x|y)$ is defined as*

$$p(x|y) = \frac{p(x,y)}{p(y)}. \tag{60}$$

From this definition the following theorem easily follows.

Theorem 10 (Bayesian factorization).

Each probability can be factored into

$$p(x) = p(x_1) \prod_{i=2}^{n} p(x_i | pa_i). \tag{61}$$

Proof. By definition of conditional probabilities we have

$$p(x) = p(x_1) \prod_{i=2}^{n} p(x_i | x_1, \cdots, x_{i-1}). \tag{62}$$

Let $pa_i \subset \{x_1, \cdots, x_{i-1}\}$. If x_i and $\{x_1, \cdots, x_{i-1}\} \setminus pa_i$ are conditionally independent given pa_i, we can simplify $p(x_i | x_1, \cdots, x_{i-1}) = p(x_i | pa_i)$.

pa_i are called the parents of variable X_i. This factorization defines a directed graph. In the context of graphical models the graph and the conditional probabilities are called a Bayesian network [13,7]. The factorization is used by the FDA.

FDA

STEP 0: Set $t \Leftarrow 0$. Generate $N \gg 0$ points randomly.
STEP 1: Selection.
STEP 2: Compute the conditional probabilities $p^s(x_i | pa_i, t)$ using the selected points.
STEP 3: Generate a new population according to

$$p(x, t+1) = \prod_{i=1}^{n} p^s(x_i | pa_i, t).$$

STEP 4: If termination criteria is met, FINISH.
STEP 5: Set $t \Leftarrow t + 1$. Go to STEP 2.

FDA can be used with an exact or an approximate factorization. It is not restricted to Bayesian factorization. FDA uses *finite samples* of points to estimate the conditional distributions. Convergence of FDA to the optimum will depend on the size of the samples.

If the factorization does not contain conditional marginal distributions, but only marginal distributions, FDA can be theoretically analyzed. The difference equations of the marginal distributions are of the form given in (16) [15].

The amount of computation of FDA depends on the size of the population (N) and the number of variables used for the factors. There exist many problems where the size of the factors is bounded by k independent from n. In this case FDA is very efficient [16]. But for the function *BigJump* an exact factorization needs a factor of size n. Then the amount of computation of FDA is exponential in n. We have seen before that for *BigJump*, UMDA will already find the global optimum. Thus an exact factorization is not a necessary condition for convergence. But it is necessary if we want to be sure that the optimum is found.

8 Finite Populations

In finite populations, convergence of UMDA or FDA can only be probabilistic. Since UMDA is a specialized FDA algorithm, it is sufficient to discuss FDA. This section is extracted from [16].

Definition 9. Let ϵ be given. Let $P_{conv}(N)$ denote the probability that FDA with a population size of N converges to the optima. Then the critical population size is defined as

$$N^*(\epsilon) = \min_{N} P_{conv}(N) \geq 1 - \epsilon. \tag{63}$$

If FDA with a finite population does not convergence to an optimum, then at least one gene is fixed to a wrong value. The probability of fixation is reduced if the population size is increased. We obviously have for FDA

$$P_{conv}(N_1) \leq P_{conv}(N_2), \quad \text{for } N_1 \leq N_2.$$

The critical question is: how many sample points are necessary to reasonably approximate the distribution used by FDA? A general estimate from Vapnik [29] can be a guideline. One should use a sample size which is about 20 times larger than the number of free parameters.

We discuss the problem with a special function called *Int*. $Int(x)$ gives the integer value of the binary representation:

$$Int(n) = \sum_{i=1}^{n} 2^{i-1} x_i. \tag{64}$$

The fitness distribution of this function is not normally distributed. The function has 2^n different fitness values. We show the cumulative fixation probability in table 4 for $Int(16)$. The fixation probability is larger for stronger selection. For a given truncation selection the maximum fixation probability is at generation 1 for very small N. For larger values of N, the fixation probability increases until a maximum is reached and then decreases again. This behaviour has been observed for many fitness distributions.

Boltzmann selection with $\beta = 0.01$ is still very strong for the fitness distribution given by $Int(16)$. For $N = 700$, the largest fixation probability is still at the first generation. Therefore the critical population size for Boltzmann selection with $\beta = 0.01$ is very high ($N^* > 700$). For truncation selection with $\tau = 0.25$ we have $N^*(0.1) \leq 80$.

t	$\tau = 0.25$ $N = 30$	$\tau = 0.5$ $N = 30$	$\tau = 0.25$ $N = 80$	$\tau = 0.5$ $N = 60$	Boltzm. $N = 500$	Boltzm. $N = 700$
1	0.0955	0.0035	0.0	0.0	0.2520	0.0885
2	0.4065	0.0255	0.0025	0.0095	0.2980	0.1110
3	0.5955	0.1040	0.0165	0.0205	0.3180	0.1275
4	0.6880	0.2220	0.0355	0.0325	0.3295	0.1375
5	0.7210	0.3270	0.0575	0.0490	0.3385	0.1455
6	0.7310	0.4030	0.0695	0.0630	0.3435	0.1510
7	0.7310	0.4470	0.0740	0.0715	0.3505	0.1555
8	0.7310	0.4705	0.0740	0.0780	0.3530	0.1565
9	0.7310	0.4840	0.0740	0.0806	0.3555	0.1575

Table 4. Cumulative fixation probability for $Int(16)$. Truncation selection versus Boltzmann selection with $\beta = 0.01$

Because Boltzmann selection in finite populations critically depends on a good annealing schedule, we normally run FDA with truncation selection. This selection method is a good compromise. It has an important property, which we formulate as an empirical law. It has been confirmed by many numerical experiments.

Empirical law. *Let ϵ be reasonable small, e.g., $\epsilon = 0.1$. Then the number of generations to converge to the optimum remains constant for $N \geq N^*(\epsilon)$*

$$GEN_e(N^*(\epsilon)) = GEN_e(N) = GEN_e(N = \infty) \quad \text{for } N \geq N^*(\epsilon). \tag{65}$$

Truncation selection has a free parameter, the truncation threshold τ. It seems obvious that the smaller the threshold τ, the larger N^* has to be. But numerical experiments have shown that there exists a threshold τ_{min} which leads to a minimal N^*_{min}. This means that N^* also increases for very low selection. The reason for this phenomenon is genetic drift. Slow selection leads to a large number of generations which increases the probability of gene fixation. This problem was first investigated by Mühlenbein and Schlierkamp-Voosen [21] for *OneMax* and GAs.

We denote the critical population size for given τ by $N^*(\epsilon, \tau)$. Because ϵ is fixed, we omit ϵ and write just $N^*(\tau)$. For *Int* we have approximately computed $N^*(\tau)$ by a Markov chain analysis. The Markov model was simplified, therefore we formulate the result as a conjecture.

Conjecture 1. *Let $\tau_k = 2^{-k}$. For FDA with fitness function Int, the critical population size $N^*(\tau)$ is approximately given by*

$$N^*(\tau_k) \approx N^*(\tau_1) * 2^{\frac{k-1}{2}} \quad \text{for } k \geq 1.$$

If $N^*(\tau)$ has been determined, then an optimal truncation threshold τ_{opt} can be computed. This threshold gives the minimum number of function evaluations *FE*.

Definition 10. The optimum truncation threshold τ_{opt} is defined by

$$\tau_{opt} = \min_\tau FE(\tau) = \min_\tau GEN_e(\tau) * N^*(\tau). \tag{66}$$

In general τ_{opt} is different from τ_{min} which needs the minimal population size. The following result follows from $k > 1$ from the above conjecture.

Empirical Law. *For Int the optimal truncation threshold τ is contained in the interval $[0.125, 0.4]$.*

Proof. Part of the result follows from the approximate formulas. For $\tau = 2^{-k}$ using the critical population size, we obtain

$$FE \approx \frac{n}{k} \times N^*(\tau_1) \times 2^{\frac{k-1}{2}} \propto \frac{1}{\sqrt{\tau} \log(1/\tau)} \quad \text{for } k \geq 1. \tag{67}$$

The minimum is at $k = 0.5(1 + \sqrt{17})$.

This short discussion indicates the difficulty of the critical population size problem. In principle, UMDA depends only on one parameter, the critical population size. But this size depends on the function to be optimized. Numerical estimates are very difficult. The finite size problem is discussed in a different context in the next section. There we introduce an algorithm which computes a good factorization from search points.

9 LFDA – Learning a Bayesian Factorization

Computing the structure of a Bayesian network from data is called learning. Learning gives an answer to the question: *given a population of selected points $M(t)$, what is a good Bayesian factorization fitting the data?* The most difficult part of the problem is to define a quality measure also called scoring measure.

A Bayesian network with more arcs fits the data better than one with less arcs. Therefore a scoring metric should give the best score to the minimal Bayesian network which fits the data. It is outside the scope of this tutorial to discuss this problem in more detail. The interested reader is referred to the two papers by Heckerman and Friedman et al., both in [13].

For Bayesian networks two quality measures are most frequently used: the *Bayes–Dirichlet* (BDe) score and the *minimal description length* (MDL) score. We concentrate on the MDL principle. This principle is motivated by universal coding. Suppose we are given a set D of instances, which we would like to store. Naturally, we would like to conserve space and save a compressed version of D. One way of compressing the data is to find a suitable model for D that the encoder can use to produce a compact version of D. In order to recover D we must also store the model used by the encoder to compress D. Thus the total description length is defined as the sum of the length of the compressed version of D and the length of the description of the model. The MDL principle postulates that the optimal model is the one that minimizes the total description length.

In the context of learning Bayesian networks, the model is a network B describing a probability distribution p over the instances appearing in the data. Several authors have approximately computed the MDL score. Let $M = |D|$ denote the size of the data set. Then MDL is approximately given by

$$\mathrm{MDL}(B, D) = -\mathrm{ld}(P(B)) + M \cdot H(B, D) + \tfrac{1}{2} PA \cdot \mathrm{ld}(M) \tag{68}$$

with $\mathrm{ld}(x) := \log_2(x)$. $P(B)$ denotes the prior probability of network B, $PA = \sum_i 2^{|pa_i|}$ gives the total number of probabilities to compute. $H(B, D)$ is defined by

$$H(B, D) = -\sum_{i=1}^{n} \sum_{pa_i} \sum_{x_i} \frac{m(x_i, pa_i)}{M} \mathrm{ld} \frac{m(x_i, pa_i)}{m(pa_i)} \tag{69}$$

where $m(x_i, pa_i)$ denotes the number of occurrences of x_i given configuration pa_i. $m(pa_i) = \sum_{x_i} m(x_i, pa_i)$. If $pa_i = \emptyset$, then $m(x_i, \emptyset)$ is set to the number of occurrences of x_i in D.

The formula has an interpretation which can be easily understood. If no prior information is available, $P(B)$ is identical for all possible networks. For minimizing, this term can be left out. $0.5PA \cdot \text{ld}(M)$ is the length required to code the parameter of the model with precision $1/M$. Normally one would need $PA \cdot \text{ld}(M)$ bits to encode the parameters. However, the central limit theorem says that these frequencies are roughly normally distributed with a variance of $M^{-1/2}$. Hence, the higher $0.5\,\text{ld}(M)$ bits are not very useful and can be left out. $-M \cdot H(B, D)$ has two interpretations. First, it is identical to the logarithm of the maximum likelihood ($\text{ld}(L(B|D))$). Thus we arrive at the following principle:

Choose the model which maximizes $\text{ld}(L(B|D)) - \frac{1}{2}PA \cdot \text{ld}(M)$.

The second interpretation arises from the observation that $H(B, D)$ is the conditional entropy of the network structure B, defined by PA_i, and the data D. The above principle is appealing, because it has no parameter to be tuned. However, the formula has been derived under many simplifications. In practice, one needs more control about the quality versus complexity tradeoff. Therefore we use a weight factor α. Our measure to be maximized is called *BIC*:

$$BIC(B, D, \alpha) = -M \cdot H(B, D) - \alpha PA \cdot \text{ld}(M). \qquad (70)$$

This measure with $\alpha = 0.5$ was first derived by Schwarz [28] as the *Bayesian information criterion (BIC)*. To compute a network B^* which maximizes *BIC* requires a search through the space of all Bayesian networks. Such a search is more expensive than to search for the optima of the function. Therefore the following greedy algorithm has been used. $k_{\max}$ is the maximum number of incoming edges allowed.

BN(α, $k_{\max}$)

STEP 0: Start with an arc-less network.
STEP 1: Add the arc (x_i, x_j) which gives the maximum increase of $BIC(\alpha)$ if $|PA_j| \leq k_{\max}$ and adding the arc does not introduce a cycle.
STEP 2: Stop if no arc is found.

Checking whether an arc would introduce a cycle can be easily done by maintaining for each node a list of parents and ancestors, i.e., parents of parents, etc. Then $(x_i \to x_j)$ introduces a cycle if x_j is ancestor of x_i.

The BOA algorithm of Pelikan [24] uses the BDe score. This measure has the following drawback. It is more sensitive to coincidental correlations implied by the data than the MDL measure. As a consequence, the BDe measure will prefer network structures with more arcs over simpler networks [3]. The *BIC* measure with $\alpha = 1$ has also been proposed by Harik [10]. But Harik allows only factorizations without conditional distributions. This distribution is only correct for separable functions.

Given the *BIC* score we have several options to extend FDA to LFDA which learns a factorization. Due to limitations of space we can only show results of an algorithm which computes a Bayesian network at each generation using algorithm

BN(0.5, k_{max}). FDA and LFDA should behave fairly similarly, if LFDA computes factorizations which are in probability terms very similar to the FDA factorization. FDA uses the same factorization for all generations, whereas LFDA computes a new factorization at each step which depends on the given data M.

We have applied LFDA to many problems [16]. The results are encouraging. Here we only discuss the functions introduced in section 6. We recall that UMDA finds the optimum of *BigJump* and *Saw*. UMDA uses univariate marginal distributions only. Therefore its Bayesian network has no arcs.

Function	n	α	N	τ	Succ.%	SDev
OneMax	30	UMDA	30	0.3	75	4.3
	30	0.25	100	0.3	2	1.4
	30	0.5	100	0.3	38	4.9
	30	0.75	100	0.3	80	4.0
	30	0.25	200	0.3	71	4.5
BigJump(30,3,1)	30	UMDA	200	0.3	100	0.0
	30	0.25	200	0.3	58	4.9
	30	0.5	200	0.3	96	2.0
	30	0.75	200	0.3	100	0.0
	30	0.25	400	0.3	100	0.0
Saw(32,2,0.5)	32	UMDA	50	0.5	71	4.5
	32	UMDA	200	0.5	100	0.0
	32	0.25	200	0.5	41	2.2
	32	0.5	200	0.5	83	1.7
	32	0.75	200	0.5	96	0.9
	32	0.25	400	0.5	84	3.7
Deceptive-4	32	UMDA	800	0.3	0	0.0
	32	FDA	100	0.3	81	3.9
	32	0.25	800	0.3	92	2.7
	32	0.5	800	0.3	72	4.5
	32	0.75	800	0.3	12	3.2

Table 5. Numerical results for different algorithms, LFDA with BN(α, 8)

Table 5 summarizes the results. For LFDA we used three different values of α, namely $\alpha = 0.25, 0.5, 0.75$. The smaller α, the less penalty for the size of the structure. Let us discuss the results in more detail. $\alpha = 0.25$ gives by far the best results when a network with many arcs is needed. This is the case for *Deceptive-4*. Here a Bayesian network with three parents is optimal. $\alpha = 0.25$ performs badly on problems where a network with no arcs defines a good search distribution. For the linear function *OneMax*, *BIC*(0.25) has only a success rate of 2%. The success rate can be improved if a larger population size N is used. The reason is as follows. *BIC*(0.25) allows denser networks. But if a small population is used, spurious correlations may arise. These correlations have a negative impact for the search distribution. The problem can be solved by using a larger population. Increasing the value from $N = 100$ to $N = 200$ increases the success rate from 2% to 71% for *OneMax*.

For *BigJump* and *Saw* a Bayesian network with no arcs is able to generate the optimum. An exact factorization requires a factor with n parameters. We used the heuristic BN with $k_{max} = 8$. Therefore the exact factorization cannot be found. In all these cases $\alpha = 0.75$ gives the best results. $BIC(0.75)$ enforces smaller networks. But $BIC(0.75)$ performs very badly on *Deceptive-4*. Taking all the results together $BIC(0.5)$ gives good results. This numerical results supports the theoretical estimate.

The numerical result indicates that control of the weight factor α can substantially reduce the amount of computation. For Bayesian networks we have not yet experimented with control strategies. We have intensively studied the problem in the context of neural networks [33].

10 Conclusion

We have shown that GAs can be approximated by an algorithm which keeps the population in linkage equilibrium. This algorithm, called UMDA, transforms the discrete optimization problem max $f(x)$ into a continuous one defined by max $\tilde{W}(p_1, \ldots, p_n)$, where $0 \le p_i \le 1$ is a univariate marginal distribution. With proportionate selection UMDA performs gradient ascent on $\tilde{W}$.

UMDA solves difficult multimodal optimization problems. This explains the success of genetic problems in practical problems. Of course, there are functions where UMDA fails. These functions are defined by highly correlated variables. In these cases search distributions using multivariate distributions and conditional marginal distributions have to be used. This is done by the algorithm FDA. If the factorization cannot analytically be computed, it has to be estimated from the set of selected points. This leads to a synthesis problem: finding a good factorization for a search distribution defined by a finite sample. This problem is addressed by the algorithm LFDA. It uses Bayesian networks to represent the distribution. For Bayesian networks numerically efficient algorithms have been developed which compute a network which best explains the data. LFDA computes a Bayesian network which minimizes the *BIC*.

The computational effort of both FDA and LFDA is substantially higher than that of UMDA. Thus UMDA should be the first algorithm to be tried in practice. All three algorithms are designed for unconstrained optimization problems. We believe that they can be extended to constrained optimization problems in an easier way than GAs. A first step has already been made in [20].

References

1. H. Asoh and H. Mühlenbein. On the mean convergence time of evolutionary algorithms without selection and mutation. In Y. Davidor, H.-P. Schwefel, and R. Männer, editors, *Proceedings of the 3rd Conference on Parallel Problem Solving from Nature*, LNCS 866, pages 88–97. Springer-Verlag, Berlin Heidelberg New York, 1994.
2. S. Baluja and R. Caruana. Removing the genetics from the standard genetic algorithm. In A. Prieditis and S. Russell, editors, *Proceedings of the 12th International Conference on Machine Learning*, pages 38–46. Morgan Kaufmann, San Francisco, 1995.
3. R. R. Bouckaert. Properties of Bayesian network learning algorithms. In R. Lopez de Mantaras and D. Poole, editors, *Proceedings of the Tenth Conference on Uncertainty in Artificial Intelligence*, pages 102–109. Morgan Kaufmann, San Francisco, 1994.
4. F. B. Christiansen and M. W. Feldman. Algorithms, genetics and populations: the schemata theorem revisited. *Complexity*, 3:57–64, 1998.
5. M. Dorigo and G. Di Caro. The ant colony optimization meta-heuristic. In D. Corne, M. Dorigo, and F. Glover, editors, *New Ideas in Optimization*. MacGraw–Hill, New York, 1999.
6. R. Feistel and W. Ebeling. *Evolution of Complex Systems. Self-Organization Entropy and Development*. Kluwer, Dordrecht, 1989.
7. B. J. Frey. *Graphical Models for Machine Learning and Digital Communication*. MIT Press, Cambridge, 1998.
8. H. Geiringer. On the probability theory of linkage in Mendelian heredity. *Annals of Math. Stat.*, 15:25–57, 1944.
9. D. E. Goldberg. *Genetic Algorithms in Search, Optimization and Machine Learning*. Addison–Wesley, Reading, MA, 1989.
10. G. Harik. Linkage learning via probabilistic modeling in the ecga. Technical Report IlliGal 99010, University of Illinois, Urbana-Champaign, 1999.
11. J. Hofbauer and K. Sigmund. *Evolutionary Games and Population Dynamics*. Cambridge University Press, Cambridge, 1998.
12. J. H. Holland. *Adaptation in Natural and Artificial Systems*. University of Michigan Press, Ann Arbor, MI, 1975/1992.
13. M. I. Jordan. *Learning in Graphical Models*. MIT Press, Cambridge, 1999.
14. M. Mitchell, J. H. Holland, and S. Forrest. When will a genetic algorithm outperform hill climbing? *Advances in Neural Information Processing Systems*, 6:51–58, 1994.
15. H. Mühlenbein and T. Mahnig. Convergence theory and applications of the factorized distribution algorithm. *Journal of Computing and Information Technology*, 7:19–32, 1999.
16. H. Mühlenbein and T. Mahnig. FDA – A scalable evolutionary algorithm for the optimization of additively decomposed functions. *Evolutionary Computation*, 7(4):353–376, 1999.
17. H. Mühlenbein. Evolution in time and space – the parallel genetic algorithm. In G. Rawlins, editor, *Foundations of Genetic Algorithms*, pages 316–337. Morgan Kaufmann, San Francisco, 1991.
18. H. Mühlenbein. The equation for the response to selection and its use for prediction. *Evolutionary Computation*, 5(3):303–346, 1997.
19. H. Mühlenbein, M. Gorges-Schleuter, and O. Krämer. Evolution algorithms in combinatorial optimization. *Parallel Computing*, 7:65–88, 1988.
20. H. Mühlenbein, T. Mahnig, and A. Rodriguez Ochoa. Schemata, distributions and graphical models in evolutionary optimization. *Journal of Heuristics*, 5:215–247, 1999.

21. H. Mühlenbein and D. Schlierkamp-Voosen. The science of breeding and its application to the breeder genetic algorithm. *Evolutionary Computation*, 1:335–360, 1994.
22. H. Mühlenbein and H.-M. Voigt. Gene pool recombination in genetic algorithms. In J. P. Kelly and I. H. Osman, editors, *Metaheuristics: Theory and Applications*, pages 53–62. Kluwer Academic, Norwell, 1996.
23. T. Nagylaki. *Introduction to Theoretical Population Genetics*. Biomathematics, Vol. 21. Springer-Verlag, Berlin Heidelberg New York, 1992.
24. M. Pelikan, D. E. Goldberg, and E. Cantu-Paz. BOA: The Bayesian optimization algorithm. Technical Report IlliGal 99003, University of Illinois, Urbana-Champaign, 1999.
25. M. Peschel and W. Mende. *Predator-Prey-Model: Do We Live in a Volterra World?* Akademie-Verlag, Berlin, 1986.
26. A. Prügel-Bennet and J. L. Shapiro. An analysis of a genetic algorithm for simple random Ising systems. *Physica D*, 104:75–114, 1997.
27. L. M. Rattray and J. L. Shapiro. Cumulant dynamics of a population under multiplicative selection, mutation and drift. *Theoretical Population Biology*. To be published, 1999.
28. G. Schwarz. Estimating the dimension of a model. *Annals of Statistics*, 7:461–464, 1978.
29. V. Vapnik. *Statistical Learning Theory*. Wiley, New York, 1998.
30. H.-M. Voigt. *Evolution and Optimization*. Akademie-Verlag, Berlin, 1989.
31. M. Vose. *The Simple Genetic Algorithm: Foundations and Theory*. MIT Press, Cambridge, 1999.
32. S. Wright. Random drift and the shifting balance theory of evolution. In K. Kojima, editor, *Mathematical Topics in Population Genetics*. Springer-Verlag, Berlin Heidelberg New York, 1970.
33. Byoung-Tak Zhang, P. Ohm, and H. Mühlenbein. Evolutionary induction of sparse neural trees. *Evolutionary Computation*, 5:213–236, 1997.

Properties of Fitness Functions and Search Landscapes

L. Kallel[1], B. Naudts[2], and C. R. Reeves[3]

[1] CMAP – UMR CNRS 7641
  Ecole Polytechnique
  Palaiseau 91128, France
  E-mail: *kallel@cmapx.polytechnique.fr*

[2] Departement Wiskunde-Informatica
  Universiteit Antwerpen (RUCA)
  Groenenborgerlaan 171
  B-2020 Antwerpen, Belgium
  E-mail: *bnaudts@ruca.ua.ac.be*

[3] School of Mathematical and Information Sciences
  Coventry University
  Coventry, UK
  E-mail: *C.Reeves@coventry.ac.uk*

Abstract. This tutorial is an introduction to the study of properties of fitness functions and search landscapes in the context of predicting the difficulty of search problems for genetic algorithms and, more generally, for stochastic iterative algorithms. Central to this topic is the Walsh transform, which presents a view on the fitness function in terms of the interactions between the variables of this function. The first part of this tutorial introduces the Walsh decomposition of fitness functions, and its relation to epistasis variance. Exchanging the fitness function for the fitness landscape, the second part discusses two important consequences of putting a topology on the search space: the modality and the ruggedness. Methods to estimate properties of landscapes are discussed. The last part moves away from property learning of general fitness functions and landscapes, and instead focuses on properties of classes of fitness functions originating from well-known NP-hard search problems. Important issues here are the factorization of the joint probability distribution of the fitness values and the engineering of interactions to achieve a highly multimodal, symmetric fitness landscape resembling a needle-in-a-haystack problem.

Keywords
Walsh transformation, landscape decomposition, landscape modality and ruggedness, interaction and symmetry, hard search problems

1 Introduction

It is a challenging task to develop a sound theory for genetic algorithms (GAs), as they contain several probabilistic operators such as mutation, recombination (also termed crossover), and one of many different selection schemes. There are actually few theoretical results that are of practical use to optimize a given search problem. This fact stimulates numerical investigation of the GA behavior on some reference problems, in order to determine features or properties of fitness landscapes that make a problem difficult.

Consider a fitness function $f : E \to \mathbb{R}$ to be maximized. In the context of evolutionary algorithms, one is interested in characterizing some properties of the function f that might reflect, explain, or allow one to predict the behavior of the GA when searching for the point of E that maximizes f.

A first approach to the study of f is to decompose f in some functional basis. Section 2 presents the Walsh decomposition and its relation to epistasis variance. However, the choice of basis matters, and the resulting coefficients are not necessarily meaningful for predicting the GA behavior.

A second approach consists of studying the landscape of f, given a topology on E. If the topology is induced by some modification operator of the GA (for example, k-bit-flip mutation), then one might expect to capture landscape properties of influence to the GA. Along this line, section 3 starts by giving a formal definition of fitness landscapes, and presents the graph representation of a landscape. A first important aspect of a fitness landscape is its *modality*: the number of local optima and the sizes of their basins of attraction. This is the concern of section 5. A second aspect is its *ruggedness*. Ruggedness properties can be captured by some statistics (the correlation functions) computed from a random walk on the landscape. Section 6 reviews some theoretical results about correlation functions pointing out the lack of a systematic numerical exploitation of the available theory, for once. To finish the discussion of landscapes, section 7 comments on the methods commonly used to estimate the properties of a given landscape, and gives some references for sample sizing and rigorous estimation of modality and ruggedness.

Many NP-hard search problems are defined in terms of interactions between the variables of the problem, which allows a direct computation of all Walsh coefficients. More particularly, many problems like the graph coloring and clique problem consists only of interactions between pairs of variables, allowing for a graphical representation of the interactions. Section 8 recalls how algorithms like the factorized distribution algorithm use the underlying interaction structure to optimize efficiently relatively difficult search problems. Section 9 discusses the interaction structures of easy and difficult search problems, and their relationship to the fitness functions they induce. The SK-model from statistical physics, the binary constraint satisfaction problem, and the graph coloring problem serve as the three example problem classes.

2 Decomposition of Functions

Suppose we have a function f defined for a variable x, it is natural to ask if this function can be decomposed into a superposition of simpler functions

$$f(x) = \sum_j w_j \zeta_j(x)$$

where the ζ_j ideally should have some meaning that can be related to properties of the function f. In the case of GAs, we normally have x represented as a vector (often a binary string), which we shall denote by $\boldsymbol{x}$, and its components by x_i. Occasionally it will be convenient to abuse our notation and write x or $\boldsymbol{x}$ as seems most appropriate. The most well-known set of functions for binary strings are the *Walsh* functions, defined as follows:

$$\psi_j(\boldsymbol{x}) = \prod_{i=1}^{\ell} (1 - 2x_i)^{j_i}$$

where $\boldsymbol{j}$ is the binary vector representing the integer j. There are several equivalent definitions: one such is

$$\psi_j(\boldsymbol{x}) = \xi(\boldsymbol{x} \wedge \boldsymbol{j})$$

where $\wedge$ is the bitwise AND operator, and ξ is the parity function

$$\xi(\boldsymbol{y}) = \begin{cases} +1 & \text{if } \sum_i y_i \text{ is even} \\ -1 & \text{if } \sum_i y_i \text{ is odd.} \end{cases}$$

For example, if $\boldsymbol{x} = (110)$ and $j = 5$, which means that $\boldsymbol{j} = (101)$, we find that $\boldsymbol{x} \wedge \boldsymbol{j} = (100)$, with odd parity, so that

$$\psi_5(110) = -1.$$

For strings of length ℓ, there are 2^ℓ Walsh functions, so that the full decomposition is

$$f(\boldsymbol{x}) = \sum_{j=0}^{2^\ell - 1} w_j \psi_j(\boldsymbol{x}).$$

For computational purposes, it is convenient to note that the Walsh functions can also be written as

$$\psi_j(\boldsymbol{x}) = 1 - 2[(\sum_{i=1}^{\ell} j_i x_i) \bmod 2].$$

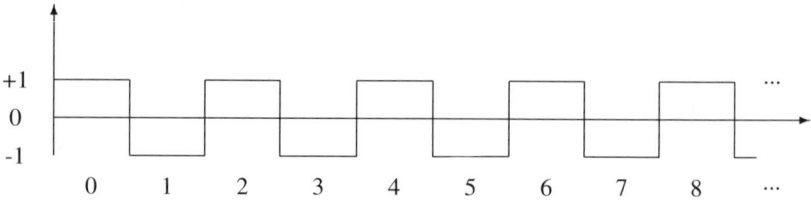

Figure 1. The Walsh function $\psi_1(x)$

Such a decomposition is, of course, a close analogy of what is done for functions of real numbers, where it is known as a Fourier decomposition, and entails the superposition of trigonometric functions, of which Walsh functions are a discrete analogue, forming a set of rectangular waveforms. For example, the function $\psi_1(x)$ is as shown in figure 1.

The Walsh coefficients $\{w_j\}$ can be obtained from the values of $f(x)$ by means of the Walsh transform

$$w_j = \frac{1}{2^\ell} \sum_{x=0}^{2^\ell-1} f(x)\psi_j(x)$$

and as in the case of Fourier transforms, there are computational algorithms that can speed up the process. A wealth of detail on such matters can be found in [2][1].

2.1 Experimental Design

Another slant on Walsh coefficients can be obtained from the statistical discipline of experimental design (ED). It is customary to break down the value of a function into *effects* of different orders:

$$f(x) = \text{constant} + \sum_{i=1}^{\ell}(\text{effect of allele at gene } i)$$
$$+ \sum_{i=1}^{\ell-1}\sum_{j=i+1}^{\ell}(\text{interaction between alleles at gene } i \text{ and gene } j)$$
$$+ \ldots$$
$$+(\text{interaction between alleles at gene } 1, \text{gene } 2, \ldots, \text{gene } \ell)$$

---

[1] It should be noted that the Walsh literature has different notions of numbering the Walsh functions. That used in this tutorial follows Goldberg [15] rather than the standard order used in [2].

In conventional experimental design, the above model would actually be written in parametric form. For example, the model for a vector of 3 components would be as follows:

$$f(pqr) = \mu + \alpha_p + \beta_q + (\alpha\beta)_{pq} + \gamma_r + (\alpha\gamma)_{pr} + (\beta\gamma)_{qr} + (\alpha\beta\gamma)_{pqr} \quad (1)$$

where (pqr) represent the allele values that are instantiated at each locus. In the case of binary strings, the coefficients obtained are precisely the same as the Walsh coefficients (see [43] for details). It turns out that the Walsh coefficients have a simple interpretation: the number of 1s in the binary representation of their indices shows the order of the interaction it represents, while the positions of those 1s indicate which genes are involved in the interaction. For example, consider the coefficient w_{13} in a 4-bit problem: the binary equivalent of 13 is 1101, so this represents a 3-factor interaction, and the factors participating in this interaction (reading – conventionally in the Walsh literature – from right to left) are numbers 1, 3, and 4, i.e., w_{13} measures what ED would call the interaction $(\alpha\gamma\delta)$.

2.2 Epistasis Variance

The concept of epistasis has been borrowed from biology, and refers to the effect on chromosome fitness of a combination of alleles which is not merely a linear function of the effects of the individual alleles. It is meant to convey the difficulty of a function in the sense of how *non-linear* it is. The assumption is that the more epistatic a problem is, the harder it may be for a GA to find its optimum. Davidor [5,6] was the first to explore a means of capturing this idea by calculating a quantity we shall call η, but his approach is somewhat naive, and it is clear even from [5,6] that it leads to some difficulties, the reason for which is explained in [43,42].

Davidor's 'variance' measure, as modified by Reeves and Wright [42], can be written as

$$\eta = \frac{\sum(f - \sum \text{linear effects})^2}{\sum(f - \bar{f})^2} \quad (2)$$

where $\bar{f}$ is the mean value of all strings in the Universe, the outer sum is over all strings, and the inner sum over the subscripts denoted by each string in turn.

In fact, Davidor used only the numerator of this quantity; however, it is clearly crying out for some form of normalization. As discussed in [44], from an ED perspective the normalization is natural – to divide by the overall sum of squared deviations from the mean. The value η as given above is then translation-invariant, and has the simple interpretation that it measures the fraction of the total 'variance' not explained by linear effects. Thus, we might expect that a value near to 0 expresses the fact that the function is close to linear, while a value near to 1 indicates a function that is highly non-linear.

Standard ED theory (for example, [18] or [32]) further shows that sums of squared deviations can be additively partitioned, a result that we can show in this context as follows:

Proposition 1. *Davidor's variance measure can be written as the sum of the squared interaction effects, i.e.,*

$$\eta = \frac{\sum (\text{interaction effect})^2}{\sum (f - \bar{f})^2} \tag{3}$$

where the sum is over all strings.

Note (1). An immediate corollary of this result is as stated in [42]: to every function f with a given value of η, there will be many functions whose decomposition differs from f only in the sign of the interaction effects. Clearly such functions cannot be distinguished by Davidor's variance measure.

Note (2). The idea inherent in the definition of η can be extended to epistasis of a particular order:

$$\eta_k = \frac{\sum (f - \sum \text{effects})^2}{\sum (f - \bar{f})^2}$$

where the inner sum is over all interactions of order $\leq k$. It follows that $\eta = \eta_1 \geq \eta_2 \geq \ldots \geq \eta_\ell = 0$. In other words, the sequence of η values measures the effect of successively better approximations to the function v. It is precisely this idea that underlies the statistical technique of Analysis of Variance [43].

2.3 Caveats

In [42] the importance of the *signs* of the interaction effects is pointed out. If the sign of an interaction is such that it reinforces the message of its associated main effects, there is no problem. We call such interactions *benign*. On the other hand, if the interaction acts counter to the joint influence of its associated main effects, we have *malign* interactions – essentially what the GA community commonly calls deception. Figure 2, taken from [42], demonstrates this point.

Thus, epistasis variance starts with a severe handicap: the value η cannot distinguish between cases where epistasis helps and where it harms, so its value cannot easily be interpreted as indicative of problem difficulty. A high value of η would certainly indicate small main effects, but if the interactions had the 'right' signs, it could still be a trivial problem to solve.

Even if we ignore this question, there is a further difficulty. The above analysis is predicated on knowledge of the complete Universe. In reality, if we knew the Universe, it would be the end of story: there would be no point in an optimization algorithm. Naturally we don't really know the Universe, but we may be dealing with an enormous search space, so the temptation is clearly simply to take a random sample and hope that η is still in some sense meaningful.

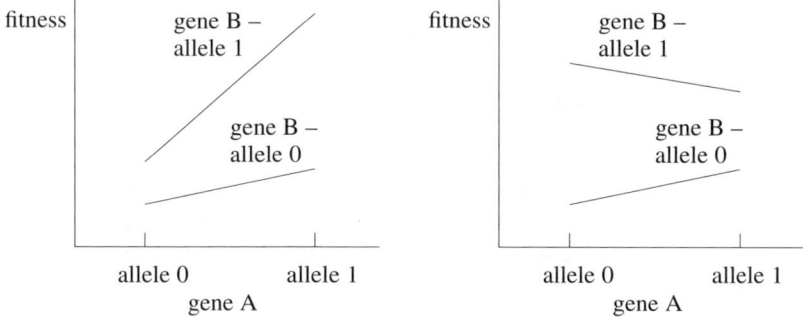

Figure 2. Benign and malign interactions. In the left diagram, the best allele for each gene is 1; there is epistasis, since the joint effect of having A and B set at 1 exceeds the sum of the individual main effects, but the interaction merely reinforces the main effects. However, on the right, the interaction has a malign influence: the best allele for both A and B is 1, but it is better to set gene A at 0.

Unfortunately, this will depend not only on the size of the sample, but also on its composition. In [42] it is shown that in principle it is not possible to estimate the value of the effects independently of each other. All that we can do is to estimate some linear combination of the effects, a combination that depends upon the composition of the sample used. The tutorials [42] and [44] give an extended treatment of these issues.

Moreover, there is yet one more difficulty, at least if we wanted to use this information directly to calculate the optimal solution. Suppose we eventually achieve reliable estimates of all the non-negligible effects; if we assume maximization as the goal, we would like to be able to assign + signs to all the positive effects. Unfortunately, this may not be possible if epistasis is malign – giving a + sign to two main effects, for example, might then entail a negative influence for their associated interaction. Essentially, we have a satisfiability problem to solve – an assignment of Boolean values has to be found that will maximize some weighted sum of 'clauses' – and of course such problems are NP-hard. As we will see in section 9, even an exact Walsh analysis for a problem of bounded epistasis (i.e., an upper limit on the number of factors that can interact) is insufficient to guarantee inferring the global optimum in polynomially bounded time. Restricting to second-order interactions is sufficient to show this.

Therefore, it is important to keep in mind that having total interaction knowledge is not equivalent to solving the optimization problem: it is merely a different view on the same problem.

3 Landscapes

The last section shows that the magnitude and sign of the interactions (or Walsh coefficients) are both important quantities that will affect the properties of functions that we try to optimize by means of evolutionary algorithms. Other suggestions have been made as to how to measure epistasis (e.g., in [37]), and an extended survey can be found in [44]. However, a factor that is ignored in most attempts to measure epistasis is the way in which the search algorithm interrelates with the fitness function.

Epistasis *per se* says nothing about how the fitness values are topologically related to each other: do they form a smooth progression building towards a solitary peak in the 'landscape' (Mount Fuji), or do they form a spiky pattern of many isolated peaks (Dolomites)? In fact, this will depend not only on the fitness function, but also on the search algorithm. Jones [26] has clearly demonstrated that the 'landscape' observed for a particular function is an artifact of the algorithm used or, more particularly, the neighborhood induced by the operators it employs.

3.1 Definition of a Landscape

Given a topology on the search space E, *local optima* are defined as follows. A point $m \in E$ is a local optimum of the function f if there exists a neighborhood N of m such that

$$\forall x \in N, \ f(m) - f(x) \geq 0.$$

If this topology is specified by means of a distance, then we can define a landscape Λ for the function f as a triple $\Lambda = (E, f, d)$, where d denotes a distance measure $d : E \times E \to \mathbb{R}^+ \cup \{\infty\}$ for which is required that

$$d(s, t) \geq 0$$
$$d(s, t) = 0 \Leftrightarrow s = t$$
$$d(s, u) \leq d(s, t) + d(t, u)$$

for all $s, t, u \in E$. Note that we do not need to specify the representation explicitly (for example, binary or Gray code), since this is assumed to be implied in the description of f. We have also decided, for the sake of simplicity, to ignore questions of search strategy and other matters in the definition of a landscape, unlike the more comprehensive definition of Jones [26], for example.

This definition says nothing about how the distance measure arises. In fact, for many cases a 'canonical' distance measure can be defined. Often, it is symmetric, i.e., $d(s, t) = d(t, s)$ for all $s, t \in E$, so that d also defines a *metric* on E. This is clearly a nice property, although it is not essential.

3.2 Neighborhood Structure

In the context of evolutionary algorithms, the *neighborhood relation* in E is often induced by some stochastic operator μ that maps $x \in E$ into $y \in E$ with some probability $p(x, y) = \mathbb{P}(y = \mu(x))$. The neighborhood of x is then the set of points of E that can be obtained by one application of μ to x

$$N(x) = \{y \in E, \mathbb{P}(y = \mu(x)) \neq 0\}.$$

What we have called a 'canonical distance measure' is typically related to the neighborhood structure. The canonical distance measure d_μ induced by μ can be defined as follows:

$$d_\mu(s, t) = 1 \Leftrightarrow t \in N(s).$$

The distance between non-neighbors is then the length of the shortest path between them, if one exists.

Note that such a path always exists if μ is an irreducible operator, in the sense that for all x and y of E, there exists a finite sequence of applications of μ that map x into y. This condition of irreducible operator ensures, for example, that the search space can be split into a partition of basins of attraction.

On the other hand, other distances have been suggested to take into account the probability of visiting different neighbors of x. Ristad and Yanilos [45] propose to relate the distance between two strings to the probability that a random sequence of operator applications transforms a string into another, or to the probability of the most likely sequence that transforms a string into another.

We are now ready to define the notion of a *basin of attraction* induced by a neighborhood relation and some local search algorithm. The basin of attraction of a local optimum m^j is the set of points $x^1, \ldots, x^k$ of the search space such that a local search algorithm starting from x^i, with $1 \leq i \leq k$, ends in the local optimum m^j.

Among the possible algorithms, we present the steepest ascent algorithm which selects the best neighbors after the entire neighborhood is examined. An alternative algorithm, the so-called first improvement algorithm, accepts the first favorable neighbor as soon as it is found. Note that extra free parameters are required for the latter: the order in which the neighborhood is searched needs to be determined.

3.3 Examples of Landscapes Related to Different Operators

If E is the binary hypercube $\mathbf{Z}_2^\ell$, the bit-flip (BF) operator can be defined as

$$\phi(k) : \mathbf{Z}_2^\ell \to \mathbf{Z}_2^\ell : \begin{cases} z_k \mapsto 1 - z_k \\ z_i \mapsto z_i & \text{if } i \neq k \end{cases}$$

where z is a binary vector of length ℓ. It is clear that the distance metric induced by ϕ is the well-known Hamming distance. Thus we could describe this landscape as a Hamming landscape (with reference to its distance measure), or as the BF landscape (with reference to the operator). Another useful operator is the CX operator:

$$\gamma(k) : \mathbf{Z}_2^\ell \to \mathbf{Z}_2^\ell : \begin{cases} z_i \mapsto 1 - z_i & \text{for } i \geq k \\ z_i \mapsto z_i & \text{otherwise.} \end{cases}$$

This is closely related to the one-point crossover operator frequently used in GAs. (For that reason, it has been named [20] the complementary crossover or CX operator). Further, if the vertices of the binary hypercube are re-ordered by transforming to a Gray code, it is easy to show that the neighbors of a point in Gray-coded space under BF are identical to those in the original binary-coded space under CX. This is an example of an *isomorphism* of landscapes.

In the following, unless specified explicitly, we take examples from the binary search space $E = \{0, 1\}^\ell$, and consider landscapes related to the single-bit-flip mutation operator (applied to a string of E, it flips one bit uniformly chosen in $\{1, ..\ell\}$).

4 Decomposition of Landscapes

4.1 Graph Representation of Landscapes

Neighborhood structures are clearly just another way of defining a graph Γ, which can be described by its $(n \times n)$ *adjacency matrix* A. The elements of A are given by $a_{ij} = 1$ if the indices i and j represent neighboring vectors, and $a_{ij} = 0$ otherwise. For example, the graphs induced by the bit-flip ϕ and the CX operator γ on binary vectors of length 3 have the $(2^3 \times 2^3)$ adjacency matrices

$$A_\phi = \begin{bmatrix} 0 & 1 & 1 & 0 & 1 & 0 & 0 & 0 \\ 1 & 0 & 0 & 1 & 0 & 1 & 0 & 0 \\ 1 & 0 & 0 & 1 & 0 & 0 & 1 & 0 \\ 0 & 1 & 1 & 0 & 0 & 0 & 0 & 1 \\ 1 & 0 & 0 & 0 & 0 & 1 & 1 & 0 \\ 0 & 1 & 0 & 0 & 1 & 0 & 0 & 1 \\ 0 & 0 & 1 & 0 & 1 & 0 & 0 & 1 \\ 0 & 0 & 0 & 1 & 0 & 1 & 1 & 0 \end{bmatrix} \quad A_\gamma = \begin{bmatrix} 0 & 1 & 0 & 1 & 0 & 0 & 0 & 1 \\ 1 & 0 & 1 & 0 & 0 & 0 & 1 & 0 \\ 0 & 1 & 0 & 1 & 0 & 1 & 0 & 0 \\ 1 & 0 & 1 & 0 & 1 & 0 & 0 & 0 \\ 0 & 0 & 0 & 1 & 0 & 1 & 0 & 1 \\ 0 & 0 & 1 & 0 & 1 & 0 & 1 & 0 \\ 0 & 1 & 0 & 0 & 0 & 1 & 0 & 1 \\ 1 & 0 & 0 & 0 & 1 & 0 & 1 & 0 \end{bmatrix},$$

where the vectors are indexed in the usual binary-coded integer order (i.e., (000), (001), etc). It is simply demonstrated that permuting the rows and columns of A_γ so that they are in the order 0, 1, 3, 2, 6, 7, 5, 4 reproduces the adjacency matrix A_ϕ – another way of demonstrating the isomorphism mentioned earlier. In other words

$$P^{-1} A_\phi P = A_\gamma$$

where P is the associated permutation matrix of the binary-to-Gray transformation, so that the eigenvalues and eigenvectors of the two matrices are the same.

4.2 Laplacian Matrix

The *graph Laplacian* Δ is defined as

$$\Delta = A - D$$

where D is a diagonal matrix such that d_{ii} is the degree of vertex i. Usually, these matrices are vertex-regular and $d_{ii} = k \; \forall i$, so that

$$\Delta = A - kI.$$

The significance of this matrix is found in its related eigensystem. From the eigenvectors $\{\varphi_i\}$, f can be expanded as

$$f(x) = \sum_i a_i \varphi_i(x).$$

Stadler and Wagner [51] call this a 'Fourier expansion'. Usually, the eigenvalues are not simple, and this sum can be further partitioned into a sum

$$f(x) = \sum_p \beta_p \tilde{\varphi}_p(x)$$

over the distinct eigenvalues of Δ. The functions $\tilde{\varphi}_p(x)$ are called 'elementary landscapes'. The corresponding values

$$|\beta_p|^2 = \sum |a_k|^2$$

(where the sum is over the coefficients that correspond to the pth distinct eigenvalue) form the *amplitude spectrum*, which expresses the relative importance of different components of the landscape.

In the case of the BF landscape, the eigenvectors turn out to be the Walsh functions, apart from a trivial change of scale (and in some cases, of sign). The distinct eigenvalues correspond to sets of Walsh coefficients of different orders, and the amplitude spectrum is exactly the set of components of the 'epistasis variance' that we used to compute $\{\eta_k\}$ above. The effect of the permutation inherent in mapping from the BF landscape to the CX landscape is to re-label some of the vertices of the graph, and hence some of the Walsh coefficients. Thus some coefficients that previously referred to linear effects now refer to interactions, and vice-versa.

Recombination operators are more difficult to analyze, but Stadler and Wagner [51] have shown that the eigenvectors associated with certain recombination operators are again the Walsh functions, so that the coefficients of the expansion are again the Walsh coefficients, although grouped in a different way in order to compute the amplitude spectrum. Table 1 gives an example of the differences in groupings for 3 different operators; note that the important factor for the crossover landscape is the *separation* between the outermost single-bits.

Index	binary coding	BF	CX	crossover	Index	binary coding	BF	CX	crossover
0	0000	0	0	0	8	1000	1	2	1
1	0001	1	1	1	9	1001	2	3	4
2	0010	1	2	1	10	1010	2	4	3
3	0011	2	1	2	11	1011	3	3	4
4	0100	1	2	1	12	1100	2	2	2
5	0101	2	3	3	13	1101	3	3	4
6	0110	2	2	2	14	1110	3	2	3
7	0111	3	1	3	15	1111	4	1	4

Table 1. Illustration of the different groupings of the Walsh coefficients associated with the BF, CX and recombination landscapes

We can see that the linear Walsh coefficients (and hence the linear component of epistasis variance) are the same in both the BF and the crossover landscapes, but not for the CX landscape, where the re-labelling gives a different grouping. However, the coefficients in the crossover landscape do not form a natural grouping in terms of interactions, and consequently the different components of variance for the recombination landscape do not have a simple interpretation as due to interactions of a particular order.

4.3 Consequences

The above analysis has shown how we can, in principle, identify certain properties of the function that we wish to optimize, and even extend this to the landscape that is created when a particular search operator is employed. It has also demonstrated that in the case of binary strings, the Walsh coefficients are of fundamental importance. Thus it appears that comparing the amplitude spectra of different landscapes is feasible and potentially informative, so that we can choose a 'good' landscape over which to search.

However, there are still some problems in attempting to measure landscape properties while making no assumptions as to the nature of our function f. First, calculation of the amplitude spectrum of a landscape faces the same problem as that of epistasis variance: it cannot distinguish between Walsh coefficients of opposite signs, so that it does not fully capture the nature of such properties as the number and type of local optima that are present. Second, in practice we will only have estimates of the Walsh coefficients, and obtaining reliable estimates may be computationally expensive.

There are alternative steps that can be taken. On the one hand, we could preserve some generality while trying to find measures that reflect more precisely the landscape properties that make the search harder or easier – such as the number of local optima, and the associated sizes of their basins of attraction. On the other hand, we can place restrictions on the type of function that we wish to consider and explore their properties in the hope that we can learn more about them in the restricted domain. In the rest of this tutorial we shall discuss some ideas that follow both these lines of inquiry.

5 Modality of Landscapes

Early attempts to characterize landscapes [17,16,9] propose criteria based on isolation (needle-in-a-haystack) and multimodality. It seems clear that these criteria certainly contribute to problem difficulty for EAs. However, there exist examples showing that multimodality on its own is neither necessary nor sufficient for making a landscape difficult to search.

5.1 Isolation

Consider the case of a needle-in-a-haystack, which is strictly positive in one arbitrary string, and zero elsewhere. There is no information to guide a search algorithm towards the optimum, and no search algorithm can perform better than random search, which requires 2^ℓ samples on average (2^ℓ is the size of the search space). On the other hand, it is well known that a random walk of single-bit-flip mutation also requires on average 2^ℓ steps to hit the optimum once. This result has been improved by [13] showing that a random walk using c/ℓ-mutation (each bit is flipped independently with probability c/ℓ) requires on average $2^\ell/(1 - e^{-c})$ iterations to hit the optimum. Hence, c/ℓ-mutation, having a maximal neighborhood size, performs significantly worse than single-bit-flip mutation on the needle-in-a-haystack.

In practice, misleading information can even be worse than the absence of information. For example, the global optimum can be 'isolated' due to the presence of local optima with very large basins of attraction in the landscape. Examples of such cases have been studied in detail in Garnier and Kallel's contribution to this volume (page 343).

Note that Rogers and Prügel-Bennett (in this volume, page 207) consider a population trapped in a non-optimal basin, compute the time required for the algorithm to produce a point in the optimal basin of attraction, and investigate the influence of crossover, mutation, and selection on the flow of the population's distribution in the landscape.

5.2 Multimodality: Number of Local Optima

One important characteristic of a landscape is its number of local optima, as they are obstacles for local search algorithms. Some authors [39,50] also conjecture that there is a close relation between ruggedness (nearest neighbor correlation, see section 6 for a detailed discussion) and the number of local optima: fewer local optima probably result in a larger correlation and easier optimization.

In the extreme and intractable case of a random fitness landscape ($f(x)$ are i.i.d.[2] random variables) there are on average $2^\ell/\ell$ local optima in the binary space $E = \{0, 1\}^\ell$. The following examples from the literature show that a landscape having a maximal number of local optima, can be alternatively easy or difficult for GAs, depending on the mutation rate.

An Easy Multimodal Function. Horn and Goldberg [22] construct an easy maximally multimodal function, defined for $x \in \{0, 1\}^\ell$ as follows:

$$f(x) = \mathbf{1}_{\text{odd}(\sum_{i=1}^\ell x_i)} \sum_{i=1}^\ell x_i. \tag{4}$$

(**1** denotes the indicator function.) This function has a maximal possible number of $2^{\ell-1}$ local optima, and each basin of attraction (w.r.t. the single-bit-flip neighborhood and steepest ascent local search) has a size of less than ℓ. Experiments show that a GA without crossover, using single-bit-flip mutation, easily finds the optimum.

Horn and Goldberg's Function Can Present Difficulties. On the other hand, Vose and Wright [56] prove that for functions similar to f defined above, a crossover-only GA with a large population will be trapped in a suboptimal point. The authors conclude that a fully non-deceptive function can be difficult for a crossover-only GA because of an exponential number (in terms of string length ℓ) of stable suboptimal attractors. Experiments show that the function f is still difficult for GAs with crossover and a low mutation rate. Note however that one would come to the same conclusion for the one-max function: in the neighborhood relation induced by crossover, the one-max function obviously presents many stable traps for populations.

This simple multimodal landscape above illustrates the importance of comparing algorithms with similar operators if one wants to develop a sound theory of landscapes. One cannot compare apples and oranges. In the following, we stick to GAs that use a significant amount of local-neighborhood mutation.

5.3 Unimodality

The previous section shows that multimodality does not necessarily cause problems for GAs. This section shows the other extreme: Hamming landscapes with only one

---
[2] independent and identically distributed

local optimum (unimodal functions) can present serious difficulties for both hill-climbers and GAs.

Convergence Time on Unimodal Functions. One of the easiest functions one could think of is probably the one-max function, where fitness and distance to the optimum are proportional: the closer we come to the optimum the better the fitness.

The one-max functions has been given a great deal of attention in the literature. One can find at least two reasons for this fact: first, these studies provide lower bounds on convergence time which in turn allows one to quantify the notion of quick and slow convergence for a given algorithm. Second, theoretical investigation is very amenable because the fitness function can be written as a function of the distance to the optimum.

Consider a (1+1)-evolution strategy and a one-max landscape. Assuming that the starting point x^0 is such that $d(x^0, x^f) = k$, then the expectation of the first hitting time of the optimum $\bar{T}$ is given by [14]:

$$\mathbb{E}_k[\bar{T}] = \ell \sum_{j=1}^{k} j^{-1} \stackrel{\ell \geq k \gg 1}{\simeq} \ell(\ln k + C) \quad \text{for single-bit-flip mutation}$$

$$\mathbb{E}_k[T] \stackrel{\ell \geq k \gg 1}{\simeq} \ell \left(\frac{e^c}{c} \ln k + R(c) \right) \quad \text{for } c/\ell\text{-mutation}$$

where $C \simeq 0.58$ is Euler's constant, and $R(c)$ a function of c. Hence, on the one-max problem, convergence time is linear in the problem size for both mutation operators, but strangely enough, single-bit-flip mutation is a factor of e^c/c quicker than c/ℓ-mutation.

At the other end of the spectrum of unimodal functions one finds longpath functions, for which convergence can take much more time than for the one-max function. This issue is addressed below for the same mutation operators.

Long k-path problems [23] have been introduced to show that unimodal problems can cause difficulties for single-bit-flip hill-climbers because of an exponential path length. By construction, such a problem requires an exponential convergence time for hill-climbers and population-based GAs using single-bit-flip mutation. Efforts were also devoted to extend this result to some complex GA instances showing that population-based GAs with $1/\ell$-mutation spend a super-polynomial time (w.r.t. to the problem size ℓ) [12] to optimize a sub-family of long k-path problems (for $k > \sqrt{\ell}$ and $\ell \gg 1$). Furthermore, [27] proposes a more general construction of fitness functions for which the trajectory of a mutation-only GA is predictable and its convergence is super-polynomial. Experiments show that this result is also valid for GAs with crossover, in the sense that the best individual so far necessarily visits a user-defined ordered sequence of subspaces of the search space before it hits the optimum.

The overall conclusion is that GAs can take a very long time to optimize unimodal functions; mutation and crossover operators that potentially allow *shortcuts* (jumps to a point at a much higher position on the unimodal path) are of no use on some unimodal landscapes. These landscapes are possibly not very realistic. However, we now have a better idea of the variety of behavior of the GA on unimodal functions: the convergence time ranges from linear to super-polynomial for both single-bit-flip and $1/\ell$-mutation, and the convergence of GAs with both mutation and crossover can be very slow and predictable.

These facts, in turn, lead us to separate completely the notion of multimodality from that of the easiness of an optimization task by means of GAs.

5.4 Multimodality: Number and Sizes of Basins of Attraction

Let us come back to the general case of functions with multiple local optima. While section 5.2 only considered the number of basins, this section concentrates on the size of the basins of attraction. The distribution of the basin sizes gives a lot more information about the landscape than the mere number does.

An example of well studied multimodal functions are the NK-landscapes [28]. After extensive numerical investigation, the author concludes that the greater the K value, the more local optima there are, and the more rugged and difficult the landscape is. Besides, for low K values, the fitter the local optima the larger their basins and the closer to the global optimum. Hence, the author suggests that the distribution of basin sizes, coupled with information on the height of peaks in the basins, allows one to form an idea of the landscape difficulty.

Note that before being able to investigate the modality properties of an unknown landscape, we need a reliable (and not too expensive) sampling method. It is therefore important to characterize the notion of a representative sample of a given landscape. Section 7 discusses this issue.

6 Landscape Correlations and Ruggedness

This section presents a broad overview of empirical correlation functions of landscapes and their potential use in practice to gather some information about the landscape.

Consider a simple random walk $\{x^0, x^1, \ldots\}$ on the fitness landscape, which induces a time series $\{f(x^0), f(x^1), \ldots\}$. Weinberger [57] has suggested using this time series to gather some information about the fitness landscape itself. The autocorrelation function of this time series is defined as

$$r(s) = \frac{\mathbb{E}_t[f(x^{t+s})f(x^t)] - \mathbb{E}_t[f(x^t)]^2}{\mathbb{E}_t[f(x^t)^2] - \mathbb{E}_t[f(x^t)]^2}.$$

Most of the empirical correlation functions which have been computed so far are decaying exponentials of the form

$$r(s) = \exp(-s/\tau)$$

where τ is the correlation length. Numerically, many authors neglect the deviations of $r(s)$ from the exponential form and estimate τ by some interpolation from $r(\tau) = 1/e$ or simply by setting $\tau = -\frac{1}{ln(r(1))}$.

The correlation length allows one to reduce the information contained in the correlation functions to single number. The value of τ determines the rate at which the fitness of a point of the landscape is forgotten as the walk moves away from this point.

6.1 Exploiting Correlation Functions of Landscapes

In the following, we present a number of results that exploit the correlation length to find some information about the landscapes.

Under the assumption that the fitness values of points of the landscape are jointly Gaussian distributed[3], we have the following result. If the fitness of a given point x is f_x, the distribution of fitnesses f_y of neighboring points on the landscape is Normal [40] with mean and variance given by

$$\mathbb{E}[f_y \mid \text{fitness of } x \text{ is } f_x] = \mu + r(1)(f_x - \mu)$$
$$\text{Var}[f_y \mid \text{fitness of } x \text{ is } f_x] = \sigma^2(1 - r^2(1))$$

where μ and σ are the marginal fitness mean and variance of x. (By the hypothesis of statistical isotropy, these are the same for all points.)

This result is very useful for estimating the order of the fitness difference between neighbors

$$\Delta f \simeq \sigma\sqrt{1 - r^2(1)}.$$

A second noticeable feature of correlation functions is that $r(s) = \lambda^s$ if f is an *elementary* landscape, but the practical use of this result is not straightforward. We refer to [48] for a possible exploitation of the result by decomposing an arbitrary landscape into a weighted sum of elementary landscapes with increasing ruggedness.

Furthermore, as mentioned above, it is a common conjecture that the more rugged the landscape is, the more local optima it contains. However, there is no systematic method for relating the number of local optima to the correlation structure of the landscape. Nevertheless, such relations have been estimated for particular landscapes such as the spin glasses and the NK-landscapes on sequences of length ℓ over an alphabet of size α.

---

[3] This assumption is valid for many problems where the fitness function is defined as the sum of independent quantities (due to the central limit theorem). This is usually the case with, for example, NK-landscapes and the travelling salesman problem.

Finally, note that the correlation length can simply be used to set the mutation strength so that starting from x, one can mainly reach points within a radius of τ from x.

The use of landscape correlations is not really common practice in the GA literature, although they seem very promising. Up to now, there has been no routine procedure tested on some benchmark of problems which can guide the choice of the optimizer according to the landscape's ruggedness (obtained by an estimation of the correlation length).

7 Sample sizing, Rigorous Estimation of Landscape Modality, and Ruggedness

While the previous sections concentrate on reviewing some studied properties of landscapes, this section comments on methods commonly used to identify such properties. The easiest way to ensure that a landscape has a given property is to construct it fully. This is the case for some of the examples cited above. Unfortunately, in practice one has to detect such properties in unknown landscapes, given some non-exhaustive sample (or random walk) of the search space. Exhaustive sampling is often out of computational reach. Usually, the sample size is decided by a user-defined criterion, and little attention is paid to find the reliability of the result with respect to the sample size.

In order to overcome these drawbacks, a methodology has been developed by Garnier and Kallel (this volume, page 343). It shows how one can estimate the minimal uniform sample size that ensures at least one visit to each basin of attraction of a given landscape. Note that the size of such a sample is fully determined by the knowledge of the size of the smallest basin of attraction of the landscape. If the smallest basin has a normalized size of α, then the probability that a uniform sample of size N visits this basin is $1 - (1 - \alpha)^N$; hence a sample size around $N = 1/\alpha$ ensures that all basins are visited with a high probability.

The methodology proposed by Garnier and Kallel allows one to estimate the distribution of the sizes of the basins of a given landscape. As a consequence, it gives an idea of the uniform sample size required to visit each basin at least once. In order to find these estimates, a random sampling of the landscape and a local steepest ascent search are performed. This allows one to determine the number of points sampled in the same basin, and to use these data to estimate the distribution of basin sizes. Estimations are then given with an associated confidence level which is not user-defined but dependent on the landscape. We refer to their paper for more details.

Analogously, when investigating the ruggedness properties, one also has to choose the length of the random walk used to derive the correlation function of the landscape. This issue is addressed by Hordijk [21] who shows how one can assess the reliability of the estimated ruggedness computed on a random walk.

8 Factorization of the Joint Probability Distribution

Some search algorithms operate directly on the fitness distribution of a population, bypassing all genetic operators. Two nice examples are UMDA and FDA ([33], in this volume, page 135) and [35]). Although this tutorial is not explicitly concerned with these algorithms, some of their concepts will be used in the next section. Therefore we will briefly recall them.

Let $p(x, t)$ be a probability distribution over the search space at time t, with $x = x_1 x_2 \ldots x_\ell \in \{0, 1\}^\ell$. In the infinite population model, fitness proportionate selection changes these probabilities according to

$$p^s(x, t) = p(x, t) \frac{f(x)}{\bar{f}(t)}$$

where $\bar{f}(t) = \sum p(x, t) f(x)$ is the average fitness. It is easy to see that the system

$$p(x, t+1) = p^s(x, t)$$

converges to a distribution where only optimal elements have a non-zero probability of occurrence – ideal for optimization. However, real populations are not infinite but restricted in size to a small subset of the (finite) search space.

To ensure that the selection operator does not cause premature convergence to the population's fittest element, the less fit members of the population need to be replaced by new strings. To construct these new strings, a GA incorporates genetic operators like mutation and crossover.

An alternative way of generating new strings is to sample from an approximation of the fitness distribution based on the fitness values of the current population. Starting off with a large enough population and a well-chosen model for the fitness distribution, the iteration of selection and sampling constitutes a usable search algorithm.

UMDA approximates the distribution by univariate marginal distributions (whence its name), which are equivalent to first-order schemata. New points are generated according to

$$p(x, t+1) = \prod_{i=1}^{\ell} p^s(x_i, t)$$

where $p^s(x_i, t)$ denotes the probability after selection of the first-order schema $\#\ldots\#x_i\#\ldots\#$. In relation to the previous sections, we can easily say that this algorithm will perform well on problems with linear effects and/or benign interactions only. Indeed, a malign interaction cannot be accounted for because of the univariate factorization of the distribution.

More complicated factorizations can replace the univariate one of the UMDA. The route to the *factorized distribution algorithm* (FDA, [35]) goes via a conceptual algorithm with Boltzmann selection, similar to simulated annealing. Its selection step is defined as

$$p^s(x,t) = p(x,t) \frac{v^{f(x)}}{\sum_y p(y,t) v^{f(y)}} \tag{5}$$

with $v \geq 1$. Mühlenbein and Mahnig show that when the initial distribution is a Boltzmann distribution

$$p(x,0) = \frac{u^{f(x)}}{\sum_y p(y,t) u^{f(y)}}$$

and new points are sampled according to (5), then

$$p(x,t) = \frac{w^{f(x)}}{\sum_y p(y,t) w^{f(y)}}$$

with $w = u \cdot v^t$, i.e., only the basis of the distribution changes over time. This conceptual algorithm is now turned into the actual algorithm FDA by factorizing the Boltzmann distribution $p(x, t)$.

Let $S_n = \{s_1, \ldots, s_n\}$ be a set of sets of variables, and define for each $i \in \{1, \ldots, n\}$ the sets d_i, b_i and c_i, satisfying

$$d_i = \bigcup_{j=1}^{i} s_j \qquad b_i = s_i \setminus d_{i-1} \qquad c_i = s_i \cap d_{i-1} \qquad d_0 = \emptyset.$$

If these sets also satisfy $b_i \neq \emptyset$ for all i, $d_n = \{x_1, \ldots, x_\ell\}$ and

$$\forall i \geq 2, \exists j < i \cdot c_i \subseteq s_j \tag{6}$$

then $p(x, t)$ can be factorized as (we omit the t)

$$p(x) = p(\Pi_{b_1} x) p(\Pi_{b_2} x | \Pi_{c_2} x) p(\Pi_{b_3} x | \Pi_{c_3} x) \ldots p(\Pi_{b_n} x | \Pi_{c_n} x). \tag{7}$$

Property (6) is called the running intersection property [30], and is a sufficient condition for the factorization to be exact, and hence for the FDA to converge to the global optima given a large enough population size.

We will see in the coming sections how the running intersection property and the form of exact factorizations can be used for problem difficulty prediction.

9 Predicting Problem Difficulty from Interaction Information

Many NP-hard optimization problems are essentially defined by the interactions between the components of the problem. The computation of the Walsh transformation of the search space then becomes a simple activity.

Consider, for example, the class of functions of the form

$$f : \{-1, 1\}^\ell \to \mathbb{R} : \sigma_0\sigma_1\ldots\sigma_{\ell-1} \to - \sum_{0 \leq i < j < \ell} J_{ij}\sigma_i\sigma_j - \sum_i h\sigma_i \quad (8)$$

where the parameters J_{ij} and h take real values. This NP-hard problem class is clearly defined by terms of interacting pairs of variables and terms depending on one variable only. It takes some algebra to compute the Walsh coefficients from the J_{ij} and the h, but algorithmically this transformation can be done in polynomial time [52]. When the J_{ij} are taken as independent random variables with zero mean and variance $1/\ell$, the problem is referred to as the Sherrington–Kirkpatrick (SK) model in spin-glass theory [47,31]. We will use the term SK-problem to indicate the search problem of optimizing a function of the form (8) with arbitrary parameters.

It is the aim of this section to give some first hints of how to construct indicators of problem difficulty based on interaction information rather than samples of the search space. We will restrict this to problems where only interactions between pairs of variables occur. This has the advantage that the interaction information can be split into two separate entities: the interaction graph showing which string positions are linked to each other, and the interaction coefficients which specify how two string positions are linked. In our example of the SK-problem, the vertices of the graph then represent the variables, and edges are placed between variables i and j when $J_{ij} \neq 0$. Their labels carry the interaction coefficients J_{ij}.

In many problems, the interaction coefficients are orthogonal to the interaction graph when problem difficulty is concerned. This can easily be seen by considering two extreme cases. When the graph of a problem instance is simple, the problem difficulty will be low, regardless of the interaction coefficients. For example, the instance with $J_{ij} = 0$ except when $j = i + 1$ can be solved in linear time, regardless of the values of the J_{ij}. When the interaction coefficients are trivially chosen (e.g., all J_{ij} set to 1), the form of the graph becomes unimportant.

To characterize difficulty, one must know the common interaction features of hard problem instances – hard independently of the algorithm chosen. We postulate that such instances must show a search landscape consisting of many local optima, typically an exponential number of them, and only one or a few global optima hidden between the local optima. Part of the process of creating indicators of problem difficulty is therefore to study how interactions have to be designed to achieve such a needle-in-a-haystack like landscape.

Before dealing with hard problem instances, we briefly discuss two other search problems and study their simplest instances.

9.1 Two Other Search Problems

In this section we discuss the graph coloring problem and the binary constraint satisfaction problems. Our choice is motivated by the extreme positions these problems occupy with respect to their amount of interaction coefficients. They will also allow us to elaborate on some differences between decision and optimization problems.

Graph Coloring. The famous problem of graph coloring consists of partitioning the vertices of a graph $G = (V, E)$ into k color classes, under the single constraint that two adjacent vertices cannot belong to the same color class. If such a partitioning exists, the graph is said to be k-colorable. Two versions of the problem exist: the determination of the smallest k such that the graph is still k-colorable, and the decision version to determine whether a graph is k-colorable or not.

Clearly, the graphical structure is the only determining factor; interaction coefficients are absent. Instances can be generated randomly by tossing a coin for each possible edge separately. At a given probability, called the threshold, the instances of the decision problem for k-colorability become empirically hard to solve (e.g., [19]). One proves that a phase transition from solvable to non-solvable occurs at this threshold, and hence the average number of solutions moves through one. Characteristics of instances near the threshold will be discussed in section 9.6.

Fewer theoretical results are known about the optimization version. Famous benchmarks exist [25] and the competition for ever better heuristics continues (e.g, [11]), but they are of less immediate use to our problem. The SK-problem is in a similar position: it is an optimization problem for which no phase transition in a edge density parameter is known. However, attempts were made in statistical physics to find the building blocks of hard problem instances of the SK-model. We will find in section 9.6 that these building blocks are conceptually similar to the characteristics of instances of the decision problem near the threshold.

Binary Constraint Satisfaction. A binary constraint satisfaction problem (e.g., [55]) consist of a set of variables $\{x_i\}$, a domain D_i corresponding to each variable, and a subset $C_{ij} \subset D_i \times D_j$ for each pair of variables (i, j), with $i < j$. The set C_{ij} represents a *constraint* when it is non-empty; the elements of C_{ij} are called *conflicts*. The goal of a CSP is to assign to the variables $\{x_i\}$ a value from their domain D_i in such a way that all constraints are satisfied. A constraint C_{ij} is satisfied if and only if $(x_i, x_j) \notin C_{ij}$, otherwise it is *violated*. A search landscape is generated by the *standard penalty function*, which simply counts the number of constraints violated by one assignment, and the Hamming metric.

It is obvious to see that the decision version of the graph coloring problem is a special case of binary constraint satisfaction: simply set the domain sizes to k, let a constraint correspond to each edge, and set the conflicts of all constraints to avoid equal values.

The random generation of CSP instances (e.g., [49]) typically depends on two parameters, once the number of variables n and the fixed domain size m has been specified: the *connectivity* or *constraint density* p_1 and the *tightness* p_2. The connectivity indicates the probability that a couple of variables (x_i, x_j) will be constrained; more formally, to generate a random CSP, $\binom{n}{2}$ i.i.d. Bernouilli experiments with success rate p_1 are performed. Success indicates that the couple is constrained; failure means no constraint is present between the variables. The tightness sets the probability of a conflict being present in a constraint. For each constrained pair of variables, m^2

i.i.d. Bernouilli experiments with success rate p_2 take place, with success indicating the presence of a conflict.

Empirically, one detects a phase transition from solvable to unsolvable when connectivity and tightness are increased. Close to the phase transition, algorithms find it hard to prove or disprove whether solutions exist. This region of interesting instances, with an expected number of solutions close to 1, is called the *mushy region*.

Contrary to the graph coloring problem, characteristics of instances in the mushy region will have to include properties of conflict matrices, which are harder to describe. We have not found any theoretical work on this subject in the literature.

9.2 Simple Interaction Graphs

The simplest interaction graph is one without edges, indicating that the fitness function is fully separable over all variables, and that the problem is solvable in linear time. As mentioned earlier in this tutorial, epistasis variance measures the distance between a function and the first-order functions, which form a subclass of the fully separable functions. Algorithms like UMDA [33] and PBIL [1] are guaranteed to solve problems with this type of interaction graph.

Second easiest is a one-dimensional chain of interactions, as depicted below:

$$\bullet \text{———} \bullet \text{———} \bullet \text{———} \cdots \text{———} \bullet \text{———} \bullet \qquad (9)$$

Using a *divide-and-conquer* technique, problems with this type of graph can be optimized deterministically in $O(na^2)$, with n the number of variables and a the size of the alphabet. Note that the problem difficulty does not increase when the chain becomes a closed loop, i.e., when the leftmost variable also interacts with the rightmost. Although the interaction content of this chain does not affect the running time of the divide-and-conquer algorithm, we will show in section 9.5 that simulated annealing and other stochastic algorithms can fail to optimize such problems efficiently. Note also that the MIMIC algorithm [7] is guaranteed to solve problems with a chain graph.

Time complexity does not change when the chain is replaced by a tree structure. In this case, the Boltzmann probability distribution can factorized into

$$p(x) = p(x_j) \prod_{i \neq j} p(x_i | x_{\text{pa}_i}) \qquad (10)$$

where pa_i is the parent of i in the tree structure. Only loops or cycles can slow down the running time, making the factors larger (i.e., contain more variables) to satisfy the running intersection property.

At this point, we can make a first suggestion to improve epistasis variance as a measure of interaction: construct an algorithm to measure how close the structure of the interaction graph is to a tree structure.

9.3 Simple Interaction Coefficients

Away from the mushy region in the binary constraint satisfaction decision problem, the values of the interaction coefficients make the problem easy to solve. On the unsolvable side, i.e., where the tightness is relatively high, the $\mathcal{O}(n^2)$ *arc consistency checker* (e.g., [55]) can decide unsolvablility for a large proportion of instances.

A simple example shows why this is possible. Assume three variables x_1, x_2, and x_3, with domains $D_i = D = \{0, 1, 2\}$, and suppose only the combinations $(1, 0)$ and $(2, 0)$ between (x_1, x_2) and the combinations $(1, 1)$, $(1, 2)$, $(2, 0)$ between (x_2, x_3) do not generate a conflict. Looking at the interaction between x_1 and x_2, we see that x_2 must have the value 0. But according to the interaction (x_2, x_3), it should have values 1 or 2 ... We can immediately decide that no solution exists.

On the other side of the mushy region, conflicts are rare, and many constraints can be removed simply because they do not impose a real constraint on the values of the variables. A simple extension of the arc consistency checker will quickly decide the majority of instances not too close to the mushy region [46].

The improvement on epistasis variance, introduced in the previous section, will clearly suffer from the same weakness as epistasis variance itself: by only looking at the structure of the interactions, it ignores the interaction coefficients.

9.4 Characterizations of the Interaction Graph

Nevertheless, characterizing the interaction graph is a feasible task and a fair amount of literature is available. The theory of graphical models and Bayesian networks [41,30,10], already touched on in section 8, provides a lot of useful material.

The Euclidean Dimension. A first, crude way of characterizing the complexity of interaction graphs is by their Euclidean dimension or coordination number. As discussed above, instances with one-dimensional interaction graphs, where variables only interact with their neighbors on a line, are optimizable in linear time.

Depending on the type of problem, two-dimensional interaction graphs pose problems or do not. The coloring of two-dimensional rectangular graphs is trivial. But as soon as interaction coefficients are introduced, simple algorithms fail to work. In the case of binary CSP, for example, a phase transition from solvable to unsolvable is observed when varying the tightness, with instances in the mushy region being fairly hard to tackle [46]. NP-completeness proofs are unknown to the authors.

Analytically, problems on two-dimensional lattices are in a way still tangible. For example, the transfer matrix technique from statistical physics [29] allows some analytical calculations on two-dimensional SK-problems. As soon as the dimension becomes three or more, though, this opportunity is lost. It is unknown, and not necessarily true, that problem classes become harder with their Euclidean dimension increasing. Again, we do not know of any NP-completeness proof of the subproblem of the SK-problem on three- (or higher-)dimensional lattices.

Measuring the Euclidean dimension of a graph, if it exists, is an NP-complete problem in itself – although the only useful measure would be to verify whether the graph can be embedded in a two-dimensional lattice or not. Although the Euclidean dimension seems to be less powerful and practical as a practical predictor of graph complexity, it is very useful for the construction of graphs, especially then for dimensions 2 and 3.

Factorization Complexity. The convergence behavior of the FDA is studied in detail in [34,36]. When an exact factorization of the joint probability distribution is known, and the factors are of manageable size, the FDA can find the optimum very efficiently. Hard problems therefore require a factorization with large factors. Of course, due to the orthogonality of interaction graph and interaction coefficients, we have to keep in mind that a such a factorization may also hide a simple problem.

Only tree-structured interaction graphs can yield a factorization where each factor contains only one variable, possibly conditioned on one other variable. Cycles in the interaction graph necessarily result in larger factors due to the running intersection property. Consider, for example, the interaction graph

$$\begin{array}{ccc} x_1 & \text{———} & x_2 \\ | & & | \\ x_4 & \text{———} & x_3 \end{array}$$

The factorization

$$p(x) = p(x_1)p(x_2 \mid x_1)p(x_3 \mid x_1)p(x_4 \mid x_2, x_3)$$

does not satisfy the running intersection property: the variables x_2 and x_3 occur together in the last factor, but not in the other factors. Note that to store this factorization, $2+4+4+8 > 16$ values are required when working with a binary alphabet. Exact factorizations, like

$$p(x) = p(x_1)p(x_2 \mid x_1)p(x_3, x_4 \mid x_1, x_2)$$
$$p(x) = p(x_1, x_2)p(x_3, x_4 \mid x_1, x_2)$$

necessarily contain a factor of size 4: two variables conditioned on two others. The joint probability distribution has the same size as the largest factor, which makes it pointless to factorize this single quadrangle.

The size of the largest factor in a factorization could therefore be an indicator of problem difficulty at the 'easy-to-intermediately difficult' level, with the caveat that the problem may turn out to be simple even when a hard one is predicted. A two-dimensional lattice with identical interaction coefficients, as an instance of the SK-problem, serves as a good example. It is easy to show that any exact factorization of such a lattice requires factors of size $\sqrt{n}$, with n the total number of variables.

The computation of the factorization of a search problem is similar to performing a Walsh transformation, and theoretically requires knowledge of the whole search space. Greedy approximate algorithms exist, however, and the FDA can be extended to a learning FDA (LFDA) which learns the factorization while optimizing the problem ([33], in this volume, page 135).

Other Complexity Measures. From the material above we can conclude that in hard problem instances, the number of interactions which really contribute to problem difficulty (call them 'significant' for a moment) is neither extremely small nor extremely large. It is therefore justifiable to look at the way the number of significant interactions grows with the number of variables. This law might not follow the laws of Euclidean dimensions, in which case we speak of a fractal dimension of the structure of significant interactions. In papers from neuroscience ([53], for example), one finds that loosely connected clusters of dense interactions provide the highest complexity.

9.5 Properties of Hard Search Landscapes

Earlier on we postulated that intrinsically hard problem instances, i.e., instances hard for all realistic search algorithms, are in a way equivalent to the needle-in-a-haystack problem: they contain a super-polynomial number of local optima, amongst which only a few global optima are hidden.

By its very definition, *symmetry* is an important property of such instances: the more symmetry in the search landscape, the more areas which will seem to be identical to a search algorithm. Given the relatively low number of free variables in an NP-complete optimization problem, such symmetry is necessary to generate the high number of local optima.

The standard one-dimensional Ising model $H = -J \sum_{i=0}^{n-1} \sigma_i \sigma_{i+1}$, with $J > 0$, $\sigma_i \in \{-1, +1\}$ and $\sigma_n \equiv \sigma_0$, which is a subproblem of the SK-problem, is a perfect example to demonstrate how symmetry in the interactions can be the origin of a highly symmetric search landscape featuring an exponential amount of local optima for a single-bit-flip neighborhood [3]. A gentle introduction to the model and its symmetry can be found in [38] and in Van Hoyweghen's contribution to this volume, page 423. In brief, the combination of spin-flip symmetry (applying $\sigma_i \leftarrow -\sigma_i$ for all i does not change the fitness value), rotational symmetry (applying $\sigma_{i+1} \leftarrow \sigma_i$ for all i does not change the fitness value), and the loosest possible coupling between the variables, results in a very flat and symmetric fitness landscape where fitness improvements are only possible when the number of domains (sequences of variables with the same spin value) in the string decreases. The following sequence of bit-flips shows the merging of two domains, decreasing the energy by $2J$ when the

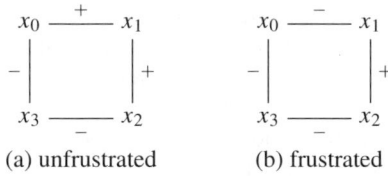

Figure 3. An unfrustrated and a frustrated loop on a two-dimensional rectangular lattice. The frustrated one can be recognized by the negative product of the signs.

two walls collide:

```
+ + +|− − −|+ + +
+ + + +|− −|+ + +
+ + + +|−|+ + + +
+ + + + + + + + +
```

The vertical bars, called domain walls, separate the domains.

Local search algorithms based on a single-bit-flip neighborhood, like ordinary simulated annealing, will spend a long time searching for one of the two optima. This is entirely due to the fact that they only process local information which results in a random walk of the domain walls. If one extends the model a little bit to allow control parameters for each interaction, as in $H = -\sum_{i=0}^{n-1} J_i \sigma_i \sigma_{i+1}$, an effect called 'pinning' can occur which causes the single bit-flip hill-climbers to spend an exponential (in the number of variables) amount of time to reach one of the two optima [24].

This model shows that (a) symmetry in the interactions can create a symmetric search landscape, and that (b) the combination of a symmetric search landscape and a large amount of local optima *per se* is insufficient to create hard instances, since the optima of the extended model are found in linear time, as discussed in section 9.2. The reason for the latter can be found at the interaction level: none of the interactions are in competition with each other.

9.6 Symmetry Due to Frustration

How do we engineer the interaction graph and coefficients to obtain a highly multimodal, hard-to-search landscape? This section gives some first hints to this open problem.

Let us start in the context of the SK-problem, with an example of an unfrustrated and a frustrated loop on a two-dimensional rectangular lattice [54,8], depicted in figure 3. In order for an interaction to decrease the energy (and increase the fitness), its variables should have equal values in the case of the label + and different values in case of the label −. In figure 3(a), the assignments $(-1, -1, -1, +1)$

and $(+1, +1, +1, -1)$ satisfy all interactions. This situation is impossible in figure 3(b): at least one interaction will never contribute. Clearly the interactions are in competition with each other; we say that this loop is frustrated. The product of the labels in the loop is used to distinguish between the two cases. One easily sees that a loop of length ℓ generates a landscape with 2ℓ different optima for the single-bit-flip neighborhood, each at the same energy level. While the factor two is due to spin-flip symmetry, the factor ℓ is due to the rotational symmetry of the unsatisfied interaction.

A situation very much like this can be found in the graph coloring problem. A graph G is called *critical* when it is not k-colorable, but becomes k-colorable when one of its edges is removed. The following example is a critical graph for the 3-color problem:

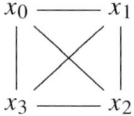

No assignment exists which does not violate at least one of the constraints. However, the following subgraphs are 3-colorable:

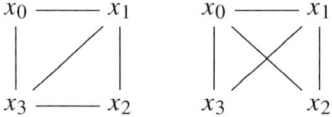

Using the standard penalty function, one can speak of a fairly symmetric search landscape with $k|E|$ local optima. Note that critical graphs need not necessarily be complete subgraphs.

Culberson and Gent [4] found that instances of the 4-colorability problem near the threshold contain critical subgraphs of size not smaller than $\mathcal{O}(\ell)$, i.e., linear in the number of variables. As a result, fast algorithms which, for example, search for 5-cliques to disprove colorability will be of no use. Culberson and Gent conjecture that no intelligent technique can exist to solve these instances, which turns them into what we called intrinsically hard search problems. Note that one needs to identify these large critical subgraphs for both the decision version and the optimization version of graph coloring: for the former to decide uncolorability, for the latter to prove that one additional color is justified.

The idea of critical graphs can be generalized to the binary constraint satisfaction problem. Recall that we can embed a graph coloring instance by making a constraint for each edge, setting the conflicts along the diagonal. If a critical graph is embedded in this way, the corresponding CSP has no solution unless at least one conflict is removed. Note that removing one conflict instead of removing the whole constraint breaks the color symmetry.

Translating the idea of large critical graphs back to the SK-problem is nontrivial, and is part of ongoing research. In [8] we find one caveat: the two-dimensional triangular lattice with only negative interactions is fully frustrated, yet it contains no phase transition near low temperatures. This indicates that the optimization problem is not difficult, which one can indeed easily show. The most probable cause of this easiness is that the graph is far too regular.

9.7 Assessing a Construction

In the process of engineering hard search problems, a means of assessing new constructions is needed. The estimation of the number and sizes of the basins of attraction, as described in this tutorial and in detail on page 343 of this volume, is the first technique that comes to mind. Experimental results are not yet available, however.

When dealing with decision problems, an alternative is readily available: run as many search algorithms as possible to test newly constructed instances. If none of the algorithms can decide whether there is a solution faster than the best algorithms perform in the mushy region, then the created instance is of the same level of difficulty as the hardest decision problems. This technique is used in [46].

In the context of optimization, exhaustive studies of the search landscape for small problem sizes seem to be the only alternative to modality measurements.

Acknowledgments

Bart Naudts is a Postdoctoral Fellow of the Fund for Scientific Research – Flanders, Belgium (F.W.O.).

References

1. S. Baluja and R. Caruana. Removing the genetics from the standard genetic algorithm. In A. Prieditis and S. Russel, editors, *Proceedings of the 12th International Conference on Machine Learning*, pages 38–46. Morgan Kaufmann, San Francisco, 1995.
2. K. G. Beauchamp. *Walsh Functions and Their Applications*. Academic Press, London, 1975.
3. A. J. Bray and M. A. Moore. Metastable states in spin glasses. *J. Phys. C: Solid St. Phys.*, 13:469–476, 1980.
4. J. Culberson and I. P. Gent. Well out of reach: why hard problems are hard. Technical Report 13, APES, June 1999.
5. Y. Davidor. Epistasis variance: a viewpoint on representations, GA hardness, and deception. *Complex Systems*, 4:369–383, 1990.
6. Y. Davidor. Epistasis variance: a viewpoint on GA-hardness. In G. J. E. Rawlins, editor, *Foundations of Genetic Algorithms*, pages 23–35. Morgan Kaufmann, San Francisco, 1991.
7. J. S. De Bonet, C. L. Isbell, and P. Viola. MIMIC: Finding optima by estimating probability densities. In M. Mozer, M. Jordan, and T. Petsche, editors, *Advances in*

Neural Information Processing Systems 9, pages 424–431. MIT Press, Boston, MA, 1997.
8. K. H. Fisher and J. A. Hertz. *Spin Glasses*. Cambridge University Press, Cambridge, 1991.
9. S. Forrest and M. Mitchell. Relative building-block fitness and the building-block hypothesis. In L. D. Whitley, editor, *Foundations of Genetic Algorithms 2*, pages 109–126. Morgan Kaufmann, San Francisco, 1993.
10. B. J. Frey. *Graphical Models for Machine Learning and Digital Communication*. MIT Press, Boston, MA, 1998.
11. P. Galinier and J.-K. Hao. Hybrid evolutionary algorithms for graph coloring. *Journal of Combinatorial Optimization,* to appear.
12. J. Garnier and L. Kallel. Statistical distribution of the convergence time for longpath problems. Technical Report 378, CMAP, Ecole Polytechnique, April 1998.
13. J. Garnier, L. Kallel, and M. Schoenauer. Rigourous results of the first hitting times for binary mutations. *Evolutionary Computation*, 7(2):173–203, 1999.
14. Josselin Garnier, Leila Kallel, and Marc Schoenauer. Rigorous results of the first hitting times of some Markov chains. Technical Report 389, CMAP, Ecole Polytechnique, June 1998.
15. D. E. Goldberg. Genetic algorithms and Walsh functions. Part I: a gentle introduction. *Complex Systems*, 3:129–152, 1989.
16. D. E. Goldberg. *Genetic Algorithms in Search, Optimization and Machine Learning*. Addison–Wesley, Reading, MA, 1989.
17. D. E. Goldberg. Making genetic algorithms fly: a lesson from the Wright brothers. *Advanced Technology for Developers*, 2:1–8, February 1993.
18. K. Hinkelmann and O. Kempthorne. *The Design and Analysis of Experiments*. John Wiley, New York, 1994.
19. T. Hogg, B. A. Huberman, and C. P. Williams. Editorial: Phase transitions and the search problem. *Artificial Intelligence*, 81:1–15, 1996.
20. C. Höhn and C. R. Reeves. The crossover landscape for the *onemax* problem. In J. Alander, editor, *Proceedings of the 2nd Nordic Workshop on Genetic Algorithms and their Applications*, pages 27–43. University of Vaasa Press, Vaasa, Finland, 1996.
21. W. Hordijk. A measure of landscapes. *Evolutionary Computation*, 4(4):336–360, 1996.
22. J. Horn and D. E. Goldberg. Genetic algorithms difficulty and the modality of fitness landscapes. In L. D. Whitley and M. D. Vose, editors, *Foundations of Genetic Algorithms 3*, pages 243–269. Morgan Kaufmann, San Francisco, 1995.
23. J. Horn, D. E. Goldberg, and K. Deb. Long path problems. In Y. Davidor, H.-P. Schwefel, and R. Männer, editors, *Proceedings of the 3rd Conference on Parallel Problem Solving from Nature*, volume 866 of *LNCS*, pages 149–158. Springer-Verlag, Berlin Heidelberg New York, 1994.
24. J. Jäckle, R. B. Stinchcombe, and S. Cornell. Freezing of nonequilibrium domain structures in a kinetic Ising model. *J. Stat. Phys.*, 62:425–433, 1991.
25. D. S. Johnson, C. R. Aragon, L. A. McGeoch, and C. Schevon, editors. *Proceedings of the 2nd DIMACS Implementation Challenge*, volume 26 of *DIMACS Series in Discrete Mathematics and Theoretical Computer Science*, American Mathematical Society, 1996.
26. T. Jones. *Evolutionary Algorithms, Fitness Landscapes and Search*. PhD thesis, The University of New Mexico, 1995.
27. L. Kallel and B. Naudts. Candidate longpaths for the simple genetic algorithm. In W. Banzhaf and C. R. Reeves, editors, *Foundations of Genetic Algorithms 5*, pages 27–44. Morgan Kaufmann, San Francisco, 1999.

28. S. A. Kauffman. Adaptation on rugged fitness landscapes. In *Lectures in the Sciences of Complexity*, volume I of *SFI studies*, pages 619–712. Addison–Wesley, Reading, MA, 1989.
29. H. A. Kramers and G. H. Wannier. Statistics of the two-dimensional ferromagnet. Part I. *Phys. Rev.*, 60:252–262, 1941.
30. S. L. Lauritzen. *Graphical models*. Clarendon Press, Oxford, 1996.
31. M. Mezard, G. Parisi, and M. A. Virasoro. *Spin Glass Theory and Beyond*. World Scientific, Singapore, 1987.
32. D. C. Montgomery. *Design and Analysis of Experiments*. John Wiley, New York, 1997.
33. H. Mühlenbein and T. Mahnig. Evolutionary algorithms: from recombination to search distributions. This volume, pages 135-173.
34. H. Mühlenbein and T. Mahnig. Convergence theory and applications of the factorized distribution algorithm. *J. Comput. Inf. Technol.*, 7:19–32, 1999.
35. H. Mühlenbein, T. Mahnig, and A. Ochoa Rodriguez. Schemata, distributions and graphical models in evolutionary optimization. Technical Report, RWCP Theoretical Foundations GMD Laboratory, 1999.
36. H. Mühlenbein and Th. Mahnig. FDA – a scalable evolutionary algorithm for the optimization of additively decomposed functions. *Evolutionary Computation*, 7(4):353–376, 1999.
37. B. Naudts and L. Kallel. A comparison of predictive measures of problem difficulty in evolutionary algorithms. *IEEE Transactions on Evolutionary Computing*, 4(1):16, 2000.
38. B. Naudts and J. Naudts. The effect of spin-flip symmetry on the performance of the simple GA. In A. E. Eiben, Th. Bäck, M. Schoenauer, and H.-P. Schwefel, editors, *Proceedings of the 5th Conference on Parallel Problem Solving from Nature*, volume 1498 of *LNCS*, pages 67–76. Springer-Verlag, Berlin Heidelberg New York, 1998.
39. R. Palmer. Optimization on rugged landscapes. In A. S. Perelson and S. A. Kauffman, editors, *Molecular Evolution on Rugged Landscapes: Proteins, RNA and the Immune System*, pages 3–25. Addison–Wesley, Reading, MA, 1991.
40. A. Papoulis. *Probability, Random Variables and Stochastic Processes*. McGraw–Hill, New York, 1965.
41. J. Pearl. *Probabilistic Reasoning in Intelligent Systems*. Morgan Kaufmann, San Francisco, 1988.
42. C. Reeves and C. Wright. Epistasis in genetic algorithms: an experimental design perspective. In L. J. Eshelman, editor, *Proceedings of the 6th International Conference on Genetic Algorithms*, pages 217–230. Morgan Kaufmann, San Francisco, 1995.
43. C. Reeves and C. Wright. An experimental design perspective on genetic algorithms. In L. D. Whitley and M. D. Vose, editors, *Foundations of Genetic Algorithms 3*, pages 7–22. Morgan Kaufmann, San Francisco, 1995.
44. C. R. Reeves. Predictive measures for problem difficulty. In *Proceedings of the 1999 Congress on Evolutionary Computation*, pages 736–743. IEEE Press, New York, 1999.
45. E. S. Ristad and P. N. Yanilos. Learning string edit distance. Technical Report CS-TR-532-96, Princeton University, 1996.
46. L. Schoofs and B. Naudts. Empirical comparison of search algorithms near the mushy region of a binary constraint satisfaction problem. Technical Report, Department of Mathematics and Computer Science, University of Antwerp, Belgium, 2000.
47. D. Sherrington and S. Kirkpatrick. Solvable model of a spin-glass. *Phys. Rev. Lett.*, 35:1792–1796, 1975.

48. V. Slavov and N. Nikolaev. Genetic algorithms, fitness sublandscapes and subpopulations. In W. Banzhaf and C. R. Reeves, editors, *Foundations of Genetic Algorithms 5*, pages 199–218. Morgan Kaufmann, San Francisco, 1999.
49. B. M. Smith and M. E. Dyer. Locating the phase transition in binary constraint satisfaction problems. *Artificial Intelligence*, 81:155–181, 1996.
50. P. Stadler. Towards a theory of landscapes. In R. Lopez-Pena, R. Capovilla, R. Garcia-Pelayo, H. Waelbroeck, and F. Zertouche, editors, *Complex Systems and Binary Networks*, volume 461 of *Lecture Notes in Physics*, pages 77–173. Springer-Verlag, Berlin Heidelberg New York, 1995.
51. P. F. Stadler and G. P. Wagner. Algebraic theory of recombination spaces. *Evolutionary Computation*, 5:241–275, 1998.
52. D. Suys. *A Mathematical Approach to Epistasis*. PhD thesis, Department of Mathematics and Computer Science, University of Antwerp, Belgium, 1998.
53. G. Tononi, G. M. Edelman, and O. Sporns. Complexity and coherency: integrating information in the brain. *Trends in Cognitive Sciences*, 2(12):474–484, December 1998.
54. G. Toulouse. Theory of the frustration effect in spin glasses: I. *Communications on Physics 2*, pages 115–119, 1977. Also in [31].
55. E. Tsang. *Foundations of Constraint Satisfaction*. Academic Press, New York, 1993.
56. M. D. Vose and A. H. Wright. Stability of vertex fixed points and applications. In L. D. Whitley and M. D. Vose, editors, *Foundations of Genetic Algorithms 3*, pages 103–114. Morgan Kaufmann, San Francisco, 1995.
57. E. D. Weinberger. Correlated and uncorrelated fitness landscapes and how to tell the difference. *Biological Cybernetics*, 63:325–336, 1990.

A Solvable Model of a Hard Optimisation Problem

A. Rogers and A. Prügel-Bennett

Department of Electronics and Computer Science
University of Southampton
Southampton SO17 1BJ, UK
E-mail: *a.rogers@ecs.soton.ac.uk*

Abstract. The dynamics of a genetic algorithm (GA) on a model of a hard optimisation problem are analysed using a formalism which describes the changing fitness distribution of the GA population under ranking selection, uniform crossover, and mutation. The time to solve the optimisation problem is calculated in a closed form expression which enables the effect of the various GA parameters – population size, mutation rate, and selection scheme – to be understood.

Keywords
Fitness distribution, macroscopic dynamics, convergence time, hard problem, finite population effects

1 Introduction

Much of the theoretical analysis of genetic algorithms (GAs) has necessarily been performed on simple simple fitness landscapes such as one-max [4,12,6,7] and the royal road functions of Mitchell, Holland and Forest [3]. Whilst giving some insight into the behavior of GA, they are not representative of the types of problems to which GA are often applied. GA are shown to perform poorly on these problems compared to other techniques such as simple hill climbing and little knowledge regarding the setting of the various GA parameters is gained.

The simple GA models studied usually omit the crossover operator in their analysis. Empirical comparisons, however, suggest that the performance of a GA is qualitatively different with and without crossover [5]. Thus, drawing general conclusions from the analysis is often misleading.

In this paper we analyse a previously published model of a hard optimisation problem [10] – the 'Basin with a Barrier' fitness landscape. This model has some of the features of a real-world problem whilst still being amenable to analysis. It is one of the few test problems where a GA is shown to have a distinct advantage over other conventional search methods such as simulated annealing [8].

The dynamics of a generational GA using ranking selection [1], uniform crossover [11], and mutation are calculated accurately on this model by describing the effect that each genetic operator has on several macroscopic variables which describe the population distribution. Closed form expressions can be derived for the end point of evolution and the number of function evaluations required to solve the problem. In this paper, we use these results to study the effect that the various GA parameters have on how the problem space is searched and show how the optimal mutation rate is obtained. It is hoped that by studying this problem, some insight into real-world problems is gained.

2 The Basin with a Barrier

Discussions of what characterises a hard optimisation problem often result in a debate regarding the validity of certain measures of problem difficulty. Despite this debate, there is some consensus as to what features make a problem space hard to search. There may be many local minima. These local minima may be separated by a fitness barrier from even fitter solutions, resulting in the need for non-local search steps. If these local minima occupy the majority of the search space it may take a long time to generate the moves necessary to fall into the basin of attraction of the global minimum.

The 'Basin with a Barrier' attempts to model these types of landscapes [10]. It has some of the features of a hard optimisation problem but is still amenable to analysis.

The landscape consists of a large local minimum separated from the global minimum by a potential barrier. We consider a series of L spins whose value may be 1 or -1 and consider the total 'magnetisation', M, of the string

$$M = \sum_{i=1}^{L} S_i \quad \text{where} \quad S_i = \{-1, 1\}. \tag{1}$$

The potential, which we are trying to minimize, is a function of this magnetisation and is given generically as

$$V(M) = \begin{cases} (M - M_l)^2 + V_l & \text{if } M \leq M_b \\ 0 & \text{if } M > M_b. \end{cases} \tag{2}$$

In describing the landscape we must also consider the entropy of the system, S. Entropy describes the number of possible states which the system may occupy. Clearly there are only two states with the maximum or minimum magnetisation, whilst the majority of possible states represent magnetisations somewhere in between.

The entropy of the system is such that the number of states in the global minimum is much smaller than that in the local minimum, whilst the maximum entropy state is some distance from both the local and global minima. Figure 1 shows the landscape schematically.

We propose that whilst being a toy problem, this model holds some of the features seen in combinatorial optimisation problems. We expect random search to produce

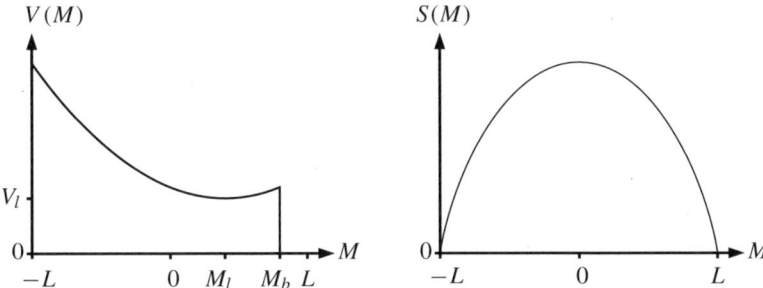

Figure 1. Diagram of potential and entropy for the 'Basin with a Barrier' fitness landscape

poor solutions most of the time and we expect these poor solutions to occupy the majority of the problem space. We expect good solutions to be near one another in problem space but they may be separated by non-local moves.

3 Modelling the GA

The GA we consider is a finite population generational GA using ranking selection [1] with either roulette wheel or stochastic universal sampling (SUS) [2]. Mutation is applied with a fixed small probability that each spin mutates at each generation and uniform crossover [11] is applied between pairs.

The technique we use to model the GA is that developed by Prügel-Bennett, Shapiro and Rattray. Several macroscopic variables are used to describe the magnetisation distribution of the population. In this analysis, we make a Gaussian approximation to the distribution and describe it by the first two cumulants – the mean of the distribution, K_1, and the variance of the distribution, K_2 – and a measure of the degree of replication of individuals within the population, C. By combining these variables, we can also calculate the correlation of the population, q. These parameters describe the ensemble average of many finite populations and thus predict the mean behavior of a finite population.

We make a number of approximations as we wish to capture the full dynamics of the GA but not at the cost of an overcomplex model. The complete analysis including the derivation of the results used here and justification underlining the approximations made are published elsewhere [8]. In this paper we use the results of this analysis and use them to study the effect of changing the various GA parameters.

As each genetic operator – selection, mutation, and crossover – acts on the population, we obtain a set of coupled equations which describe the changing state of the macroscopic variables – K_1, K_2 and C – in terms of the GA parameters – the selection strength, MAX, the mutation rate, γ, the population size, P, the string length, L, and a measure of the rate of genetic drift inherent in the selection scheme, r. By iterating these, we can describe the full dynamics of the GA.

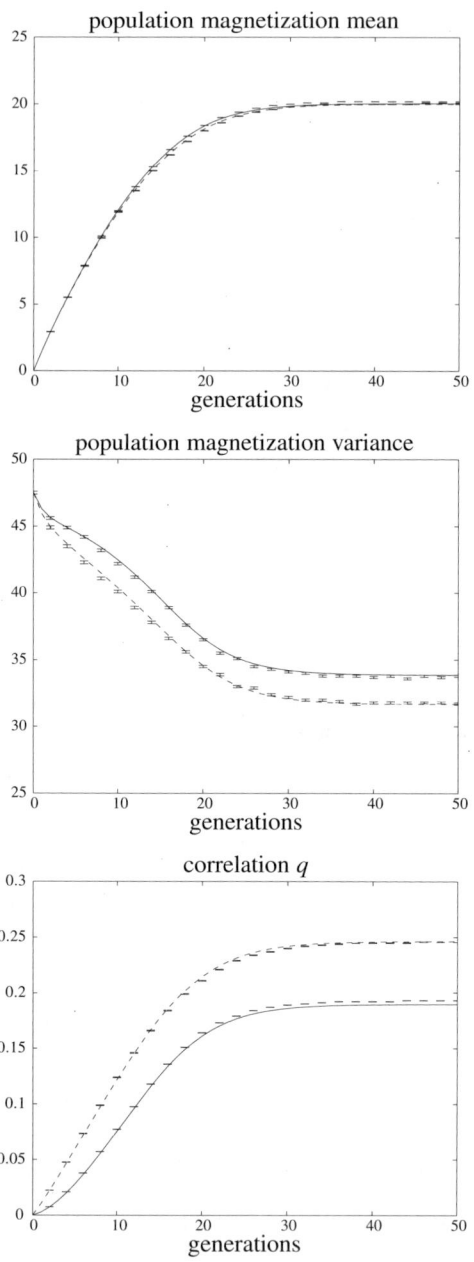

Figure 2. Comparison of theoretical and simulation results for the first two cumulants and correlation for a GA with ranking selection, mutation, and crossover. Roulette wheel selection (*dashed line*) and SUS (*solid line*) are shown. Parameters used were $L = 48$, $\gamma = 1/48$, $P = 100$, MAX $= 1.4$, and $M_l = L/2$.

Selection

$$K_1 \to K_1 + (\text{MAX} - 1)\sqrt{\frac{K_2}{\pi}}\,\text{erf}\left(\frac{M_l - K_1}{\sqrt{K_2}}\right)$$

$$K_2 \to r\left[1 - \frac{2(\text{MAX} - 1)}{\pi}\exp\left(-\frac{(M_l - K_1)^2}{K_2}\right)\right.$$
$$\left. - \frac{(\text{MAX} - 1)^2}{\pi}\text{erf}^2\left(\frac{M_l - K_1}{\sqrt{K_2}}\right)\right]K_2$$

$$C \to 1 - r(1 - C)$$

Mutation

$$K_1 \to \Gamma K_1$$
$$K_2 \to \Gamma^2 K_2 + L\left(1 - \Gamma^2\right)$$
$$C \to \frac{\Gamma^2\left(K_1^2 - L^2\right)}{\Gamma^2 K_1^2 - L^2}C \quad \text{where} \quad \Gamma = 1 - 2\gamma$$

Crossover

$$K_1 \to K_1$$
$$K_2 \to \frac{K_2}{2} + \frac{L}{2}(1 - q) \quad \text{where} \quad q = C + (1 - C)\frac{K_1^2}{L^2} \qquad (3)$$

Besides giving an accurate description of the dynamics, these expressions provide an intuitive idea of how the GA is searching the fitness landscape.

- Selection acts to focus the population onto areas of improved fitness and thus increases the mean magnetisation. In doing so, however, it reduces the variance of the population leading to a smaller area of the landscape being sampled and increasing the correlation of the population.
- Mutation will generate new population members around the selected area but will act to push the population back to the maximum entropy state, thereby increasing the variance and decreasing the correlation and mean.
- Crossover does not effect the mean but forces the variance towards a *natural* value defined by the correlation of the population, q. This acts to restore variance to the population lost through selection.

By considering an initial population whose spins are assigned randomly – $K_1 = 0$, $K_2 = L$, and $C = 0$ – we can iterate these equations to predict the dynamics of the GA. Figure 2 show the theory predictions compared to simulation results from repeated runs of a real GA using roulette wheel selection and stochastic universal sampling. The simulation data is averaged over 10 000 runs and uses the parameters $L = 48$, $\gamma = 1/48$, $P = 100$, MAX $= 1.4$, and $M_l = L/2$. The figures show very good agreement between theory and simulations.

4 Equilibrium Point

The conflicting forces of selection, mutation, and crossover result in the population evolving to an equilibrium state where the magnetisation mean, variance, and correlation no longer change from generation to generation. Although at the macroscopic level the population is in equilibrium, the GA is continually generating new population members and sampling the area of problem space described by the macroscopic variables. We can solve these expressions as a set of simultaneous equations by simply making the demand that after all the three operator – selection, mutation, and crossover – the macroscopic variables remain unchanged. Doing this numerically is easy, but in order to derive closed form expressions for the equilibrium point several approximations are made. In the case when the equilibrium distribution has a significant overlap with the local minimum, a simple approximation can be made for the error term [1] in the selection term

$$\mathrm{erf}(x) \approx x \quad \text{where} \quad x = \frac{M_l - K_1}{\sqrt{K_2}}.$$

The response of the magnetisation mean to selection simplifies significantly in this case and is no longer dependent on the variance of the distribution. We can then simply solve for the equilibrium value of K_1 to give

$$K_1^* \approx \frac{(\mathrm{MAX} - 1)\,\Gamma M_l}{\sqrt{\pi}\,(1 - \Gamma) + (\mathrm{MAX} - 1)\,\Gamma}. \tag{4}$$

This approximation holds in most cases except when the mutation rate is so high and the selection strength so weak that the population does not move far from the initial position. The equilibrium correlation is determined by the balance of mutation and selection and the equilibrium mean. We can simply solve for the equilibrium value of C

$$C^* = \frac{x\,(1 - r)}{1 - xr} \tag{5}$$

where

$$x = \frac{\Gamma^2 \left(K_1^{*2} - L^2\right)}{\Gamma^2 K_1^{*2} - L^2}. \tag{6}$$

---

[1] Where erf(x) represents the standard error function.

The equilibrium correlation is thus given by

$$q^* \approx C^* + (1 - C^*) \frac{K_1^{*2}}{L^2}. \tag{7}$$

To find the equilibrium variance, we again assume that the equilibrium distribution overlaps the local minimum, and we can approximate the response to selection of the variance given in (3) as

$$K_2 \to r \left[1 - \frac{2(\text{MAX} - 1)}{\pi} \right] K_2. \tag{8}$$

Solving the remaining expressions as a set of simultaneous equations thus gives the equilibrium variance as

$$K_2^* \approx \frac{L(1 - \Gamma^2) + L(1 - q^*)}{2 - \Gamma^2 r [1 - 2(\text{MAX} - 1)/\pi]}. \tag{9}$$

5 First Passage Time

The picture which emerges from this analysis is of a population evolving towards the local minimum and reaching an equilibrium point close to it. How close it is to the minimum depends on the mutation rate and selection strength. The variance of the population at this equilibrium point depends on the various GA parameters in a rather complicated way.

We are interested in how long it takes for mutation and crossover to generate an individual which is at the extremes of the distribution and is in the global minimum. Clearly this will be related to the population size and mean and variance of the population magnetisation distribution at its equilibrium point.

As we are describing the magnetisation distribution as a Gaussian, the probability of finding any one population member with a magnetisation less than the barrier, M_b, is simply given by

$$p = \Phi(x) \qquad \text{where} \qquad x = \frac{M_b - K_1^*}{\sqrt{K_2^*}} \tag{10}$$

and $\Phi(x)$ represents integration of a unit Gaussian from negative infinity to x. We can approximate this expression to

$$p \approx 1 - \frac{\exp(-x^2/2)}{x\sqrt{2\pi}}. \tag{11}$$

The probability of finding one member above the barrier, and thus in the global minimum in any generation, is $1 - p^P$ and since p is small this can be approximated

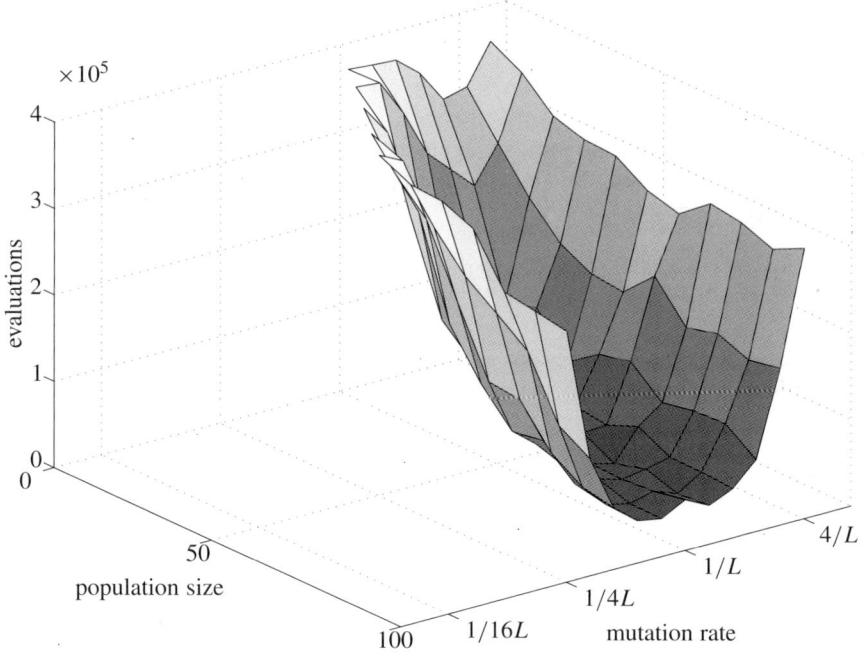

Figure 3. Simulation results to solve 'Basin with a Barrier' using a GA with stochastic universal sampling, mutation, and uniform crossover. Parameters used were MAX $= 2$, $L = 48$, $M_l = L/2$, and $M_b = 7L/8$.

to $1 - Pp$. Thus, the expected time in terms of function evaluations, n, is given by $P/(1 - Pp)$. Using the above result gives

$$n \approx x\sqrt{2\pi} \exp(x^2/2). \tag{12}$$

The most significant factor here is the exponential dependence on the equilibrium mean and variance

$$n \propto \exp\left(\frac{(M_l - K_1)^2}{2K_2}\right). \tag{13}$$

The second interesting point is the absence of any population size dependence. Once the population has reached the equilibrium point, the number of evaluations required to find a solution is independent of the population size.

6 Simulation Results

The time to solve a typical size 'Basin with a Barrier' problem was found for a range of population sizes and mutation rates. The results are shown in figure 3 and are the results of averaging over 100 runs.

The results show a very clear optimal mutation rate with some population size dependence particularly when the population is small.

7 Theoretical Analysis

Having derived expressions for the time to solve the problem and understanding how each of the genetic operators influence the evolution of the population, we can consider what effect GA parameters such as mutation rate and population size have on search. Figure 4 shows the results of solving the analytical solutions for the time to solve the problem simulated in figure 3. The theoretical analysis very accurately predicts the time required and shows the same influence of mutation rate and population size.

7.1 Population Size

As seen in the analysis of the first passage time, the size of the population does not enter our expressions in (12) for the time required for our population in equilibrium to produce a solution in the global minimum. The population size does, however, affect this time indirectly by changing the final end point distribution of the population. With small populations the stochastic nature of the selection operator becomes significant and must be accounted for to accurately describe the dynamics and end point distribution.

Population size enters our model as an extra factor, r, caused by the stochastic nature of the selection scheme. It is less than one and tends to one as the population size goes to infinity. It causes a further reduction in population variance than that predicted for an infinite population and also occurs as a factor increasing population correlation at each selection step. A small population thus experiences a faster reduction in population variance and reaches an equilibrium magnetisation with less variance. This clearly has a strong effect on the time to solve the problem.

What signifies a small population can be found by calculating how r changes with population size. The factor is given by

$$r = \frac{P - \langle n^2 \rangle}{P - 1}. \tag{14}$$

The term $\langle n^2 \rangle$ is the expected square of the number of times any population member is selected and is a function of the population size, selection strength and the

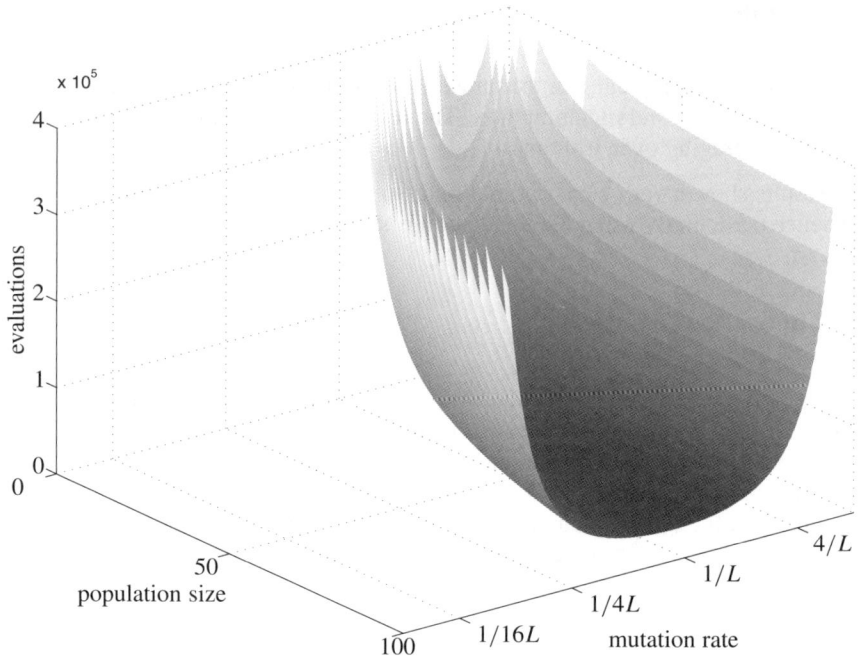

Figure 4. Theoretical results to solve 'Basin with a Barrier' using a GA with stochastic universal sampling, mutation, and uniform crossover. Parameters used were MAX = 2, $L = 48$, $M_l = L/2$, and $M_b = 7L/8$.

selection scheme used – roulette wheel or stochastic universal sampling. It has been calculated in another paper [8] and has been used by the authors in a comparison of various selection schemes [9]. Figure 5 shows the results of these against population size.

Clearly the curve approaches unity – the infinite population limit – very quickly. For small population sizes, however, the deviation from unity is large. This feature explains the relative lack of dependence of population size on the time to solve the problem for large population sizes and the very rapid decline in performance as the population size becomes small.

In this analysis, we have assumed cases where the initial dynamics are comparatively short in comparison to the time spent at equilibrium, whilst waiting for the mutation and crossover operators to generate a solution in the global minimum. If we consider a more realistic problem consisting of a cascade of barriers, it is clear that this initial dynamic phase favors a smaller population as it requires less function evaluations to move the population to its new equilibrium point.

This suggests an optimum population size which is a balance between the need to maintain the speed at which the population can move around the landscape, whilst

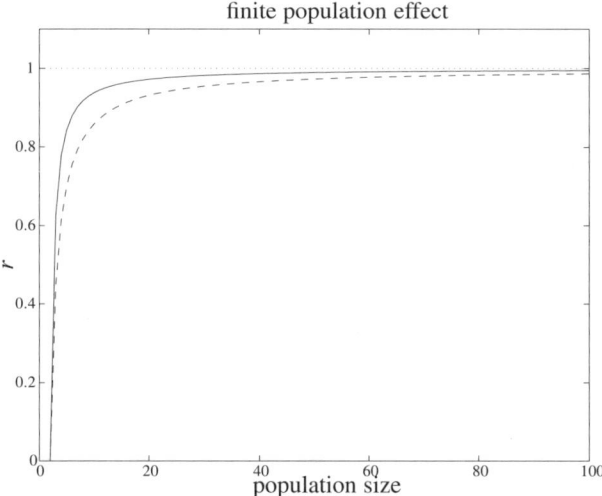

Figure 5. Finite population effect for roulette wheel (*dashed line*) and stochastic universal sampling (*solid line*)

not being so small that the area of the landscape being searched is significantly reduced by finite population effects.

7.2 Selection Scheme

Stochastic universal sampling [2] was suggested by Baker as an alternative to roulette wheel selection. Baker noted that whilst on average any individual is expected to be selected a certain number of times depending on its fitness within the population, the stochastic nature of roulette wheel selection allows anywhere between 0 and P copies to be selected. This is the source of convergence of a finite population due to stochastic effects – genetic drift – and what we are calculating in this formalism when we calculate r.

Ranking selection allows us to calculate r for both stochastic universal sampling and roulette wheel selection [8]. In figure 5 both selection schemes are plotted. For roulette wheel selection, the finite population effects are much larger over the entire range of population sizes.

In figure 2, showing the evolving dynamics of the GA, stochastic universal sampling reaches an equilibrium distribution with greater variance and less correlation than the equivalent roulette wheel case. This is a direct cause of the reduction in finite population effects achieved through using stochastic universal sampling. In the test problem shown in this paper, this difference in variance leads to a halving of the number of function evaluations required to solve the problem.

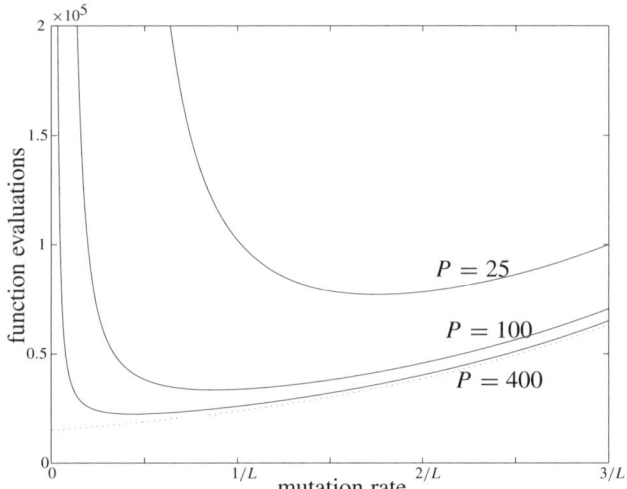

Figure 6. Theoretical results to solve 'Basin with a Barrier' with varying mutation rate at three different population sizes. $P = 25, 100$, and 400. The dotted line is the infinite population response. Parameters used were MAX $= 2$, $L = 48$, $M_l = L/2$, and $M_b = 7L/8$.

7.3 Mutation Rate

Perhaps the most significant feature of the surfaces plotted is the strong dependence on mutation rate, with extremely poor performance outside the optimal range. Understanding this feature involves the interplay of all the effects previously discussed.

Mutation has been shown to increase the variance of the final population equilibrium distribution but also to move the mean of the distribution away from the global minimum back towards the maximum entropy state. The second effect is most significant and mutation has a detrimental effect on performance. Increasing the mutation rate increases the number of function evaluations required to solve the problem.

In an infinite population, the optimum mutation rate would thus be zero. However, we are dealing with finite populations and must consider the correlation of the population caused by selection. With no mutation, the correlation of the population will increase very rapidly limiting the crossover's ability to restore variance to the population. This will result in a very small equilibrium variance which searches a very small area of the problem space and thus takes a very long time to reach the global minimum.

A balance is achieved when mutation is large enough to prevent the correlation of the population but not so large as to disrupt the search.

Figure 6 shows theoretical results for the time to solve the 'Basin with a Barrier' with varying mutations rates. Three different population sizes are shown along with the infinite population case as a dotted line. Clearly as P increases, the finite popula-

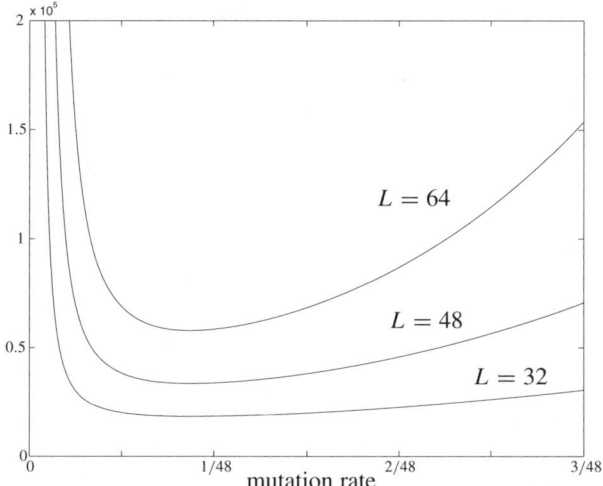

Figure 7. Theoretical results to solve 'Basin with a Barrier' with varying mutation rate at three string lengths – $L = 32, 48$, and 64. Vertical scales for $L = 32$ and $L = 64$ are multiplied and divided by ten, respectively, to enable easy comparison. Parameters used were MAX $= 2$, $P = 100$, $M_l = L/2$, and $M_b = 7L/8$.

tion effects decrease and the finite populations approach the infinite case. However, as the mutation rate becomes small, the correlation of the population through finite population effects becomes significant.

The optimum mutation rate is seen to be dependent on the population size. A large population suffers less correlation due to finite population effects and thus will not suffer the same correlation until the mutation rate is very small. For this particular problem we find that the optimum mutation rate is approximately proportional to $1/\sqrt{P}$.

As the effects discussed here are independent of the string length, we find that the optimum mutation rate is also independent of the string length. Figure 7 shows the results of varying the length of the string for various mutation rates. As expected after performing this analysis, we see no dependence on the optimum mutation rate with string length.

8 Conclusions

The 'Basin with a Barrier' problem is a caricature of a real-world optimisation problem. Unlike simpler models such as one-max and royal road functions, it has local and global minima and thus shows some of the features of a hard optimisation problem.

We have been able to study the effects of all the GA parameters and say definitively how these parameters affect search on this model problem. In this case, we have seen mutation to be a disruptive force. Unlike crossover, it has no knowledge of which parts of the strings are shared by many population members and thus disrupts parts of the string which are beneficial to fitness.

Crossover has been shown to be the dominant search operator on this landscape. Indeed without crossover the GA performs many orders of magnitude worse. By mixing those parts of the strings which are not identical, it is able to produce new population members without disrupting what has already been gained. However, in the absence of mutation, selection very rapidly produces a highly correlated population which prevents crossover from operating. Thus we require a minimum level of mutation to overcome this correlation without disrupting the search.

At larger population sizes, the increase in correlation of the population is slower and thus the optimum mutation rate is lower. This optimum is seen to be independent of the length of the string.

Relating this work to real problems, however, is still some way off. However, the techniques developed here have enabled a model problem to be analysed and definite statements made about the influence of parameters. In this way it represents a first step towards understanding the influences these parameters may have on real-world problems.

References

1. J. Baker. Adaptive selection methods for genetic algorithms. In J.J. Grefenstette, editor, *Proceedings of the First International Conference on Genetic Algorithms*, pages 101–111. Lawrence Erlbaum, Hillsdale, 1985.
2. J. Baker. Reducing bias and inefficiency in the selection algorithm. In J.J. Grefenstette, editor, *Proceedings of the Second International Conference on Genetic Algorithms*, pages 14–21. Lawrence Erlbaum, Hillsdale, 1987.
3. M. Mitchell, J. Holland, and S. Forrest. When will a genetic algorithm outperform hill climbing? In J. Cowan, G. Tesauro, and J. Alspector, editors, *Advances in Neural Information Processing Systems*, pages 51–58. Morgan Kaufmann, San Francisco, 1994.
4. A. Nix and M. D. Vose. Modeling genetic algorithms with markov chains. *Annals of Mathematics and Artificial Intelligence*, 5:79–88, 1991.
5. G. Ochoa, I. Harvey, and H. Buxton. On recombination and optimal mutation rates. In *Proceedings of the Genetic and Evolutionary Computation Conference, GECCO 1999*, volume 1, pages 488–495. Morgan Kaufmann, San Francisco, 1999.
6. A. Prügel-Bennett and J. L. Shapiro. An analysis of genetic algorithms using statistical mechanics. *Phys. Rev. Lett.*, 72(9):1305–1309, 1994.
7. A. Prügel-Bennett and J. L. Shapiro. The dynamics of a genetic algorithm for simple random Ising systems. *Physica D*, 104:75–114, 1997.
8. A. Rogers and A. Prügel-Bennett. The dynamics of a genetic algorithm on a model hard optimization problem. *Complex Systems*, 11(6):437–464, 2000.

9. A. Rogers and A. Prügel-Bennett. Genetic drift in genetic algorithm selection schemes. *IEEE Transactions on Evolutionary Computation*, 3(4):298–303, 1999.
10. J. L. Shapiro and A. Prügel-Bennett. Genetic algorithms dynamics in two-well potentials with basins and barriers. In R. K. Belew and M. D. Vose, editors, *Foundations of Genetic Algorithms 4*, pages 101–139. Morgan Kaufmann, San Francisco, 1997.
11. G. Syswerda. Uniform crossover in genetic algorithms. In *Proceedings of the Third International Conference on Genetic Algorithms*, pages 2–9. Morgan Kaufmann, San Francisco, 1989.
12. M. D. Vose. Modelling simple genetic algorithms. In D. Whitley, editor, *Foundations of Genetic Algorithms 2*, pages 63–74. Morgan Kaufmann, San Francisco, 1992.

Bimodal Performance Profile of Evolutionary Search and the Effects of Crossover

M. Oates, J. Smedley, D. Corne and R. Loader

BT Laboratories
Martlesham Heath, Suffolk, UK
E-mail: *moates@srd.bt.co.uk*

Abstract. Tunable performance profiles for evolutionary search on instances of the *adaptive distributed database management problem* have previously been plotted and published by the authors. This demonstrates a bimodal feature of convergence time with respect to population size and mutation rate. Preliminary results on other problems (one-max, De Jong functions, etc.) led to the tentative conclusion that the features of the complex profile discovered could indeed be generic, and four key hypotheses were presented. These covered the effects of problem complexity and evaluation limit on optimal and non-optimal mutation rates. This paper expands significantly on these results looking in more detail at the one-max and royal staircase problems, and demonstrates the effect of various rates of crossover on the performance profile of evolutionary search. Crucially, these results continue to demonstrate the bimodal feature and show that reduced levels of crossover extend the influence of the bimodal region to higher population sizes. A study of the coefficient of variation of convergence time shows importantly that this can be at a minimum at an optimal mutation rate which can also deliver consistent results in a minimum number of evaluations.

Keywords
Bimodal response, one-max, royal staircase, convergence time, crossover, optimal mutation rate, population size

1 Introduction

When considering the use of evolutionary algorithms (EA) [3–5] in real-time industrial control problems, it is crucial to have a high degree of confidence that the algorithm will consistently find good solutions. It is not sufficient to have an algorithm that finds excellent solutions some of the time, but occasionally fails by a wide margin. Equally it is important to know that the algorithm will find good results within a fixed (ideally minimal) amount of time and is robust to the changing problem environment. A good example of such an application is the *adaptive distributed database management problem* (ADDMP) [10,9,8] which has been extensively studied by the authors. This problem considers how best to distribute user workload over a number of potential distributed servers, by using a performance model of the servers and communication links, and an optimiser which iteratively tries to find the optimum user distributions as workload changes. It has been shown [11,9,8] that a suitably tuned evolutionary algorithm can outperform more conventional optimisers (hill-climbers, greedy algorithms, simulated annealing, etc.) on this task. To find the optimal tuning parameters for the EA (i.e., population size, mutation rate, evaluation limit, etc.), extensive tests were performed to determine a *performance profile* for the EA [12,14]. This allowed the EA to run for up to 20 000 evaluations noting the best solution it could find and the evaluation number this was first found at. This was repeated 50 times and the mean values plotted against a population size/mutation rate pairing which varied from 2 members to 100 members in steps of 2 and 0% mutation per gene, through $10^{-5}\%$ mutation, then doubling at each step up to 83% mutation. Thus each point represented the results of 1 million evaluations and the whole plot represented some 1.25×10^9 evaluations. Similar tests were also run at population sizes of 10 to 500 members in steps of 10.

These results presented a bimodal feature in the plot of the number of evaluations taken to first find the best solution in up to 20 000 evaluations. This is in line with recent results from studies by Deb and Agrawal ([2] and personal communications) and theoretical studies by Rogers and Prügel-Bennett (in this volume, page 207). In these plots the number of evaluations was seen to rise as mutation rate increased, as the EA was able to increasingly avoid premature convergence on poor fitness solutions.

As mutation was increased further, a point was reached where the EA began to find the global optimum solution after a high number of evaluations. This produced the first peak in our bimodal feature. As mutation was increased further, the number of evaluations taken to find the global optimum decreased until an optimum mutation rate was reached, allowing the EA to find the global optimum solution in a minimum number of evaluations. On problems studied so far, this mutation rate is typically seen to correspond to around one divided by the chromosome length, a result previously seen by others [1,2,6] and referred to as the $1/L$ rule. As mutation was increased further still, the number of evaluations started to increase again until at too high a level of mutation, the EA could no longer reliably find the global optimum. This is the second peak. At even higher mutation rates the EA degenerated into random search.

The performance plot also showed that at zero and low levels of mutation, the number of evaluations 'exploited' by the EA rose linearly with population size; a surprising result given the complexity of the evolutionary search process. Obviously at low population sizes this represents premature convergence on exceedingly unfit solutions. When looking at the plot of the number of times (out of the 50 runs) that the EA failed to find the global optimum, termed the *error surface*, a distinct trough was seen in the middle to higher range of mutation values. This trough included the value of mutation which induced the minimum number of required evaluations as discussed previously. The trough was seen to have a high degree of independence from population size. At lower mutation rates, the error also reduced with increased population size as the increased diversity available to crossover allowed more effective exploration of the search space by the crossover operator.

These results were obtained using a generational breeder style GA [7] employing 50% elitism and uniform selection of parents from within the highest ranked half of each generation. However, it is important to note that very similar results were also obtained from a simple three way tournament selection EA, i.e., a steady state EA. The EA employed uniform crossover [15], but again similar results have now been seen with one-point and two-point crossover. The representation was a natural, non-binary one, where (given a ten server node scenario), ten gene positions contained alleles in the range one through ten. Uniformly distributed allele replacement mutation was then employed.

These results were presented and discussed in previous publications [12,14] together with preliminary results from similar investigations into other standard test problems (e.g., one-max, De Jong functions, etc.). These experiments gave similar performance profiles, however, it was noted that the particular rates of mutation at which the features occurred varied with problem complexity, evaluation limit, and customisation of crossover operator [8,13,14], and as a result, four generic hypotheses were proposed:

1. The 'optimum' rate of mutation consistently finding best solutions in the minimum number of evaluations, shows a high degree of independence from population size. This is also true of the lower rate of mutation at which a locally maximum number of evaluations is exploited.
2. The 'optimum' rate of mutation for minimum number of evaluations reduces slightly with problem difficulty.
3. The 'locally worst' mutation rate reduces slightly with improved suitability of the crossover operator.
4. The 'optimum' rate of mutation reduces slightly with increased evaluation limit. The 'locally worst' mutation rate reduces significantly with increased evaluation limit.

In the above experiments the EA was operating with a crossover probability of unity followed by per gene mutation at the specified rate. In this paper we report on the effects of different rates of crossover on two standard binary representation test problems – one-max (unitation problem) and the royal staircase. In the former, fitness is measured by the number of ones occurring in the chromosome (here of length 50). With the latter, fitness is equal to the number of consecutive ones in the chromosome (again of length 50) which must start with the first gene position. This problem has also been studied by van Nimwegen and Crutchfield [17,16] using a mutation-only EA as the problem does not necessarily lend itself to useful exploitation of normal crossover operators, as will be shown later.

The remainder of this paper is set out as follows. Section 2 gives more details on the test problems and the method employed. Section 3 describes the new results presented in this paper, and section 4 summarises these results and presents our conclusions.

2 Method

For the one-max problem, the 50% elitist breeder EA is allowed around 5000 evaluations, with the generational limit adjusted according to population size. For a given combination of population size, mutation rate, and uniform crossover probability, 50 runs are performed, each starting with a randomly generated population. A binary representation is used with a string length of 50. With the higher cardinality representation used on the ADDMP, a given mutation rate was used to uniformly generate one of 10 possible alleles, 9 of which would represent a change. Thus, to compare this to a traditional guaranteed flip mutation rate, would mean reducing the published rate by only 10%. However, in the case of our binary one-max problem, the effective rate of guaranteed flip mutation is reduced by 50%, as our stated mutation rate has a 50% chance of replacing the allele with its original value. This should be borne in mind when comparing results with other studies. Groups of runs are performed with fixed mutation rates varying from 0% mutation per gene, through 10^{-5}% mutation then doubling at each step up to 83% mutation. Population sizes vary from 2 to 100 members, in steps of 2.

For each run the EA notes the fitness of the best solution found so far and the evaluation number at which this was first found. This evaluation number was termed the *convergence time* in [12,14] and although this may be considered a controversial definition, the authors now have a considerable body of unpublished data supporting it. Our experiments have demonstrated a marked and rapid reduction in population diversity around this number of evaluations, particularly for 'ideally tuned' EAs.

The mean of the convergence times across the 50 runs each with the same combination of population size and mutation rate on the one-max problem is then plotted to produce a performance surface as shown in figure 1. The number of runs (out of 50) which fail to find the global optimum (in this case a chromosome of all ones) is also plotted as an error surface as shown in figure 2. These two plots are from EAs using a uniform crossover probability of unity followed by mutation at the specified rate. As will be shown later, this probability is changed in subsequent experiments looking at crossover probabilities of 85%, 70%, 50% and 10%.

For later experiments with the 'harder' royal staircase problem (also length 50), the EAs are allowed 20 000 evaluations, and results are shown over a larger population range – 10 to 500 in steps of 10.

Finally, plots are shown using the coefficient of variation of the convergence time of each of the runs in a group. This is the standard deviation of the 50 values divided by the mean which gives an indication of how 'stable' the evolutionary process is in terms of the variability of run time to find best solutions.

3 Results

Figure 1 shows the performance profile for our EA with 100% probability of crossover on the one-max problem. The linear feature relating convergence time to population size at low mutation rates is clearly visible in the left-hand side of the plot. Also, the bimodal feature is clearly present particularly at low population sizes. The 'locally worst' mutation rate which requires a high number of evaluations, is at 0.04% whilst the 'optimum' rate is between 2.6% and 5.2% mutation at low population sizes. As population size increases, the bimodal feature is seen to have a reducing effect, becoming insignificant above population sizes of around 90. This reduction, combined with the rising floor of the rest of the profile as population size increases, induces a slight reduction in the 'optimum' mutation rate at higher population sizes. At around population sizes of 80, the 'optimum' mutation rate has reduced to around 1.3%, and beyond 90 members, the mutation rate inducing a minimal convergence time becomes any value below about 2.6%.

This profile can only be correctly interpreted in conjunction with the associated error surface plot of figure 2 which shows the number of times out of the 50 runs that the EA failed to find the global optimum. Clearly, where the EA has too little diversity available to it from both too small a population size and too low a mutation rate (rear left-hand side), the EA completely fails to find the global optimum. At low population sizes, as mutation rate is increased, a critical mutation rate is reached at 0.3% whereby the EA is suddenly able to consistently find the global optimum with 100% success rate. This continues up to mutation rates of 10%, beyond which reliability rapidly deteriorates. This region shows a marked independence from population size and represents the EA working in hill-climber mode, where recombination from crossover is not the dominant operator. At very high rates of mutation, here above 10%, the EA degenerates into random search. This trough of good performance also coincides with the trough seen in figure 1, and suggests a region where the EA can be said to be 'optimally tuned' i.e., consistently delivering global optimum solutions in a minimum number of evaluations. It can be seen to demonstrate a high degree of independence from population size, and this will explored in more detail in later plots.

By contrast, the left-hand side of figure 2 shows how the ability of the EA to find the global optimum improves with increased population size. Above 30 members, the EA starts to occasionally find the global optimum, improving until at above 60 members, failure rates are below 10%. Results beyond 100 members, not shown here, demonstrate that this region also flattens out to give consistently good performance above population sizes of about 150. Here, a high degree of independence is shown from the mutation rate, with no significant difference in the range 0% to 0.1% mutation. The EA can here be said to be operating in recombination mode, following the classical mechanisms of genetic algorithms according to the schema theorem [4]. It is important to note, however, that the number of evaluations used by the EA at a population size of 100 with 0% mutation is around 900 evaluations (some 18 generations). In the low population trough at 'optimal' mutation this value is around 500 evaluations for a wide range of population sizes (2–40 members). However, as will be shown later, performance in this region is not necessarily as stable, with a higher degree of variability of evaluations needed across the 50 runs despite the lower mean value.

Crossover	M_p	M_t	P_m	S_e	P_s	P_e	M_l	M_r
100%	0.04%	2.6-5.2%	64	8.2	24	62	0.3%	10%
85%	0.04%	2.6-5.2%	72	8.8	24	70	0.3%	10%
70%	0.04%	2.6-5.2%	74	9.9	28	80	0.3%	10%
50%	0.04%	2.6-5.2%	92*	11.8	36	>100	0.3%	10%
10%	0.04%	2.6-5.2%	>100	16.2	92	≫100	0.3%	10%

Table 1. Feature variation with varying crossover rates

These results are consistent with those seen on 'easy' variants of the ADDMP, whilst 'harder' variants show that the troughs in both convergence time and error plots become much tighter. Further, on 'harder' problems, the improved performance in error profile at low mutation rates with high populations does not manage to achieve such consistently good performance as the 'optimal' mutation trough.

These experiments were then repeated with EAs which were given lower rates of crossover probability. Specific runs were performed at 85%, 70%, 50% and 10% probability of crossover, followed by mutation at the specified rate. Figure 3 shows the performance profile for our EA with only 10% probability of crossover. As would be expected, the slope relating convergence time to population size is now steeper, as crossover is performed 10 times less frequently. The bimodal feature now extends for the fully visualised range of population sizes, showing as a complete ridge with the 'optimal' mutation trough again present for the whole range of population sizes. Importantly, the value of the 'locally worst' mutation rate has not been changed by the reduced level of crossover, nor has the 'optimum' mutation rate at which convergence time is quickest. Figure 4 shows the error surface for the EA with 10% crossover probability, which clearly shows the continued existence of the 'optimum' mutation trough but, as one would expect, the reduced levels of crossover have prevented the EA from finding the global optimum even at population sizes of 100. The mutation rates defining the 'optimal' performance trough remain unchanged.

From these figures we define eight key values of mutation rate and population size. The values of these performance landscape parameters are then given in table 1 for the experiments with other mutation rates. M_p is defined to be the 'locally worst' mutation value of the first bimodal peak seen on the convergence plots. M_t is defined to be the 'optimal' mutation value of the bottom of the trough on the convergence plots. P_m is defined to be the population size along the first bimodal peak ridge at which the number of evaluations utilised is at a minimum. S_e is defined as the slope of the linear feature relating population size to evaluations exploited. P_s is defined to be the population size at low mutation rates at which the EA is first able to find the global optimum in at least 1 of the 50 runs taken from the error surface plot. P_e is the population size, at low mutation rates, at which the EA first succeeds in finding the global optimum at least 45 times out of 50. M_l is the lower bound mutation rate at which, given moderate population sizes, the EA first manages to find the global optimum 50 times out of 50. M_r is the upper bound mutation rate at which the EA manages to find the global optimum 50 times out of 50.

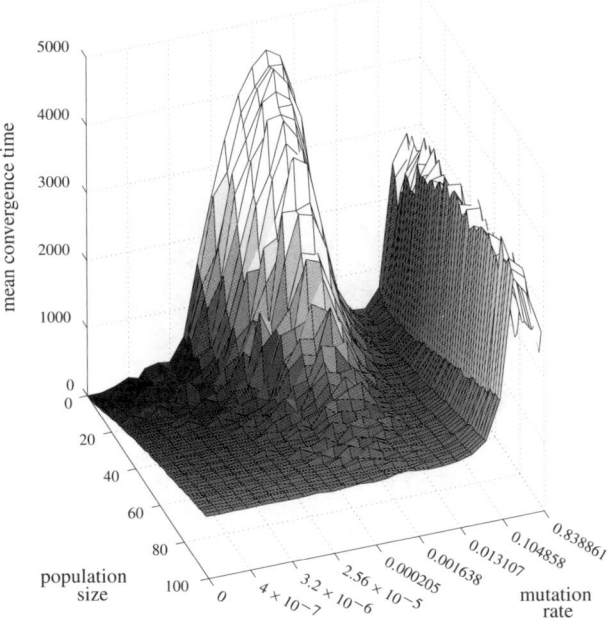

Figure 1. Mean convergence time for one-max problem with 100% crossover

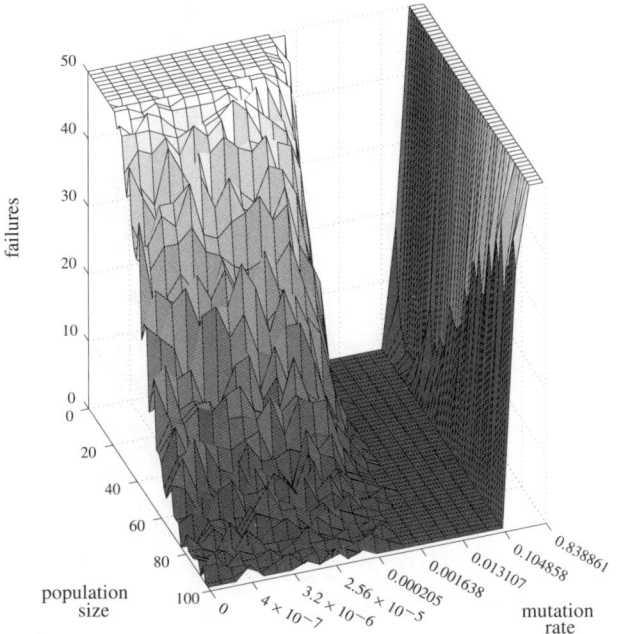

Figure 2. Error surface for one-max problem with 100% crossover

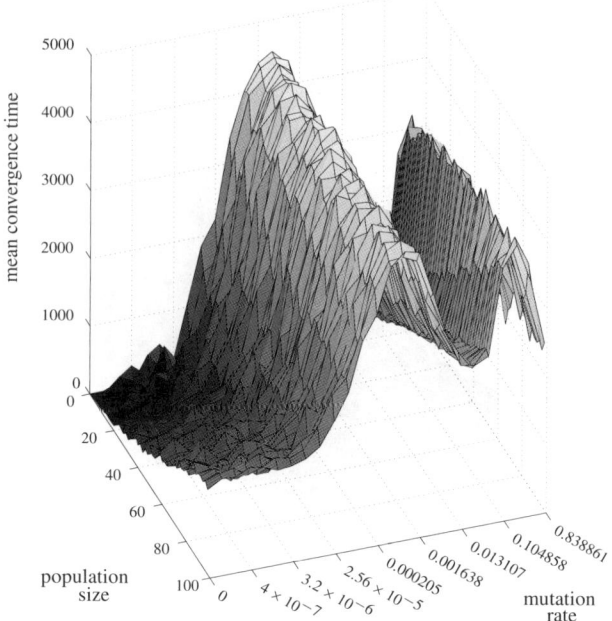

Figure 3. Mean convergence time for one-max problem with 10% crossover

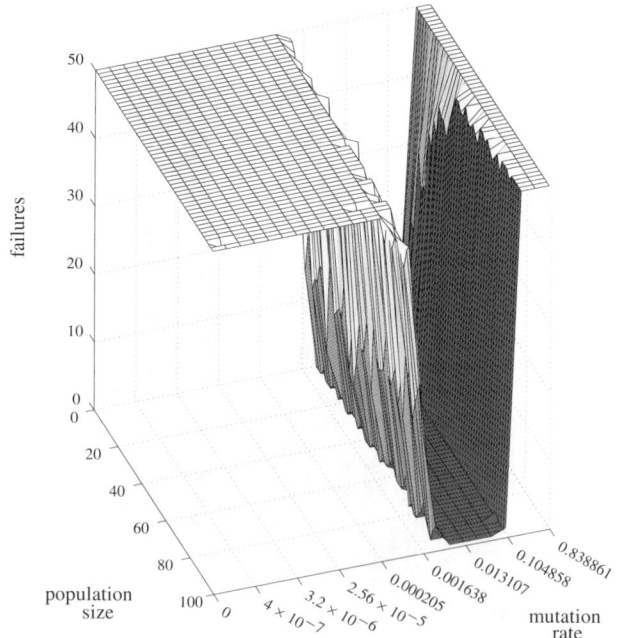

Figure 4. Error surface for one-max problem with 10% crossover

From this table it is clear that the different rates of crossover examined have no significant affect on any mutation-related feature. The slope of the linear feature (S_e) at very low mutation rates, clearly increases with reduced rates of crossover, as do the points on the error surface plot giving the population size at which the EA first finds the global optimum and the point where it finds it 90% of the time. Given that all three of these features are related to the crossover's ability to process schemata in the absence of any significant mutation, this is perhaps not at all surprising.

Figure 5 shows the relative slopes of the linear feature at 0% mutation, relating convergence time to population size at different crossover rates. Here it can clearly be seen that reduced levels of crossover increase the slope of the feature. However, it can also be seen that lower rates of crossover make the steeper plot subject to more noise. It is also noticeable that the projected slope of the plot for 10% mutation would meet the population size axis at a value considerably larger than the other two plots. This demonstrates a 'dead' region, where with too small a population size and too low a rate of crossover, the evolutionary search stalls immediately with insufficient diversity to make any progress beyond the members of the initial population. Taken to an extreme, with 0% crossover (and 0% mutation), this 'dead' region would dominate the plot as the search would be entirely restricted to the precise members of the initial population. Indeed, further experiments at lower rates of crossover (3% and 1%) indicate that the linearity of this feature becomes ever more erratic and the slope begins to reduce again. Arguably, with no mutation, one might expect a direct relation between the relative slopes of the linear features and the percentage of time crossover is allowed to operate, (i.e., 1/10 the crossover rate leading to 10 times steeper slope). However, this does not seem to be the case and is probably due to the fact that at low crossover probabilities, there is a high probability that reproduction will simply copy an unchanged member of the breeding pool. Given the 50% elitist strategy, this will rapidly cause premature convergence in the absence of any mutation.

Figure 6 shows the relative plots of mean convergence time against population size for the 'locally worst' mutation rate (0.04%), i.e., the ridge feature, for experiments with 100%, 50% and 10% probability of crossover. Clearly, the plot for 100% crossover can be seen to fall rapidly reaching a minimum at a population size of 64 from which the curve begins to rise again in line with the background linear feature seen in figure 1. At 50% probability of crossover, it can be seen that the rate of descent is much slower reaching a plot minimum at a population size of 92, however experiments to extend this feature reveal that this minimum is actually more like 110 before the curve begins to rise again. At only 10% probability of crossover the rate of descent is seen to be significantly slower, with only a minor reduction being evident over the population size range shown. Indeed the effect of change of crossover rate on the maximum slopes of these curves is much more in direct relation to the changes in crossover rate (i.e., a factor of 2 and 10 respectively), reflecting the fact that mutation is still an active force here.

Figure 7 shows the relative plots of mean convergence time against population size around the 'optimal' mutation value of 2.6%, for experiments with crossover rates of 100%, 50%, and 10%. At only 10% crossover the plot has a high slope value

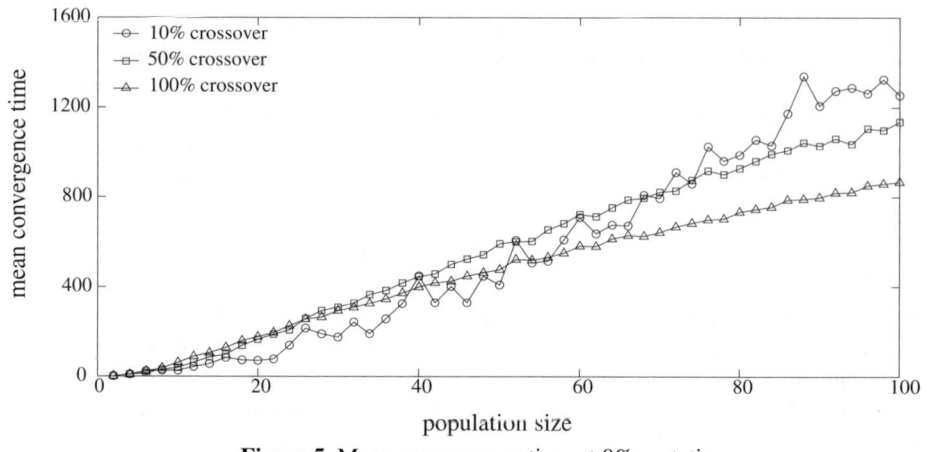

Figure 5. Mean convergence time at 0% mutation

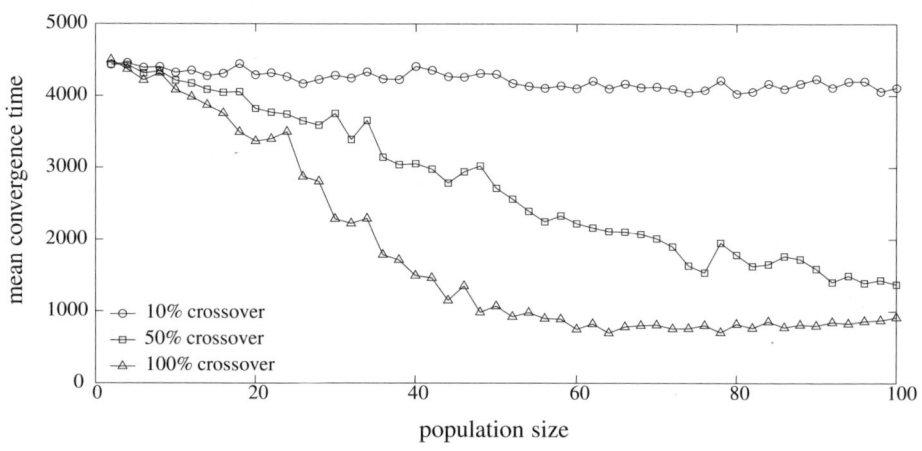

Figure 6. Mean convergence time at 'locally worst' rate of mutation

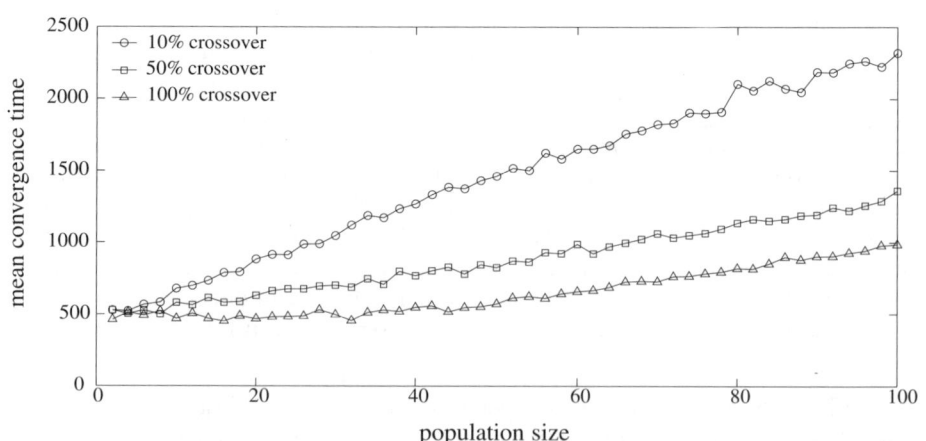

Figure 7. Mean convergence time at 'optimum' rate of mutation

(higher than that at 0% mutation) and has no significant 'dead' region close to the axes. However, as the probability of crossover is increased, the relative slopes of this feature reduce and a significant 'dead' region is introduced in which population size (and hence generation count) has no influence on the number of evaluations exploited. The growth of this 'dead' region of population independence with increased crossover is particularly interesting as it demonstrates the advantages of an EA employing both crossover and mutation, over a mutation-only process. It is important to bear in mind that these plots of convergence time are all within the trough of zero failures on the error surface and hence all runs are finding the global optimum with a 50 out of 50 run success rate. Interestingly, the relative slopes of the 50% and 100% crossover runs are not significantly different, however, the extent of the 'dead' region is what distinguishes the two plots. This low dependence on population size is in line with recent studies from the Santa Fe Institute from work by van Nimwegen and Crutchfield [17,16] who have recently proposed theories on population size dependency and independency on mutation-only searches when looking at different instances of the tunable royal staircase problem.

In an attempt to see how stable and repeatable the performance of the EA was, for all points on the convergence time plot of figure 1 (showing the mean of 50 runs), a plot of the coefficient of variation was also plotted (see figure 8). This shows the standard deviation of the 50 runs at each point, divided by the mean. This gives an indication as to the degree of variability in the convergence time required at different mutation rates and population sizes. Here, it can clearly be seen that the regions of lowest variation (less than 10%) are to be found at high population sizes and low mutation rates. This is perhaps not surprising as this is the region of EA operating with least dependence on purely random processes (although it is accepted that crossover itself is still dependent on random values). However, it must be borne in mind that this region of low variation at high population size is also in a region of relatively high mean convergence time (see figure 1). Given that the variation appears to have 'bottomed out' at a population size of around 80, perhaps this suggests an ideal crossover based region in which to operate, as evaluations required for convergence is lower here than at higher population sizes. It must be remembered, however, that from figure 2, we are still in a region in which the EA occasionally fails to find the global optimum. It is also worth pointing out that for any given moderately sized population, the variation is seen to have a locally worst rate of variance at mid-range mutation rates. This further emphasises the concept of two distinctly different modes of EA operation as mentioned earlier. These being either recombination mode at high population sizes and very low mutation rates, or hill-climber mode utilising significant mutation (circa $1/L$) with moderate population size – although as has been shown, crossover can still contribute to improve performance in this region.

What is of significant interest is the low trough in variation at lower population sizes and moderate rates of mutation (between 2.6% and 10% mutation). Above a population size of 40 this variation very rarely exceeds 20%, and above 60 members is nearer 10%. This trough coincides with our troughs of both minimum evaluations needed to converge, and zero failure rate, indicating a region which consistently

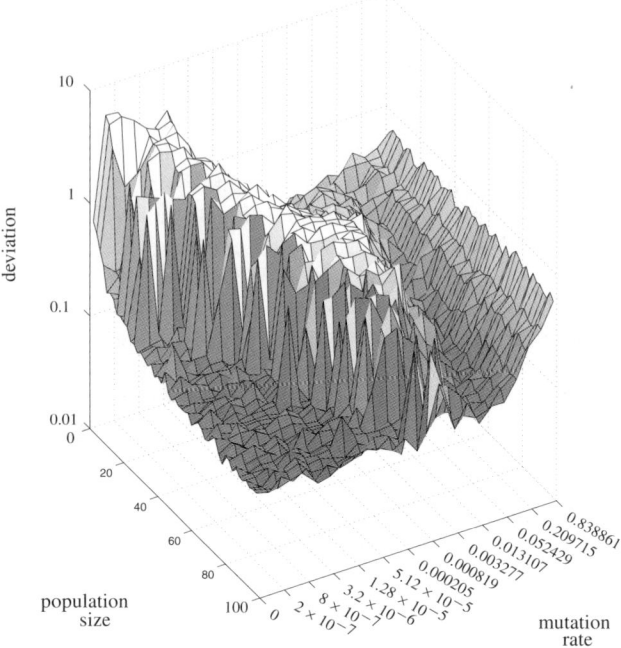

Figure 8. Convergence variation for one-max

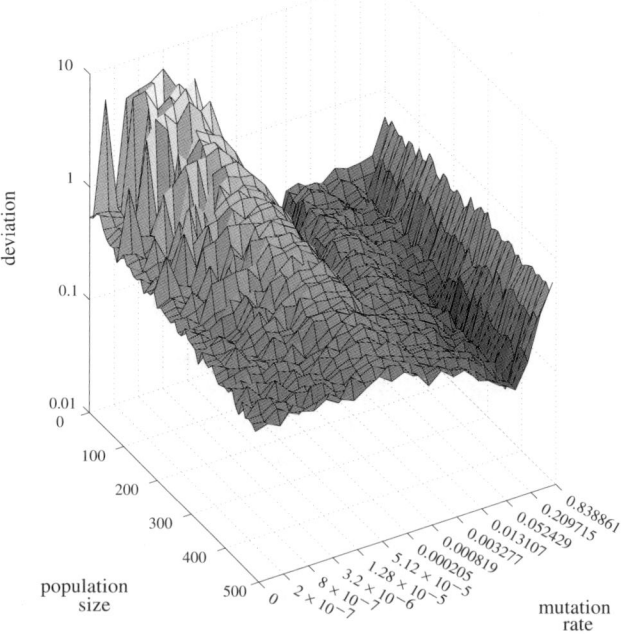

Figure 9. Convergence variation for royal staircase

gives good performance, in a minimum number of evaluations with a minimum variance in run time – surely an ideally tuned EA!

To further emphasise this point, another plot of the coefficient of variation is shown in figure 9, taken from an ongoing study into performance on tunable instances of the royal staircase problem described earlier. Even in its simplest form (grouping factor of only 1), the problem can be shown to be 'harder' for EA than one-max [17]. For this experiment, the EA was allowed up to 20 000 evaluations and population sizes of 10 to 500 in steps of 10 were investigated. Figure 9 shows, in comparison to figure 8, that although the variation in convergence time is still reducing with increased population size, at low mutation rates this has still not dropped below 10% variation by population sizes of 500, being near a typical value of 15%. In contrast, the 'optimal' mutation trough has achieved less than 10% variation by a population size of only 450, albeit only in a very tight mutation range. However, the data implies that this mutation rate range is increasing with increased population size and this will be investigated in due course. The dip in figure 9 at medium-range mutation rates with very low population sizes corresponds to the peak of the 'locally worst' mutation rate. Here, the variation is artificially depressed as the mean convergence time approaches the limit of 20 000 evaluations, hence limiting the scope for variation.

Finally, it is worth pointing out the difference in mean convergence time at typical points representing our two different modes of EA operation. At the 'optimal' mutation trough with variation of around 15% on figure 9 (population size of 140, mutation rate of 5.2%) the mean is only 6510 evaluations compared to a typical value of 12 000 evaluations at 15% variation with a population size of 500 and 0% mutation. Once again, an EA with the correct mutation rate is seen to outperform a crossover dependent EA with an arbitrarily low choice of mutation rate.

4 Conclusions

These results continue to clearly demonstrate the existence of the linear and bimodal features in the tunable performance profile of evolutionary search on two further problems. They add weight to the hypothesis that these features are generic to the characterisation of the process.

The rate of crossover is seen to have little effect on the rates of mutation at which certain features occur, but does affect the slope of the linear feature linking convergence time to population size at very low mutation rates. However this is not in itself a simple proportional relationship. Rates of crossover are also seen to affect the 'depth of penetration' of the bimodal feature as population size is increased. The rate at which this effect is changed, however, does seem more in line with the relative level of crossover. Crossover rates are also seen to have a significant effect on the ability of the EAs to find the global optimum at low mutation rates, but this is not at all surprising.

The population size independent region at optimum mutation (the 'dead' region) is also seen to be significantly extended by increased crossover and probably represents a region of operation for the EA in which it can be said to be optimally tuned – particularly on this problem.

Not surprisingly, the EA shows increased stability of run-time with increased population size, but moderate amounts of mutation can cause this to deteriorate. Importantly, a critical mutation rate seems to exist that does give minimal variation, minimal convergence time, and minimum error, which if predictable could have important implications for EA parameter value determination.

Further work is under way exploring other problems and representations to establish whether these results are truly generically applicable to evolutionary search.

References

1. T. Bäck. *Evolutionary Algorithms in Theory and Practice*. Oxford University Press, 1996.
2. K. Deb and S. Agrawal. Understanding Interactions Among Genetic Algorithm Parameters. In W. Banzhaf and C. Reeves, editors, *Foundations of Genetic Algorithms 5*, pages 265–286. Morgan Kaufmann, San Francisco, 1998.
3. D. Goldberg. *Genetic Algorithms in Search Optimisation and Machine Learning*. Addison–Wesley, Reading, MA, 1989.
4. J. H. Holland. *Adaptation in Natural and Artificial Systems*. University of Michigan Press, Ann Arbor, MI, 1975.
5. Z. Michalewicz. *Genetic Algorithms + Data Structures = Evolution Programs*. Springer-Verlag, Berlin Heidelberg New York, 3rd edition, 1996.
6. H. Mühlenbein. How Genetic Algorithms Really Work: I. Mutation and Hillclimbing. In R. Manner and B. Manderick, editors, *Proceedings of the 2nd Int'l Conf. on Parallel Problem Solving from Nature*, pages 15–25. Elsevier, Amsterdam, 1992.
7. H. Mühlenbein and D. Schlierkamp-Voosen. The Science of Breeding and its application to the Breeder Genetic Algorithm. *Evolutionary Computation*, 2(3):335–360, 1994.
8. M. Oates. Autonomous Management of Distributed Information Systems using Evolutionary Computing Techniques. In *Computing Anticipatory Systems*, pages 269–281, 1998.
9. M. Oates and D. Corne. Investigating Evolutionary Approaches to Adaptive Database Management Against Various Quality of Service Metrics. In *Proceedings of the 5th Int'l Conf. on Parallel Problem Solving from Nature*, LNCS 1498, pages 775–784. Springer-Verlag, Berlin Heidelberg New York, 1998.
10. M. Oates and D. Corne. QoS-based GA Parameter Selection for Autonomously Managed Distributed Information Systems. In *Proceedings of the 1998 European Conference on Artificial Intelligence*, pages 670–674. IEEE Press, 1998.
11. M. Oates, D. Corne, and R. Loader. Investigating Evolutionary Approaches for Self-Adaptation in Large Distributed Databases. In *Proceedings of the 1998 IEEE Int'l Congress on Evolutionary Computation*, pages 452–457. IEEE Press, New York, 1998.
12. M. Oates, D. Corne, and R. Loader. Investigation of a Characteristic Bimodal Convergence-time/Mutation-rate Feature in Evolutionary Search. In *Proceedings of*

the 1999 IEEE Int'l Congress on Evolutionary Computation, volume 3, pages 2175–2182. IEEE Press, New York, 1998.
13. M. Oates, D. Corne, and R. Loader. Skewed Crossover and the Dynamic Distributed Database Problem. In *Artificial Neural Nets and Genetic Algorithms*, pages 280–287. Springer-Verlag, Vienna New York, 1999.
14. M. Oates, D. Corne, and R. Loader. Variation in Evolutionary Algorithm Performance Characteristics on the Adaptive Distributed Database Management Problem. In *Proceedings of the Genetic and Evolutionary Computation Conference, GECCO 1999*, pages 480–487. Morgan Kaufmann, San Francisco, 1999.
15. G. Syswerda. Uniform Crossover in Genetic Algorithms. In *Proceedings of the Third International Conference on Genetic Algorithms*, pages 2–9. Morgan Kaufmann, San Francisco, 1989.
16. E. van Nimwegen and J. Crutchfield. Optimizing Epochal Evolutionary Search: Population-Size Dependent Theory. Technical Report 98-10-090, Santa Fe Institute, 1998.
17. E. Van Nimwegen and J. Crutchfield. Optimizing Epochal Evolutionary Search: Population-Size Independent Theory. Technical Report 98-06-046, Santa Fe Institute, 1998.

Evolution Strategies in Noisy Environments – A Survey of Existing Work

D. V. Arnold

Department of Computing Science XI
University of Dortmund
44221 Dortmund, Germany
E-mail: *arnold@ls11.cs.uni-dortmund.de*

Abstract. Noise is a factor present in almost all real-world optimization problems. While it can potentially improve convergence reliability in multimodal optimization by preventing convergence towards merely local optima, it is generally detrimental to the velocity with which an optimum is approached. Evolution strategies (ES) form a class of evolutionary optimization procedures that are believed to be able to cope quite well with noise. A number of theoretical results as well as empirical findings regarding the influence of noise on the performance of ES can be found in the literature. The purpose of this survey is to summarize what is known regarding the behavior of ES in noisy environments and to outline directions for future research.

Keywords
Evolution strategy, fitness noise, progress rate

1 Introduction

Noise in optimization problems can stem from sources as different as possible systematic errors of measurement in experiments, numerical inaccuracies in computations, stochastic sampling or simulation procedures, or from interaction with users, to name but a few. It is a well-established fact that, usually, reliability and efficiency are two conflicting goals in optimization, and that while the presence of noise has the potential to improve reliability by preventing the optimization process from getting stuck in a merely local optimum, it has generally negative effects with regard to the efficiency of the process. In particular, reduced convergence velocity and the occurrence of a residual optimum location error are two consequences that have been observed in many experiments. Altogether, the effects of noise on optimization algorithms are worth further investigation.

Evolutionary algorithms (EA) in general, and therefore evolution strategies (ES) in particular, are optimization procedures that are generally believed to work well in noisy environments. This hope is fostered by the observation that in many applications EA do seem to be able to cope with noise and yield satisfactory results. However, there are hardly more than a handful of studies dealing explicitly with the effects of noise on the convergence behavior of EA, most of those that do exist are empirical, and it is questionable whether their results can easily be generalized. An improved theoretical understanding of the effects of noise on evolutionary processes is hoped to pave the road both to improved algorithms and to additional insight regarding optimal parameter settings.

Theoretical research into ES emphasizes a view of evolutionary optimization procedures as dynamical systems and focuses on the computation of local performance measures to arrive at conclusions regarding the convergence behavior in simple settings. As of now, theoretical results regarding the potential increase in convergence reliability do not exist. This survey summarizes the state of knowledge regarding the loss in performance that comes along as a consequence of noise and points out the remaining holes. Section 2 provides a short introduction to ES and their analysis in general. Section 3 deals with ES in noisy environments. Section 4 concludes with a brief summary and directions for future research.

2 Analysis of the $(\mu/\rho \overset{+}{,} \lambda)$-ES

This section introduces $(\mu/\rho \overset{+}{,} \lambda)$-ES and the ideas commonly applied to their analysis. Note, however, that the mechanism for the self-adaptation of strategy parameters is left out here despite its relevance for practical applications. For an introduction to self-adaptation and a more extensive introduction to $(\mu/\rho \overset{+}{,} \lambda)$-ES see [7] and the references therein.

2.1 The $(\mu/\rho \overset{+}{,} \lambda)$-ES

The $(\mu/\rho \overset{+}{,} \lambda)$-ES is a biologically inspired algorithm for the optimization of arbitrary fitness functions. It is commonly employed in situations where little is known about the fitness function, where the fitness function has characteristics that make the use of other, more traditional optimization algorithms fruitless, or where factors such as a dynamically changing fitness landscape or noise complicate the optimization process.

The $(\mu/\rho \overset{+}{,} \lambda)$-ES uses a population of μ individuals that move towards regions of increasingly higher fitness in search space as time progresses. In every generation, λ offspring individuals are created by means of recombination and mutation. Using (,) selection the population of the next generation is composed of those μ individuals from the pool of offspring with the highest fitness, and using the (+) variant the μ fittest individuals from the union of the offspring pool and the parent population are selected.

The offspring individuals are created independently from each other. For the creation of a new individual, ρ parents are selected randomly from the population. The genetic information of these parents is combined by either intermediate or by dominant recombination. Using intermediate recombination the genetic information of the parents is averaged, requiring the computation of the 'center of mass' of the parent individuals. Using dominant recombination involves random sampling from the parent individuals for each component of the offspring individual to be generated. Subsequently, the result of recombination is subject to mutation. For this purpose, normally distributed disturbances with mean zero and standard deviation σ are added to its components. In what follows, σ is referred to as mutation strength. The determination of optimal settings for σ is one of the central goals of ES theory. In practical applications correlated mutations can sometimes improve performance of ES, but as more strategy parameters have to be introduced this complicates the issue of mutation strength control and usually precludes theoretical analyses.

In practical applications, a nearly optimal mutation strength can be obtained and sustained by means of self-adaptation. For that purpose, the mutation strength is made part of the genetic information of the individuals and subjected to recombination, mutation, and selection along with the object parameters. As self-adaptation renders the task of analyzing ES considerably more complex and is theoretically understood only in the most simple cases, it will not be considered here.

2.2 Performance Analyses of $(\mu/\rho \, {}^+_, \lambda)$-ES

Theoretical performance analyses of ES usually employ very simple fitness models so as to make it possible to obtain analytical results. The most commonly used fitness model assumes a spherically symmetric objective function

$$f(y) = g(\|y - \hat{y}\|), \quad y \in \mathbb{R}^n \tag{1}$$

where g is a real-valued, strictly increasing function of a single parameter. Usually, the parameter space dimension n is assumed to be large, and in general the results of the analytical investigations reported below are approximations that are exact in the limit $n \to \infty$. Note that the goal here is minimization, and that therefore fitness is inversely related to objective function values. An overview of other fitness models can be found in [7].

Most commonly, the performance of ES is quantified by computing the expected change in distance to the optimum $\hat{y}$ from one generation to another, that is, the progress rate

$$\varphi = \mathrm{E}\left[r^{(t)} - r^{(t+1)}\right] \tag{2}$$

where $r^{(t)}$ is the average distance of the population to the optimum at time t. A positive progress rate indicates progress towards the optimum. Alternatively, sometimes progress is not measured in object variable space, but in the space of objective function values. The corresponding performance measure, the quality gain, is the expected change in fitness from one generation to another.

Note that progress rate as well as quality gain are local progress measures describing microscopic aspects of the evolution. However, in some highly symmetric cases such as a spherical fitness landscape or an infinite ridge, these local measures can be sufficient for capturing the global dynamics of ES. In particular, for the spherical fitness model, introducing normalized mutation strength $\sigma^* = \sigma n/r$ and computing the normalized progress rate $\varphi^* = \varphi n/r$ as a function of σ^*, the dynamics of the optimization process are described by

$$r^{(t+1)} = r^{(t)} \left[1 - \frac{\varphi^*(\sigma^*)}{n} \right] \qquad (3)$$

ascertaining linear convergence order if the mutation strength is appropriately adjusted.

3 ES in Noisy Environments

Fitness noise in investigations of the performance of ES is commonly modeled by an additive, bias-free, normally distributed noise term. That is, it is assumed that selection is based on information provided by the noisy objective function

$$f_{\sigma_\epsilon}(\mathbf{y}) = f(\mathbf{y}) + \sigma_\epsilon \mathcal{N} \qquad (4)$$

where $\mathcal{N}$ is a standard normally distributed random variable. If the noise stems from a variety of independent sources, a justification for this approach can be found in the central limit theorem of probability theory. As a consequence of the presence of noise it is possible that inferior individuals are selected to become part of the population while superior ones die out. If the standard deviation of fitness values in the offspring pool is small compared to σ_ϵ the population performs a random walk in parameter space. Noise strength σ_ϵ may be constant throughout the search space or vary relative to $f(\mathbf{y})$.

For the spherical fitness model introduced above, it is useful to introduce normalized noise strength

$$\sigma_\epsilon^* = \sigma_\epsilon \frac{n}{rf'} \qquad (5)$$

where $f' = df/dr$. To dynamically investigate the effect of noise on convergence velocity, most often constant normalized noise strength σ_ϵ^* is assumed. A residual optimum location error occurs if the noise strength σ_ϵ does not vanish for $r \to 0$. Its amount can be computed by equating the progress rate with zero and solving for r.

An important question a theoretical investigation of evolutionary optimization in noisy environments can help to answer is whether resampling or increasing the population size or the number of offspring yields superior results. This question has long been discussed in the realm of genetic algorithms (see for example Fitzpatrick and Grefenstette [8]). Sampling the objective function k times at one location effectively reduces the noise strength by a factor of $\sqrt{k}$. Knowing the effects of population sizing and of noise strength on the performance of ES, explicit recommendations can be given that hopefully generalize beyond the simple fitness models used in the analyses. The following sections summarize known results regarding the behavior of selected ES in noisy environments.

3.1 The (1 + 1)-ES

In (1 + 1)-ES the single individual forming the population generates one offspring individual at a time which replaces the parent if and only if it has a higher fitness. The progress rate formula

$$\varphi^*_{1+1}(\sigma^*, \sigma^*_\epsilon) = \frac{\sigma^*}{\sqrt{2\pi}} \frac{1}{\sqrt{1+2\varsigma^2}} \exp\left(-\frac{1}{2}\left(\frac{\sigma^*/2}{\sqrt{1+2\varsigma^2}}\right)^2\right) \\ - \frac{\sigma^{*2}}{2}\left[1 - \Phi\left(\frac{\sigma^*/2}{\sqrt{1+2\varsigma^2}}\right)\right] \tag{6}$$

where $\varsigma = \sigma^*_\epsilon/\sigma^*$ denotes the 'noise to signal ratio', has been derived by Beyer [1] and is valid for sphere functions with $n \gg 1$ if the parent is reevaluated in every generation. The term in square brackets in (6) is the success probability $P_{success}$, i.e., the probability that the offspring is selected to replace its parent.

Figure 1 illustrates the local performance of the (1 + 1)-ES for spherical fitness functions of infinite parameter space dimension and is obtained through numerical evaluation of (6). Part (a) of the figure shows the progress rate φ^*_{1+1} as a function of mutation strength σ^* for several values of the noise strength σ^*_ϵ, revealing the existence of a noise strength-dependent mutation strength $\hat{\sigma}^*_{1+1}$ for which the progress rate attains a maximum $\hat{\varphi}^*_{1+1}$. Parts (b) through (d) of the figure display the dependence of optimal mutation strength $\hat{\sigma}^*_{1+1}$, maximal progress rate $\hat{\varphi}^*_{1+1}$, and the success probability $\hat{P}_{success}$ in case of maximal progress as functions of noise strength σ^*_ϵ, respectively. The optimal mutation strength $\hat{\sigma}^*_{1+1}$ increases with increasing noise strength from its value of 1.224 for the noise-free case. Note, however, that this is so only in the case of infinite parameter space dimension. For finite n there exists a noise strength above which no positive progress gain can be achieved with any mutation strength and consequently it is best to avoid being driven away from the optimum by using zero mutation strength. Progress rate and success probability in case of optimally adjusted mutation strength decrease monotonously from their values of 0.202 and 0.270 for the noise-free case to zero as the noise strength increases. It is expected that this behavior is qualitatively the same for the finite but sufficiently high n. The relatively weak dependence of $\hat{P}_{success}$ on σ^*_ϵ for not too high noise strength is exploited in success probability based mutation strength adaptation schemes such as Rechenberg's 1/5-success rule.

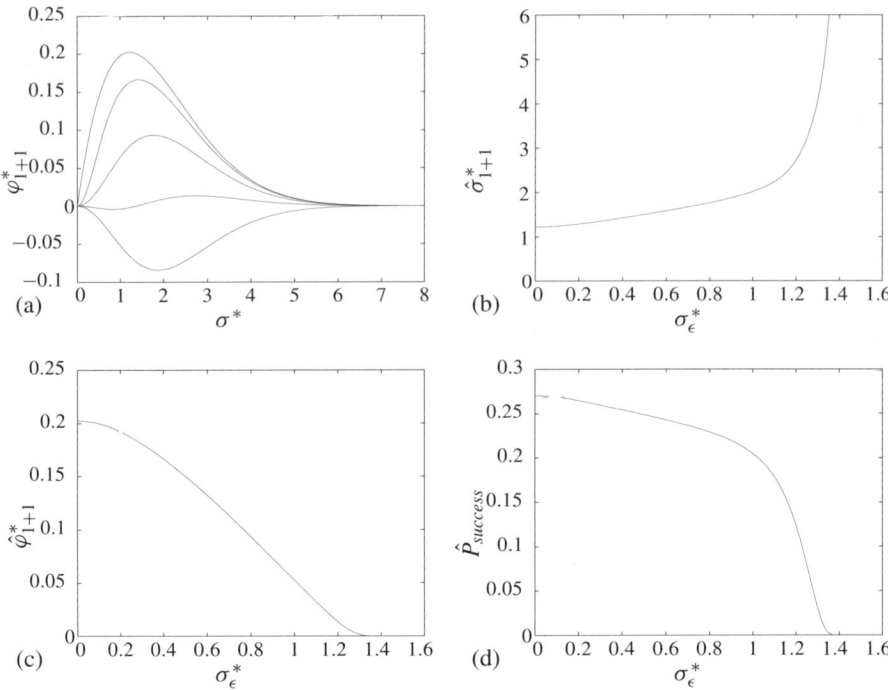

Figure 1. Local performance of the (1+1)-ES for sphere functions with $n \gg 1$. (a) Progress rate φ^*_{1+1} as a function of mutation strength σ^* for, from top to bottom, noise levels $\sigma^*_\epsilon = 0.0, 0.4, 0.8, 1.2,$ and 1.6. (b) Optimally adjusted mutation strength $\hat{\sigma}^*_{1+1}$ as a function of noise strength σ^*_ϵ. (c) Progress rate $\hat{\varphi}^*_{1+1}$ in case of optimally adjusted mutation strength as a function of noise strength σ^*_ϵ. (d) Success probability $\hat{P}_{success}$ in case of optimally adjusted mutation strength as a function of noise strength σ^*_ϵ.

3.2 The $(1, \lambda)$-ES

$(1, \lambda)$-ES differ from $(1 + 1)$-ES in that λ offspring individuals are generated at a time – naturally allowing for parallel processing – the best of which replaces the parent even if it is inferior. The progress rate formula

$$\varphi^*_{1,\lambda}(\sigma^*, \sigma^*_\epsilon) = \frac{c_{1,\lambda}\sigma^*}{\sqrt{1+\varsigma^2}} - \frac{\sigma^{*2}}{2} \tag{7}$$

has been given both by Beyer [1] and by Rechenberg [11] and is valid for sphere functions with $n \gg 1$. The progress coefficient $c_{1,\lambda}$ captures the influence of selection and can be obtained through numerical integration. For large λ it has been shown that

$$c_{1,\lambda} = \mathcal{O}\left(\sqrt{\log \lambda}\right) \tag{8}$$

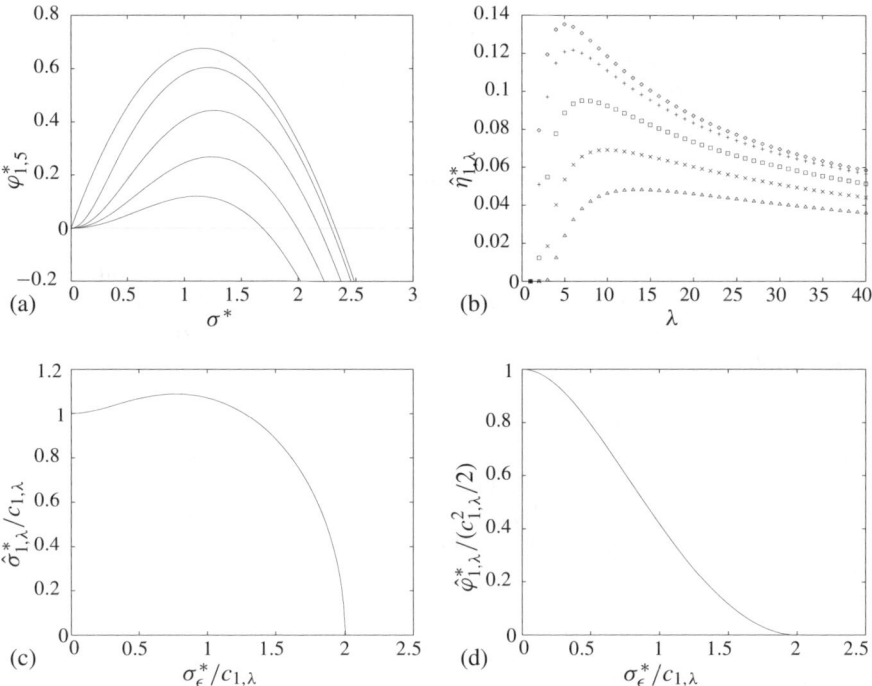

Figure 2. Local performance of the $(1, \lambda)$-ES for sphere functions with $n \gg 1$. (a) Progress rate $\varphi_{1,5}^*$ of the $(1, 5)$-ES as a function of mutation strength σ^* for, from top to bottom, noise levels $\sigma_\epsilon^* = 0.0, 0.4, 0.8, 1.2,$ and 1.6. (b) The optimal fitness efficiency $\hat\eta_{1,\lambda}^* = \hat\varphi_{1,\lambda}^*/\lambda$ as a function of λ for, from top to bottom, noise strength $\sigma_\epsilon^* = 0.0, 0.4, 0.8, 1.2,$ and 1.6. (c) Dependence of optimal mutation strength $\hat\sigma_{1,\lambda}^*$ on noise strength σ_ϵ^*. Note the scaling of the axes. (d) Dependence of the progress rate in case of optimally adjusted mutation strength $\hat\varphi_{1,\lambda}^*$ on noise strength σ_ϵ^*. Note the scaling of the axes.

and that therefore the gain that can be expected from increasing the number of offspring does not scale very favorably with λ.

It is well known that in the absence of noise optimal progress $\hat\varphi_{1,\lambda}^* = c_{1,\lambda}^2/2$ is achieved for mutation strength $\hat\sigma_{1,\lambda}^* = c_{1,\lambda}$, and that the $(1, 5)$-ES is the most fitness efficient $(1, \lambda)$-ES in that it generates the largest progress per offspring. Furthermore, from (8) it is clear that the speedup that can be achieved by increasing the number of offspring is merely logarithmic in λ.

Figure 2 illustrates the local performance of the $(1, \lambda)$-ES for spherical fitness functions of infinite parameter space dimension and is obtained through numerical evaluation of (7). Part (a) of the figure shows the progress rate $\varphi_{1,5}^*$ as a function of mutation strength σ^* for several values of the noise strength σ_ϵ^*. In contrast to the $(1 + 1)$-ES the progress rate approaches negative infinity for $\sigma^* \to \infty$. Part (b) reveals that the $(1, 5)$-ES is not necessarily the most efficient strategy in noisy environments, and that the optimal number of offspring increases with increasing noise strength. In addition, it can be seen that for high noise strength the efficiency decreases only slowly with increasing number of offspring, making the choice of λ a rather uncritical one provided that it is large enough.

As shown by Rechenberg [11], a careful analysis of (7) reveals that, normalizing noise strength and optimal mutation strength by dividing by the value of optimal mutation strength in the absence of noise, the optimal mutation strength $\hat{\sigma}_{1,\lambda}^*$ depends on the noise strength σ_ϵ^* in a manner that is independent of λ. Part (c) of figure 2 shows that dependence, illustrating that the optimal mutation strength initially increases with increasing noise strength so as to increase the signal strength to make it discernible from the noise, and then rapidly decreases to zero to avoid being driven away from the optimum.

Similarly, normalizing the progress rate in case of optimally adjusted mutation strength by dividing by its value in the absence of noise, $\hat{\varphi}_{1,\lambda}^*$ depends on the appropriately normalized noise strength in a manner that is independent of λ as well. Part (d) of figure 2 shows that dependency being qualitatively similar to the corresponding dependency for $(1+1)$-ES. For $\sigma_\epsilon^* \geq 2c_{1,\lambda}$ no positive progress rate is possible.

3.3 Rescaled Mutations

Following a suggestion by Ostermeier [11] the convergence properties of ES on a spherical fitness model in the presence of fitness noise can be improved by the method of rescaled mutations. To rescale mutations, an additional strategy parameter κ is introduced. The decision whether an offspring individual survives is not made on the basis of the individual's fitness, but on the basis of the fitness of the virtual individual resulting from a κ-fold increase of mutation strength. That is, if the mutation vector of the individual is z, the virtual individual has mutation vector κz. For large values of κ the large virtual mutations have the effect of increasing the 'signal' strength much more than the effective noise strength. Consequently, the ability to make good selections is much improved.

The method has been analyzed theoretically only for $(1, \lambda)$-ES on sphere functions with $n \gg 1$. The progress rate formula

$$\varphi_{1,\lambda}^*(\kappa, \sigma^*, \sigma_\epsilon^*) = \frac{c_{1,\lambda}\sigma^*}{\sqrt{1+(\zeta/\kappa)^2}} - \frac{\sigma^{*2}}{2} \tag{9}$$

has been found by both Rechenberg [11] and Beyer [5]. Evidently, rescaling mutations effectively reduces the noise strength by a factor κ. Infinite κ therefore completely eliminates noise. Note, however, that this is true only in the limit $n \to \infty$, and that in finite dimensional spaces κ cannot be indefinitely increased for optimal performance. A progress rate law for finite n has been presented by Beyer [6].

3.4 The (μ, λ)-ES

(μ, λ)-ES are simple multi-parent strategies without recombination. Offspring is generated by mutation of randomly selected parent individuals. In the absence of noise, Rechenberg [11] contends that the progress rate law for (μ, λ)-ES on sphere functions with $n \gg 1$ reads

$$\varphi^*_{\mu,\lambda}(\sigma^*) = c_{\mu,\lambda}\sigma^* - \frac{\sigma^{*2}}{2} \tag{10}$$

where the progress coefficients $c_{\mu,\lambda}$ can be obtained either by averaging over a large number of ES runs or by an analytical approximation obtained from calculations carried out by Beyer [2]. Comparison of the $c_{\mu,\lambda}$ with the $c_{1,\lambda}$ reveals that for a given λ the strategy with $\mu = 1$ is always the most efficient one. Without noise present, increasing the population size beyond one does not have beneficial effects.

However, Rechenberg [11] notes that this is no longer the case if fitness noise is present. The progress law for (μ, λ)-ES in the presence of noise reads

$$\varphi^*_{\mu,\lambda}(\sigma^*, \sigma^*_\epsilon) = k_{\mu,\lambda}(\sigma^*, \sigma^*_\epsilon)c_{\mu,\lambda}\sigma^* - \frac{\sigma^{*2}}{2} \tag{11}$$

where experimental evidence shows that $k_{\mu,\lambda}$ can far exceed the factor $1/\sqrt{1+\varsigma^2}$ in the progress rate law for $(1, \lambda)$-ES. Currently, there is no satisfactory theoretical explanation for the increase in performance resulting from the increase in population size which would allow for the computation of optimal population size or mutation strength.

3.5 The $(\mu/\mu, \lambda)$-ES

Introducing recombination into ES brings a factor for which Beyer [3] has coined the term "genetic repair" into play. Recombination tends to decrease the harmful parts of mutations while retaining the beneficial ones, thereby making it possible to use much higher mutation strengths. For intermediate recombination with $\rho = \mu$, the progress rate law for sphere functions without noise present as formulated both by Rechenberg [11] and by Beyer [3] reads for $n \to \infty$

$$\varphi^*_{\mu/\mu,\lambda}(\sigma^*) = c_{\mu/\mu,\lambda}\sigma^* - \frac{\sigma^{*2}}{2\mu} \tag{12}$$

where the $(\mu/\mu, \lambda)$ progress coefficients $c_{\mu/\mu,\lambda}$ can be obtained by numerical integration as outlined by Beyer. Beyer also gives the relationship

$$c_{\mu/\mu,\lambda} = \mathcal{O}\left(\sqrt{\log\frac{\lambda}{\mu}}\right) \tag{13}$$

for the asymptotic growth of the progress coefficients. Maximal progress $\hat{\varphi}^*_{\mu/\mu,\lambda} = \mu c^2_{\mu/\mu,\lambda}/2$ is achieved for mutation strength $\hat{\sigma}^* = \mu c_{\mu/\mu,\lambda}$.

While no theoretical results exist for the noisy case, empirical evidence presented by Nissen and Propach [9] suggests that, as for multi-parent ES without recombination, using dominant recombination, progress rates far exceeding those of single-parent strategies can be achieved. Part of this speedup results from the use of higher mutation strengths which lead to a relative increase of signal strength as compared to noise strength. How the performance of the algorithm compares with that of the algorithm without recombination in noisy environments is an open question.

4 Summary and Outlook

To summarize, the effects of fitness noise on ES are quite well understood only for the spherical fitness model and the most simple strategies. In particular, analytical results exist regarding the $(1 + 1)$-ES where the parent is reevaluated every generation and for the $(1, \lambda)$-ES. The method of rescaled mutations has been analyzed only for the $(1, \lambda)$-ES.

The list of things to look into in the future is much longer than this. First of all, it is desirable to extend the theoretical investigations to other forms of ES. The $(1 + 1)$-ES without reevaluating the parent is a natural candidate for future analysis. More importantly, the investigation of the local performance of multi-parent strategies is still pending. It is hoped that a theoretical explanation for the observed performance advantage of multi-parent strategies over single-parent ones yields valuable insight into the working principles of ES. Moreover, variants of the basic strategies may be worth further thought in the presence of noise. For example, while Rudolph [12] has shown that in the absence of noise Cauchy distributed mutations have no performance advantage over normally distributed ones it is not immediately obvious that this holds in noisy environments as well. Second, finding theoretical results for fitness models other than the sphere, such as the infinite ridge [10], which have inherently different characteristics, should prove interesting. Third, of immense practical importance is the investigation of the mechanism of self-adaptation for strategy parameters such as the mutation strength. While some theoretical results exist for the noise-free case, knowledge regarding self-adaptation with noise present is at best rudimentary. Finally, there is the challenge of understanding the effects of noise in multimodal environments where it can ideally serve to improve convergence reliability.

Acknowledgments

Support under grant Be 1578/6-1 by the Deutsche Forschungsgemeinschaft (DFG) is gratefully acknowledged. Thanks also to Leila Kallel for useful comments on the final version of this paper.

References

1. H.-G. Beyer. Toward a theory of evolution strategies: some asymptotical results from the $(1 \stackrel{+}{,} \lambda)$-theory. *Evolutionary Computation*, 1(2):165–188, 1993.
2. H.-G. Beyer. Toward a theory of evolution strategies: the (μ, λ)-theory. *Evolutionary Computation*, 2(4):381–407, 1995.
3. H.-G. Beyer. Toward a theory of evolution strategies: on the benefit of sex – the $(\mu/\mu, \lambda)$-theory. *Evolutionary Computation*, 3(1):81–111, 1995.
4. H.-G. Beyer. *Zur Analyse der Evolutionsstrategien*. Habilitationsschrift, Universität Dortmund, 1996.
5. H.-G. Beyer. Mutate large, but inherit small! On the analysis of rescaled mutations in $(\tilde{1}, \tilde{\lambda})$-ES with noisy fitness data. In A. E. Eiben, T. Bäck, M. Schoenauer, and H.-P. Schwefel, editors, *Parallel Problem Solving from Nature V*, volume 1498 of LNCS, pages 109–118. Springer-Verlag, Berlin Heidelberg New York, 1998.
6. H.-G. Beyer. Evolutionary algorithms in noisy environments: theoretical issues and guidelines for practice. *Computer Methods in Applied Mechanics and Engineering*, 186(2–4):269–294, 2000.
7. H.-G. Beyer and D. V. Arnold. Theory of evolution strategies – a tutorial. This volume.
8. J. M. Fitzpatrick and J. J. Grefenstette. Genetic algorithms in noisy environments. In P. Langley, editor, *Machine Learning*, pages 101–120. Kluwer, Dordrecht, 1998.
9. V. Nissen and J. Propach. Optimization with noisy function evaluations. In A. E. Eiben, T. Bäck, M. Schoenauer, and H.-P. Schwefel, editors, *Parallel Problem Solving from Nature V*, volume 1498 of LNCS, pages 159–168. Springer-Verlag, Berlin Heidelberg New York, 1998.
10. A. I. Oyman, H.-G. Beyer, and H.-P. Schwefel. Where elitists start limping: evolution strategies at ridge functions. In A. E. Eiben, T. Bäck, M. Schoenauer, and H.-P. Schwefel, editors. *Parallel Problem Solving from Nature V*, volume 1498 of LNCS, pages 34–43. Springer-Verlag, Berlin Heidelberg New York, 1998.
11. I. Rechenberg. *Evolutionsstrategie '94*. Frommann-Holzboog, Stuttgart, 1994.
12. G. Rudolph. Local convergence rates of simple evolutionary algorithms with Cauchy mutations. *IEEE Transactions on Evolutionary Computation*, 1(4):249–258, 1997.

Cyclic Attractors and Quasispecies Adaptability

J. E. Rowe

Department of Computer & Information Science
De Montfort University
Milton Keynes MK7 6HP, UK
E-mail: *jrowe@dmu.ac.uk*

Abstract. Using the techniques presented in a previous paper, it is possible to identify cyclic attractors (in the infinite population limit) for the simple genetic algorithm (with zero crossover) when the fitness function is periodic with respect to time. These techniques are applied here to study the adaptability of a quasispecies as the fitness function changes to favour another genotype. The effects of different mutation rates are examined, illustrating the 'error threshold' effect, that too high a mutation rate leads to a general collapse of the quasispecies and an inability to track the optimum. Experiments with very low mutation rates illustrate the inability of the population to evolve away from the initial 'master sequence' (in which the population clusters around a single genotype) in the face of adverse selective pressure. This result is not reflected in theoretical calculations, indicating that there is a *metastable* cyclic attractor that dominates for low mutation rates.

Keywords
Infinite population model, quasispecies model, population cyclic attractors

1 Introduction

In a previous paper [5] a technique was presented for calculating cyclic attractors, in the infinite population limit, of the simple genetic algorithm (GA) with zero crossover when the fitness function varies periodically with time. The method extends the dynamical systems model due to Vose [7,8] by representing *sequences* of populations as vectors in an abstract space. Operators on this space are then constructed representing the effects of proportional selection, S, and mutation, M, on the sequence as a whole. A further operator, R, is introduced which gives the effect of a single time step. The eigensystem of the combined operator RMS is then analysed to give information about a periodic attractor for the system. The period of the attractor is the same as that of the fitness function. This technique is further described in section 2.

The chief concern of the current paper is to apply this technique to the study of quasispecies adaptability [2]. A typical experiment in the literature is, for example [3], which involves a fitness function where one genotype, the *master sequence* has a high fitness value, while all other genotypes have an equal, low value. The experiment is concerned with the ability of an evolving population to stay on and around the master sequence as the mutation rate increases, forming a *quasispecies*. It can be shown that beyond a certain level, called the *error threshold*, the quasispecies collapses and falls away from the master sequence. In this paper, we aim to experiment with this idea further, by looking at what happens if the fitness function changes from favouring one master sequence (say, all zeros) to another (say, all ones). The quasispecies idea would indicate that when mutation is too high or too low, the population will fail to track either master sequence. However, there ought to be an optimal mutation rate for which quasispecies are relatively stable, but adaptable enough to track the change to a new master sequence when it arises.

To test this idea we define a periodic fitness function that alternately favours one master sequence then another. This is analysed using the method outlined to give the cyclic attractors of the system. It is indeed predicted that high mutation rates lead to virtually no tracking of the master sequences, while modest mutation rates do allow successful adaptation. However, with very low mutation rates, the theoretical model predicts still further successful adaptation, something which is not seen in practice. This means that there is a *metastable* attractor in operation, brought into effect by the fact that the population has a finite size [6,4].

2 Predicting Cyclic Attractors

The dynamical systems model represents populations as vectors $(p_0, \ldots, p_{n-1})$, where n is the size of the search space. In this vector, p_j is the proportion of the population taken up by the chromosome corresponding to j. The set of vectors

$$\Lambda = \left\{ (x_0, \ldots, x_{n-1}) : \sum_{j=0}^{n-1} x_j = 1, x_j \geq 0 \right\}$$

is known as the *simplex*. Selection, mutation, and crossover are operators that map the current population, p, to the expected next population, $\mathcal{G}p$, which also defines the multinomial distribution of next populations. As the population size tends to infinity, the actual next population tends to $\mathcal{G}p$. For this reason, it is sometimes referred to as the *infinite population* model.

This model is adapted in [5] to the case where the fitness function has period τ. The fitness of individual k at time t is $f(k, t)$. Let us think of k as being a *genotype*. It has a different fitness at different time steps. We can therefore think of k as having a different *phenotypic expression* at different time steps. Let z_{t_k} denote the phenotype of individual k at time step t. It is to be understood that z_{t_k} becomes $z_{(t+1 \bmod \tau)_k}$ at the next time step.

Let $x \in \mathbb{R}^{\tau n}$. We shall index such vectors in the following unusual manner

$$x = (x_{0_0}, \ldots x_{0_{n-1}}, \ldots, x_{t_k}, \ldots x_{(\tau-1)_{n-1}}).$$

When all the entries of x are non-negative and

$$\sum_{k=0}^{n-1} x_{t_k} = 1$$

for all $t = 0, \ldots, \tau - 1$, then x can be thought of as representing a set or sequence of populations, one for each time step within a single period. We will say then that x is *normal*.

A norm is required on this vector space. We will use the absolute norm

$$\|x\| = \sum_{t,k} |x_{t_k}|. \tag{1}$$

Notice that when x is normal, then $\|x\| = \tau$. We will further denote

$$\tilde{x} = \frac{x}{\|x\|}.$$

We now define the operators on this space which relate to proportional selection, bitwise mutation, and the action of a single time step. Define a $\tau n \times \tau n$ diagonal matrix S with

$$S_{t_k, t_k} = f(k, t)$$

where the indexing of the matrix S follows the same pattern as for vectors.

Define a $n \times n$ matrix U, where $U_{i,j}$ is the probability that individual j mutates into individual i. We now define a $\tau n \times \tau n$ matrix M as a block diagonal matrix with τ copies of U on the diagonal

$$M = \begin{pmatrix} U & & & \\ & U & & \\ & & \ddots & \\ & & & U \end{pmatrix}.$$

Define

$$R_{a_i, b_j} = [(i = j) \wedge (a = b + 1 \bmod \tau)] \tag{2}$$

where $[expr]$ evaluates to 1 if $expr$ is true and zero otherwise. This means that R is a kind of 'skewed' version of the identity matrix. In block form it is

$$R = \begin{pmatrix} & I & \\ I & & \end{pmatrix}.$$

The matrix RMS has all non-negative entries, so there is a non-negative eigenvector v with eigenvalue λ

$$RMSv = \lambda v.$$

Let

$$w_t = (v_{t_0}, \ldots, v_{t_{n-1}})$$

and

$$y_t = (0, \ldots, 0, v_{t_0}, \ldots, v_{t_{n-1}}, 0, \ldots, 0)$$

so that

$$v = \sum_{t=0}^{\tau-1} y_t.$$

Notice that $\tilde{w}_t$ is a population. Its average fitness is

$$\bar{f}(\tilde{w}_t) = \bar{f}(\tilde{y}_t) = \sum_{k=0}^{n-1} f(k, t) \tilde{w}_{t_k}.$$

The main result of [5] is the proof of the following.

Theorem 1. *If v is a non-negative eigenvector of RMS and*

$$y_t = (0, \ldots, 0, v_{t_0}, \ldots, v_{t_{n-1}}, 0, \ldots, 0)$$

for each t, then

$$\frac{RMS\tilde{y}_t}{\bar{f}(\tilde{w}_t)} = \tilde{y}_{t+1}. \tag{3}$$

So, given the population $\tilde{w}_t$, a single generation of the GA produces expected next population $\tilde{w}_{t+1}$. Therefore we have a cyclic attractor of the GA

$$\tilde{w}_0, \ldots, \tilde{w}_{\tau-1}.$$

In summary, the method for finding the cyclic attractor is as follows.

1. List all the possible phenotypic states in order: $z_{0_0}, \ldots, z_{\tau-1_{n-1}}$.
2. Let S be a diagonal matrix with fitnesses $f(k, t)$ on the diagonal.
3. Define matrix U by $U_{i,j}$ = probability of mutating j to i.
4. Let M be the block diagonal matrix with τ copies of U on the diagonal.
5. Let R be the 'right-shift' matrix in (2).

6. Calculate the eigensystem of RMS. Find any non-negative eigenvector v and its corresponding eigenvalue λ.
7. Normalise v, turning into a sequence of τ population vectors (w_t).

The sequence (w_t) then gives a cyclic fixed-point for the infinite population GA. The average fitnesses can be calculated directly, or via the normalisation factors

$$\bar{f}(\tilde{y}_t) = \lambda \frac{\|y_{t+1}\|}{\|y_t\|}$$

where λ is the calculated eigenvalue (see [5] for further details).

3 Theoretical Results Concerning Quasispecies

We wish to study what happens to a quasispecies that is centred on a particular master sequence, when the environment changes to favour a new master sequence. The quasispecies idea would indicate that when mutation is too low, there will not be enough adaptability for the population to track the change. When the mutation rate is too high, there will be no clearly defined quasispecies at all. However, for some moderate level of mutation, there should be sufficient quasispecies definition plus the adaptability necessary to track the change to another master sequence.

To examine this situation, we will have a fitness function which alternates between two master sequences. Initially, the bitstring (of length 5) comprising all zeros will be the master sequence, with a fitness of 2. All other bitstrings will score 1. After five time steps, however, the bitstring comprising all ones will be the new master sequence, scoring 2, with all other strings scoring 1. This will remain for a further five time steps, when the master sequence will revert to being all zeros. We therefore have a periodic fitness function, illustrated in Table 1.

Following our method, we construct the diagonal matrix S that contains the entries of this table reading from left to right, top to bottom. The mutation matrix U is defined for different mutation rates for functions of unitation in [4]. $U_{i,j}$ gives the probability of mutating from unitation class j to unitation class i. The matrix M is then built as a block diagonal matrix with copies of U on the diagonal. The matrix R is given by (2). Using a suitable software package[1] the eigenvectors of RMS, with varying mutation rates, are calculated, and normalized to give the cyclic attractors. For simplicity, we will just show the expected proportions of each master sequence during the period of the cycle, rather than the whole population.

Figure 1 shows the proportions of each master sequence expected at each time step of one cycle, for a relatively high mutation rate of 0.1. It can be seen that a master sequence will occupy between only three and twelve percent of the population. The mutation rate is too high for the population to cluster around the optimum. Indeed,

---

[1] In this paper, eigensystems were calculated using Mathematica 3.0 (Wolfram Research) running on a PC under Windows 95.

	UNITATION
TIME	0 1 2 3 4 5
0	2 1 1 1 1 1
1	2 1 1 1 1 1
2	2 1 1 1 1 1
3	2 1 1 1 1 1
4	2 1 1 1 1 1
5	1 1 1 1 1 2
6	1 1 1 1 1 2
7	1 1 1 1 1 2
8	1 1 1 1 1 2
9	1 1 1 1 1 2

Table 1. Fitness values for changing master sequence example

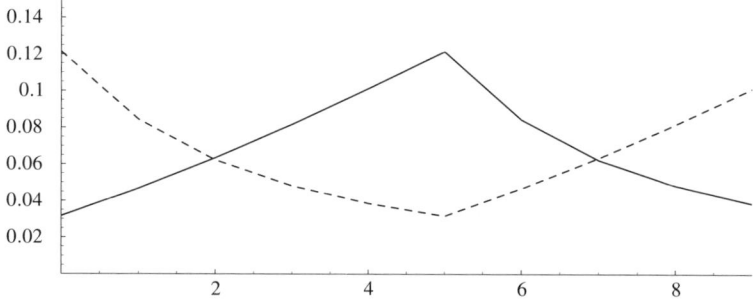

Figure 1. Proportions of the master sequences during the fixed-point cycle, for high mutation rate, 0.1, plotted against time. All zeros shown as a solid line, all ones shown as a dotted line.

looking in detail at the composition of the population at each time step reveals a roughly normal distribution centred around unitation values 2 and 3. This is in accordance with what we would expect from the quasispecies idea: there is too much adaptability, preventing the population from clustering around any single genotype. That is, there is no coherent *quasispecies definition*.

Figure 2 shows the theoretical results for a more modest mutation rate of 0.04. Now the current master sequence occupies up to forty percent of the population. We have much better quasispecies definition plus the adaptability for the population to track the change in master sequence. However, figure 3 shows this trend continuing for the still smaller mutation rate of 0.01. Contrary to this result, we actually expect to see a lack of adaptability with increased definition. If this happens in practice (see next section) then this means that the predicted attractor is no longer dominating the system. There must therefore be a metastable attractor that has come into play.

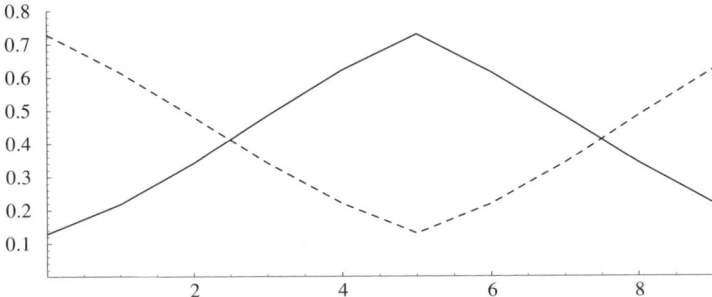

Figure 2. Proportions of the master sequences during the fixed-point cycle, for moderate mutation rate, 0.04, plotted against time. All zeros shown as a solid line, all ones shown as a dotted line.

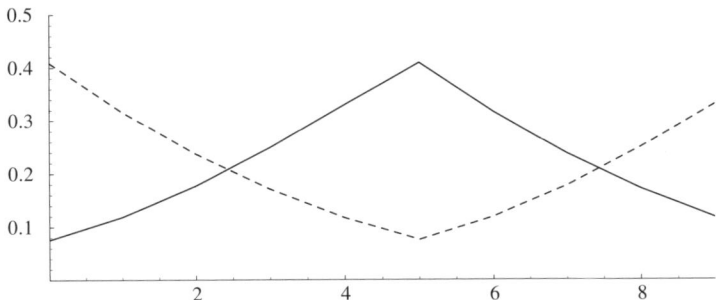

Figure 3. Proportions of the master sequences during the fixed-point cycle, for low mutation rate, 0.01, plotted against time. All zeros shown as a solid line, all ones shown as a dotted line.

4 Experimental Results

Experiments were run on the periodic fitness function described above. A population size of 1000 was used, and the proportional selection scheme was implemented using the efficient stochastic universal sampling algorithm [1]. For each mutation rate, ten runs were made of 500 generations each. The last twenty generations were examined and the proportions of the master sequence of all zeros was recorded and averaged over the ten runs. For the high mutation rate of 0.1, and the moderate rate of 0.04, the experimental results fit the theoretical prediction very well (see figures 4 and 5). However, as anticipated in the previous section, when the mutation rate gets as low as 0.01, the population can no longer track the change in master sequence. What happens in practice depends precisely on which master sequence was the first to be encountered when the GA starts. If it is all zeros, then the majority of the population will stay centred on this master sequence, even when the environment changes. The proportion of the sequence of all ones is correspondingly low. If, however, the all-ones sequence is the first to be encountered the situation is reversed. Results from a typical run are shown in figure 6.

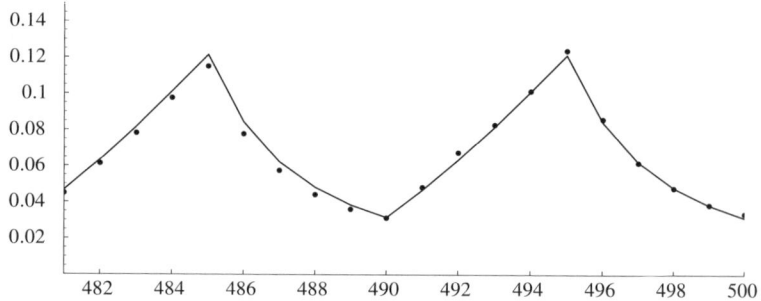

Figure 4. Proportions of the sequence 000...0 during generations 481–500, for high mutation rate, 0.1, averaged over 10 runs. The solid line gives the theoretical prediction.

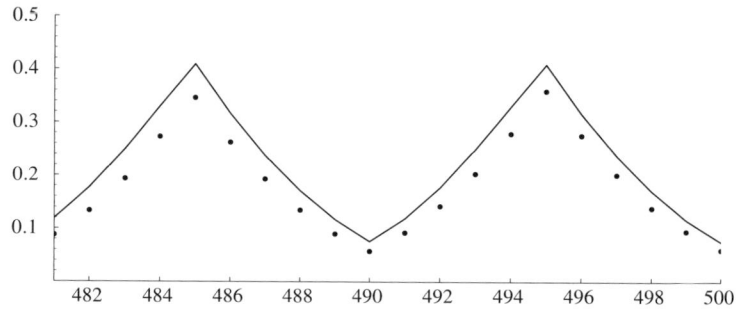

Figure 5. Proportions of the sequence 000...0 during generations 481–500, for moderate mutation rate, 0.04, averaged over 10 runs. The solid line gives the theoretical prediction.

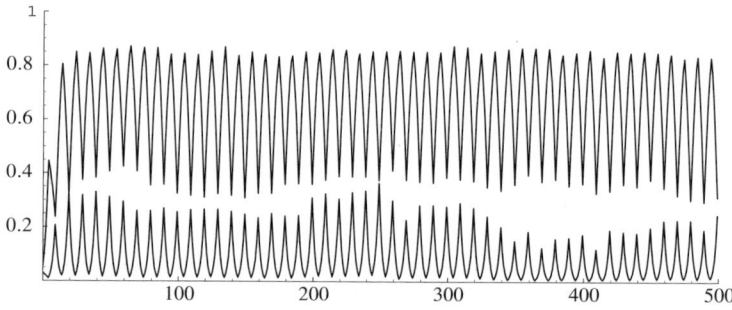

Figure 6. Proportions of the sequences 000...0 (higher curve) and 111...1 (lower curve) during a typical run of 500 generations, for low mutation rate, 0.01

5 Discussion

For moderate to high values of mutation, the theoretical predictions agree with the experiments and also with intuition guided by the quasispecies idea. That is, moderate mutation levels allow a balance of quasispecies definition (a large proportion of the population gathers around the current master sequence) with adaptability (the ability to track a change in optimum). Higher mutation rates lead to a collapse of quasispecies definition (that is, there is no discernible clustering around a single genotype). The population stays distributed normally between the two master sequences. However, for low mutation rates, we would expect high quasispecies definition and low adaptability. This is observed in practice: the population stays centred around the first master sequence throughout the run, despite changes in the environment. However, this is not the result given by the theoretical model. Therefore, the population has centred a metastable cycle. One possible source of metastable cycles would be other eigenvectors of RMS. It is possible that such an eigenvector could correspond to a sequence of populations just outside the simplex. In this case, because of continuity considerations, actual populations near one element of the sequence will be mapped to another population near the next one of the sequence.

Examining the other eigenvectors of RMS shows that there is one such sequence. However, it does not correspond to what is observed. Rather it corresponds to a cycle which is very similar to that observed in the high mutation rate case. Further thought and experimentation is required here, to complete the link between the quasispecies model and the dynamical systems theory.

References

1. J. E. Baker. Reducing bias and inefficiency in the selection algorithm. In J. J. Grefenstette, editor, *Proceedings of the Second International Conference on Genetic Algorithms*, pages 14–21. Lawrence Erlbaum, Hillsdale, NJ, 1987.
2. M. Eigen, J. McCaskill, and P. Schuster. The molecular quasi-species. In I. Prigogine and S. A. Rice, editors, *Advances in Chemical Physics LXXV*, pages 149–263. John Wiley, New York, 1989.
3. G. Ochoa and I. Harvey. Recombination and error thresholds in finite populations. In W. Banzhaf and C. R. Reeves, editors, *Foundations of Genetic Algorithms 5*, pages 245–264. Morgan Kaufmann, San Francisco, 1998.
4. J. E. Rowe. Population fixed-points for functions of unitation. In W. Banzhaf and C. R. Reeves, editors, *Foundations of Genetic Algorithms 5*, pages 69–84. Morgan Kaufmann, San Francisco, 1998.
5. J. E. Rowe. Finding attractors for periodic fitness functions. In *Proceedings of the Genetic and Evolutionary Computation Conference, GECCO 1999*, pages 557–563. Morgan Kaufmann, San Francisco, 1999.
6. E. van Nimwegen, J. P. Crutchfield, and M. Mitchell. Finite populations induce metastability in evolutionary search. *Physics Letters A*, 229(2):144–150, 1997.
7. M. D. Vose and A. H. Wright. Stability of vertex fixed points and applications. In L. D. Whitley and M. D. Vose, editors, *Foundations of Genetic Algorithms 3*, pages 103–113. Morgan Kaufmann, San Francisco, 1994.
8. M. D. Vose and A. H. Wright. Simple genetic algorithms with linear fitness. *Evolutionary Computation*, 2(4):347–368, 1995.

Genetic Algorithms in Time-Dependent Environments

C. Ronnewinkel[1,2], C.O. Wilke[1], and T. Martinetz[1]

[1] Institut für Neuro- und Bioinformatik
Medizinische Universität Lübeck
Seelandstraße 1a
D-23569 Lübeck, Germany

[2] Institut für Neuroinformatik
Ruhr-Universität Bochum
D-44780 Bochum, Germany
E-mail: *ronne@neuroinformatik.ruhr-uni-bochum.de*

Abstract. The influence of time-dependent fitnesses on the infinite population dynamics of simple genetic algorithms (GAs) without crossover is analyzed. Based on general arguments, a schematic phase diagram is constructed that allows one to characterize the asymptotic states in dependence on the mutation rate and the time scale of changes. Furthermore, the notion of *regular* changes is raised for which the population can be shown to converge towards a *generalized* quasispecies. Based on this, error thresholds and an optimal mutation rate are approximately calculated for a generational GA with a moving needle-in-the-haystack landscape. The phase diagram thus found is fully consistent with our general considerations.

Keywords
Time-dependent fitness landscape, quasispecies, error threshold, optimal mutation rate

1 Introduction

Genetic algorithms (GAs) as special instances of evolutionary algorithms have been established during the last three decades as optimization procedures, but mostly for static problems (see [2] for an overview and [3] for an in-depth presentation of the field). In view of real-world applications, such as routing in data-nets, scheduling, robotics, etc., which include essentially dynamic optimization problems, there are two alternative optimization strategies. On the one hand, one can take snapshots of the system and search "offline" for the optimal solutions of the static situation represented by each of these snapshots. In this approach, the algorithm is restarted for every snapshot and solves the new problem from scratch. On the other hand, the optimization algorithm might re-evaluate the real, current situation in order to reuse information gained in the past. In this case, the algorithm works "online". As can be argued from the analogies to natural evolution, evolutionary algorithms seem to be promising candidates for "online" optimization [2,4]. The re-evaluation of the situation or environment then introduces a *time-dependency* of the fitness landscape. This time-dependency occurs as external to the algorithm's population and does not emerge from coevolutive interactions. Coevolutive interactions as an alternative source of time-dependency in the fitness landscape are not within the scope of this work.

In recent years, many different methods and extensions of standard evolutionary algorithms for the case of time-dependent fitnesses have been analyzed on the basis of experiments (see [4] for a review) but only seldom on the basis of *theoretical* arguments (see [12,15]). To take a step towards the direction of a better theoretical understanding of "online" evolutionary algorithms, we will study the effects of simple time-dependencies of the fitness landscape on the dynamics of GAs (without crossover), or more generally, of populations under mutation and probabilistic selection. As we will see, it is possible to characterize the asymptotic states of such a system for a particular class of dynamic fitness landscapes that is introduced below. The asymptotic state forms the basis on which it can be decided whether the population is able to adapt to, or track, the changes in the fitness landscape. Our mathematical formalism applies to GAs as well as to biological self-replicating systems, since the analyzed GA model and Eigen's quasispecies model [7–9] in the molecular evolution theory (see [1] for a recent review) are very similar. Hence, all introduced concepts for GAs are valid and relevant in analogous form for molecular evolutionary systems.

In the following section, we will introduce the model to be analyzed and show the correspondence to the quasispecies model. Then, we will introduce the mathematical framework, based on which we will formally characterize the asymptotic state as a fixed-point. After presenting the main concepts, we will proceed with the construction of a phase diagram that allows one to characterize the order found in the asymptotic state for different parameter settings. Finally, a moving needle-in-the-haystack (*NiH*) landscape is analyzed and its phase diagram, including the optimal mutation rate, is calculated.

2 Mathematical Framework

In order to study the influence of a time-dependent fitness landscape on the dynamics of a GA, we consider GAs to be discrete dynamical systems. A detailed introduction to the resulting dynamical systems model is given by Rowe [13] in this volume, page 31. Here, we will only briefly introduce the basic concepts and the notations we use within the present work.

The GA is represented as a generation operator $G_t^{(m)}$ acting on the space Λ_m of all populations of size m for some given encoding of the population members. If we choose the members i to be encoded as bit-strings of length l, this state space is given by

$$\Lambda_m = \{(n_0, \ldots, n_{2^l-1})/m \mid \sum_i n_i = m, n_i \in \mathbb{N}_0\}$$

where n_i denotes the number of bit-strings in the population equal to the binary representation of $i \in \{0, \ldots, 2^l - 1\}$.

The generation operator maps the present population onto the next generation,

$$x(t+1) = G_t^{(m)}[x(t)].$$

This is achieved by applying a sampling procedure that draws the members of the next generation's population $x(t+1)$ according to their expected concentrations $\langle x(t+1) \rangle \in \Lambda_\infty$ which are defined by the mixing scheme [13,17] and the selection scheme. For an infinite population size, the sampling acts like the identity resulting in

$$G_t^{(\infty)} x(t) = x(t+1) = \langle x(t+1) \rangle.$$

Hence, $G_t := G_t^{(\infty)}$ represents, in fact, the mixing and selection scheme. For finite population size, $\langle x(t+1) \rangle \in \Lambda_\infty$ is approximated by the sampling process to obtain $x(t+1) \in \Lambda_m$. The deviations thereby possible become larger with decreasing m and distort the finite population dynamics as compared to the infinite population case. This results in fluctuations and epoch formation as shown in [13,17,16]. In the following, we will consider the infinite population limit, because it reflects the exact flow of probabilities for a particular fitness landscape. In a second step, the fluctuations and epoch formation introduced by the finiteness of a real population can be studied on the basis of that underlying probability flow.

The generation operator is assumed to decompose into a separate mutation and a separate selection operator, like

$$G_t = M \cdot S(t) \tag{1}$$

where the selection operator $S(t)$ contains the time-dependency of the fitness landscape. Crossover is not considered in this work.

Inspired by molecular evolution, and also by common usage, we assume that the mutation acts like flipping each bit with probability μ. If we set the duration of one generation to 1, μ equals to the mutation rate. The mutation operator then takes on the form

$$M = \begin{pmatrix} 1-\mu & \mu \\ \mu & 1-\mu \end{pmatrix}^{\otimes l}, \quad \text{i.e.} \quad M_{ij} = \mu^{d_H(i,j)}(1-\mu)^{l-d_H(i,j)}$$

where $\otimes$ denotes the Kronecker (or canonical tensor) product and $d_H(i,j)$ denotes the Hamming distance of i and j.

To keep the description analytically tractable, we will focus on fitness-proportionate selection

$$S(t) \cdot x = F(t) \cdot x / \langle f(t) \rangle_x, \quad \text{where } F(t) = \text{diag}(f_0(t), \ldots, f_{2^l-1}(t))$$
$$\text{and } \langle f(t) \rangle_x = \sum_i f_i(t) x_i = \|F(t) \cdot x\|_1.$$

This will already provide us with some insight into the general behavior of a GA in time-dependent fitness landscapes.

Since the GA corresponding to (1) applies mutation to the current population and selects the new population with complete replacement of the current one, it is called a *generational* GA. In addition to generational GAs, *steady state* GAs with a two-step reproduction process are also in common use: first, a small fraction γ of the current population is chosen to produce $m\gamma$ mutants according to some heuristics. Second, another fraction γ of the current population is chosen to get replaced by those mutants according to some other heuristics (see [6,11,5] and references therein). We can include steady state GAs into our description in an approximate fashion by simply bypassing a fraction $(1-\gamma)$ of the population into the selection process without mutation, whereas the remaining fraction γ gets mutated before it enters the selection process. The generation operator then reads

$$G_t = \left[(1-\gamma)\mathbb{1} + \gamma M\right] S(t). \tag{2}$$

By varying γ within the interval $]0, 1]$, we can interpolate between steady state behavior for $\gamma \ll 1$ and generational behavior for $\gamma = 1$. Equation 2 is only an approximation of the true generation operator for steady state GAs because the heuristics involved in the choice of the mutated and replaced members are neglected. But in the next section, the heuristics are expected to play a minor role for our general conclusion on an inertia of steady state GAs against time variations.

At this point, we want to review briefly the correspondence of our GA model with the quasispecies model, extensively studied by Eigen and coworkers [7–9] in the context of molecular evolution theory (see also [14] in this volume, page 251). The quasispecies model describes a system of self-replicating entities i (e.g., RNA-, DNA-strands) with replication rates f_i and an imperfect copying procedure such that mutations occur. For simplicity reasons, the overall concentration of molecules in the system is held constant by an excess flow $\Phi(t)$. In the above notation, the continuous model reads

$$\dot{x}(t) = [M \cdot F(t) - \Phi(t)]\, x(t) \tag{3}$$

where the flux needs to equal the average replication, $\Phi(t) = \langle f(t) \rangle_{x(t)}$, in order to keep the concentration vector $x(t)$ normalized. This model might then be discretized via $t \to t/\delta t$, which unveils the similarity to a steady state GA:

$$x(t+1) = \left[(1 - \delta t\, \langle f(t) \rangle_{x(t)})\mathbb{1} + \delta t\, M \cdot F(t)\right] x(t) \quad \text{for } \delta t \ll 1. \tag{4}$$

By comparison with (2), we can easily read off that $\gamma = \delta t\, \langle f(t) \rangle_{x(t)} =: \gamma_{x(t)}$. This means a low (respectively, high) average fitness leads to a small (respectively, large) replacement – a property that is not wanted in the context of optimization problems, which GAs are usually used for, because one does not want to remain in a region of low fitness for a long time. Another difference to steady state GAs is the fact that in the continuous Eigen model, selection only acts on the mutated fraction of the population – although this leads only to subtle differences in the dynamics of steady state GAs and the Eigen model.

Equation 3 is commonly referred to as 'continuous Eigen model' in the literature, because of the continuous time, and (4) is simply its discretized form which can be used for numerical calculations. Nonetheless, the notion 'discrete Eigen model' is seldom used for (4) but it is often used for the generational GA

$$x(t+1) = [M \cdot S(t)] x(t) \tag{5}$$

in the literature. This stems from the identical asymptotic behavior of (4) and (5) for static fitness landscapes. However, there are differences for time-dependent fitness landscapes, as we will see in the following two sections.

3 Regular Changes and Generalized Quasispecies

In the case of a static landscape, the fixed-points of the generation operator, which are in fact stationary states of the evolving system (if contained within Λ_m, see [13]), can be found by solving an eigenvalue problem, because of

$$x = Gx \iff MFx = \langle f \rangle_x x \,. \tag{6}$$

Let λ_i and v_i denote the eigenvalues and eigenvectors of MF with descending order $\lambda_0 \geq \cdots \geq \lambda_{2^l-1}$ and $\|v_i\|_1 = 1$. For $\mu \neq 0, 1$ the Perron–Frobenius theorem assures the non-degeneracy of the eigenvector v_0 to the largest eigenvalue and, moreover, it assures $v_0 \in \Lambda_\infty$. Often, v_0 is called the Perron vector. After a transformation to the basis of the eigenvectors $\{v_i\}$ it can be straightforwardly shown that $x(t)$ converges to v_0 for $t \to \infty$. The population represented by v_0 was called the 'quasispecies' by Eigen, because this population does not consist of only a single dominant genotype, or string, but consists of a particular stable mixture of different genotypes.

Let us now consider time-dependent landscapes. If the time-dependency is introduced simply by a single scalar factor, like

$$F(t) = F \rho(t) \quad \text{with } \rho(t) \geq 0 \text{ for all } t$$

it immediately drops out of the selection operator for GAs. For the continuous Eigen model, we note that the eigenvectors of $F(t)$ and F are the same and that $\lambda_i(t) = \lambda_i \rho(t)$. Since $\rho(t) \geq 0$, which is necessary to keep the fitness values positive, the order of the eigenvalues remains, such that $MF(t)$ will show the same quasispecies v_0 as MF. In contrast to that special case, a general, individual time-dependency of the string's fitnesses does indeed change the eigenvalues and eigenvectors of $MF(t)$ compared to MF. For an arbitrary time-dependency the Perron vector is constantly changing, and therefore, we cannot even define a unique asymptotic state. However, this problem disappears for what we call *regular* changes. After having established a theory for such changes, we can then take into account more and more non-regular ingredients. What do we mean by "regular change"? We define it heuristically in the following way: a regular change is a change that happens with fixed duration τ and obeys some deterministic rule that is the same for all change cycles. Let us express the latter more formally and make it more clear what we mean by "same rule of change". Within a change cycle, we allow for an arbitrary time-dependency of the fitness, up to the restriction that two different change cycles must be connected by a permutation of the sequence space. Thus, if the time-dependency is chosen for one change cycle, e.g., the first change cycle starting at $t = 0$, it is already fixed for all other cycles, apart from the permutations. We will represent permutations π from the permutation group $\mathfrak{S}_{2^l}$ of the sequence space as matrices

$$(P_\pi)_{ij} = \delta_{\pi(i),j} \quad \text{for } i, j \in \{0, \ldots, 2^l - 1\}.$$

The permutations of vectors x and matrices A are obtained by

$$(P_\pi x)_i = x_{\pi(i)} \quad \text{and} \quad (P_\pi A P_\pi^T)_{i,j} = A_{\pi(i),\pi(j)}$$

where P_π^T denotes the transpose of P_π with the property $P_\pi^T = P_{\pi^{-1}} = P_\pi^{-1}$.

In reference to the first change cycle, we define the fitness landscape $F(t)$ as being *single-time-dependent*, if and only if for each change cycle $n \in \mathbb{N}_0$ there exists a permutation $\pi_n \in \mathfrak{S}_{2^l}$, such that for all cycle phases $\varphi \in \{0, \ldots, \tau - 1\}$

$$P_n F(\varphi + n\tau) P_n^T = F(\varphi) \quad \text{(in short: } P_n := P_{\pi_n}).$$

We will call each permutation P_n a *jump-rule*, or simply *rule*, which connects $F(\varphi + n\tau)$ and $F(\varphi)$. To make predictions about the asymptotic state of the system, we need to relate the generation operators of different change cycles to each other. This is readily achieved if the permutations P_n *commute* with the mutation operator M. The condition for this being the case is that for all i, j

$$M_{ij} = M_{\pi_n(i),\pi_n(j)} \quad \text{or equivalently} \quad d_H(i, j) = d_H(\pi_n(i), \pi_n(j)).$$

Thus, the Hamming distances $d_H(i, j)$ need to be *invariant* under the permutations P_n. Geometrically this means that the fitness landscape gets "translated" or "rotated" by those permutations without changing the neighborhood relations. Then, we find for arbitrary $n \in \mathbb{N}_0$ and $\varphi \in \{0, \ldots, \tau - 1\}$

$$G_{\varphi+n\tau} = P_n^T G_\varphi P_n. \tag{7}$$

To study the asymptotic behavior of the system, it is useful to accumulate the time-dependency of a change cycle by introducing the τ-generation operators

$$\Gamma_{n\varphi} := G_{\tau-1+\varphi+n\tau} \cdots G_{\varphi+n\tau} \quad \text{for all } n \in \mathbb{N}_0$$

where $\varphi \in \{0, \ldots, \tau - 1\}$ in the following always denotes the phase within a cycle. Because of (7), all these operators are related to $\Gamma_{0\varphi}$ by

$$\Gamma_{n\varphi} = P_n^T \Gamma_{0\varphi} P_n.$$

This property allows us to write the time evolution of the system in the form

$$x(\varphi + n\tau) = P_{n-1}^T \Gamma_{0\varphi} P_{n-1} \cdots P_1^T \Gamma_{0\varphi} P_1 \Gamma_{0\varphi} x(\varphi). \tag{8}$$

Let us consider the special case of a single rule P being applied at the end of each change cycle, which results in $P_n = (P)^n$, e.g., imagine a fitness peak that moves at a constant "velocity" through the string space. We will see below that for those cases it is possible to identify the asymptotic state with a quasispecies in analogy to static fitness landscapes. Because of that, we can now define the notion of *regularity* of a fitness landscape formally in the following manner:

A time-dependent fitness landscape $F(t)$ is *regular*, if and only if: (i) the fitness landscape is *single-time-dependent*; (ii) there exists some rule $P \in \mathfrak{S}_{2^l}$ which is applied at the end of each cycle such that $P_n = (P)^n$; and (iii) the rule P *commutes* with the mutation operator M.

In this case, we get with $PP^T = \mathbb{1}$ the time evolution

$$x(\varphi + n\tau) = \left(P^T\right)^n \left(P\Gamma_{0\varphi}\right)^n x(\varphi). \tag{9}$$

To proceed, it is useful to permute the concentrations compatible with the rule of the fitness landscape. By this, concentrations are measured in reference to the fitness landscape structure of the start cycle $n = 0$. We will denote those concentrations by $x'(t)$ and they are related to the concentrations $x(t)$ by

$$\begin{aligned} x'(\varphi + n\tau) &= (P)^n\, x(\varphi + n\tau) \\ &= (P\Gamma_{0\varphi})^n\, x(\varphi) \quad \text{and} \quad x'(\varphi) = x(\varphi). \end{aligned} \tag{10}$$

For example, if there is no time-dependency within the cycles, some x'_i will for all cycles measure the concentration of the highest fitness string, independent of its current position in string space. Thus, $x'(t)$ evolves in a fitness landscape with periodic change, which can also be seen from the second line of (10). In analogy to the static case (6), the calculation of fixed-points of $x'(t)$ is equivalent to an eigenvalue problem,

$$x'(t + \tau) = x'(t) \iff P\widetilde{\Gamma}_{0\varphi}\, x'(t) = \|P\widetilde{\Gamma}_{0\varphi}\, x'(t)\|_1\, x'(t)$$

where $\widetilde{\Gamma}_{0\varphi}$ is the *unnormalized* τ-generation operator obtained from the accumulation of the *unnormalized* generation operators $\widetilde{G}_{\psi+\varphi} = MF(\psi + \varphi)$ with $\psi \in \{0, \ldots, \tau - 1\}$.

The corresponding periodic quasispecies v_0 can be calculated for all phases φ of the change cycle from the Perron vector v_0 of $P\widetilde{\Gamma}_{00}$ in the following way:

$$x'(\varphi + n\tau) \xrightarrow{n \to \infty} v_0(\varphi) = G_{\varphi-1} \cdots G_0\, v_0 \quad \text{for } \varphi \in \{0, \ldots, \tau - 1\}. \tag{11}$$

To find the asymptotic states of the concentrations $x(t)$, we simply need to invert (10)

$$x(\varphi + \nu\tau) = \left(P^T\right)^\nu x'(\varphi + \nu\tau) \quad \text{for } \nu \in \{0, \ldots, \eta - 1\} \tag{12}$$

where $\eta := \operatorname{ord} P$ is the order of the group element $P \in \mathfrak{S}_{2^l}$.

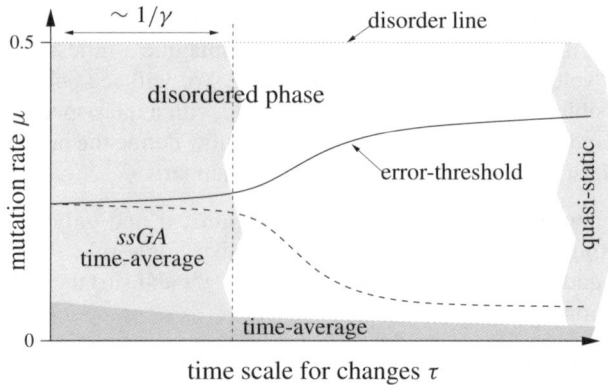

Figure 1. Schematic phase diagram: time-average regions due to low mutation (*dark gray*) and large inertia (*light gray, left*), quasi-static region for slow changes (*light gray, right*)

The essential reason for the existence of asymptotic states for $x(t)$ lies in the finiteness of the permutation group $\mathfrak{S}_{2^l}$. Because $P^\eta = \mathbb{1}$, we find directly from (9) the asymptotic state

$$x(\varphi + \tilde{n}\eta\,\tau) = (P\Gamma_{0\varphi})^{\eta\tilde{n}} x(\varphi) \xrightarrow{\tilde{n}\to\infty} v_0(\varphi)$$

where $v_0(\varphi)$ is the same as in (11), because $(P\widetilde{\Gamma}_{0\varphi})^\eta$ and $P\widetilde{\Gamma}_{0\varphi}$ have the same eigenvectors, in particular, the same Perron vector. Moreover, we get

$$x(\varphi + (\nu + \tilde{n}\eta)\tau) \xrightarrow{\tilde{n}\to\infty} (P^{\mathrm{T}})^\nu v_0(\varphi) \quad \text{for } \nu \in \{0,\ldots,\eta-1\} \qquad (13)$$

which is the same result as yielded by (11) and (12). In the limit of long strings $l \to \infty$, ord P is not necessarily finite anymore. If ord $P \to \infty$ for $l \to \infty$, then the asymptotic states (13) for $x(t)$ do not exist, but (11) still holds. Hence, a quasispecies exists even in the limit $l \to \infty$ if measured in reference to the structure of the fitness landscape.

In conclusion, equations (11) and (13) represent the *generalized* quasispecies for the class of *regular* fitness landscapes which includes as special cases static and periodic fitness landscapes. In fact, the simplest case of a *regular* change is a periodic variation of the fitness values $f_i(t) = f_i(t + \tau)$ because *no* permutations are involved ($P = \mathbb{1}$) and hence $x'(t) = x(t)$ for all t. The quasispecies was generalized for this case already in [18] and – using a slightly different formalism – in [12]. In section 5, we will study a more complicated example.

4 Schematic Phase Diagram

To get an intuitive feeling for the typical behavior of steady state GAs and generational GAs, let us consider some special lines in the plane spanned by the mutation rate μ and the time scale for changes τ, as shown in figure 1. The mutation operator represents only for $\mu < 1/2$ a copying procedure with occurring errors, whereas for $\mu > 1/2$ it systematically tends to invert strings, i.e., it resembles an inverter with occurring errors. Since mutation should introduce *weak* modifications to the strings, we will consider only $\mu \leq 1/2$.

Disorder Line. For $\mu = 1/2$, the Perron vector of $MF(t)$ is always

$$v_0^T = (1, \ldots, 1)/2^l.$$

The population will therefore converge towards the disordered state. Because of the continuity of M in μ, we already enter a disordered phase for $\mu \approx 1/2$.

Time-Average Region. For $\mu = 0$, the mutation operator is the identity. We find as time evolution simply the product average over the fitness of the evolved time steps:

$$x(t+\tau) = \left[\prod_{\varphi=t}^{t+\tau-1} S(\varphi)\right] x(t)$$

$$= \tilde{F}(t+\tau, t)\, x(t)/\|\ldots\|_1 \text{ with } \tilde{F}(t+\tau, t) = \prod_{\varphi=t}^{t+\tau-1} F(\varphi).$$

Since diagonal operators commute, the order in which the $F(\varphi)$ get multiplied does not make any difference. For the case of a τ-periodic landscape, $\tilde{F} = \tilde{F}(t+\tau, t) = \tilde{F}(\tau, 0)$ is *independent* of t. The quasispecies is then a linear superposition of the eigenvectors of the largest eigenvalue of the product averaged fitness landscape $\tilde{F}$ – there might be more than one such eigenvector, since $\tilde{F}$ is diagonal and the Perron-Frobenius theorem does not apply. Because of the continuity of M in μ the dynamics are already governed for $0 < \mu \ll 1$ by the product average $\tilde{F}$. Analogous conclusions apply to those non-periodic landscapes for which by choosing a suitable time scale τ a meaningful average $\tilde{F}(t+\tau, t)$ can be defined.

For steady state GAs, γ is small and we find to first order in $\tau\gamma$:

$$x(t+\tau) = (1 - \tau\gamma)\tilde{F}(t+\tau, t)$$

$$+ \tau\gamma\left(\frac{1}{\tau}\sum_{\varphi=0}^{\tau-1} S(t+\tau)\cdots \underbrace{M}_{\varphi\text{-th factor from left}} \cdots S(t)\right) + \mathcal{O}((\tau\gamma)^2).$$

If $\tau\gamma \ll 1$ holds, the time evolution is governed by $\tilde{F}(t+\tau, t)$. For changes on a time scale τ, we find time-averaged behavior if $\tau \ll 1/\gamma$. Thus, the width of the time-average region is proportional to $1/\gamma$. A detailed analysis of the effect of the different positions of the mutation operator M within the $\tau\gamma$-term, which is otherwise an arithmetic time-average, has not yet been carried out.

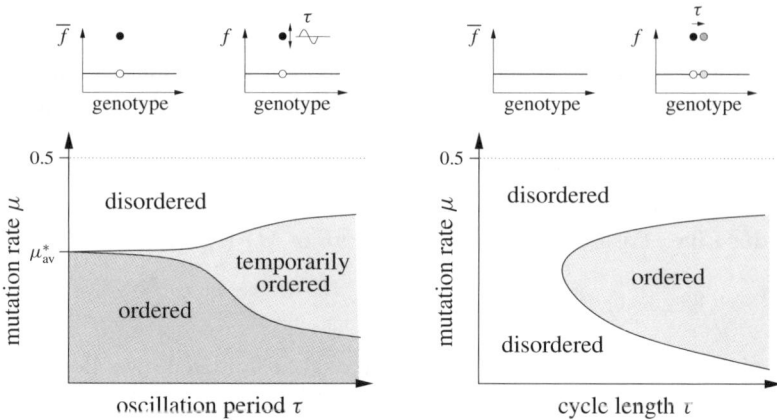

Figure 2. Phase diagrams for *(left)*: needle-in-the-haystack with oscillating height at frequency $\omega = 2\pi/\tau$, *(right)*: needle-in-the-haystack that jumps after τ time steps to a randomly chosen nearest neighbor.

Quasi-Static Region. If the changes happen on a very large time scale τ compared to the average relaxation time ($\sim 1/\langle\lambda_0-\lambda_1\rangle$) the quasispecies grows almost without noticing the changes. Thus, in the quasi-static region all quasispecies that might be expected from the static landscapes $\tilde{F} = F(t)$ will occur at some time during one cycle τ.

Wilke et al. raise in [19] the schematic phase diagram of the continuous Eigen model, which exhibits the same time-average phases as that for steady state GAs. Their result is in perfect agreement with two recently, explicitly studied time-dependent landscapes. First, Wilke et al. studied in [18] a needle-in-the-haystack (*NiH*) landscape with oscillating, τ-periodic fitness of the needle, i.e.,

$$f_0(t) > f_1 = \cdots = f_{2^l-1} = 1 \quad \text{and} \quad f_0(t) = \sigma \exp\{\varepsilon \sin(2\pi t/\tau)\}.$$

The continuous model was represented for $\delta t \to 0$ as (4) and the periodic quasispecies (11) was calculated. Figure 2 *(left)* shows the resulting phase diagram. For small τ, the error threshold is given by the one of the time-averaged landscape, whereas for large τ, the error threshold oscillates between minimum and maximum values corresponding to $\min_t f_0(t)$ and $\max_t f_0(t)$, as expected in the quasi-static regime. Second, Nilsson and Snoad studied in [10] a moving *NiH* that jumps randomly to one of its nearest neighbor strings every τ time steps. The time-average of this landscape over many jump cycles is a totally flat or neutral landscape, which explains the extension of the disordered phase to small μ and small τ as is shown in figure 2 *(right)*. In the quasi-static region, order is expected because the needle stays long enough at each position for a quasispecies to grow. Hence, we can understand the existence of the observed and calculated phase diagrams in figure 2 from simple arguments. In fact, they are special instances of the general schematic phase diagram depicted in figure 1.

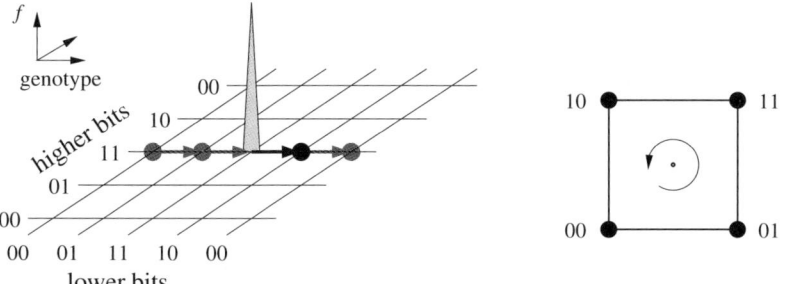

Figure 3. A regularly moving needle-in-the-haystack for string length $l = 4$. In *(left)*, the *solid arrow* represents the next jump to happen, whereas the *gray and solid arrows* all together represent the jumps that happen one after the other under the rule P of rotating the two lower bits as shown in *(right)* with rotation angle $\pi/2$ at every jump.

In the following, we will consider regularly moving *NiH*s and derive the infinite population behavior of a generational GA in such landscapes. This is interesting, since generational GAs should be considered to adapt faster to changes compared to steady state GAs, as the missing time-average region of generational GAs for small τ suggests. To clarify whether a different phase diagram compared to figure 2 *(right)* emerges for generational GAs with moving *NiH*, we will calculate the phase diagram including the optimal mutation rate that maximizes a lower bound for the concentration of the needle string in the population.

5 Generational GA and a Moving *NiH*

In this section, we want to analyze quantitatively the asymptotic behavior of a generational GA with *NiH* that moves *regularly* in the sense of section 3 to one of its l nearest neighbors every τ time steps. At the end, we will also be able to comment on the case of a *NiH* that jumps *randomly* to one of its nearest neighbors.

A simple example of a *NiH* that moves regularly to nearest neighbors is shown in figure 3 *(left)*. Each jump corresponds to a $\pi/2$-rotation of the four-dimensional hypercube $\{0, 1\}^4$ along the 1100 axis, i.e., the lower two bits are rotated as shown in figure 3 *(right)*. We will call the set of strings $\{P^n i \mid n \in \mathbb{N}\}$ which is obtained by applying the same rule $P \in \mathfrak{S}_{2^l}$ over and over to some initial string $i \in \{0, 1\}^l$, the *orbit of i under P*. The period length 4 of the orbit shown in figure 3 *(left)* originates from the rotation angle $\pi/2$ and hence is independent of the string length l. The orbits of such rotations will always be restricted to only four different strings. For reasons that will become clear below, we are looking for *regular* movements of the needle that are *not* restricted to such a small subspace of the string space. Instead, the needle is supposed to move 'straight away' from previous positions in string space. Since a complete classification and analysis of all possible *regular* movements for given string length l and jump distance d is out of the scope of

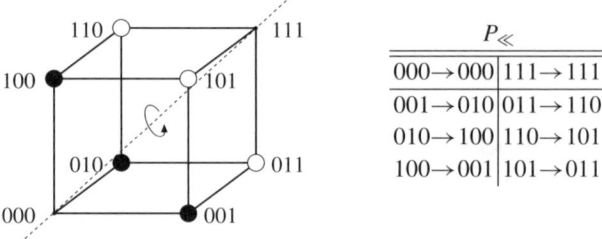

Figure 4. The equivalence of a $2\pi/3$-rotation along the 111 axis and a cyclic 1-bit left-shift, denoted by $P_{\ll}$, for string length $l = 3$

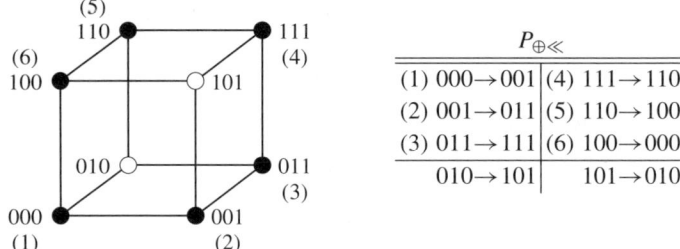

Figure 5. The orbit of 000 under $P_{\oplus\ll}$ (*black dots*) for string length $l = 3$. The numbers (1), ..., (6) show the order in which the strings are visited by the needle, starting from 000.

this work, we will simply give an example of a rule $P \in \mathfrak{S}_{2^l}$ that generates such movements: the composition of a cyclic 1-bit left-shift, which we denote by $P_{\ll}$, and an exclusive-or with $0\cdots01$, which we denote by $P_{\oplus}$.

For string length $l \leq 3$, $P_{\ll}$ corresponds to a $2\pi/l$ rotation along the $1\cdots1$ axis as can be seen in figure 4. Moreover, the orbit of $0\cdots0$ under $P_{\oplus\ll} = P_{\oplus} \circ P_{\ll}$ is shown in figure 5 also for $l = 3$. For arbitrary string length l, it is more difficult to visualize the action of $P_{\ll}$ and hence of $P_{\oplus\ll}$. But, it is easily verified that starting from all zeros $0\cdots0$, the string with $n \leq l$ ones $0\cdots01\cdots1$ will be reached after exactly n jumps. Moreover, the orbit of $0\cdots0$ under $P_{\oplus\ll}$ has the period length $2l$. In the limit of long strings $l \to \infty$, this periodicity is broken because the needle never (i.e., after an infinite number of jumps) returns to all zeros $0\cdots0$, but – as we have shown in (11) using (10) – there still exists an asymptotic quasispecies.

How does our simple GA behave with a *NiH* that moves according to $P_{\oplus\ll}$? In figure 6, two typical runs of a generational GA with a *NiH* like that are depicted. The setting (m, l, f_0, τ) was kept fixed but two different mutation rates μ were chosen. In the case of figure 6 *(right)*, the mutation rate is 'too high' to allow the population to track the movement. The concentration of the future needle string (solid line) cannot grow much within one jump cycle resulting in a decreasing initial condition (bullet) for the growth of the needle concentration (dotted line) in the next

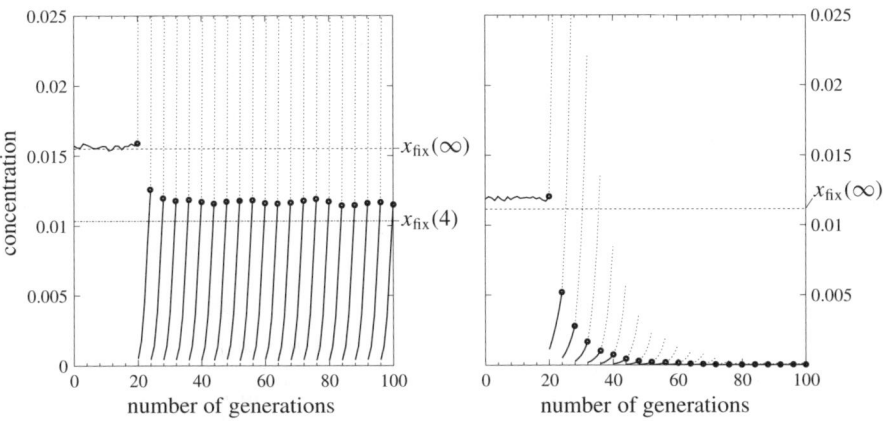

Figure 6. Run of a generational GA with *regularly* moving needle-in-the-haystack. The parameter setting was $m = 1\,000\,000$, $l = 20$, $f_0 = 5$, $\tau = 4$, *(left)*: $\mu = 0.022$, *(right)*: $\mu = 0.055$. In both cases the system evolved for 100 generations (not shown) without any jumps occurring, in order to let a typical quasispecies grow around the initial needle string. In generation 20 the first jump happened and afterwards every $\tau = 4$ generations. *Solid line*: $x_1(n,t)$, *dotted line*: $x_0(n,t)$, *bullet*: jump $- x_0(n+1, 0) = x_1(n, \tau)$.

cycle. The population loses the peak – in this case after ≈ 90 generations. It might happen that the population finds the needle again by chance (or better, the moving needle jumps into the population), but the population will not be able to stably track the movement. In contrast to that, the mutation rate was chosen to maximize the concentration of the future needle string at the end of each jump cycle (bullets) in figure 6 *(left)*.

Since, in that case, the best achievable initial condition is given to each jump cycle, the movement of the needle is tracked with the highest possible stability for the given setting (m, l, f_0, τ). As can be expected from figure 6, and which is affirmed by further experiments, the bullets keep on fluctuating around an average value for $n \to \infty$ which is for the infinite population given by the quasispecies (11). In the following, we are going to model that system with some idealizations and we will calculate a lower boundary for this average value.

We adopt the viewpoint of permuting the concentration vector compatible to the movement of the needle as we have done implicitly in figure 6 and formally in the definition of $\boldsymbol{x}'(t)$ in (10), but we drop the primes henceforth. The concentration of the needle string within jump cycle n is denoted by $x_0(n, \varphi)$ and the concentration of the string the needle will move to with the $(n+1)$-th jump (i.e., the future needle string in jump cycle n) is denoted by $x_1(n, \varphi)$. The initial cycle prior to which *no* jump has occurred is $n = 0$. Within a cycle, the time or generation is counted as phase $\varphi \in \{0, \ldots, \tau\}$. Two succeeding cycles are connected by the (approximated)

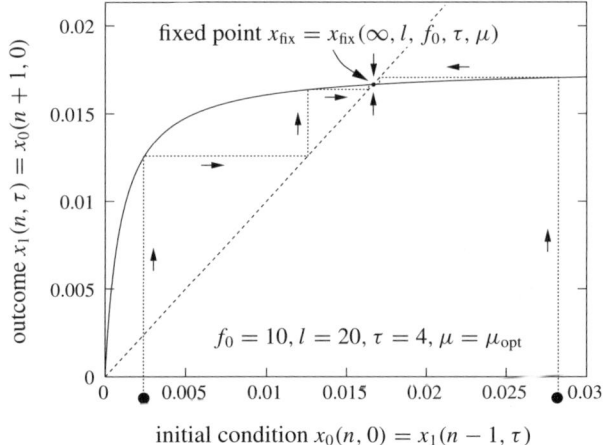

Figure 7. The fixed-point which is reached by an infinite population for $n \to \infty$

rule of change

$$x_0(n+1, 0) = x_1(n, \tau) \quad \text{and} \quad x_1(n+1, 0) \approx 0. \tag{14}$$

The second relation is an approximation which is made to simplify the coming calculations, but it holds only if the needle jumps onto a string which has not been close to one of the previous needle positions. Otherwise, the future needle string could already be present with a concentration significantly larger than $1/2^l \approx 0$. In figure 6, we have chosen the rule $P_{\oplus \ll}$ to get experimental data for a case in which this assumption is fulfilled. Later on we will see that we can still make useful comments about cases in which that approximation is partly broken.

If we plot $x_0(n+1, 0) = x_1(n, \tau)$ against $x_0(n, 0)$, we get an intuitive picture for the system's evolution towards the quasispecies. The concentration $x_0(n, 0)$ converges for $n \to \infty$ towards a fixed-point

$$x_{\text{fix}} := \lim_{n \to \infty} x_0(n, 0)$$

as shown in figure 7 for a finite value of x_{fix}. Obviously, this fixed-point depends on the full setting $x_{\text{fix}} = x_{\text{fix}}(m, l, f_0, \tau, \mu)$. Since we are especially interested in the effects of various cycle lengths τ and mutation rates μ, we keep (m, l, f_0) fixed, such that $x_{\text{fix}} = x_{\text{fix}}(\tau, \mu)$.

In the remaining of this section, we will calculate $x_0(n+1, 0) = x_1(n, \tau)$ in dependence on $x_0(n, 0)$, which is the solid curve in figure 7, for arbitrary parameter settings. From this knowledge, we will construct the phase diagram. Since we stay within one jump cycle, we drop n to take off some notational load.

5.1 Derivation of the Fixed-Point Concentrations

To calculate $x_1(\tau)$, it is sufficient to take only x_0 and x_1 into account, because the assumed initial condition is $x_1(0) \approx 0$, such that the main growth of x_1 is produced by the mutational flow from the needle. Moreover, we assume μ to be small enough such that terms proportional to μ^2 can be neglected. This means we restrict ourselves to the case in which the system is mainly driven by one-bit mutations. Without normalization, the evolution equations then read

$$\begin{aligned} y_0(t+1) &= (1-\mu)^l \; f_0 \, y_0(t) + \{\mu(1-\mu)^{l-1} \, y_1(t)\} \\ y_1(t+1) &= \mu(1-\mu)^{l-1} f_0 \, y_0(t) + (1-\mu)^l \; y_1(t) \end{aligned} \qquad (15)$$

where y_i denote unnormalized concentrations in contrast to the normalized concentrations x_i.

For $f_0(1-\mu) \gg \mu$, which is always the case for large enough f_0, we can further neglect the back-flow $\{\cdots\}$ from the future needle string compared to the self-replication of the current needle string. The solution of (15) is then given by

$$y_0(t) = \left[(1-\mu)^l f_0\right]^t y_0(0)$$
$$y_1(t) = \kappa_t(\mu) \, y_0(0) + (1-\mu)^{lt} y_1(0)$$

$$\text{with} \begin{cases} \kappa_t(\mu) = \mu(1-\mu)^{lt-1} \alpha_t \\ \alpha_t = \sum_{v=1}^{t} f_0^v = f_0 \frac{f_0^t - 1}{f_0 - 1}. \end{cases}$$

The coefficient $\kappa_t(\mu)$ measures the growth of $y_1(t)$ starting from the initial condition $y_1(0) \approx 0$, $y_0(0) \neq 0$. As long as $y_0(t) + y_1(t) \ll 1$, this already gives a good approximation for the concentrations $x_0(t)$ and $x_1(t)$. But, in general, this approximation breaks down for large t, because of the exponential growth of $y_0(t)$. We need to normalize our solution, which can be done by

$$x(t) = y(t) / \langle f \rangle_0 \cdots \langle f \rangle_{t-1}, \quad \text{where } \langle f \rangle_t = (f_0 - 1)x_0(t) + 1. \qquad (16)$$

By expressing the fitness averages in terms of $y_0(t)$, we find, after solving a simple recursion

$$\begin{aligned} \langle f \rangle_0 \cdots \langle f \rangle_{t-1} &= 1 + (f_0 - 1)\left[\sum_{v=0}^{t-1}(1-\mu)^{lv} f_0^v\right] x_0(0) \\ &= 1 + (f_0 - 1)\beta_t(\mu) x_0(0) \end{aligned}$$

$$\text{where } \beta_t(\mu) = \frac{\tilde{f}^t - 1}{\tilde{f} - 1} \text{ and } \tilde{f} = (1-\mu)^l f_0.$$

Finally, we arrive at the normalized concentrations

$$x_0(t) = \left[(1-\mu)^l f_0\right]^t x_0(0) \Big/ [1 + (f_0 - 1)\beta_t(\mu)x_0(0)]$$
$$x_1(t) = \left[\kappa_t(\mu) x_0(0) + (1-\mu)^{lt} x_1(0)\right] \Big/ [1 + (f_0 - 1)\beta_t(\mu)x_0(0)].$$

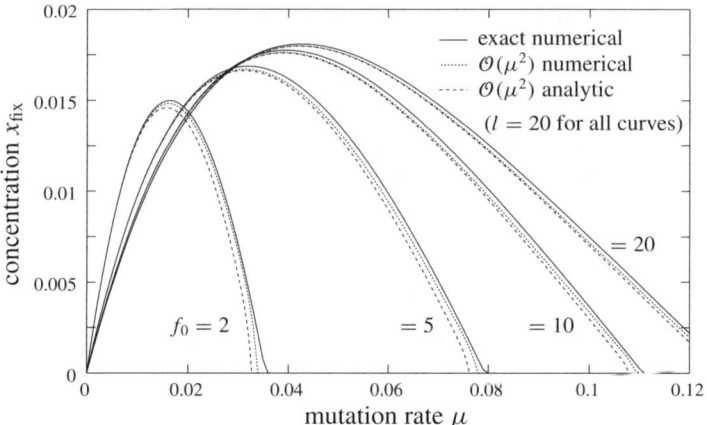

Figure 8. Comparison of the exact numerical and the $\mathcal{O}(\mu^2)$ calculation for different values of the needle fitness f_0

The asymptotic state can now be calculated by using the initial condition $x_1(0) \approx 0$, $x_0(0) \neq 0$ and demanding $x_1(\tau) = x_0(0)$. It is then easily verified that for the fixed-point

$$x_{\text{fix}}(\tau, \mu) = \frac{\kappa_\tau(\mu) - 1}{(f_0 - 1)\beta_\tau(\mu)}. \tag{17}$$

5.2 Consistency in the Quasi-Static Limit

How can we test the quality of the approximate result (17)? For large cycle lengths τ, we enter the quasi-static regime, where we can approximate the population at the end of each cycle by the quasispecies of the corresponding static landscape. Figure 8 shows a comparison of the exact numerical calculations of the quasispecies ($\tau \to \infty$) and the $\mathcal{O}(\mu^2)$ calculations ($\tau = 100$). In the numerical $\mathcal{O}(\mu^2)$ calculation, the back-flow from the future needle string to the current needle string is included. Overall, we find the error threshold and the maximum of the fixed-point concentration well represented. This also suggests that the deviation of the $\mathcal{O}(\mu^2)$ approximation from the exact values should be small for smaller τ, because those deviations add up for $\tau \to \infty$ by the iterative procedure.

How do the calculated fixed-point concentrations compare to simulations with large finite populations? In figure 6, the values of $x_{\text{fix}}(\infty, \mu)$ and $x_{\text{fix}}(4, \mu)$ are shown. For $\tau \to \infty$, the deviation from the average $\langle x_1(n, \varphi) \rangle$ (in generations 0–20) is in fact the same as what can be read off in figure 8. The deviation of $x_{\text{fix}}(4, \mu)$ from the average value $\langle x_0(n, 0) \rangle$ in generations 24, 28, ..., 100 is significantly larger. This is caused by the neglect of all other strings' contributions apart from the

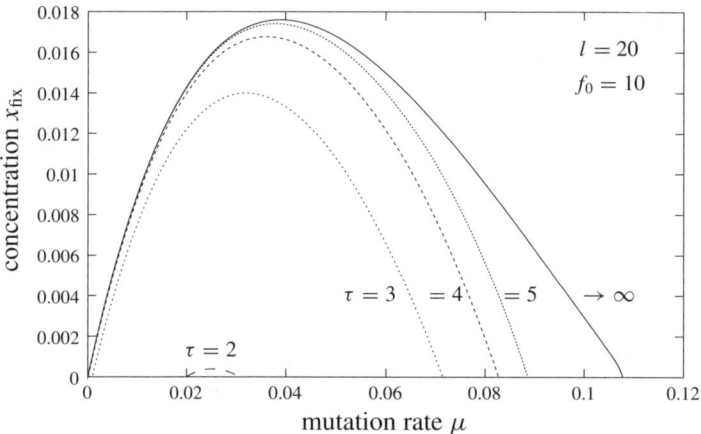

Figure 9. Fixed-point concentration $x_{\text{fix}}(\tau, \mu)$ for different values of τ. For faster changes, the fixed-point concentration rapidly drops down.

current needle string's contribution to the flow onto the future needle string. These neglected contributions increase the average fixed-point concentration measured in the experiment in comparison to the calculated value $x_{\text{fix}}(\tau, \mu)$. But even though there are deviations, we conclude that the approximately calculated value is always a lower bound for the exact value. In the next section, we will use this observation to derive an expression for the mutation rate that maximizes the average fixed-point concentration.

5.3 Phase Diagram

In figure 9, the fixed-point values $x_{\text{fix}}(\tau, \mu)$ are shown for small cycle lengths τ. For the shown parameter setting, the region with $x_{\text{fix}}(2, \mu) > 0$ is extremely small. We notice that there are two error thresholds, one for 'too low' mutation rates, $\mu_{\text{th}<}$, and one for 'too high' mutation rates, $\mu_{\text{th}>}$. The intuition behind that was already given in section 4. For too low mutation rates the population becomes slow and evolves in the averaged, flat landscape, whereas for too high mutation rates the usual transition to the disordered phase takes place. In the following we will calculate the phase diagram starting from (17).

Error Thresholds. The error thresholds are given by

$$x_{\text{fix}}(\tau, \mu) = 0 \quad \Longleftrightarrow \quad \kappa_\tau(\mu) = 1. \tag{18}$$

This is the same condition as one would get using only unnormalized concentrations $y_i(t)$. Since $y_i(t) \approx 0$ near the error thresholds, the neglect of the normalization is not critical for the calculation of the error thresholds themselves, whereas it is important for the optimal mutation rate and, of course, for the fixed-point concentration. Since (18) cannot be solved for μ in closed form, we write down the corresponding recursion relation that converges, for a suitable starting value of μ, to the solution of (18) in the limit $k \to \infty$,

$$\mu_{\text{th}<}^{(k)} = 1 \Big/ \alpha_\tau \left(1 - \mu_{\text{th}<}^{(k-1)}\right), \qquad \mu_{\text{th}<}^{(0)} = 0$$

$$\mu_{\text{th}>}^{(k)} = 1 - \left(1 \Big/ \alpha_\tau \mu_{\text{th}>}^{(k-1)}\right)^{1/(l\tau-1)}, \qquad \mu_{\text{th}>}^{(0)} = 1 - f_0^{-1/l} =: \mu_{\text{th}}^\infty.$$

For $\mu_{\text{th}<}$, a good starting value is 0, since $\mu_{\text{th}<} \approx 0$ anyway. For $\mu_{\text{th}>}$, the approximate value for the error threshold of the static (i.e., $\tau \to \infty$) landscape μ_{th}^∞ can be chosen, which is obtained by calculating the fixed-point [using (15) and (16)]

$$x_0(t+1) = x_0(t) \iff x_{\text{fix}}^\infty = \frac{(1-\mu)^l f_0 - 1}{f_0 - 1}$$

setting it to zero and solving for μ.

Optimal Mutation Rate. In order to track changes with the best achievable stability for a given setting (m, l, f_0, τ), the lowest possible concentration (infimum of) $x_0(n, \varphi)$ needs to be maximized, because a low concentration might result in the loss of the needle string in a finite population. Since for infinite populations $x_0(n, \varphi)$ is monotonously increasing with φ it is sufficient to maximize $x_0(n, 0)$. Moreover, we derived above that $x_0(n, 0)$ approaches the fixed-point value $x_{\text{fix}}(\tau, \mu)$ for $n \to \infty$. For finite populations, we expect similar behavior, but the strict monotony of $x_0(x, \varphi)$ in φ will be destroyed by fluctuations and also the fixed-point value itself will fluctuate around some average value $\langle x_{\text{fix}} \rangle$ as can be seen in figure 6. However, the *safest* way to avoid any loss of the needle string is still to maximize the average fixed-point value $\langle x_{\text{fix}} \rangle$. In this sense, we define the *optimal mutation rate* μ_{opt} as the one that maximizes $\langle x_{\text{fix}} \rangle$. In the previous section 5.2, we noted that our approximated infinite population value $x_{\text{fix}}(\tau, \mu)$ represents a lower bound for $\langle x_{\text{fix}} \rangle$, where the maxima of the two curves are expected to coincide for fixed τ. Thus, μ_{opt} can be obtained by maximization of $x_{\text{fix}}(\tau, \mu)$.

We can derive an expression for the optimal mutation rate μ_{opt} from

$$\frac{\partial x_{\text{fix}}}{\partial \mu}(\tau, \mu_{\text{opt}}) = 0.$$

If we neglect the μ dependence of $\beta_\tau(\mu)$ in (17), which corresponds to the approach in [10], we simply find $\mu_{\text{opt}}^{\text{NS}}(\tau, l) = 1/l\tau$. Because of $\mu_{\text{opt}}^{\text{NS}} \xrightarrow{\tau \to \infty} 0$, this result is inconsistent with the quasi-static limit, because μ_{opt} should approach the value for which the concentration of 1-mutants in the quasispecies of the corresponding static *NiH* landscape is maximized. We conclude that the μ dependence of $\beta_\tau(\mu)$ cannot be neglected for the correct optimal mutation rate, which we are going to calculate now.

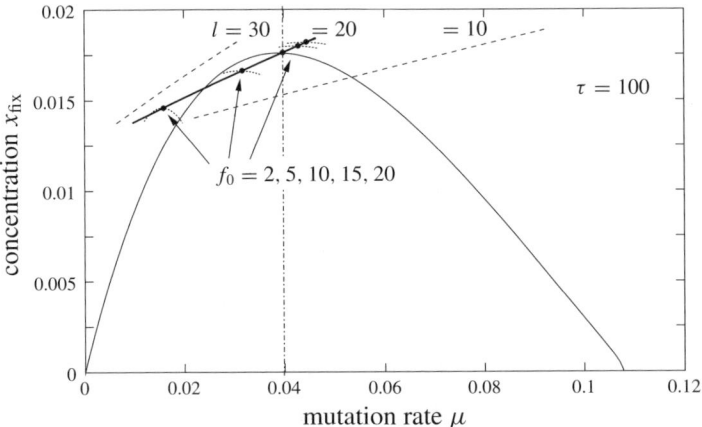

Figure 10. The optimal mutation rate $\mu_{\text{opt}}^\infty(f_0, l)$ from (19) in dependence on needle height f_0 and string length l

For $\alpha_\tau \gg 1$, which is the case for $\tau \gg 1$ and $f_0 > 1$, or $\tau \approx 1$ and $f_0 \gg 1$, we can neglect the -1 in the numerator of $x_{\text{fix}}(\tau, \mu)$ and take only α_τ into account for the calculation of $\partial x_{\text{fix}}/\partial \mu$. After some algebra, we find

$$\mu_{\text{opt}} = \frac{(\tilde{f}^\tau - 1)(\tilde{f} - 1)}{l(\tilde{f}^{\tau+1} - (\tau+1)\tilde{f} + \tau)}, \quad \text{where } \tilde{f} = f_0(1 - \mu_{\text{opt}})^l.$$

Since $\tilde{f} = \tilde{f}(\mu_{\text{opt}})$, this equation cannot be solved in a closed form for μ_{opt}. However, for $\tau \to \infty$ the equation simplifies to

$$\mu_{\text{opt}}^\infty = \begin{cases} (\tilde{f} - 1)/l\tilde{f} & : \tilde{f} > 1 \\ 0 & : \tilde{f} \le 1. \end{cases}$$

In the case $\tilde{f} > 1$, we find

$$(1 - l\mu_{\text{opt}}^\infty)(1 - \mu_{\text{opt}}^\infty)^l = 1/f_0.$$

By approximating $(1 - \mu)^l \approx (1 - l\mu)^2$, we get a cubic equation. The real root of that equation is approximately[1] given by (see also figure 10)

$$\mu_{\text{opt}}^\infty(f_0, l) \approx \mu_+ \left[1 + \frac{(l-1)\mu_+(1 - l\mu_+)}{3l(l-1)\mu_+^2 - 2\mu_+(3l-1) + 4}\right],$$

$$\text{with } \mu_+ = \frac{1}{l}\left[1 + f_0^{-1/2}\right]. \quad (19)$$

---

[1] A more detailed explanation and analysis of the used approximation will be presented elsewhere.

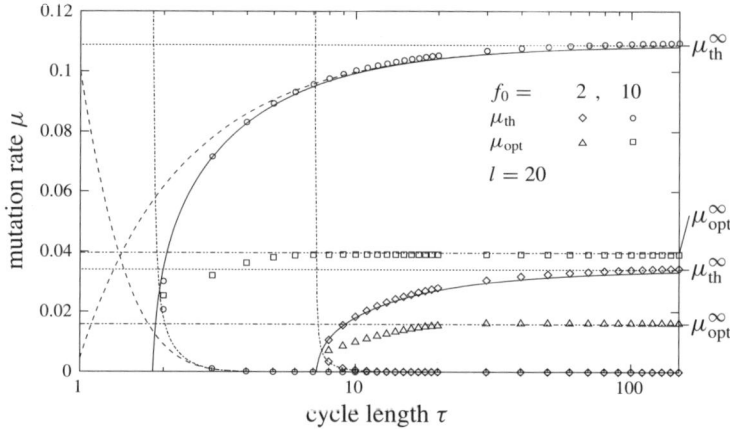

Figure 11. The calculated phase diagram for a generational GA with stochastically moving needle-in-the-haystack; two settings are shown: $f_0 = 2, 10$, both for $l = 20$.

Resulting Phase Diagram. From the above, we are able to plot the phase diagram for our model as shown in figure 11. Two settings are plotted. For $f_0 = 2$ (or 10, resp.) the diamonds (or circles, resp.) are the numerically obtained error thresholds. The solid and dash-dotted lines are $\mu_{\text{th}<}^{(5)}$ and $\mu_{\text{th}>}^{(5)}$. To show the convergence property of $\mu_{\text{th}<,>}^{(k)}$, $\mu_{\text{th}<,>}^{(0)}$ are plotted for $f_0 = 10$ as dashed lines. Obviously, the corrections needed to the chosen starting values increase for smaller τ, such that more iterations are needed to describe the error thresholds correctly for small τ. The expressions $\mu_{\text{th}<,>}^{(5)}$ are already a good approximation for the given settings. Representing the quasi-static limit, μ_{th}^∞ is plotted as dotted lines and gets consistently approached by $\mu_{\text{th}>}(\tau)$ for $\tau \to \infty$. Furthermore, μ_{opt}^∞ is plotted as dash-dot-dotted lines. The numerically measured values for $\mu_{\text{opt}}(\tau)$ are shown for $f_0 = 2$ (or 10, resp.) as triangles (or squares, resp.). They approach μ_{opt}^∞ very quickly and coincide with it already for $\tau \approx 20$ (respectively, 10).

We conclude that the above quantitative description is in good agreement with the numerical observations and approaches the quasi-static region in a consistent way. Moreover, the phase diagram fits well into the general one raised in section 4. Even in the considered case of a generational GA, we find – depending on the parameter setting – a time-averaged phase for very small τ. The time-averaged phase broadens for small f_0.

5.4 Stochastically Moving *NiH*

Up to now, we analyzed a regularly moving *NiH*, for example, with the rule $P_{\oplus\ll}$. What happens if the *NiH* is allowed to move to a *randomly* picked nearest neighbor, as is shown in figure 12 for $l = 4$? Two typical runs of a generational GA with this

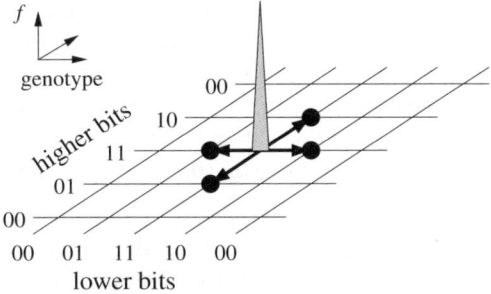

Figure 12. A *stochastically* moving needle-in-the-haystack for string length $l = 4$. The needle is allowed to jump to one of its nearest neighbors which is chosen at random.

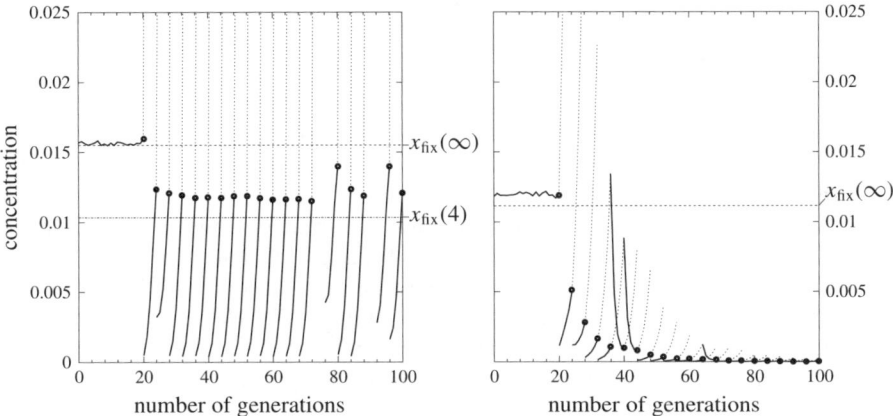

Figure 13. Run of a generational GA with *stochastically* moving needle-in-the-haystack. The parameter setting in *(left)* and *(right)* were the same as in figure 6 *(left)* and *(right)*.

fitness landscape are depicted in figure 13. The setting (m, l, f_0, τ) was chosen to be the same as in figure 6 which allows for a direct comparison of the GA's behavior for regularly and stochastically moving *NiH*s. The overall behavior is similar. For large mutation rates, the population loses the needle string, whereas the moving needle is tracked stably for mutation rates close to the above defined optimal mutation rate. In addition, strong fluctuations in the values of $x_1(n, 0)$ (lower ends of solid lines) as well as $x_0(n + 1, 0) = x_1(n, \tau)$ (bullets) occur in the *stochastic* case. These result from *back-jumps*. If, at the end of the current cycle, the needle jumps back to the string it has been to in the previous cycle, then $x_1(n, 0) = x_0(n - 1, \tau)$ is significantly larger than zero. This can be seen in figure 13 *(right)* at generations 36, 40, and 64 and also in figure 13 *(left)* at generations 72 and 88 (the gaps in figure 13 *(left)* correspond to x_1, x_0 being much larger than 0.025). If no back-jumps occur, as in generations 24–72 in figure 13 *(left)*, the system with stochastic *NiH* behaves nearly indistinguishably from the one with regularly moving *NiH*. Since back-jumps always increase the concentrations of the needle string in the very next occurring jumps, the above calculated fixed-point $x_{\text{fix}}(\tau, \mu)$ is still a lower bound. Thus, our previous notion of optimal mutation rate remains applicable to the stochastically moving *NiH* although the assumption $x_1(n, 0) \approx 0$ from (14) is not always fulfilled.

Nilsson and Snoad [10] did their analysis of the continuous Eigen model (3) with stochastic *NiH* in a similar way as we did above. In analogy to their calculation for the continuous Eigen model, we find for a generational GA the optimal mutation rate $\mu_{\text{opt}}^{\text{NS}}(\tau, l) = 1/l\tau$ which is inconsistent with the quasi-static limit (see section 5.3). The reason is the missing normalization (16) in the work of Nilsson and Snoad. Furthermore, they could not derive an expression for the fixed-point concentration $x_{\text{fix}}(\tau, \mu)$ for that same reason.

5.5 Jumps of Larger Distance

To conclude this section about the behavior of generational GAs with different kinds of *NiH*s that move to *nearest* neighbors, let us briefly discuss jumps of Hamming distance d *larger* than one. Obviously, the analytical calculations get more complicated, because the $\mathcal{O}(\mu^2)$-approximation is not sufficient anymore as it connects only nearest neighbors. To describe jumps of a larger distance, the concentrations of some intermediate sequences need to be taken into account, so that we have to solve a time evolution much more complicated than (15). Hence, we cannot make simple statements for finite τ. On the other hand, the system approaches the quasi-static region for large τ and it is characterized by $\mu_{\text{th}<,>}^{\infty}$ and $\mu_{\text{opt}}^{\infty}$ as we have seen in figure 11.

The exact quasispecies for $\tau \to \infty$ is shown in figure 14. The plotted values are error class concentrations, in order to make the higher error classes visible at all. Each k-mutant has a concentration of $\tilde{x}_k / \binom{l}{k}$ in the quasispecies state, because for a *NiH* the mutant's fitness depends only on its Hamming distance to the needle and therefore all $\binom{l}{k}$ k-mutants have the same concentration in the quasispecies. For finite populations, this is only true on average, because the asymptotic state is distorted by fluctuations. But in the following, we assume that the quasispecies is still representative for the average distribution of the population in the asymptotic state. Then, the optimal mutation rate, in the sense of section 5.3, for jumps of distance d is by definition the position of the maximum of $\tilde{x}_d$. For $d \geq l/2$, optimal mutation rate and error threshold become identical. Although $\tilde{x}_d$ is maximized for mutation rates close to the error threshold it amounts, as do all other concentrations, to only $\approx 1/2^l$, which leads to an approximately random drift for finite populations. On the other hand, the chance of tracking the needle decreases even further for small mutation rates because then the concentration $\tilde{x}_d$ becomes even smaller. In this sense, the quasispecies distribution, which is centered on the needle string, is useless for tracking the next jump if $d \geq l/2$. This also suggests – in agreement with the experimental findings of Rowe [14] – that finite populations, for low mutation

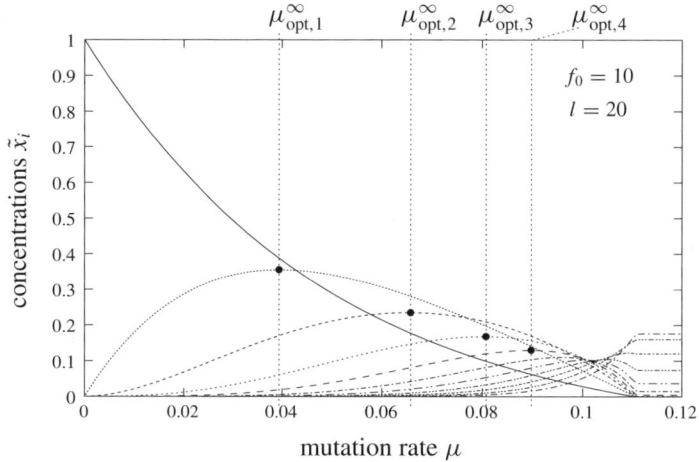

Figure 14. The quasispecies for the static *NiH* in dependence on the mutation rate μ. The concentrations $\tilde{x}_i$ of the ith error class for $i \in \{0, \ldots, \lfloor l/2 \rfloor\}$ are depicted. The optimal mutation rates for jumps of Hamming distance $d = 1, 2, 3, 4$ are shown as *dotted lines*.

rates, are unable to track large jumps – in particular in the extreme case $d = l$. Only for jumps of $d < l/2$ does the corresponding error class concentration $\tilde{x}_d$ show a concentration maximum significantly above $1/2^l$. From the heights of the concentration maxima, we see that the difficulty of tracking the changes increases with the Hamming distance d of the jumps. Vice versa, the advantage a population gets after a jump from its structure prior to the jump decreases with increasing jump distance d. In addition, a mutation rate which is simultaneously optimal for more than one distance cannot be found.

6 Conclusions and Future Work

On the basis of general arguments, the phase diagrams of population-based mutation and probabilistic selection systems like the above generational GA, steady state GA, and Eigen model in time-dependent fitness landscapes can be easily understood. The notion of regular changes allows for an exact calculation of the asymptotic state in the sense of a generalized, time-dependent quasispecies. For a generational GA with *NiH* that moves regularly to nearest neighbors, the quasispecies can be straightforwardly calculated under simplifying assumptions. The result is a lower bound for the exact quasispecies. With that lower bound, we have constructed the phase diagram in the infinite population limit. This phase diagram is in agreement with the one raised from first principles.

In order to improve our quantitative analysis, we need to weaken our assumptions. In particular, we have to overcome the restriction of taking into account only the flow from the current towards the future needle string. The presence of other contributions to the flow has to be modeled in some way. Another future step could be an investigation of the fluctuations that are introduced by the finiteness of realistic populations (discreteness of Λ_m) around the quasispecies. This would lead to a lower boundary for the population size above which the needle string is not lost due to those fluctuations.

A generalization of the analysis to non-regularities, like the occurrence of more than a single jump rule, can be achieved by averaging the time evolution (8) for $n \to \infty$ according to each rule's probability of being applied. A similar averaging procedure will be necessary if fluctuations of the cycle length τ are present. Finally, an extension of our description to broader, more realistic peaks, as well as GA models including crossover and other selection schemes, are important topics for future work.

References

1. E. Baake and W. Gabriel. Biological evolution through mutation, selection, and drift: an introductory review. *Ann. Rev. Comp. Phys.*, 7:203–264, 2000.
2. T. Bäck, U. Hammel and H.-P. Schwefel. Evolutionary computation: comments on the history and current state. *IEEE Transactions on Evolutionary Computation*, 1(1):3–17, 1997.
3. T. Bäck, D. B. Fogel and Z. Michalewicz, editors. *Handbook of Evolutionary Computation*. IOP Publishing, Bristol, 1997.
4. J. Branke. Evolutionary algorithms for dynamic optimization problems, a survey. Technical Report 387, AIFB University Karlsruhe, 1999.
5. J. Branke, M. Cutaia and H. Dold. Reducing genetic drift in steady state evolutionary algorithms. In W. Banzhaf et al., editors, *Proceedings of GECCO 1999*, pages 68–74. Morgan Kaufmann, San Francisco, 1999.
6. K. DeJong and J. Sarma. Generation gaps revisited. In L. D. Whitley, editor, *Foundations of Genetic Algorithms 2*, pages 19–28. Morgan Kaufmann, San Francisco, 1993.
7. M. Eigen. Self-organization of matter and the evolution of biological macromolecules. *Naturwissenschaften*, 58:465–523, 1971.
8. M. Eigen and P. Schuster. *The Hypercycle – a Principle of Natural Self-Organization.* Springer-Verlag, Berlin Heidelberg New York, 1979.
9. M. Eigen, J. McCaskill and P. Schuster. The molecular quasispecies. *Adv. Chem. Phys.*, 75:149–263, 1989.
10. M. Nilsson and N. Snoad. Error thresholds on dynamic fitness-landscapes. Working Paper 99-04-030, Santa Fe Institute, 1999.
11. A. Rogers and A. Prügel-Bennett. Modeling the dynamics of a steady state genetic algorithm. In W. Banzhaf and C. Reeves, editors, *Foundations of Genetic Algorithms 5*, pages 57–68. Morgan Kaufmann, San Francisco, 1998.
12. J. E. Rowe. Finding attractors for periodic fitness functions. In W. Banzhaf et al., editors, *Proceedings of the Genetic and Evolutionary Computation Conference, GECCO 1999*, pages 557–563. Morgan Kaufmann, San Francisco, 1999.
13. J. E. Rowe. The dynamical systems model of the simple genetic algorithm. This volume, pages 31–57.

14. J. E. Rowe. Cyclic attractors and quasispecies adaptability. This volume, pages 251–259.
15. L. M. Schmitt, C. L. Nehaniv and R. H. Fujii. Linear analysis of genetic algorithms. *Theoretical Computer Science*, 200(1–2):101–134, 1998.
16. E. van Nimwegen, J. P. Crutchfield and M. Mitchell. Statistical dynamics of the Royal-Road genetic algorithms. *Theoretical Computer Science*, 229(1-2):41-102, 1999.
17. M. D. Vose. *The Simple Genetic Algorithm – Foundations and Theory*. MIT Press, Cambridge, MA, 1999.
18. C. O. Wilke, C. Ronnewinkel and T. Martinetz. Molecular evolution in time-dependent environments. In D. Floreano, J.-D. Nicoud and F. Mondada, editors, *Proceedings of the European Conference on Artificial Life 1999*, pages 417–421. Springer-Verlag, Berlin Heidelberg New York, 1999.
19. C. O. Wilke and C. Ronnewinkel. Dynamic fitness landscapes in the quasispecies model. *Physics Reports,* Elsevier, in press.

Statistical Machine Learning and Combinatorial Optimization

A. Berny

IRIN, Université de Nantes
2 rue de la Houssinière
BP 92208
44322 Nantes cedex 3, France
E-mail: *Arnaud.Berny@ectia.ec-nantes.fr*

Abstract. In this work we apply statistical learning methods in the context of combinatorial optimization, which is understood as finding a binary string minimizing a given cost function. We first consider probability densities over binary strings and we define two different statistical criteria. Then we recast the initial problem as the problem of finding a density minimizing one of the two criteria. We restrict ourselves to densities described by a small number of parameters and solve the new problem by means of gradient techniques. This results in stochastic algorithms which iteratively update density parameters. We apply these algorithms to two families of densities, the Bernoulli model and the Gaussian model. The algorithms have been implemented and some experiments are reported.

Keywords
Learning a Gibbs distribution, free energy minimization, gradient descent

1 Introduction

In this work, we apply statistical learning methods in the context of combinatorial optimization, which is understood as finding a binary string minimizing a given cost function. We transform the initial search problem over binary strings into a search problem over binary strings probability distributions. The solutions of the latter are mixtures of Dirac distributions charging optimal solutions of the former. However, the search for a probability distribution rather than for a string does not reduce the complexity. In order to make this model tractable, we decide to restrict ourselves to distributions completely determined by a small – compared to the size of the binary strings space – number of parameters. In this paper we consider two families of probability distributions, the Bernoulli model and the Gaussian model.

The search criterion in the space of binary strings is the cost function of the problem itself. We define two additional criteria for the search over distributions. The first one is simply the expectation of the cost function relative to a distribution. The second one is the Kullback–Leibler divergence between a distribution and the Gibbs distribution at a given temperature. Finally, we replace the search for a minimum over the set of binary strings, which is discrete and finite, by the search for a minimum of one of the two criteria, which are differentiable functions relative to distribution parameters. In order to minimize one of the two criteria, we use gradient dynamical systems acting on distribution parameters. There are some well-known limitations to this approach, namely, the fact that it is subject to local minima. But we do not simulate this dynamical system. We rather turn to a stochastic approximation of it, enabling the system to escape local minima to some extent. We set the initial conditions of the algorithm such that, at first, it behaves like a random search. While relaxing, it is biased towards binary strings with lower cost. We are aware of the fact that this approach is likely to be defeated by deceptive functions (such as those exhibiting Hamming-isolated optimal solutions). It is intended to demonstrate the link between statistical learning procedures and combinatorial optimization.

The paper is organized as follows. Section 2 is a derivation of the first algorithm from the principles of statistical physics, the Gibbs distribution and the Kullback–Leibler divergence. In section 3 we study the second algorithm which relies on the concept of reinforcement learning. In section 4 we consider the choice of Bernoulli random variables and in section 5 the choice of the Gaussian law to take into account second-order statistics. Section 6 is a summary of the main equations, including dynamics and distributions. Finally, in section 7 and section 8, we apply the above ideas to two combinatorial optimization problems, the four peaks problem and the iterated prisoner's dilemma problem, and report some experiments and results.

2 Free Energy Minimization

This section first introduces some concepts drawn from statistical physics, in particular, the Gibbs distribution which models a system in thermo-dynamical equilibrium at a given temperature. Such a statistical framework has been introduced into many fields. One can cite combinatorial optimization [9], machine learning [14,10] or image processing [6]. However, relaxation methods (without memory) have prevailed in the field of combinatorial optimization. In the remainder of the paper we focus on the application of learning or adaptive algorithms to combinatorial optimization. We use the Gibbs distribution as a reference distribution and try to approximate or learn it with a target distribution. This is motivated by the fact that the Gibbs distribution charges low cost states at low temperatures. The comparison between both distributions relies on the Kullback–Leibler divergence which gives a statistical criterion to be minimized. The process of minimizing the criterion is then achieved by means of a gradient dynamical system which is finally approximated by a stochastic discrete-time dynamical system.

2.1 Statistical Framework

We first give some mathematical notations. Let X be a finite set of n binary variables X_i taking their values in the hypercube $\mathcal{S} = \{0, 1\}^n$. Let x_i be the value of variable

X_i. The problem consists in minimizing a function $H : \mathcal{S} \to \mathbb{R}$ called the cost function of the problem.

In order to plunge an optimization problem into a statistical framework we use the Gibbs distribution [6]. The Gibbs distribution maps each value of the cost function onto a probability defined by

$$p_T^*(x) = \frac{1}{Z} \times \exp\left(-\frac{H(x)}{T}\right)$$

where $T > 0$ is the analogue of a temperature and $Z = \sum_{y \in \mathcal{S}} \exp(-H(y)/T)$ is a function which only depends on T. Z is sometimes called the *partition function*. The Gibbs distribution p_T^* has the fundamental property that when $T \to 0$ it converges to a limit distribution p^* which uniformly charges optimal solutions. Put another way, if $U = \{x \in \mathcal{S} / \forall y \in \mathcal{S}, H(x) \leq H(y)\}$ is the set of all optimal solutions, then

$$p^*(x) = \begin{cases} 1/|U| & \text{if } x \in U \\ 0 & \text{otherwise} \end{cases}$$

where $|U|$ is the cardinal of U [6].

2.2 Optimization as Learning

The theoretical distribution p_T^* is an implicit information source. The Metropolis algorithm allows one to sample from p_T^* by building a Markov chain whose stationary distribution is precisely p_T^* [8]. We adopt a different point of view in this work. It is clear from the statistical framework (see page 288) that if we approximate or learn the theoretical distribution p_T^* with a target distribution p then low cost states will be more likely than high cost states. In that sense optimization can be seen as a learning task.

We need some sort of distance between the Gibbs distribution and the target distribution to evaluate the quality of approximation. For this purpose, we use the Kullback–Leibler (KL) divergence [10] which is well-suited to exponential functions. The KL divergence between p and p_T^* is defined by

$$\mathcal{D}(p, p_T^*) = -\sum_{x \in \mathcal{S}} p(x) \times \log \frac{p_T^*(x)}{p(x)}.$$

It has, in particular, the following properties:

1. $\mathcal{D}(p, p_T^*) \geq 0$;
2. $\mathcal{D}(p, p_T^*) = 0$ if and only if $p = p_T^*$.

Replacing p_T^* by its expression, we find

$$\mathcal{D}(p, p_T^*) = \sum_{x \in \mathcal{S}} p(x) \times \left(\log Z + \frac{H(x)}{T} + \log p(x)\right).$$

Since Z is a function which only depends on the temperature T and since

$$\sum_{x \in \mathcal{S}} p(x) = 1$$

we simplify the expression

$$\mathcal{D}(p, p_T^*) = \log Z + \frac{1}{T}\left(\sum_{x \in \mathcal{S}} p(x)H(x) + T \times \sum_{x \in \mathcal{S}} p(x) \log p(x)\right)$$

$$= \log Z + \frac{1}{T}(E - TS)$$

where we have introduced some thermo-dynamical quantities:

- $E = \sum_{x \in \mathcal{S}} p(x)H(x)$ is the *internal energy* of the system,
- $S = -\sum_{x \in \mathcal{S}} p(x) \log p(x)$ is its *entropy*.

The quantity $F = E - TS$ is called the *free energy* of the system. When operating at constant temperature, minimizing the KL divergence, i.e., the learning task, reduces to minimizing the free energy of the system.

2.3 Dynamics

Recall that we want to minimize the free energy F of the system. The corresponding search space is the set of probability distributions over $\mathcal{S}$. However, $2^n - 1$ positive numbers are needed to describe a probability distribution. So we restrict the search to probability distributions described by a reduced number v of parameters $\theta = (\theta_1, \ldots, \theta_v)^t$, where $\cdot^t$ denotes transpose. We write such a distribution $p(\cdot, \theta)$. We also require $p(x, \cdot)$ to be differentiable with respect to its parameters for each $x \in \mathcal{S}$ so that the free energy F is itself differentiable, although the initial space $\mathcal{S}$ is discrete. The problem may now be restated as finding a vector $\theta^* \in \mathbb{R}^v$ such that $F(\theta^*) = \min_{\mathbb{R}^v} F$. In order to yield a solution, we apply gradient techniques.

We start with a dynamical system defined by the ordinary differential equation

$$\frac{d\theta}{dt} + \frac{\partial F}{\partial \theta} = 0$$

where $\partial F/\partial \theta$ is the gradient vector $(\partial F/\partial \theta_1, \ldots, \partial F/\partial \theta_n)^t$. We exchange the sum over the state space and the gradient

$$\frac{d\theta}{dt} + \sum_{x \in \mathcal{S}} \left(H(x)\frac{\partial p}{\partial \theta}(x) + T \times \frac{\partial (p \log p)}{\partial \theta}(x)\right) = 0.$$

We finally obtain

$$\frac{d\theta}{dt} + \sum_{x \in \mathcal{S}} (H(x) + T(1 + \log p(x)))\frac{\partial p}{\partial \theta}(x) = 0.$$

We write $\partial p/\partial\theta = p \times \partial\log p/\partial\theta$ and we apply the stochastic approximation technique [5], which means that we replace the ordinary differential equation by a stochastic discrete time dynamical system: we just leave the weighted sum over the state space and only retain the current sample. An intermediary solution is to compute a partial sum before updating parameters. The basic update rule is then

$$\Delta\theta = -\alpha\bigl(H + T(1 + \log p)\bigr)\frac{\partial\log p}{\partial\theta} \qquad (1)$$

where $\alpha > 0$ is a small constant called the learning rate. This update rule is similar to the one derived by Sabes and Jordan in the field of unsupervised associative learning [14].

2.4 Algorithm

Assuming we know how to randomly generate solutions in $\mathcal{S}$ following a given probability distribution $p(\cdot, \theta)$, we are now ready to write the algorithm:

step 1 initialize θ;
step 2 generate $x \in \mathcal{S}$ from $p(\cdot, \theta)$;
step 3 evaluate $H(x)$;
step 4 update θ with (1);
step 5 go to step 2.

The initial parameter θ_0 should be chosen such that $p(\cdot, \theta_0)$ is uniform over $\mathcal{S}$ in order to ensure sufficient exploration before convergence. The algorithm runs until a global optimum is reached, a time bound is encountered, or the distribution is concentrated enough so that one considers that the system has converged. Both the temperature T and the learning rate α are chosen empirically.

3 Reinforcement Learning

Reinforcement learning is the second framework for mixing learning and optimization that we study. It also leads to a dynamical system and an update rule similar to the one derived in the previous section. Basic reinforcement learning involves an agent faced with a learning task and an environment. Learning is unsupervised in the sense that the agent is not taught with examples but rather makes trials evaluated by the environment which sends back to the agent a reinforcement signal. The agent modifies its behavior in order to maximize or minimize the reinforcement signal which is a scalar information. In the context of combinatorial optimization, the agent learns to produce low cost binary strings and the reinforcement signal is simply the cost function. We first review some related work on reinforcement learning and then derive an update rule from its principles. Finally, we give the corresponding algorithm.

3.1 Related Work

Williams has reviewed many reinforcement learning schemes and has abstracted them into a generic class called REINFORCE [15]. He has focused his work on tasks which involve learning with associative networks. We can also cite the work of Barto, Sutton and Anderson in the context of learning control [3].

Let us recall the REINFORCE algorithm:

$$\Delta w_{ij} = \alpha_{ij}(r - b_{ij}) \frac{\partial \log g_i}{\partial w_{ij}}$$

where r is the reinforcement signal to be maximized in this context, w_{ij} is the weight of the connection from input j to unit i, g_i is the probability of the output of unit i given the input and the weights to this unit. b_{ij} is called the baseline: it should be independent of the output of unit i. α_{ij} is the learning rate: it should depend at most on the weights w_{ij} and time. The partial derivative of the logarithm of g_i is called the eligibility of weight w_{ij}. We have already seen such a log-derivative in (1) while minimizing the free energy of the system. Williams has proved that REINFORCE algorithms are in fact gradient-following algorithms.

3.2 Reinforcement Learning for Combinatorial Optimization

In Williams' model, there are three sources of randomness:

1. the environment's choice of input to the network;
2. the network's choice of output;
3. the environment's choice of reinforcement value.

In the context of combinatorial optimization, we only keep the second source of randomness: there is no input and the reinforcement signal (the cost function) is deterministic. We will only consider stationary reinforcement signals.

We propose to modify the REINFORCE algorithm to fit our task:

$$\Delta \theta_i = -\alpha \left(H(x) - b_i \right) \frac{\partial \log p}{\partial \theta_i} \tag{2}$$

where α is the learning rate, b_i is the baseline, and $\partial \log p / \partial \theta_i$ is the eligibility. Formally, this is still a REINFORCE algorithm but we directly prove that it is gradient-following:

$$E_X[\Delta \theta_i] = - E_X \left[\alpha \left(H(x) - b_i \right) \frac{1}{p} \frac{\partial p}{\partial \theta_i} \right]$$

$$= - \sum_{x \in \mathcal{S}} \alpha \left(H(x) - b_i \right) \frac{\partial p}{\partial \theta_i}$$

$$= - \alpha \sum_{x \in \mathcal{S}} H(x) \frac{\partial p}{\partial \theta_i} + \alpha b_i \sum_{x \in \mathcal{S}} \frac{\partial p}{\partial \theta_i}$$

where we have used the fact that $\partial \log p/\partial \theta_i = 1/p \times \partial p/\partial \theta_i$ and that both α and b_i are independent of X. The second term of the sum vanishes since $\sum_{x \in S} p(x) = 1$ and $\sum_{x \in S} \partial p/\partial \theta_i = 0$. We exchange the sum and the derivation:

$$E_X[\Delta \theta_i] = -\alpha \frac{\partial}{\partial \theta_i} \sum_{x \in S} H(x) p(x)$$
$$= -\alpha \frac{\partial}{\partial \theta_i} E_X[H(x)]$$

which completes the proof. If we let $\boldsymbol{E} = E_X[H(x)]$, then (2) is a stochastic approximation of the dynamical system

$$\frac{d\theta}{dt} + \frac{\partial \boldsymbol{E}}{\partial \theta} = 0.$$

Moreover, $\boldsymbol{E}$ is the internal energy of the system as defined on page 290 in an analogy with statistical physics. Consequently, reinforcement learning is equivalent to minimizing the internal energy of the system (not its free energy).

Although the baseline b_i is not necessary to the proof, it greatly improves performance. For a discussion on this issue see [3,15]. We have studied two different possibilities for baseline which both relies on first-order linear estimation of some expectations.

The first strategy for baseline is the expectation $E_X[H(x)]$ itself. The idea behind it is that if a randomly generated solution x happens to have a cost $H(x)$ lower than the average cost $E_X[H(x)]$, then the automata should be reinforced to make this solution x more likely. The average cost $E_X[H(x)]$ is estimated with the linear recursive equation

$$b(t+1) = \gamma b(t) + (1-\gamma) H(x(t)) \qquad (3)$$

where $0 < \gamma < 1$. The baseline is independent of the component of the vector θ. This is the common approach and we will refer to it as the "expectation strategy".

The second strategy for baseline stems from the choice of minimizing the variance of $\Delta \theta_i$ which is written $V_X[\Delta \theta_i]$. We have $V_X[\Delta \theta_i] = E_X[\Delta \theta_i^2] - E_X^2[\Delta \theta_i]$ and the second term is independent of b_i, so minimizing the variance reduces to minimizing the first term, $J_i = E_X[e_i^2(H(x) - b_i)^2]$, where $e_i = \partial \log p/\partial \theta_i$ is the eligibility. J_i is a polynomial of degree 2 in b_i and its minimum is reached at

$$b_i = \frac{E_X[e_i^2 H(x)]}{E_X[e_i^2]}.$$

Both the numerator n and the denominator d are estimated with linear recursive equations:

$$n(t+1) = \gamma n(t) + (1-\gamma) e_i^2(t) H(x(t))$$
$$d(t+1) = \gamma d(t) + (1-\gamma) e_i^2(t) \qquad (4)$$

In this strategy each component has a different baseline. We will refer to it as the "minimum variance strategy".

3.3 Algorithm

The following algorithm is a variant of the one described on page 291:

step 1 initialize θ and b (expectation) or n_i and d_i for each component (minimum variance);
step 2 generate $x \in S$ from $p(\cdot, \theta)$;
step 3 evaluate $H(x)$;
step 4 update θ with (2);
step 5 update b with (3) (expectation) or n_i and d_i for each component with (4) (minimum variance);
step 6 go to step 2.

4 Bernoulli Random Variables

At this point we have seen two dynamics or update rules which operate on the parameter vector of the target or approximate distribution: free energy minimization for (1) and reinforcement learning for (2).

We now have to choose a family of distributions from which we can sample efficiently. In doing so, we may restrict the class of problems that are solvable with our approach but this is the only way to derive practical algorithms. In this section, we choose the random variables X_i as being independent. $p(x)$ then is written as the product

$$\prod_i \left(x_i p_i + (1 - x_i)(1 - p_i)\right) \tag{5}$$

where p_i is the probability that x_i equals 1. We emphasize the fact that, in the context of population genetics, the asymptotic distribution of an infinite population of binary strings evolved through crossover only is precisely of the form of (5) and that convergence to equilibrium is fast (see [13]).

The n numbers p_i cannot be the parameters of the distribution on which (1) or (2) operate since they are bounded over [0, 1] and nothing in these equations ensures that this constraint is enforced. We thus insert a nonlinear function which bounds the p_i. We choose for example a sigmoid such as

$$p_i = g(\theta_i) = \frac{1}{2}(1 + \tanh(\beta \theta_i)).$$

Hopfield has used this technique in his fully connected neural networks [7]; it is also used in multilayer perceptrons trained with back-propagation.

From the expression of $p(x)$, we compute the gradient of its logarithm:

$$\frac{\partial \log p}{\partial \theta_i} = \frac{\partial \log p}{\partial p_i} \times \frac{\partial p_i}{\partial \theta_i}$$

$$= \frac{2x_i - 1}{x_i p_i + (1 - x_i)(1 - p_i)} g'(\theta_i)$$

We take advantage of the fact that $(\tanh)' = 1 - \tanh^2$ and express $g'(\theta_i)$ as a function of p_i

$$g'(\theta_i) = 2\beta p_i (1 - p_i).$$

We substitute $g'(\theta_i)$ in the expression of the derivative

$$\frac{\partial \log p}{\partial \theta_i} = 2\beta(2x_i - 1) \frac{p_i(1 - p_i)}{x_i p_i + (1 - x_i)(1 - p_i)}$$

and finally factorize both cases $x_i = 0/1$ into the following expression:

$$\frac{\partial \log p}{\partial \theta_i} = 2\beta(x_i - p_i).$$

The mean of the random variable X_i is p_i. We then write

$$\frac{\partial \log p}{\partial \theta} = 2\beta(x - \mu) \tag{6}$$

where $\mu = (p_1, \ldots, p_n)^t$ is the mean of X.

Narendra and Thathachar [12] have previously studied similar stochastic learning automata based upon Bernoulli random variables, for which the environment is itself stochastic rather than deterministic. Baluja and Caruana [2] have proposed an algorithm which combines a population technique with selection and an update rule similar to that proposed by Narendra and Thathachar. Let us describe their algorithm.

step 1 initialize μ with $p_i = 1/2, i = 1, \ldots, n$;
step 2 generate a population of N individuals sampled from the Bernoulli distribution p with parameter μ;
step 3 evaluate each individual according to H;
step 4 select the fittest x among the population;
step 5 update μ using the fittest x with the rule

$$\mu(t + 1) = \mu(t) + \alpha(x - \mu(t)); \tag{7}$$

step 6 go to step 2.

Provided that $0 < \alpha < 1$ and $\mu(t)$ lies inside the hypercube $\mathcal{S}$ and since $x \in \mathcal{S}$ the resulting μ will lie inside this same hypercube. There is no need for a bounding function in this approach. The right part of (7) is similar to the expression of the log-derivative in (6). The main difference with our work is that the cost function H is made implicit in (7) via selection, in contrast to both (1) and (2).

5 Gaussian Probability Distribution

The choice of independent Bernoulli random variables is the simplest one and may not fit complex problems. One direction to improve algorithm performance may be to take into account second-order statistics. In the context of combinatorial optimization, it means being able to generate binary random variables following second-order statistics specifications, not estimating those statistics. In this regard, the Gaussian probability distribution is of interest and can be considered as a quadratical approximation of optimal distributions such as the Gibbs distribution p_T^* defined on page 289. In this section we will mainly compute the log-derivative of the Gaussian probability density p.

For any real-valued n-dimensional vector x, we consider the Gaussian density

$$p(x) = \frac{1}{(2\pi)^{\frac{n}{2}}} (\det C)^{\frac{1}{2}} \exp\left(-\frac{1}{2}(x - \mu)^t C(x - \mu)\right)$$

where C is the inverse of the covariance matrix, μ is the mean, and $\det C$ is the determinant of C. Since C is symmetric positive definite, it can be factorized into its positive Choleski's decomposition $C = LL^t$, where L is a lower triangular matrix with $l_{ii} > 0$. We substitute L in $p(x)$

$$p(x) = \frac{1}{(2\pi)^{\frac{n}{2}}} \times \det L \times \exp\left(-\frac{1}{2}(x - \mu)^t LL^t (x - \mu)\right).$$

Its logarithm is written

$$\log p(x) = K + \log \det L - \frac{1}{2}(x - \mu)^t LL^t (x - \mu).$$

We want to compute

- $\partial \log p(x) / \partial L$ which is the matrix of general element $\partial \log p(x) / \partial l_{ij}$ and
- $\partial \log p(x) / \partial \mu$ which is the vector of general element $\partial \log p(x) / \partial \mu_i$.

We first compute the matrix $\partial \log p(x) / \partial L$. Since L is lower triangular its determinant is equal to $\prod_i l_{ii}$. Consequently,

$$\frac{\partial}{\partial L} (\log \det L) = \mathrm{diag}\left(\frac{1}{l_{ii}}\right).$$

Using the fact that $\partial (x^t LL^t x) / \partial L = 2xx^t L$ to derive the last term of $\log p(x)$

$$\frac{\partial}{\partial L} \left((x - \mu)^t LL^t (x - \mu)\right) = 2(x - \mu)(x - \mu)^t L.$$

We finally obtain

$$\frac{\partial \log p(x)}{\partial L} = \mathrm{diag}\left(\frac{1}{l_{ii}}\right) - (x - \mu)(x - \mu)^t L.$$

We now compute the log-derivative with respect to the mean μ. We simply have

$$\frac{\partial \log p(x)}{\partial \mu} = LL^t(x - \mu).$$

An important issue is the complexity of the Gaussian approach. The complexity of update rules is proportional to n^3 (matrix product). Another time-consuming operation is the random generation of samples. In fact, we use the linear transformation $x = (L^t)^{-1} y + \mu$ to produce a sample x from $p(x)$, y being a normal Gaussian random variable. The complexity lies in the inversion of L which is proportional to n^3. This complexity is independent of that of the cost function evaluation.

There are several constraints on this model. First, the matrix L has to stay lower triangular. We simply do not update elements l_{ij} for $j > i$.

Second, the matrix L has to stay invertible or, equivalently, its diagonal elements l_{ii} have to stay positive. This constraint is not taken into account in any of the update rules (see (1) and (2)). One way to ensure that diagonal elements of L stay positive is to introduce n new parameters λ_i such that $l_{ii} = e^{\lambda_i}$ and make the dynamical system evolve λ_i instead of l_{ii}. This technique is close to that of the sigmoid function described on page 294. It has been implemented but has not improved convergence on the four peaks problem.

Finally, we deal with binary values not real values. Consequently, any outcome x of the random generator has to be mapped onto a binary value x' such that $x'_i = 1$ if $x_i > 1/2$ and $x'_i = 0$ otherwise. This mapping may introduce some distortion between second-order statistics of both sets of random variables.

Sabes and Jordan have used Gaussian units in the context of associative learning by probability matching [14]. They have chosen a linear combination of Gaussian kernels with covariance matrices equal to $\sigma^2 I$. The Gaussian law is also used to model complex mutation schemes in evolutionary strategies [1], especially in the case of self-adaptation.

6 Summary

We recall the two dynamics we have encountered:

Free Energy Minimization

$$\Delta \theta = -\alpha \big(H + T(1 + \log p) \big) \frac{\partial \log p}{\partial \theta}.$$

Reinforcement Learning

$$\Delta \theta = -\alpha (H - b) \frac{\partial \log p}{\partial \theta}.$$

Both update rules are similar and the expression $-T(1 + \log p)$ taken from the free energy minimization update rule can be identified with the baseline of the reinforcement learning update rule.

We also recall the two log-derivatives each corresponding to a different way to parameterize the target distribution p:

Bernoulli Model

$$\frac{\partial \log p}{\partial \theta} = 2\beta(x - \mu).$$

Gaussian Model

$$\frac{\partial \log p(x)}{\partial L} = \text{diag}\left(\frac{1}{l_{ii}}\right) - (x - \mu)(x - \mu)^t L$$

$$\frac{\partial \log p(x)}{\partial \mu} = LL^t(x - \mu).$$

7 Four Peaks Problem

7.1 Problem Definition

This problem is taken from [2]. The cost function to be maximized is

$$H(x) = \max(z(x), o(x)) + R(x)$$

where $z(x)$ is the number of trailing "0" in the string x and $o(x)$ is the number of leading "1". $R(x)$ is a conditional reward defined by:

$$R(x) = \begin{cases} n & \text{if } z(x) > s \text{ and } o(x) > s \\ 0 & \text{otherwise} \end{cases}$$

where s is a threshold equal to $n/10$. Let us give some examples in the case $n = 6$ and $s = 1$:

- for $x = 111111$, $R(x) = 0$ and $H(x) = 6$ (local maximum);
- for $x = 111110$, $R(x) = 0$ and $H(x) = 5$;
- for $x = 111100$, $R(x) = 6$ and $H(x) = 10$ (global maximum).

We want to maximize H, which is equivalent to minimizing $-H$. H has four maxima, among which two are global maxima and two are local maxima. Local search methods quickly become trapped into attraction basins of local maxima. Quoting [2], as s increases,

> "the basins of attraction surrounding the inferior local optima increase in size exponentially while the basins around the global optima decrease at the same rate".

Moreover H has large plateaus which bring no information to local search methods.

7.2 Evaluation with Bernoulli Random Variables

De Bonet, Isbell and Viola propose an optimization algorithm called MIMIC and compare it with three other algorithms [4]:

- population based incremental learning (PBIL) as described on page 294,
- random hill climbing and
- standard genetic algorithm.

We first briefly describe MIMIC and then compare our algorithm with those reviewed in [4].

The main idea in MIMIC is to estimate the probability distribution of solutions whose cost is below a given threshold. The threshold is then lowered and new samples are generated from the estimated distribution. The process is iterated until a good solution is found. However, 2^n numbers are needed to completely describe the probability distribution of n binary random variables. De Bonet, Isbell and Viola reduce the complexity with distributions which can be expressed as

$$p(x_{\sigma(1)}|x_{\sigma(2)})p(x_{\sigma(2)}|x_{\sigma(3)}) \ldots p(x_{\sigma(n-1)}|x_{\sigma(n)})p(x_{\sigma(n)})$$

where σ is a permutation of $1, 2, \ldots, n$. This implies a search for σ which is performed with a greedy algorithm minimizing a KL divergence.

We follow the benchmark described in [4]. We average over 1000 independent trials the number of cost function evaluations before a global maximum is found. This is different from the experimental protocol followed in [2], which consists in allowing a fixed number of cost function evaluations and averaging the best evaluations over 25 runs.

We have evaluated the reinforcement algorithm of page 294 with Bernoulli random variables. We have carried out two series of experiments, with the expectation strategy and the minimum variance one. All results are given in table 1 and table 2. The values of simulation parameters are: learning rate $\alpha = 0.5$, baseline factor $\gamma = 0.9$, sigmoid gain $\beta = 1/n$, where n is the size of the problem.

The results with the minimum variance strategy are significantly worse than those with the expectation one. At least for this problem, the minimum variance strategy is not the optimal policy. It is not clear why the expectation one is a better choice.

MIMIC and PBIL have better results than GA and RHC. MIMIC finds an optimum in approximately one-tenth the number of evaluations required by PBIL. Our algorithm performs better than PBIL but worse than MIMIC. MIMIC and PBIL show a complexity which is linear in the problem size contrary to our algorithm.

7.3 Evaluation with the Gaussian Probability Distribution

We have evaluated a reinforcement algorithm with the Gaussian probability distribution and the expectation strategy. Due to the complexity of the approach (in n^3)

Problem Size	30	40	50	60	70	80
Min.	227	1584	4351	9934	17524	38759
Mean	8065	21169	38320	69110	109160	169260
Median	5455	15454	32125	58878	99122	154270
Max.	102860	313260	754300	990030	386350	793520
Std. Dev.	9002	26089	33493	55980	51577	76104
Avg. Dev.	5277	12436	17632	29634	38839	55227

Table 1. Average time before a global maximum is found for different sizes of problem (expectation strategy).

Problem Size	30	40	50	60	70	80
Min.	206	1463	3736	11102	19179	37191
Mean	9973	28238	52066	91049	152330	229530
Median	6951	19113	42292	82019	136300	208370
Max.	117230	1220500	979850	637520	1026500	1864500
Std. Dev.	10937	56052	49568	55267	84558	114760
Avg. Dev.	6751	18799	25920	37007	57709	78550

Table 2. Average time before a global maximum is found for different sizes of problem (minimum variance strategy).

and available computational power, we only consider the case $n = 30$. Figure 1 shows the evolution of the cost function value until a global maximum is found. The case $s = 3$ is rather easy since the algorithm quickly "finds" the reward and the remaining part of the graph shows the effect of a stochastic gradient ascent. The results are given in the table below:

Min.	Mean	Median	Max.	Std. Dev.	Avg. Dev.
10149	15692	15388	21993	2751	2126

These numbers are calculated from the outcome of 100 instead of 1000 independent trials. The algorithm performs on average worse than Bernoulli random variables with expectation strategy. However, worst case analysis, i.e., the maximum number of cost function evaluations before global optimum, shows that the Gaussian model is superior to the Bernoulli model. The Gaussian algorithm has also proved to be very sensitive to the value of the learning rate: higher learning rates lead to faster convergence, if any, but are also more likely to make the algorithm diverge. Consequently, we have chosen a learning rate which ensures convergence and slows down the algorithm. The standard and average deviations of the number of cost function evaluations are significantly lower than with the Bernoulli model.

The complexity of the approach makes it rather impractical although it may use statistical dependencies between random variables better than the simpler scheme of Bernoulli random variables.

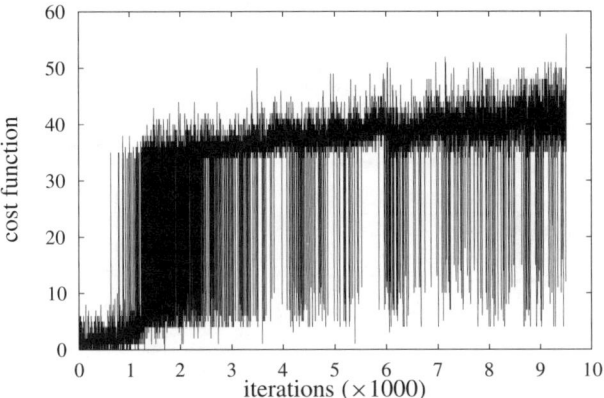

Figure 1. Evolution of the cost function value until a global maximum is reached

8 Prisoner's Dilemma

8.1 Problem Definition

The prisoner's dilemma is a game for two players where each player chooses an action between two possible actions in a synchronous and independent manner. The two actions are:

- cooperate,
- defect.

Depending on the outcome of the round, each player is credited with a payoff whose matrix is given in the following table:

	D	C
D	1	5
C	0	3

A game is a finite sequence of rounds and each player tries to maximize its accumulated score. This is known as the iterated prisoner's dilemma. The problem consists in designing a strategy in order to maximize a player's score.

We start with the encoding proposed in [11] for a genetic algorithm. Each action is coded with one bit (D/C). A strategy is defined as the way to play in the light of the 3 previous rounds (or 6 actions). A 64 bit string encodes such a strategy. For the first round, a player has to make an assumption for the 3 virtual previous rounds which can be coded with 6 more bits. Finally 70 bits are necessary to encode a full strategy.

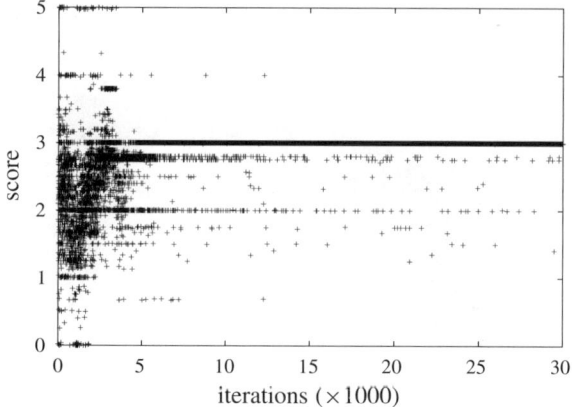

Figure 2. Prisoner's average score evolution for a reinforcement algorithm

8.2 Simulations

In the following simulations, two prisoners play against each other. They are implemented with Bernoulli random variables as seen on page 294.

The cost function to be maximized is defined as the score at the end of the game (finite sequence of rounds). The simulation algorithm is organized as a loop which decomposes as:

1. generate strategies for each player;
2. for 1000 rounds
   - generate first player's choice;
   - generate second player's choice;
   - credit each player with the corresponding payoff;
3. compute both average payoffs;
4. reinforce both automata.

We put the stress on the fact that the environment of each player (its cost function H) is not stationary on the contrary to the analysis of previous sections: the opponent evolves itself and tries to maximize its own score.

We have first implemented the reinforcement learning algorithm of page 294. Figure 2 shows the evolution of a player's score. In this simulation, both players share the same parameters. The most common limit strategy is cooperation, each player being credited with 3 points in average. The algorithm also converges sometimes to suboptimal strategies, i.e., non-cooperative ones which can be interpreted as local optima.

We have also implemented the free energy minimization algorithm of page 291. The parameters of these simulations are: learning rate $\alpha = 1$ and sigmoid gain $\beta = 0.05$.

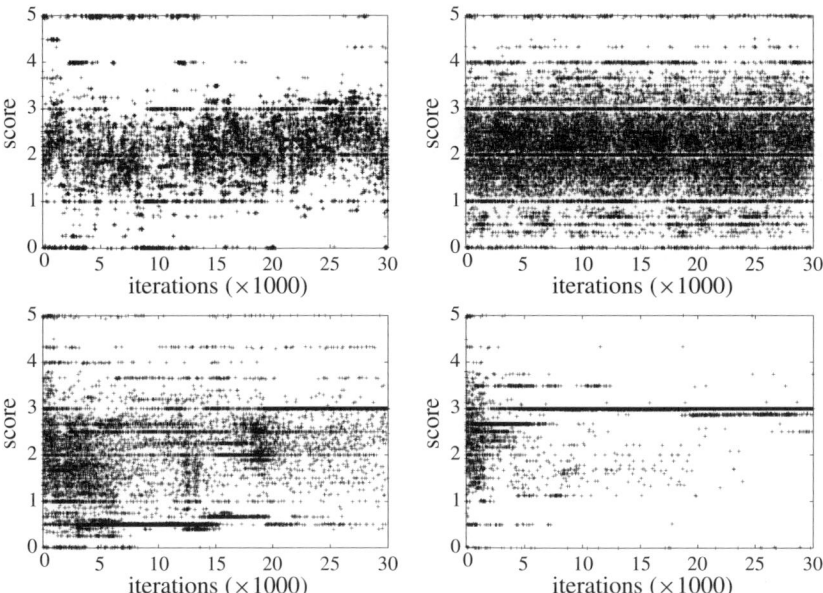

Figure 3. Prisoner's average score evolution for different values of temperature. From left to right and from top to bottom, we have $T = 1, 0.1, 0.06$, and 0.005.

Figure 3 shows the evolution of a player's score for different values of temperature. From an analogy with statistical physics, at high temperatures the system is purely random and at low temperatures the system is close to low energy states (or ground states). This is what we observe in this simulation with the exception of the first graph: high temperatures introduce a bias in (1). At the beginning of the simulation, we have $p = 1/2^n$. Let us define the quantity

$$A = H + T(1 + \log p) < H_{\max} + T(1 - n \log 2).$$

For $T > H_{\max}/(n \log 2 - 1)$ A is always negative, hence the bias and the fast "saturation" of sigmoid functions. This results in "less random" than expected behaviors. Figure 4 shows the evolution of the parameter θ_i of a Bernoulli random variable X_i (which encodes the player's response to three particular previous rounds) and figure 5 shows the evolution of the corresponding probability $p_i = g(\theta_i)$.

9 Conclusion

We have shown how to apply statistical learning schemes to combinatorial optimization problems. The approach has been divided into two parts: the target or approximate probability distribution of the random variables which is parameterized by a real vector and the dynamics which operates on the parameters and makes the target

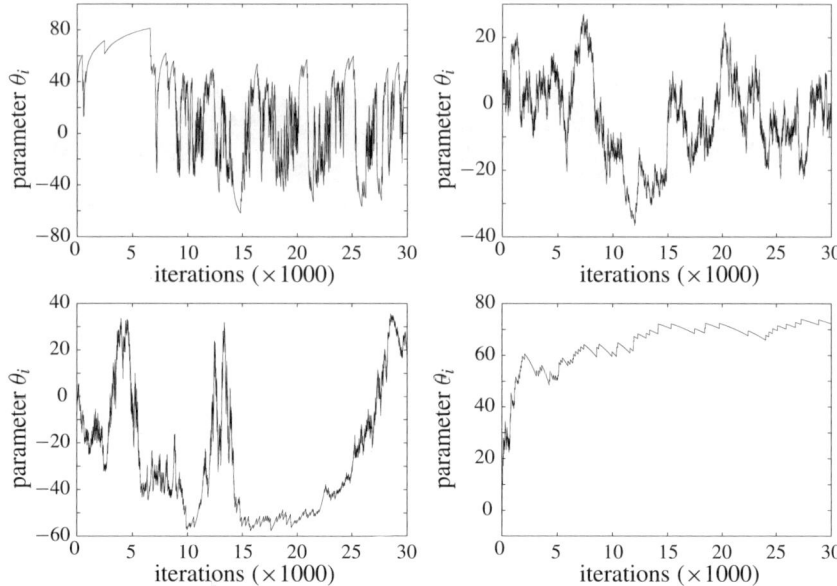

Figure 4. Evolution of the parameter θ_i of a Bernoulli random variable X_i for the same values of temperature as in figure 3

distribution converge to a limit distribution charging optimal solutions. Two families of dynamics, free energy minimization algorithms, and reinforcement learning algorithms, and two families of probability distributions, Bernoulli random variables and Gaussian random variables, have been studied. Several simulations have been carried out with both distributions and both dynamics. The algorithms have been applied to two optimization problems: the four peaks problem, for which our results have been compared to those yielded by other approaches, and the iterated prisoner's dilemma problem. In the former, the environment is stationary whereas it is not in the latter since both player strategies evolve at the same time. In the case of the free energy minimization, we have shown the influence of the temperature on the evolution of the average payoffs.

Both dynamics lead to global optima although only local optima may be found in the case of bigger or less regular problems. The free energy minimization dynamics automatically adjusts its baseline contrary to the reinforcement learning one, which requires a comparison strategy. However, the former introduces a new parameter, the temperature, that one has to choose or control with a cooling scheme. Finally, many thermo-dynamical quantities may be measured with the free energy minimization dynamics (such as the specific heat). The expectation strategy for reinforcement learning proved to be more efficient than the minimum variance one. We have no theoretical results about this fact but we believe that the latter makes the algorithm follow the trajectories of the deterministic dynamical system more closely

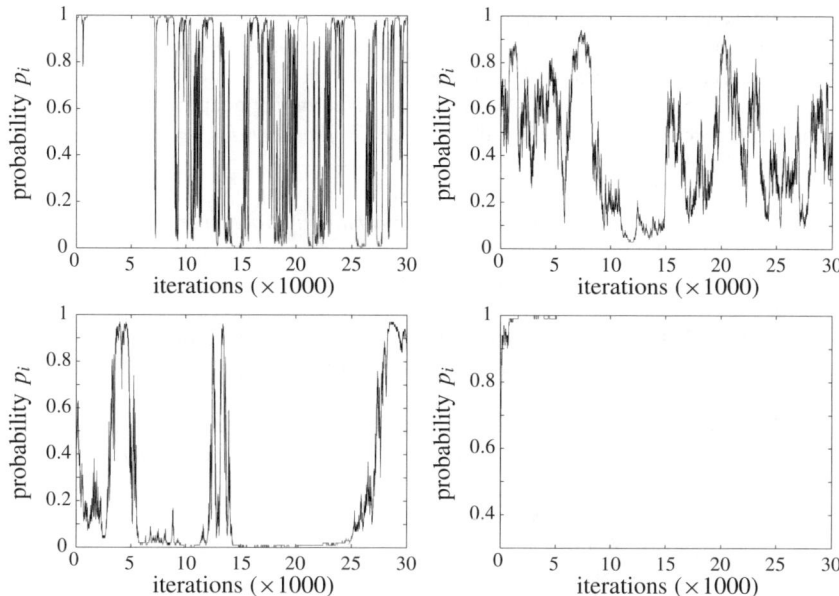

Figure 5. Evolution of the probability p_i of a Bernoulli random variable X_i for the same values of temperature as in figure 3

and therefore lowers its exploration ability.

Bernoulli random variables lead to algorithms which are faster (in terms of cost function evaluations) but less stable than their Gaussian counterparts. The Gaussian distribution seems less practical due to its time (matrix inversion) and space complexity (parameters) and requires fine-tuning of the learning rate to avoid divergence.

More experiments need to be carried out. Both dynamics should be compared directly and the overall approach should be applied to other problems. It should be instructive to compare our approach against well-known strategies for the prisoner's dilemma problem.

There are many open problems relative to combinatorial optimization by means of statistical learning. The first one is to determine the class of problems that are solvable with this approach. A strongly related issue is the regularity conditions which the cost function should fulfill.

References

1. T. Bäck, U. Hammel, and H.-P. Schwefel. Evolutionary computation: comments on the history and current state. *IEEE Transactions on Evolutionary Computation*, 1(1):3–17, 1997.

2. S. Baluja and R. Caruana. Removing the genetics from the standard genetic algorithm. In A. Prieditis and S. Russel, editors, *Proceedings of the 12th Annual Conference on Machine Learning*, pages 38–46. Morgan Kaufmann, San Francisco, 1995.
3. A. Barto, R. S. Sutton, and C. W. Anderson. Neuronlike adaptive elements that can solve difficult learning control problems. *IEEE Transactions on Systems, Man, and Cybernetics*, 13(5):834–846, 1983.
4. J. S. De Bonet, C. L. Isbell, and P. Viola. MIMIC: finding optima by estimating probability densities. In *Advances in Neural Information Processing Systems*, volume 9. MIT Press, Cambridge, MA, 1996.
5. M. Duflo. *Algorithmes Stochastiques*. Springer-Verlag, Berlin Heidelberg New York, 1996.
6. S. Geman and D. Geman. Stochastic relaxation, Gibbs distributions, and the Bayesian restoration of images. *IEEE Transactions on Pattern Analysis and Machine Intelligence*, 6(6):721–741, 1984.
7. J. J. Hopfield. Neurons with graded response have collective computational properties like those of two-state neurons. *Proceedings of the National Academy of Science*, 81:3088–3092, 1984.
8. M. Jerrum and A. Sinclair. *Approximation Algorithms for NP-Hard Problems*. Chapter: The Markov chain Monte Carlo method: an approach to approximate counting and integration. PWS, Boston, 1996.
9. S. Kirkpatrick, C. D. Gelatt, and M. P. Vecchi. Optimization by simulated annealing. *Science*, 220(4598):671–680, 1983.
10. E. Levin, N. Tishby, and S. A. Solla. A statistical approach to learning and generalization in layered neural networks. *Proceedings of the IEEE*, 78(10):1568–1574, 1990.
11. Z. Michalewicz. *Genetic Algorithms + Data Structures = Evolution Programs*, 3rd rev. and extended edition. Springer-Verlag, Berlin Heidelberg New York, 1996.
12. K. S. Narendra and M. A. L. Thathachar. *Learning Automata: an Introduction*. Prentice–Hall, Englewood Cliffs, NJ, 1989.
13. Y. Rabani, Y. Rabinovitch, and A. Sinclair. A computational view of population genetics. In *Proceedings of the 27th ACM Symposium on Theory of Computing*, pages 83–92, 1995.
14. P. N. Sabes and M. I. Jordan. Reinforcement learning by probability matching. In *Advances in Neural Information Processing Systems*, volume 8, pages 1080–1086. MIT Press, Cambridge, MA, 1995.
15. R. J. Williams. Simple statistical gradient-following algorithms for connectionist reinforcement learning. *Machine Learning*, 8:229–256, 1992.

Multi-Parent Scanning Crossover and Genetic Drift

C. A. Schippers

Leiden University
Niels Bohrweg 1
2333 CA Leiden, The Netherlands
E-mail: *adriaan@cs.leidenuniv.nl*

Abstract. Genetic drift is a well-known phenomenon from biology. Only recently has it gained attention in the field of evolutionary computation. In this article we argue that occurrence-based scanning causes a stronger than usual genetic drift. This is done in the context of a genetic algorithm based on the steady state model. To prove our claim we define three kinds of events: drift off events, drift back events, and neutral events. Drift off events constitute those events that increase the number of dominant alleles. Drift back events constitute those events that decrease the number of dominant alleles. Neutral events leave the numbers intact. We prove that in our context, with occurrence-based scanning, the probability of drift off events is always bigger than the probability of drift back events. Furthermore, we prove that this tendency to drift off amplifies itself, i.e., drifting off increases the probability of drifting off even further. Finally, we prove that the tendency to drift off gets stronger when the number of parents increases. For comparison we show that uniform scanning, another multi-parent crossover operator, does not influence genetic drift at all in this context.

Keywords
Genetic drift, multi-parent crossover, scanning crossover, steady state model

1 Introduction

In this article we study the phenomenon of genetic drift in the context of genetic algorithms (GAs). More specifically, we study the relation between multi-parent crossover and genetic drift. We focus on two multi-parent crossover operators: uniform scanning crossover and occurrence-based scanning crossover. In this study we use the steady state model for our genetic algorithm, which implies we have to do replacement selection as well as parent selection. Because we are studying genetic drift, we disable the forces opposing this phenomenon. That is, we disable selective pressure and mutation in our GA. The selective pressure is disabled by using uniformly random selection for both parent and replacement selection.

The structure of this article is as follows. In section 2 we give a short introduction to multi-parent crossover operators, and concentrate on two of them – uniform scanning crossover and occurrence-based scanning crossover. In section 3 we give an introduction to genetic drift and define it in the context of a steady state genetic algorithm. In section 4 we study the influence of uniform scanning crossover on genetic drift. We prove rigorously that there is *no* influence at all, independent of the contents of the gene pool and also independent of the number of parents. The core of this article is section 5, in which we study the influence of occurrence-based scanning crossover on genetic drift. There it is proven that occurrence-based scanning induces genetic drift. Moreover, it is proven that the tendency to drift off amplifies itself. Finally, it is proven here that increasing the number of parents magnifies these effects. Section 6 contains some considerations on the practical implications of the theoretical results in section 4 and 5. Concluding remarks follow in section 7.

This article is a spin-off of a previous publication [5] and is part of a PhD project on exploration and exploitation in evolutionary algorithms. In the context of this project, specific attention is given to multi-parent crossover operators.

2   Multi-Parent Crossover

Multi-parent crossover is an old idea [3] which has gained renewed attention in the past few years [1,2]. This article deals with a specific family of multi-parent crossover operators – scanning crossover. The basic idea, for this family of crossover operators, is that the value of each gene in the child is chosen separately. This choice is based solely on the corresponding genes in the parents. So, if we were to decide on the value of the gene in the fifth position in the child, we would scan all parents in just the fifth position and base our decision on this scan. The author is aware of three members of this family of operator: fitness-based scanning, occurrence-based scanning, and uniform scanning crossover. In this article the latter two of the three are addressed. In the case of uniform scanning crossover, for each gene one of the parents is chosen, using a uniform probability distribution. The value of the gene is copied in the corresponding gene of the child. In the case of occurrence-based scanning, each gene of the child gets the value that occurs most frequently in the corresponding genes of the parents. In case of a tie an allele is selected uniformly randomly. A more detailed and formal description of these operators can be found in Eiben, Raué and Ruttkay [1].

3   Genetic Drift

3.1   Introduction

Genetic drift is one of the two major forces that reduce the variation within a gene pool, the other one being selective pressure. Basically, it is a reduction of variation caused by the fact that most gene pools are finite and by the fact that stochastic sampling is intrinsically imprecise. If there is no mutation, genetic drift will eventually cause certain alleles to disappear from a certain position. Once this has happened the allele is forever lost and the reduction in variation is definite, unless the allele is reintroduced by mutation. Normally, genetic drift is a very weak force and it is easily countered by mutation combined with selective pressure. Mutation can introduce new alleles, and these new alleles will spread through the population because of selective pressure, that is, if they are any good.

In this study we look at genetic drift from the perspective of one gene (position), i.e., we will not look at how frequently certain combinations of alleles occur nor at what happens with these frequencies in time. Because our context is the genetic algorithm, we have only two possible alleles for a gene: 0 and 1. We define the *dominant allele*, with respect to a gene, as the most frequent allele for that gene in the population[1]. The proportion of the population having that allele in that gene (position) will be denoted by p_d. Given the above, we can now define genetic drift a bit more precisely by defining three different events in our GA: drift off events, drift back events, and neutral events. Recall that we are using the steady state model for our GA, and that genetic drift is measured in the absence of selective pressure and mutation.

Definition 1. A drift off event with respect to some gene is the event in which an individual with the non-dominant allele for that gene is replaced by an individual with the dominant allele for that gene.

Definition 2. A drift back event with respect to some gene is the event in which an individual with the dominant allele for that gene is replaced by an individual with the non-dominant allele for that gene.

Definition 3. A neutral event with respect to some gene is the event in which an individual with a certain allele for that gene is replaced by an individual with the same allele for that gene.

3.2 Genetic Drift in the Absence of Crossover

Given the definitions in section 3.1, speaking of the probability of a drift off event or a drift back event is meaningful. Our ultimate goal is to know the effect of (multi-parent) crossover on these probabilities. For this purpose we should compare the probabilities in the presence of crossover with the probabilities in the absence of crossover. If we remove the crossover, then the algorithm is reduced to a pure sampling algorithm. For each step of this steady state sampling algorithm we first choose a *parent* (uniformly randomly) from the population and make a copy of this individual. Next, we choose an individual from the population, again uniformly randomly, to be replaced by this copy. We call this replaced individual the *loser*. Because the selection of the parent is independent of the selection of the loser, the probability

---

[1] Note, that the dominant allele is not defined in case of a tie. For simplicity's sake we ignore this case and assume that the number of individuals in the population is odd.

of a drift off event with respect to a given gene, is just the product of the probability of selecting a parent with a dominant allele for that gene and the probability of selecting a loser with a non-dominant allele for that gene. The probability of a drift back event is the product of the probability that a parent with a non-dominant allele is selected and the probability that a loser with a dominant allele is selected. In both cases the probability obeys the same formula, given by

$$P_{\text{drift off}} = P_{\text{drift back}} = p_d \cdot (1 - p_d). \tag{1}$$

3.3 Genetic Drift and Crossover

In general, crossover can have an effect on genetic drift. Especially in the case of gene combinations there is an obvious effect, because crossover tends to break up old combinations and introduce new ones. But in the case of single gene genetic drift there is most often no effect. This is caused by the fact that normally, in traditional two-parent crossover, two children are generated, and if one of the children has a gene containing the allele of one of the parents, then the corresponding gene in the other child will contain the allele of the other parent. But in the case of multi-parent crossover it is quite common to generate only one child. Therefore, it might be the case that some multi-parent crossover operators do in fact change the amount of single gene genetic drift. This is actually the main topic of this article, and we elaborate on this in the next sections.

4 Uniform Scanning Crossover and Genetic Drift

In this section we show that for uniform scanning the probabilities for, respectively, drift off, drift back, and neutral events, are equal to the corresponding probabilities in the case without crossover.

Theorem 1. *In the case of uniform scanning, the respective probabilities of drift off events, drift back events, and neutral events are equal to the corresponding probabilities without crossover.*

The probability of a drift off event equals

$$P_{\text{drift off}} = (1 - p_d) \sum_{i=0}^{n} \frac{i}{n} \binom{n}{i} p_d^i (1 - p_d)^{n-i}. \tag{2}$$

The interpretation of the formula is as follows. The upper bound, n, of the iterator, is the number of parents used. The proportion of the population that contains the dominant allele is denoted by p_d. The part before the sum symbol is the probability that the loser contains the non-dominant allele. The summation itself is the probability that the child contains the dominant allele. We sum over all possible sizes of the subset of the parents containing the dominant allele. For each possible subset size (i in the equation) we multiply three things:

- the number of possible subsets of that size, $\binom{n}{i}$,
- the probability that a specific parent set, where this subset is of size i, is chosen from the population, $p_d^i(1 - p_d)^{n-i}$, and
- the probability that, given this parent set, an individual containing the dominant allele is chosen as donor for the child, $\frac{i}{n}$.

With (2) and the interpretation above, we now prove theorem 1.

Proof (of theorem 1).
Using the fact that $\frac{i}{n} = 0$ for $i = 0$ and the fact that $\frac{i}{n}\binom{n}{i} = \binom{n-1}{i-1}$ for $i \geq 1$, and doing some rewriting, we get

$$P_{\text{drift off}} = (1 - p_d)p_d \sum_{i=1}^{n} \binom{n-1}{i-1} p_d^{i-1}(1 - p_d)^{(n-1)-(i-1)}.$$

Substituting $j := i - 1$, we obtain

$$\begin{aligned} P_{\text{drift off}} &= (1 - p_d)p_d \sum_{j=0}^{n-1} \binom{n-1}{j} p_d^j (1 - p_d)^{(n-1)-j} \\ &= (1 - p_d)p_d(p_d + (1 - p_d))^{n-1} \\ &= (1 - p_d)p_d. \end{aligned} \quad (3)$$

For $P_{\text{drift back}}$ we can find a formula that is analogous to the one for $P_{\text{drift off}}$ in (2). This formula is equal to the one in (2) with $(1 - p_d)$ substituted for p_d. Making the same substitution in (3), we find

$$P_{\text{drift back}} = p_d(1 - p_d) = (1 - p_d)p_d = P_{\text{drift off}}. \quad \square$$

Corollary 1 (of theorem 1). *In the case of uniform scanning, the respective probabilities of drift off events, drift back events, and neutral events, are independent of the number of parents.*

5 Occurrence-Based Scanning Crossover and Genetic Drift

In this section we study the effect of occurrence-based scanning crossover on genetic drift. In the case of occurrence-based scanning crossover we can distinguish two cases: the case in which the number of parents is odd, and the case in which the number of parents is even. The latter is slightly more complicated, because we can have a tie, that is, half the parents have the dominant allele and half the non-dominant allele. When this happens a uniformly random selection between the two is made. For the sake of simplicity we only study the case in which the number of parents is odd. Note, that when the number of parents is large and even, the probability of a tie is small; when the number of parents is small and even, occurrence-based scanning behaves almost the same as uniform scanning (in the case of two parents they behave exactly the same).

This section is the most important section of this article; it is divided in four parts. In section 5.1 we define three functions that facilitate the expression of theorems on the probabilities of drift off, drift back, and neutral events. Next, in section 5.2, we use those functions to prove that occurrence-based scanning induces drift. Following this, we prove in section 5.3 that the tendency to drift off amplifies itself. That is, the further the population has drifted off, the higher the probability that the population will drift off even further. Finally, in section 5.4, we study the effects of using more parents.

5.1 Some Definitions

In this section we define three important functions: f, g, and h. All these functions tell us something about the probabilities of drift off, and drift back events. All of them have two parameters: the number of parents, m, and the proportion of the population that contains the dominant allele, p_d. The latter ranges over the interval $(\frac{1}{2}, 1)$.

The domain of p_d is open at the left side because the dominant allele has a majority by definition. The interval is open at the right side because a proportion of 1 would mean that all individuals contained the dominant allele. In that case it ceases to be possible to drift back, and drifting off even further is also impossible.

One could argue that the actual domain of p_d should not be a continuous interval, but a finite set of rational numbers,

$$\left\{ \frac{1}{2} + \frac{1}{N}, \frac{1}{2} + \frac{2}{N}, \ldots, \frac{1}{2} + \frac{N-1}{N} \right\}$$

where N is the size of the population. This is true, because genetic drift is only an issue when the population is finite. but we would have to consider all different possible values for N, and this would complicate things. Also, this would result in discontinuous functions f, g and h, which would complicate things even more. Therefore, we chose to model this domain with a continuous interval and use straightforward continuous extensions of the otherwise discrete functions.

Let $p \in (0, 1)$ be the proportion of individuals in the population that contain some allele, and n the number of parents in the parent set. Recall that the n parents are chosen uniformly randomly from the population. It is evident that the probability that at most $k \in \{0 \ldots n\}$ parents contain the allele, is given by

$$B(n, k, p) = \sum_{i=0}^{k} \binom{n}{i} p^i (1-p)^{n-i}. \tag{4}$$

From [4] we learn that for all $n \geq k \in \mathbb{N}$ and $p \in (0, 1)$, the following equation holds:

$$B(n, k, p) = (n - k)\binom{n}{k} \int_0^{1-p} t^{n-k-1}(1 - t)^k \, dt. \tag{5}$$

The probability that more then k parents contain the allele $(1 - B(n, k, p))$ is given by

$$1 - B(n, k, p) = B(n, n - k - 1, 1 - p). \tag{6}$$

The probability of having more than k parents containing the allele is equal to the probability that less than $n - k$ ($n - k - 1$ or less), contain another allele. Now let p_d denote the proportion of the population that contains the dominant allele. Recall that we have an odd number of parents, say $2m + 1$, with $m \in \mathbb{N}$. The probability that more that m parents (i.e., the majority) contain the dominant allele, is then given by

$$1 - B(2m + 1, m, p_d) = B(2m + 1, m, 1 - p_d)$$

$$= \sum_{i=0}^{m} \binom{2m + 1}{i}(1 - p_d)^i p_d^{2m+1-i}. \tag{7}$$

Using (4) and (7), we can now define the following three functions.

Definition 4.

$$f : \mathbb{N} \times (\tfrac{1}{2}, 1) \to \mathbb{R} : f(m, p_d) = B(2m + 1, m, 1 - p_d)(1 - p_d).$$

This function gives us the probability of a drift off event as a function of the proportion of dominant alleles p_d, when m parents are used. $B(2m + 1, m, 1 - p_d)$ is the probability that the minority of the parents has the non-dominant allele, $(1 - p_d)$ is the probability that an individual containing the non-dominant allele is chosen to be replaced. Note that because of (7) and the fact that $1 - (1 - p_d) = p_d$, the probability of a drift back event is given by $f(m, 1 - p_d)$. The function is depicted in figures 1 and 2.

Definition 5.

$$g : \mathbb{N} \times (\tfrac{1}{2}, 1) \to \mathbb{R} : g(m, p_d) = f(m, p_d) - f(m, 1 - p_d).$$

This function tells us whether there is a tendency to drift off for the case where m parents are used and the proportion of dominant alleles is p_d. If the value of this function is positive, then there is a tendency to drift off. If the value for this function is negative, then there is a tendency to drift back. We can see the value of this function as the *net* drift; it tells us something about the speed of the drift. A higher value means a higher speed. The function is shown in figure 3.

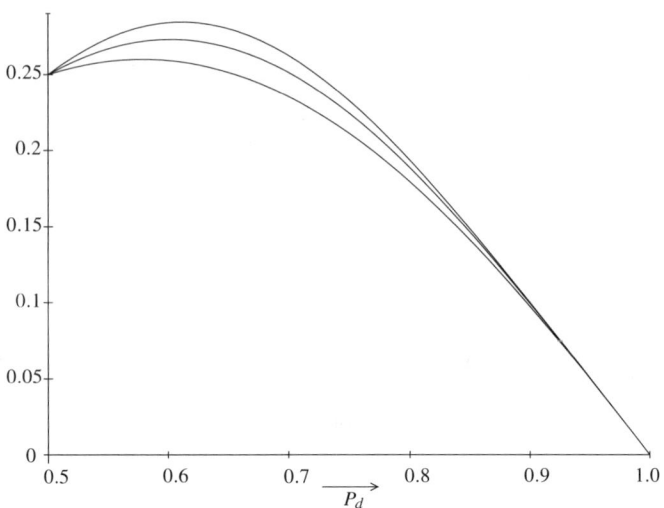

Figure 1. Probability of a drift off event, $f(m, p_d)$, for $m = 1$ (lower), $m = 2$ (middle), and $m = 3$ (upper)

Definition 6.

$$h : \mathbb{N} \times (\tfrac{1}{2}, 1) \to \mathbb{R} : h(m, p_d) = \frac{f(m, p_d)}{f(m, 1 - p_d)}.$$

This function tells us the odds of a drift off event against a drift back event, given that the event is a non-neutral event. For example, if the value of this function is 10 and we know that a non-neutral event occurred, then the odds are 10 to 1 that the non-neutral event was a drift off event. Again, m denotes the number or parents used and p_d denotes the proportion of dominant alleles in the population. The function is shown in figure 4.

5.2 Occurrence-Based Scanning Induces Drift

In this section we prove rigorously that occurrence-based scanning induces drift. More specifically, we prove that, in comparison to the case without crossover, the probability of a drift off event is bigger, and that the probability of a drift back event is smaller. We start with a lemma (lemma 1) that will not only prove to be useful in this section, but also in section 5.4. In that section we study the effect of using more parents.

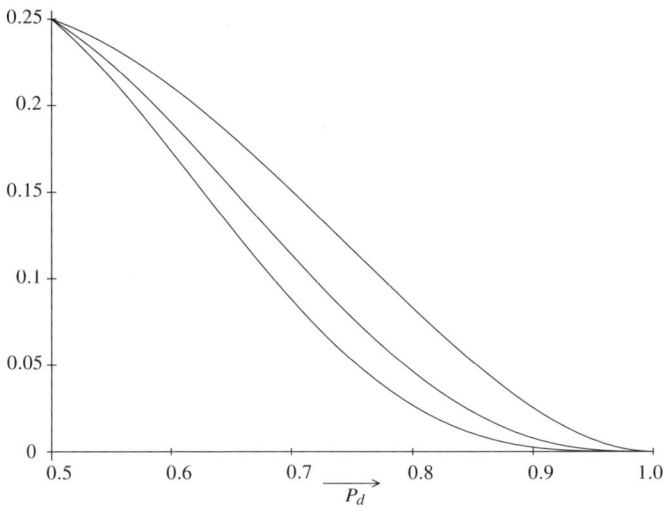

Figure 2. Probability of a drift back event, $f(m, 1 - p_d)$, for $m = 1$ (upper), $m = 2$ (middle), and $m = 3$ (lower)

Lemma 1. *For all $m \in \mathbb{N}$*

$$B(2(m + 1) + 1, m + 1, 1 - p_d) > B(2m + 1, m, 1 - p_d)$$

holds on the interval $(\frac{1}{2}, 1)$. That is, the probability that the occurrence-based scanning operator produces a child containing the dominant allele, increases when the number of parents increases.

This lemma, together with the fact that the loser is chosen independently of the child that is produced, proves that the probability of a drift off event increases when the number of parents increases.

Proof (of lemma 1).
Recall that $B(2(m+1)+1, m+1, 1-p_d)$ is the probability that drawing $2(m+1)+1$ parents from our distribution[2] gives us $m + 1$ or less non-dominant alleles. This probability can also be expressed in another way. Suppose that we first draw $2m + 1$ parents and then two more. We now have three possible events where we have $m + 1$ or less parents that contain the non-dominant allele:

1. We have $m - 1$ or less in the first $2m + 1$ parents.

---
[2] Each time a parent is chosen, a parent containing the dominant allele is chosen with probability p_d, and a parent containing the non-dominant allele is chosen with probability $1 - p_d$.

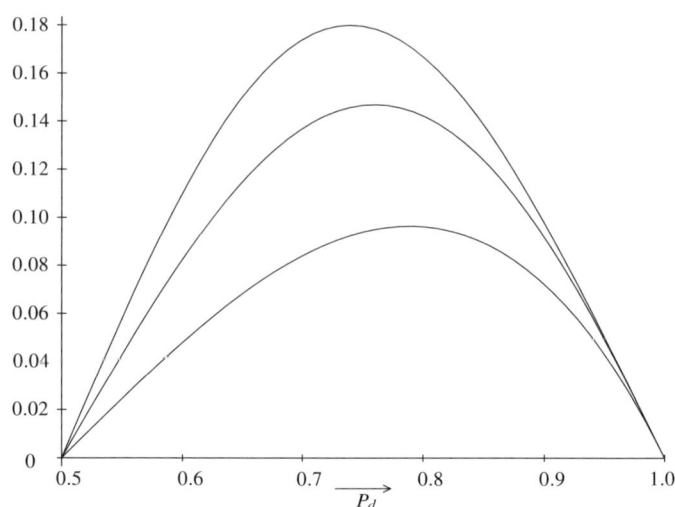

Figure 3. The net drift or drift tendency, $g(m, p_d)$, for $m = 1$ (lower), $m = 2$ (middle), and $m = 3$ (upper)

2. We have m in the first $2m + 1$, and at most one in the last two (i.e., at least one of the last parents contains the dominant allele).
3. We have exactly $m + 1$ in the first $2m + 1$ and none in the last two (i.e., the last two parents both contain the dominant allele).

Using this event partition, we get

$$B(2(m+1)+1, m+1, 1-p_d) = \sum_{i=0}^{m-1} \binom{2m+1}{i}(1-p_d)^i p_d^{2m+1-i}$$
$$+ \binom{2m+1}{m}(1-p_d)^m p_d^{m+1}\left(p_d^2 + p_d(1-p_d) + (1-p_d)p_d\right)$$
$$+ \binom{2m+1}{m+1}(1-p_d)^{m+1} p_d^m p_d^2.$$

Because $\binom{2m+1}{m+1} = \binom{2m+1}{m}$, and

$$(1-p_d)^{m+1} p_d^m p_d^2 = (1-p_d)^m p_d^{m+1} p_d(1-p_d)$$

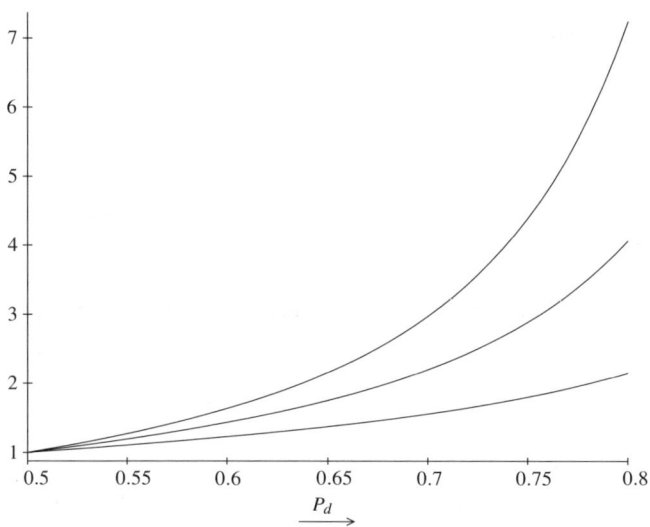

Figure 4. The odds of a drift off against a drift back event (in case of a non-neutral event), $h(m, p_d)$, for $m = 1$ (lower), $m = 2$ (middle), and $m = 3$ (upper)

we can add the last part to the next to last, and obtain

$$B(2(m+1)+1, m+1, 1-p_d) = \sum_{i=0}^{m-1} \binom{2m+1}{i}(1-p_d)^i p_d^{2m+1-i}$$
$$+ \binom{2m+1}{m}(1-p_d)^m p_d^{m+1}\left(p_d^2 + 2p_d(1-p_d) + (1-p_d)p_d\right).$$

Since $p_d \in (\frac{1}{2}, 1)$ we have $(1-p_d)p_d > (1-p_d)^2$. Therefore

$$p_d^2 + 2p_d(1-p_d) + (1-p_d)p_d > p_d^2 + 2p_d(1-p_d) + (1-p_d)^2 = 1.$$

This finally gives

$$B(2(m+1)+1, m+1, 1-p_d) > \sum_{i=0}^{m} \binom{2m+1}{i}(1-p_d)^i p_d^{2m+1-i}$$
$$= B(2m+1, m, 1-p_d). \quad \square$$

Corollary 2 (of lemma 1). *For all $m \in \mathbb{N}$*

$$B(2(m+1)+1, m+1, p_d) < B(2m+1, m, p_d)$$

on the interval $(\frac{1}{2}, 1)$. That is, the probability that the occurrence-based scanning operator produces a child containing the non-dominant allele, decreases when the number of parents increases.

Theorem 2. *For all* $m \in \mathbb{N}$

$$g(m, p_d) > 0$$

on the interval $(\frac{1}{2}, 1)$. *That is, the probability of a drift off event is bigger than the probability of a drift back event.*

Proof. We prove the theorem by induction on m. From definitions 5 and 4 it follows that

$$g(m, p_d) = B(2m + 1, m, 1 - p_d)(1 - p_d) - B(2m + 1, m, p_d)p_d.$$

Rewriting $B(2m + 1, m, p_d)$, using (7), we get

$$\begin{aligned} g(m, p_d) &= B(2m + 1, m, 1 - p_d)(1 - p_d) - (1 - B(2m + 1, m, 1 - p_d))p_d \\ &= B(2m + 1, m, 1 - p_d) - p_d \\ &= \sum_{i=0}^{m} \binom{2m+1}{i}(1 - p_d)^i p_d^{2m+1-i} - p_d. \end{aligned} \tag{8}$$

Basis: $m = 1$.

$$g(1, p_d) = -2p_d^3 + 3p_d^2 - p_d. \tag{9}$$

Clearly this function is zero for $p_d = \frac{1}{2}$ and $p_d = 1$ and positive in between those values.

Induction step: Let $m \in \mathbb{N}$ be given and assume that $g(m, p_d) > 0$. Equation (8) gives us

$$g(m + 1, p_d) = B(2(m + 1) + 1, m + 1, 1 - p_d) - p_d.$$

Lemma 1 implies that this is greater than

$$B(2m + 1, m, 1 - p_d) - p_d$$

which is, again using (8), equal to $g(m, p_d)$. Using our induction assumption we obtain $g(m + 1, p_d) > 0$. □

Corollary 3 (of theorem 2). *For all* $m \in \mathbb{N}$

$$B(2m + 1, m, 1 - p_d) > p_d$$

on the interval $(\frac{1}{2}, 1)$.

Theorem 3. *For all $m \in \mathbb{N}$*

$$f(m, p_d) > (1 - p_d)p_d.$$

That is, the probability of a drift off event when using occurrence-based scanning, is greater than the probability of a drift off event when using no crossover.

Proof. Let $m \in \mathbb{N}$ be given. Corollary 3 tells us that

$$B(2m + 1, m, 1 - p_d) > p_d.$$

By definition 4, we have

$$f(m, p_d) = B(2m + 1, m, 1 - p_d)(1 - p_d).$$

Therefore

$$f(m, p_d) > (1 - p_d)p_d. \quad \square$$

Theorem 4. *For all $m \in \mathbb{N}$*

$$f(m, 1 - p_d) < (1 - p_d)p_d.$$

That is, the probability of a drift back event when using occurrence-based scanning, is smaller than the probability of a drift back event when using no crossover.

Proof. Let $m \in \mathbb{N}$ be given. By definition 4

$$f(m, 1 - p_d) = B(2m + 1, m, p_d)p_d.$$

Using (7), we find

$$\begin{aligned} f(m, 1 - p_d) &= (1 - B(2m + 1, m, 1 - p_d))p_d \\ &= p_d - B(2m + 1, m, 1 - p_d)p_d. \end{aligned}$$

Corollary 3 tells us that $B(2m + 1, m, 1 - p_d) > p_d$. Therefore

$$f(m, 1 - p_d) < p_d - p_d^2 = (1 - p_d)p_d. \quad \square$$

5.3 Self-Amplification of the Drift Tendency

In this section we address the fact that the drift tendency, induced by occurrence-based scanning, amplifies itself. More specifically, we show that every drift off event increases the probability that the next non-neutral event will also be a drift off event. We do this by showing that the function h, as defined in definition 6, is a strictly monotonously increasing function (theorem 5). Remember that this function gives us the odds of a drift off event against a drift back event, given that we have a non-neutral event.

Theorem 5. *For all $m \in \mathbb{N}$, we have that $h(m, p_d)$ is a strictly monotonously increasing function of p_d on the interval $(\frac{1}{2}, 1)$. That is, when the proportion of dominant alleles increases, the probability that a non-neutral event will be a drift off event, increases also.*

Proof. To prove that $h(m, p_d)$ is a strictly monotonously increasing function on $(\frac{1}{2}, 1)$ (see figure 4), we must show that the derivative is positive on this interval. In this case the derivative is a very complex function. Therefore, we will prove something stronger, namely that $h(m, p_d)$ is the quotient of a strictly monotonously increasing function and a monotonously decreasing function. From this it follows that the quotient itself is strictly monotonously increasing. However, as can be seen in figures 1 and 2, definition 6 does not define $h(m, p_d)$ as a quotient of two such function. Therefore, we first rewrite this definition. The interval is split up in two parts:

$$\left(\frac{1}{2}, \frac{m+1}{2m+1}\right), \quad \text{and} \quad \left[\frac{m+1}{2m+1}, 1\right).$$

For the first part, we divide both enumerator and denominator by $(1 - p_d)^{\frac{1}{2}}$. For the second part, we divide enumerator and denominator alike by $(1 - p_d)$.

Definition 7.

$$h(m, p_d) = \frac{f(m, p_d)}{f(m, 1 - p_d)}$$

$$= \frac{B(2m+1, m, 1-p_d)(1-p_d)}{B(2m+1, m, p_d)p_d}$$

$$= \begin{cases} \frac{B(2m+1,m,1-p_d)(1-p_d)^{\frac{1}{2}}}{B(2m+1,m,p_d)(1-p_d)^{-\frac{1}{2}}p_d} & \text{if } p_d \in \left(\frac{1}{2}, \frac{m+1}{2m+1}\right) \\ \\ \frac{B(2m+1,m,1-p_d)}{B(2m+1,m,p_d)(1-p_d)^{-1}p_d} & \text{if } p_d \in \left[\frac{m+1}{2m+1}, 1\right). \end{cases}$$

Now the property does hold for both these parts. Lemmas 2, 5, 7, and 8 formalize this. The proofs of these lemmas together prove the theorem. □

To avoid forward references we start with the enumerator of the second part of definition 7, and prove that it is a strictly monotonously increasing function on the domain of the second part. We actually prove something stronger, namely that the property holds on both the domain of the first and the second part (lemma 2).

Lemma 2. *For all $m \in \mathbb{N}$, $B(2m+1, m, 1-p_d)$ is a strictly monotonously increasing function of p_d on the interval $(\frac{1}{2}, 1)$.*

Proof. Equation (5) implies

$$\frac{\partial}{\partial p_d} B(2m+1, m, 1-p_d) = (m+1)\binom{2m+1}{m}\frac{\partial}{\partial p_d}\int_0^{p_d} t^m(1-t)^m \, dt. \tag{10}$$

If we define $f(t) := t^m(1-t)^m$ and write $F(t)$ for the primitive of $f(t)$, we get

$$\begin{aligned}
\frac{\partial}{\partial p_d} &B(2m+1, m, 1-p_d) \\
&= (m+1)\binom{2m+1}{m}\frac{\partial}{\partial p_d}\int_0^{p_d} f(t)\, dt \\
&= (m+1)\binom{2m+1}{m}\frac{\partial}{\partial p_d}(F(p_d) - F(0)) \\
&= (m+1)\binom{2m+1}{m}\frac{\partial}{\partial p_d} F(p_d) \\
&= (m+1)\binom{2m+1}{m} f(p_d) \\
&= (m+1)\binom{2m+1}{m} p_d^m (1-p_d)^m.
\end{aligned} \tag{11}$$

Because $p_d > 0$, $(1-p_d) > 0$ and $m \in \mathbb{N}$ it follows that

$$\frac{\partial}{\partial p_d} B(2m+1, m, 1-p_d) > 0. \quad \square$$

We continue our proof with the enumerator of the first part of definition 7, and prove that this enumerator is a strictly monotonously increasing function on its domain (lemma 5). To accomplish this, we first introduce two extra lemmas (lemma 3 and 4), that simplify the proof.

Lemma 3. *For all $m \in \mathbb{N}$ and all $j \in \{0, \ldots, m\}$*

$$\frac{\binom{2m+1}{m-j}}{\binom{2m+1}{m}} \leq \left(\frac{m}{m+1}\right)^j.$$

Proof. Let $m \in \mathbb{N}$ be given. We prove the lemma for this m by induction on j.

Basis: $j = 0$

$$\frac{\binom{2m+1}{m-j}}{\binom{2m+1}{m}} = \frac{\binom{2m+1}{m}}{\binom{2m+1}{m}} = 1 = \left(\frac{m}{m+1}\right)^0.$$

Induction step: Let $j \in \{0, \ldots, m-1\}$ be given, and assume that

$$\binom{2m+1}{m-j} \Big/ \binom{2m+1}{m} \leq \left(\frac{m}{m+1}\right)^j.$$

Since $0 \leq j \leq m-1$, we have

$$\binom{2m+1}{m-j-1} = \binom{2m+1}{m-j} \times \frac{m-j}{m+2+j}.$$

Because $j \geq 0$

$$\frac{m-j}{m+2+j} \leq \frac{m}{m+2} \leq \frac{m}{m+1}.$$

It now follows that

$$\frac{\binom{2m+1}{m-(j+1)}}{\binom{2m+1}{m}} \leq \left(\frac{m}{m+1}\right) \frac{\binom{2m+1}{m-j}}{\binom{2m+1}{m}} \stackrel{(!)}{\leq} \frac{m}{m+1} \left(\frac{m}{m+1}\right)^j = \left(\frac{m}{m+1}\right)^{j+1}. \quad \square$$

Lemma 4. *For all $m \in \mathbb{N}$ and all $i \in \{0, \ldots, m\}$*

$$p_d \in \left(\frac{1}{2}, \frac{m+1}{2m+1}\right) \Rightarrow \frac{1}{2} \frac{\binom{2m+1}{i}}{\binom{2m+1}{m}} (1-p_d)^{i-m-1} p_d^{m-i+1} < 1.$$

Proof. Let $m \in \mathbb{N}$ be given. We first factor out $\frac{1}{2} \frac{p_d}{1-p_d}$; clearly, this is an increasing function on $(\frac{1}{2}, \frac{m+1}{2m+1})$. Substituting the supremum of the domain, $\frac{m+1}{2m+1}$, in $\frac{1}{2} \frac{p_d}{1-p_d}$ we obtain

$$\frac{1}{2} \left(\frac{m+1}{2m+1}\right) \Big/ \left(\frac{m}{2m+1}\right).$$

This is equal to $\frac{1}{2} \frac{m+1}{m}$, which has a maximal value of 1, when $m = 1$. Therefore

$$\frac{1}{2} \frac{p_d}{1-p_d} < 1.$$

We are left to prove:

$$\forall i \in \{0, \ldots, m\}: \left(\binom{2m+1}{i} \Big/ \binom{2m+1}{m}\right) \left(\frac{p_d}{1-p_d}\right)^{m-i} \leq 1.$$

Knowing that $\frac{p_d}{1-p_d}$ is smaller than $\frac{m+1}{m}$, this follows directly from lemma 3 when substituting $j := m - i$. $\quad \square$

Using the last two lemmas we can finally prove that the enumerator of the first part of definition 7, is a strictly monotonously increasing function on its domain (lemma 5).

Lemma 5. *For all* $m \in \mathbb{N}$

$$B(2m+1, m, 1-p_d)(1-p_d)^{\frac{1}{2}}$$

is a strictly monotonously increasing function of p_d on the interval $\left(\frac{1}{2}, \frac{m+1}{2m+1}\right)$.

Proof. By definition

$$\frac{\partial}{\partial p_d}\left(B(2m+1, m, 1-p_d)(1-p_d)^{\frac{1}{2}}\right)$$
$$= (1-p_d)^{\frac{1}{2}}\frac{\partial}{\partial p_d}B(2m+1, m, 1-p_d) - \frac{1}{2}(1-p_d)^{-\frac{1}{2}}B(2m+1, m, 1-p_d).$$

Using (11) and the definition of B

$$\frac{\partial}{\partial p_d}\left(B(2m+1, m, 1-p_d)(1-p_d)^{\frac{1}{2}}\right)$$
$$= (m+1)\binom{2m+1}{m}p_d^m(1-p_d)^{m+\frac{1}{2}} - \frac{1}{2}\sum_{i=0}^{m}\binom{2m+1}{i}(1-p_d)^{i-\frac{1}{2}}p_d^{2m+1-i}$$

is obtained. Factoring out the first part of the right hand side from the second part, we find

$$\frac{\partial}{\partial p_d}\left(B(2m+1, m, 1-p_d)(1-p_d)^{\frac{1}{2}}\right)$$
$$= (m+1)\binom{2m+1}{m}p_d^m(1-p_d)^{m+\frac{1}{2}}$$
$$-(m+1)\binom{2m+1}{m}p_d^m(1-p_d)^{m+\frac{1}{2}}\frac{1}{m+1}\sum_{i=0}^{m}\frac{1}{2}\frac{\binom{2m+1}{i}}{\binom{2m+1}{m}}(1-p_d)^{i-m-1}p_d^{m-i+1}.$$

Using lemma 4, we get

$$\frac{\partial}{\partial p_d}\left(B(2m+1, m, 1-p_d)(1-p_d)^{\frac{1}{2}}\right) > 0. \quad \square$$

We continue our proof with the denominator of the first part of definition 7, and prove that this denominator is a monotonously decreasing function on its domain (lemma 7). To accomplish this, we first introduce one extra lemma that simplifies the proof.

Lemma 6. *For all* $m \in \mathbb{N}$

$$\left(1 - \frac{1}{2}p_d\right) p_d^{-1} \frac{1}{m+1} \sum_0^m \frac{\binom{2m+1}{i}}{\binom{2m+1}{m}} p_d^{i-m} (1 - p_d)^{m-i} < 1$$

on the interval $\left(\frac{1}{2}, \frac{m+1}{2m+1}\right)$.

Proof. First note that $(1 - \frac{1}{2}p_d)p_d^{-1}$ is a decreasing function on $\left(\frac{1}{2}, \frac{m+1}{2m+1}\right)$ and therefore is always smaller than

$$\left(1 - \frac{1}{2}\frac{1}{2}\right)\frac{1}{2}^{-1} = \frac{3}{2}.$$

Furthermore, $p_d^{i-m}(1 - p_d)^{m-i}$ is also a decreasing function on $\left(\frac{1}{2}, \frac{m+1}{2m+1}\right)$ and therefore always smaller than

$$\frac{1}{2}^{i-m}\left(1 - \frac{1}{2}\right)^{m-i} = 1.$$

Thus, if we can show that

$$\frac{1}{m+1} \sum_0^m \binom{2m+1}{i} / \binom{2m+1}{m} \leq \frac{2}{3}$$

then we are done. We prove this by showing that

$$\sum_{i=0}^m \binom{2m+1}{i} \leq \frac{2}{3}\binom{2m+1}{m}(m+1)$$

by induction on m. Note that

$$\sum_{i=0}^m \binom{2m+1}{i} = \frac{1}{2}\sum_{i=0}^{2m+1}\binom{2m+1}{i} = \frac{1}{2}2^{2m+1} = 2^{2m}. \qquad (12)$$

Basis: ($m = 1$ and $m = 2$)

$$m = 1 \Rightarrow 2^{2m} = 2^2 = \frac{2}{3} \cdot 3 \cdot (1+1) = \frac{2}{3}\binom{3}{m}(m+1)$$

$$m = 2 \Rightarrow 2^{2m} = 2^4 < 20 = \frac{2}{3} \cdot 10 \cdot (2+1) = \frac{2}{3}\binom{5}{m}(m+1).$$

Induction step: Let $m \in \mathbb{N}$ be given and assume $\sum_{i=0}^{m} \binom{2m+1}{i} \leq \frac{2}{3}\binom{2m+1}{m}(m+1)$, then

$$2^{2(m+1)} = 4 \cdot 2^{2m}$$

$$\overset{(!)}{\leq} 4 \cdot \frac{2}{3}\binom{2m+1}{m}(m+1)$$

$$\underset{m \geq 2}{=} 4 \cdot \frac{2}{3}\binom{2m+3}{m+1} \cdot \frac{(m+1)(m+2)}{(2m+3)(2m+2)}(m+1)$$

$$= \frac{4(m+1)^2}{(2m+3)(2m+2)} \cdot \frac{2}{3}\binom{2(m+1)+1}{m+1} \cdot ((m+1)+1)$$

$$< \frac{2}{3}\binom{2(m+1)+1}{m+1}((m+1)+1). \quad \square$$

Using the last lemma we can now prove that the denominator of the first part of definition 7, is a monotonously decreasing function on its domain.

Lemma 7. *For all $m \in \mathbb{N}$, $B(2m+1, m, p_d)(1-p_d)^{-\frac{1}{2}}p_d$ is a monotonously decreasing function of p_d on the interval $\left(\frac{1}{2}, \frac{m+1}{2m+1}\right)$.*

Proof. By definition

$$\frac{\partial}{\partial p_d}\left(B(2m+1, m, p_d)(1-p_d)^{-\frac{1}{2}}p_d\right)$$

$$= (1-p_d)^{-\frac{1}{2}}p_d \frac{\partial}{\partial p_d} B(2m+1, m, p_d)$$

$$+ \left((1-p_d)^{-\frac{1}{2}} + \frac{1}{2}(1-p_d)^{-1\frac{1}{2}}p_d\right) B(2m+1, m, p_d).$$

Using (7) and (11) to compute $\frac{\partial}{\partial p_d} B(2m+1, m, p_d)$, and the definition of B to expand $B(2m+1, m, p_d)$, we find

$$\frac{\partial}{\partial p_d}\left(B(2m+1, m, p_d)(1-p_d)^{-\frac{1}{2}}p_d\right) =$$

$$- (m+1)\binom{2m+1}{m} p_d^{m+1}(1-p_d)^{m-\frac{1}{2}}$$

$$+ \left((1-p_d)^{-\frac{1}{2}} + \frac{1}{2}(1-p_d)^{-1\frac{1}{2}}p_d\right) \sum_{i=0}^{m}\binom{2m+1}{i} p_d^i(1-p_d)^{2m+1-i}.$$

The equation above is a sum of two expressions. Factoring out the first expression of the right-hand side from the second expression, and rewriting the factor before the summation symbol to $(1 - \frac{1}{2}p_d)(1 - p_d)^{-1\frac{1}{2}}$, we obtain

$$\frac{\partial}{\partial p_d}\left(B(2m + 1, m, p_d)(1 - p_d)^{-\frac{1}{2}} p_d\right) =$$

$$- (m + 1)\binom{2m + 1}{m} p_d^{m+1}(1 - p_d)^{m-\frac{1}{2}}$$

$$+ (m + 1)\binom{2m + 1}{m} p_d^{m+1}(1 - p_d)^{m-\frac{1}{2}}$$

$$\times \left(1 - \frac{1}{2}p_d\right) p_d^{-1} \frac{1}{m + 1} \sum_0^m \frac{\binom{2m+1}{i}}{\binom{2m+1}{m}} p_d^{i-m}(1 - p_d)^{m-i}.$$

Using lemma 6, we get

$$\frac{\partial}{\partial p_d}\left(B(2m + 1, m, p_d)(1 - p_d)^{-\frac{1}{2}} p_d\right) < 0. \quad \square$$

We conclude with the denominator of the second part of definition 7. We prove that this is a monotonously decreasing function on its domain (lemma 8). This concludes the proof that $h(m, p_d)$ is a strictly monotonously increasing function (theorem 5).

Lemma 8. *For all* $m \in \mathbb{N}$

$$B(2m + 1, m, p_d)(1 - p_d)^{-1} p_d$$

is a monotonously decreasing function of p_d *on the interval* $(\frac{m+1}{2m+1}, 1)$.

Proof. By definition we have

$$B(2m + 1, m, p_d)(1 - p_d)^{-1} p_d = \sum_{i=0}^{m} \binom{2m + 1}{i} p_d^{i+1}(1 - p_d)^{2m-i}. \quad (13)$$

Therefore

$$\frac{\partial}{\partial p_d}\left(B(2m + 1, m, p_d)(1 - p_d)^{-1} p_d\right)$$

$$= \frac{\partial}{\partial p_d} \sum_{i=0}^{m} \binom{2m + 1}{i} p_d^{i+1}(1 - p_d)^{2m-i}$$

$$= \sum_{i=0}^{m} \binom{2m + 1}{i}\left((i + 1)p_d^i(1 - p_d)^{2m-i} - (2m - i)p_d^{i+1}(1 - p_d)^{2m-i-1}\right)$$

$$= \sum_{i=0}^{m} \binom{2m + 1}{i}\left((i + 1)(1 - p_d) - (2m + 1)p_d\right)p_d^i(1 - p_d)^{2m-i-1}.$$

Since $i \leq m$ and $(1 - p_d) < 1$, we have

$$\big((i+1)(1-p_d) - (2m+1)p_d\big) < \big((m+1) - (2m+1)p_d\big).$$

Using $p_d \geq \frac{m+1}{2m+1}$, we get

$$\big((i+1)(1-p_d) - (2m+1)p_d\big) < 0. \quad \square$$

5.4 The Effect of Using More Parents

In this section we study the effect on genetic drift of using more parents with occurrence-based scanning crossover. This is interesting because, after all, occurrence-based scanning crossover is a multi-parent crossover operator. Basically, what we prove here is that the more parents we use, the more genetic drift is induced by occurrence-based scanning. The proofs in this section are rather short. This is caused by the fact that most of the work was done in earlier sections. This section does not only contain theorems and their proofs; it also also contains a conjecture. This is a conjecture about the effect of using more parents on the self-amplification of the drift tendency. We suspect that the proof for this conjecture is as complicated as, or even more complex than the proof for theorem 5, because it is a conjecture about the same complex derivative as in that theorem.

Theorem 6. *For all $m \in \mathbb{N}$*

$$f(m+1, p_d) > f(m, p_d)$$

on the interval $(\frac{1}{2}, 1)$. That is, the probability of a drift off event increases when the number of parents increases.

Proof. This follows directly from definition 4 (the definition of f) and lemma 1.
$\square$

Theorem 7. *For all $m \in \mathbb{N}$*

$$f(m+1, 1-p_d) < f(m, 1-p_d)$$

on the interval $(\frac{1}{2}, 1)$. That is, the probability of a drift back event decreases when the number of parents increases.

Proof. This follows directly from definition 4 (the definition of f) and corollary 2.
$\square$

Theorem 8. *For all $m \in \mathbb{N}$*

$$h(m+1, p_d) > h(m, p_d)$$

on the interval $(\frac{1}{2}, 1)$, That is, the odds of a drift off event against a drift back event, in the case of a non-neutral event, get higher when the number of parents increases.

Proof. This follows directly from lemma 1, corollary 2 and definition 6. □

Conjecture 1. *For all $m \in \mathbb{N}$*

$$\frac{\partial}{\partial p_d} h(m+1, p_d) > \frac{\partial}{\partial p_d} h(m, p_d)$$

on the interval $(\frac{1}{2}, 1)$. That is, the self-amplification of the drift off tendency increases when the number of parents increases.

This conjecture is hard to prove, because the derivative of $h(m, p_d)$ is a very complex function. For the same reason, we had such a hard time proving theorem 5. Now it is even worse, because we are saying something about how $h(m, p_d)$ changes with m, whereas theorem 5 only tells us that some property holds for all m.

6 Practical Implications

In this section we take a quick look at the practical implications of the theorems in section 5. These theorems can be summarized as follows. Occurrence-based scanning induces a stronger than usual genetic drift. This genetic drift amplifies itself. Using more parents causes the latter two effects to be stronger.

This is all well, but the theorems do not tell us anything about the significance of these effects. *Stronger* might be only a small percentage stronger, and even a hundred percent stronger genetic drift, might still be a very weak genetic drift, when compared to its opposing forces, mutation and selective pressure.

Ideally, a quantitative study would now follow. However, given the complexity of the subject, this might very well add another twenty pages to this article. So instead, we will give an example that we feel is quite representative. We use a parent set size of 7 [3], and we look at values for p_d (the proportion of the population having the dominant allele) from the set $\{0.5, 0.6, 0.7, 0.8, 0.9\}$[4].

Table 1 shows us the probability of drift off events, the probability of drift back events, and the values for the functions f, g, and h. From the table we can conclude that the effects are quite significant, even at the beginning of a run of the algorithm, when the value of p_d is small. Note, that the first three rows contain probabilities, the fourth row contains a difference between probabilities, and the fifth row contains odds.

So how about compensating for these effects by strengthening the opposing forces, mutation and selective pressure. First of all, this would mean a different mutation rate for each gene, because the frequencies by which alleles occur are different for

---

[3] We use 7, because it is an odd number and within the range that is normally used (for example in [2]).

[4] For 0.50 we take the limit from above.

p_d	0.5	0.6	0.7	0.8	0.9
$P^{p_d}_{\text{drift off}}$, i.e., $f(m, p_d)$	0.250	0.284	0.262	0.193	0.100
$P^{p_d}_{\text{drift back}}$, i.e., $f(m, 1 - p_d)$	0.250	0.174	0.088	0.027	0.002
$p_d(1 - p_d)$	0.250	0.240	0.210	0.160	0.090
$g(m, p_d)$	0.000	0.110	0.174	0.1667	0.097
$h(m, p_d)$	1.000	1.633	2.972	7.248	40.62

Table 1. Some actual values for the studied probabilities and functions. The parent set size is 7, which means that m is equal to 3. See definitions 4, 5, and 6 for an interpretation of the functions. Recall that for uniform scanning, $P_{\text{drift off}} = P_{\text{drift back}} = p_d(1 - p_d)$.

each position. The selective pressure is, at least in the case of occurrence-based scanning, based on the individual as a whole; so it cannot be handled likewise. Second, the mutation rates would have to be adjusted constantly, because the frequencies by which alleles occur change during the run. These two measures, that counter the extra genetic drift, could have serious side-effects. Especially, since we have not considered multiple gene genetic drift. In case of multiple gene genetic drift we would need yet another mutation rate for each relevant gene combination, and these would inevitably be in conflict with each other. So we must conclude that the induced genetic drift is significant and irreparable.

7 Conclusions

In this article we studied genetic drift in the context of a steady state genetic algorithm using multi-parent scanning crossover. The two main conclusions are, that uniform scanning crossover does not influence genetic drift at all in this context, and that occurrence-based scanning crossover has a negative influence on genetic drift in this context. Occurrence-based scanning induces genetic drift, and this gets worse when the number of parents increases. This is especially bad because the whole point of multi-parent crossover is that we want to use more parents.

We have shown that, in a representative case, the induced genetic drift is significant. Furthermore, we have argued that this genetic drift cannot be successfully countered by strengthening the forces that oppose genetic drift (mutation and selective pressure). Countering genetic drift in this way would have serious side-effects. But even without taking the side-effects into account, there is no way to tune the opposing forces in such a way that they counter the genetic drift of all gene combinations. To counter single gene genetic drift only, we would need a different mutation rate for each gene. Each of these mutation rates would have to adapt itself to the constantly changing strength of the genetic drift.

Further research on the subject includes: proving the conjecture that using more parents increases the self-amplification of the drift tendency, studying the effect of scanning crossover on multiple gene genetic drift, and doing the same for other multi-parent crossover operators. It would also be interesting to look at the influence of the steady state model on genetic drift.

Acknowledgments

I would like to thank the following persons: Prof. Dr. A. E. Eiben, daily supervisor of my PhD project, Dr. I. G. Sprinkhuizen-Kuyper, former colleague and first author of [5], and J. R. Floor, computer science student and a good friend of mine. Without their proofreading, corrections and helpful comments, this article would have been of a substantially lower quality. A considerable part of this article was written using Dragon Naturally Speaking, a speech recognition software package.

References

1. A. E. Eiben, P.-E. Raué, and Z. Ruttkay. Genetic algorithms with multi-parent recombination. In Y. Davidor, H.-P. Schwefel, and R. Männer, editors, *Proceedings of the 3rd Conference on Parallel Problem Solving from Nature*, volume 866 of Lecture Notes in Computer Science, pages 78–87. Springer-Verlag, Berlin Heidelberg New York, 1994.
2. A. E. Eiben and C. A. Schippers. Multi-parent's niche: n-ary crossovers on NK-landscapes. In H.-M. Voigt, W. Ebeling, I. Rechenberg, and H.-P. Schwefel, editors, *Proceedings of the 4th Conference on Parallel Problem Solving from Nature*, volume 1141 of Lecture Notes in Computer Science, pages 319–328. Springer-Verlag, Berlin Heidelberg New York, 1996.
3. D. B. Fogel, editor. *Evolutionary Computation: the Fossile Record*. IEEE Press, New York, 1998.
4. P. A. P. Moran. *An Introduction to Probability Theory*. Clarendon Press, Oxford, 1968.
5. I. G. Sprinkhuizen-Kuyper, C. A. Schippers, and A. E. Eiben. On the real arity of multiparent recombination. In W. Banzhaf, J. Daida, A. E. Eiben, M. H. Garzon, V. Honovar, M. Jakiela, and R. E. Smith, editors, *Proceedings of the 1999 Congress on Evolutionary Computation*, pages 680–686. Morgan Kaufmann, San Francisco, 1999.

Harmonic Recombination for Evolutionary Computation

F. J. Burkowski

Department of Computer Science
University of Waterloo
Waterloo, Canada
E-mail: *fjburkow@plg.uwaterloo.ca*

Abstract. When using an evolutionary algorithm to handle an optimization problem the user must immediately contend with issues related to the bit representation of a feasible solution and the choice of evolutionary operators. In this presentation we discuss research into techniques that strive to nullify the a priori decisions made when establishing the representation and bit order of a feasible solution. One objective in this research is to develop techniques that reformulate a representation so as to make it more amenable to various evolutionary operators. Aside from the obvious practical benefits that should accrue, we hope that this approach to evolutionary computation will benefit a theoretical analysis that achieves more generality by avoiding unsupported assumptions about adjacency of feasible points in the solution space. In particular, our strategy is to avoid implicit specifications of locality due to adjacency assumptions that arise from the use of particular operators. For example, the assumption of adjacency of solutions as measured via Hamming distance is a notion that is convenient for a string mutation operator but does not necessarily have anything to do with the intrinsic structure of the objective function. Naturally, some type of adjacency, explicit or implicit, has to be at work since we wish to efficiently explore a solution space of exponential size in polynomial time, but the idea is to guide this exploration by stochastic techniques that are influenced by intrinsic function properties instead of unjustified a priori assumptions.

Keywords
Discrete Fourier transform, harmonic recombination, genome representation, search locality

1 Introduction

Although there is some controversy about the use of genetic algorithms (GAs) for optimization [13], they are surprisingly effective as a heuristic strategy for tough optimization problems in a variety of applications. Nonetheless, the use of GAs is certainly an inexact science. As with many heuristic strategies, there are various parameter settings (population size, mutation rate, details associated with crossover, etc.) and so two people may implement a GA solution technique in very different ways. Over the last 15 to 20 years, various researchers have studied these issues and there is a growing body of papers that have reported on parameter settings and crossover issues. The reader may consult, for example, [10].

Despite these efforts, there are implementation issues that tend to be more of a 'black art' in that they have very few guidelines of a theoretical nature: bit string representation and choice of crossover operator.

We consider each of these in turn.

1.1 Bit String Representation

Although the search domain is typically represented by a set of bit strings, the linear order of bit positions in a string is not specified by the standard GA. Someone specifying the format of a GA bit representation involving, say three 10-bit numbers, might justifiably pause to consider several issues and options.

- In what order should I place the numbers?
- Perhaps the bits should be interleaved.
- Perhaps the bit string for one of the numbers should be reversed prior to applying crossover and then reestablished in the given order just prior to fitness evaluation.

These questions have no clear and concise resolution and are part of the representation and coding problem associated with GAs. Any a priori assumptions about the best ordering of bits leads to a particular positional bias that will affect an order sensitive operator such as one-point crossover. The reader should see [14] for a discussion about positional bias. Coding of the number (standard binary, Gray-coded or floating point) is yet another issue [2, pages 109–111].

1.2 Choice of Crossover Operator

Another aspect of positional bias is the non-uniformity of bit inheritance that is intrinsic to simple crossover. Neighboring bits within a parent are more likely to move together to a particular offspring than the bits that are further apart. This asymmetry is especially at play in one-point crossover where bits at the end of a parent will always end up in different children. One might suppose that if human DNA was circular, as in some bacteria, we would have a 'circle-centric' predisposition to see two-point crossover as natural and one-point crossover as a degenerate two-point crossover in which one of the cut points is always at a fixed position. The asymmetry of one-point crossover would then be more obvious when we were forced to make the a priori designation of the fixed cut point within the circle.

2 Theoretical Perspectives

The benefits of placing evolutionary computation, and especially GAs, in a more theoretical framework has been a recurrent research objective for more than a decade.

Various researchers have striven to get more theoretical insights into representation of feasible solutions especially with regard to interdependence among the bits in a feasible solution, hereafter referred to as a genome. While many papers have dealt with these concerns, the following references should give the reader a reasonable perspective on the more vital issues.

Liepins and Vose [21] discuss various representational issues in genetic optimization. The paper elaborates the failure modes of a GA and eventually discusses the existence of an affine transformation that would convert a deceptive objective function to an easily optimizable objective function. A later paper [30] furthers their examination of schema analysis by examining how the crossover operator interacts with schemata. A notable result here: *every function has a representation in which it is essentially the ones counting problem*. Explicit derivation of the representation is not discussed.

Manela and Campbell [22] consider schema analysis from the broader viewpoint provided by abstract harmonic analysis. Their approach starts with group theoretic ideas and concepts elaborated by [20] and ends with a discussion about epistasis and its relation to GA difficulty.

Radcliffe [24] argues that for many problems conventional linear chromosomes and recombination operators are inadequate for effective genetic search. His critique of intrinsic parallelism is especially noteworthy.

3 Objectives of This Research

Considering the importance of genome representation and epistasis, this paper describes work that strives for the following goal. Design an adaptive algorithm that attempts to derive an alternate *reduced epistasis* genome representation while the optimization search is being done.

As noted by [25, page 8], it may be possible to reduce the degree of epistasis in a problem by choosing a different coding for the genome. In attempting to derive such an encoding, our main strategy will be to generate a non-singular matrix M that will map the typical genome x to a different representation $y = Mx$. The objective function $f(x)$ then becomes $f(M^{-1}y) = g(y)$ and we hope to optimize $g(y)$ by using techniques that work well when bit dependencies are low, for example: hill-climbing, two-point crossover, or population-based incremental learning (PBIL) as described by [4] and [5]. Matrix transforms have been discussed earlier by Battle and Vose [7] who consider the notion of an invertible matrix to transform a genome x into a genome y that resides in a population considered to be isomorphic to the population holding genome x. They also note that converting a standard binary encoding of a genome to a Gray encoding is a special case of such a transformation.

In our studies we pursue techniques that attempt to derive an explicit form for such a matrix with the idea of reducing epistasis. To nullify any a priori advantage or disadvantage that may be present due to the format of the initial representation, knowledge used to construct matrix M will be derived only through evaluations of the fitness function $f(x)$. By seeing $f(x)$ as a 'black box' we are not allowed to prescribe a Gray encoding, for example, on the hunch that it will aid our optimization activity. Instead, we let the algorithm derive the new encoding which, of course, may or may not be a Gray encoding. With these 'rules of engagement' we forego any attempt to characterize those fitness functions that would benefit from the use of a Gray encoding. Instead, we aim to develop algorithms that lead to a theoretical analysis of the more general transformation, for example, the characterization of a fitness function that would guarantee the derivation of a matrix M that serves to reduce epistasis.

To emphasize some important aspects of our approach; the given genome representation is considered to be only useful for fitness evaluation. Evolutionary operators are always applied to a transformation of the genome not the genome itself. Although we start with a genome that is the usual linear bit string, our initial operations on that string will be symmetric with respect to all bits. Any future asymmetry in bit processing will arise in due course as a natural consequence of the algorithm interacting with $f(x)$. To enforce this approach, operators with positional-bias, such as single point crossover, will be avoided. For similar reasons, we eschew any analysis that deals with bit-order related concepts such as *length of schemata*.

4   Tools from Learning Theory

The publication of L. Valiant's paper [29] established a solid foundation for computational learning theory because it provided a rigorous framework within which formal questions could be posed. Subsequently, computational learning has spread to other areas of research, for example, [6] and [1] discuss the application of Valiant's PAC (probably approximately correct) model to learnability using neural nets. In this paper we borrow from the PAC model some tools that we consider to be appropriate for the study and implementation of evolutionary algorithms and GAs in particular. Learning is not new to GAs, see [8] and [26]. Other research has addressed related issues such as discovery of gene linkage in the representation [16] and adaptation of crossover operators [27], [28] and [12]. Another very interesting study can be found in [18]. Space limitations do not permit any lengthy description of the PAC learning model. The reader is urged to read [23] or [17] for introductory material. Our use of the PAC model is not to learn a target function after being given training examples, but rather to simply use some of the concepts and mathematical tools employed by PAC, in particular, those algorithms dealing with the discrete fourier transform (DFT).

5   Notation to Be Used

For our population based strategies, N members of the initial population P are randomly drawn from the set $X = \{0, 1\}^n$. A typical member of X would be

$x = (x_1, x_2, \ldots, x_n)$ where $x_i = 0$ or 1 for $i = 1, 2, \ldots, n$. The following list represents all the parameters and sets that will be required:

1. $X = \{0, 1\}^n$ (the set of all feasible solutions to the problem).
2. P = the initial population of N items randomly selected from X.
3. f mapping $X \to R$ will act as a fitness function in the usual sense.
4. $a > 0, 0 < \varepsilon < 1$, and $0 < \delta < 1$ are various parameters that will be needed later.

6 The Discrete Fourier Transform

The multidimensional discrete Fourier transform (or Walsh transform) is a very useful tool in learning theory. Given a function $f : X \to R$ we define the Fourier transform of f as

$$\widehat{f}(u) = \frac{1}{2^n} \sum_{x \in X} f(x) t_u(x)$$

where the parity functions $t_u : X \to \{-1, +1\}$ are defined for $u \in X$ as

$$t_u(x) = (-1)^{\sum_{i=1}^{n} u_i x_i}.$$

So, $t_u(x)$ has value -1 if the number of indices i at which $u_i = x_i = 1$ is odd, and 1 otherwise. We define the inner product of two functions f and g by

$$\langle f, g \rangle = \frac{1}{2^n} \sum_{x \in X} f(x) g(x).$$

Note that $\langle f, g \rangle = \mathrm{E}[fg]$ with the expectation taken over a uniform distribution of the inputs. We can define a norm on F as

$$\|f\| = \sqrt{\langle f, f \rangle} = \sqrt{\mathrm{E}[f^2]}.$$

The set $\{t_u(x)\}_{x \in X}$ is an orthonormal basis for the vector space of real-valued functions on X. We can recover f from its transform by using

$$f(x) = \sum_{u \in X} \widehat{f}(u) t_u(x).$$

This unique expansion of $f(x)$ in terms of the parity basis is its Fourier series and the sequence of Fourier coefficients is called the spectrum of $f(x)$.

Since it is often more convenient to deal with the set of indices of x where $x_i = 1$ we will define for each set $A \subseteq \{1, 2, \ldots, n\}$ the function $t_A : 2^{\{1,2,\ldots,n\}} \to \{-1, +1\}$ as

$$t_A(x) = (-1)^{\sum_{i \in A} x_i}.$$

Now we can write

$$f(x) = \sum_{A \subseteq \{1,\ldots,n\}} \widehat{f}(A) t_A(x)$$

where

$$\widehat{f}(A) = \frac{1}{2^n} \sum_{x \in X} f(x) t_A(x).$$

We will be especially interested in subsets $S \subset 2^{\{1,2,\ldots,n\}}$ such that the size $|S|$ is polynomial in n. For a given $f(x)$ and $\theta > 0$, we will say that S has the *large Fourier coefficient property* [17, page 47], if:

1. for all A such that $|\widehat{f}(A)| \geq \theta$, we have $A \in S$, and
2. for all $A \in S$, we have $|\widehat{f}(A)| \geq \frac{\theta}{\sqrt{2}}$.

Now the Fourier transform $\widehat{f}(u)$ is a sum across all $x \in X$. This involves an amount of computation that is exponential in n. To be of practical use we need a lemma from [17] that is an extension of earlier work [19].

Lemma 1. *There is an algorithm KM such that, for any function $f : X \to R$, threshold $\theta > 0$, and confidence δ with $0 < \delta < 1$, KM returns, with probability at least $1 - \delta$, a set with the large Fourier coefficient property. KM uses membership queries and runs in time polynomial in n, $1/\theta$, $log(1/\delta)$ and $\max_{x \in X} |f(x)|$.*

The large Fourier coefficient property provides a vital step in the practical use of the discrete Fourier transform. We can take a given fitness function f and approximate it with a function $f_S(x)$ defined as

$$f_S(x) = \sum_{A \in S} \widehat{f}(A) t_A(x)$$

with a set size $|S|$ that is *not* exponential in size. This is important since we do not want to search for an x giving maximum fitness by evaluating the summand an exponential number of times.

7 Harmonic Recombination

The large Fourier coefficients can be used in the creation of new genomes. Let us assume that, using Jackson's algorithm or some other technique, we have a reasonable strategy for generating a set of large Fourier coefficients along with the set of *parity strings* $\{w\}$ that represent the locations of these Fourier coefficients. In other words, we derive a set

$$W = \left\{ w : |\widehat{f}(w)| \geq \frac{\theta}{\sqrt{2}} \right\}.$$

We then extract from W a set of linearly independent w that are n in number. *These will be used as the rows of our required matrix M.* The motivation for generating M in this way rests on the following observation. Starting with: $f(x) = \sum_{u \in X} \widehat{f}(u) t_u(x)$ we replace x with $M^{-1}y$ to get

$$f(x) = f\left(M^{-1}y\right)$$
$$= g(y)$$
$$= \sum_{u \in X} (-1)^{u^T M^{-1} y} \widehat{f}(u)$$
$$= \sum_{u \in X} (-1)^{\left[(M^T)^{-1} u\right]^T y} \widehat{f}(u).$$

Now, since M^T is nonsingular it possesses a nonsingular inverse, or more noteworthy, it provides a bijective mapping of X onto X. Since u in the last summation will go through all possible n bit strings in X we can replace u with $M^T u$ and the equation is still valid. Consequently,

$$g(y) = \sum_{u \in X} (-1)^{u^T y} \widehat{f}\left(M^T u\right)$$
$$= \sum_{e \in E} (-1)^{e^T y} \widehat{f}\left(M^T e\right) + \sum_{\substack{u \in X \\ u \notin E}} (-1)^{u^T y} \widehat{f}\left(M^T u\right)$$

where E is the collection of all n bit strings that have all entries 0 with a single 1 bit. Note that when a particular e vector is multiplied by M^T we simply get one of the w in the set W and so $\widehat{f}(M^T e)$ is a large Fourier coefficient. Consequently, *if there are only a few large Fourier coefficients, we should expect the first sum in the last equation to be the more significant of the two sums.* When this is the case, a y value producing a highly fit value for $g(y)$ would be easily determined by a hill-climbing technique since the components of the first sum are independent of one another. In fact the signs of the large Fourier coefficients essentially spell out the y bit string that will maximize the first sum. For example, if all the large Fourier coefficients in this sum are negative then the value of y will be all 1's and we have the *ones counting* situation similar to that described in [30]. It should be stressed that this transformation does not 'burn any bridges' by closing off any portion of the search space. Its sole purpose is to *soften* the problem so that bit interdependence is reduced.

Note that the high correlation parity strings effectively act as *ring-sum* formulas (modulo-2-sum of monomials) that define the constraints imposed on the bit positions of genomes if they are to attain high fitness. This observation leads to the speculation that high correlation parity strings may be a reasonable or practical strategy for the specification of a type of gene linkage.

8 Working in Polynomial Time: Some Practical Considerations

The discussion in the last section gives us a characterization of the fitness function $f(x)$: *it should have large Fourier coefficients that are relatively few in number.*

The definitions of the Fourier transform and its inverse deal with summation ranges that require an exponential amount of work since $|X| = 2^n$. To be of practical use we will need to approximate the Fourier transform by deriving (as best we can) the *large* Fourier coefficients. Our current work in this direction has considered the following techniques.

1. *The KM Algorithm*: In [17, page 45], the reader will find a clear description of the Kushilevitz and Mansour algorithm for obtaining large Fourier coefficients. The primary advantage of this approach is the theoretical analysis of the technique providing certain guarantees as specified earlier in lemma 1. The analysis accounts for the sampling error due to limited population size and demonstrates that the large coefficients can be found in polynomial time. Nonetheless, the algorithm involves extensive calculation.

2. *Co-evolving Population*: This strategy is a rather novel use of a co-evolving population. We can regard a Fourier coefficient $\widehat{f}(A)$ as representing the strength of correlation between a parity function t_A and the objective function $f(x)$. We derive large coefficients by running a co-evolving population of parity strings each representing the location of a Fourier coefficient in the n-dimensional Fourier space. Like the genomes, parity strings also have a type of 'fitness' namely the absolute value of the correlation *estimated* by computing for each parity string u the value

$$\left| \sum_{x \in GP} f(x) t_u(x) \right|$$

where $x \in GP$ designates all genomes x in the genome population GP. Experiments with this approach were reasonably successful but the computation necessary to maintain the high correlation parity strings, as the population changed through evolution, was still quite time consuming.

3. *Working with Bits having High Mutual Information*: In this approach we run a heuristic procedure [9] that selects a subset of genome bits, m in number, having the highest level of mutual information [11, page 18]. To generate high correlation parity string candidates, we exhaustively compute the full set of 2^m parity strings that are 0 in all bit locations except for the specified m bits. We typically choose m so that 2^m equals the cardinality of the parity string population, a size that is conveniently made the same as the cardinality of the genome population. As before, correlation values are computed for each candidate parity string using the formula

$$\widehat{f}(A) = \frac{1}{2^n} \sum_{x \in GP} f(x) t_A(x)$$

and we then select the highest correlation strings for the matrix M. As the algorithm progresses, going through successive population generations, we compute a sequence of M matrices, each derived from a different set of bit locations (see figure 1).

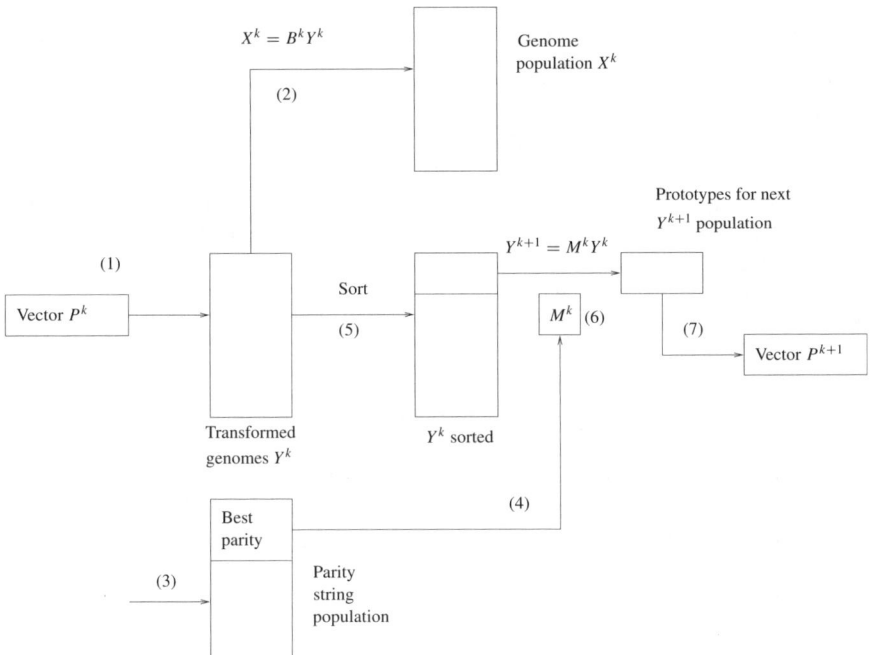

Figure 1. Data movement in step k of harmonic recombination

9 Steps in Harmonic Recombination

Referring to figure 1, our algorithm may be described as follows:

0. Initialize a probability vector P^1 to contain n values each equal to 0.5. Continue with step 1, assuming the initial value $k = 1$ and $B^1 =$ the identity matrix I.
1. Randomly create a population of N transformed genomes Y^k each with n bits. Bits in each genome follow the frequency distribution dictated by vector P^k.
2. Process the genome population:
    - Recover each genome X^k by pre-multiplying each Y^k by B^k where $B^k = B^{k-1}(M^{k-1})^{-1}$.
    - Calculate the fitness of each genome.

- Adjust each fitness value by subtracting from it the average fitness of the population.
3. Process the parity string population:
   - Generate a population of parity strings each with n bits.
   - For each parity string evaluate the strength of correlation with the transformed genomes Y^k.
4. Sort the parity strings by ascending order of the absolute value of the correlation and extract a linearly independent set of n highest correlation parity strings to form the n by n matrix M^k.
5. Sort the genomes by fitness.
6. Generate a set of prototypes:
   - Select a small subset of the most fit members of the transformed genome population.
   - For each of these genomes Y^k, generate a prototype for the next generation of transformed genomes. This is done by calculating: $Y^{k+1} = M^k Y^k$.
7. Generate the next probability vector P^{k+1}: from the set of prototypes calculate the bit frequencies in each bit location.
8. Repeat starting from step 1.

In practice we have noticed that as the algorithm progresses there is typically a gradual 'freezing up' of the Y^k representation. Various entries in the probability vector will become static holding a repeated value of 0 or 1. This is essentially a dimension reduction. When all bit positions have become frozen the algorithm is finished. The number of iterations before a complete freeze will be influenced by the number of prototypes used in the calculation of the probability vector.

9.1 Some Notes

- The *correlation value* in step 3 is an *estimate* of the Fourier coefficient corresponding to the location in Fourier space specified by the n bit parity string.
- As noted in [17] we can find the large Fourier coefficients by using a polynomial number of function evaluations, assuming $f(x)$ supports the large Fourier coefficient property. The algorithm also assumes that we are sampling $f(x)$ with x drawn from a uniform distribution. In our case, the distribution starts out uniform but then becomes non-uniform. Recalling that the Fourier coefficient is essentially an expectation evaluated over a uniform distribution, we see that we would not be evaluating the Fourier coefficient of $f(x)$, but rather $f(x)$ times $q(x)$, a function describing the non-uniform distribution. As described by [15] we can rectify this situation by redefining the norm described earlier. This involves the use of an ortho-normal set of functions first described in [3].

10 Conclusion

We contend that harmonic analysis should have significant theoretical benefits for the design of optimization algorithms. Furthermore, the Fourier spectrum of $f(x)$, its distribution of large coefficients and how they influence the location of optima should serve to quantitatively characterize functions that are compatible with these algorithms.

This paper has presented a preliminary report on evolutionary algorithms that involve an explicit use of the DFT. A possible search algorithm was described with attention drawn to some of the more challenging aspects of the implementation. Empirical results will be the subject of a future report.

Although computationally expensive, the DFT does provide a formal strategy to deal with notions such as epistasis and simple (linear) gene linkage expressible as a ring sum formula. The future value of such a theoretical study would be to see the structure of the search space expressed in terms of the spectral properties of the fitness function. Our view is that this is, in some sense, a more 'natural' expression of the intrinsic structure of the search space since it does not rely on a neighbourhood structure defined by the search operator chosen by the application programmer.

References

1. M. Anthony. Probabilistic analysis of learning in artificial neural networks: the PAC model and its variants. *http://www.icsi.berkeley.edu/~jagota/NCS/vol1.html*, 1997.
2. T. Bäck. *Evolutionary Algorithms in Theory and Practice*. Oxford University Press, New York, 1996.
3. R. R. Bahadur. A representation of the joint distribution of responses to n dichotomous items. In H. Solomon, editor, *Studies in Item Analysis and Prediction*, pages 158–168. Stanford University Press, Stanford, CA, 1961.
4. S. Baluja. *Population-Based Incremental Learning: A Method for Integrating Genetic Search Based Function Optimization and Competitive Learning*. Carnegie Mellon University, Technical Report CMU-CS-94-163, 1994.
5. S. Baluja and R. Caruna. Removing the genetics from the standard genetic algorithm. In *International Conference on Machine Learning – 12*, pages 38–46. Morgan Kaufmann, San Francisco, 1995.
6. E. B. Baum. Neural net algorithms that learn in polynomial time from examples and queries. *IEEE Transactions on Neural Networks*, 2(1):5–19, 1991.
7. D. L. Battle and M. D. Vose. Isomorphisms of genetic algorithms. In G. Rawlins, editor, *Foundations of Genetic Algorithms*, pages 242–251. Morgan Kaufmann, San Francisco, 1991.
8. R. K. Belew. When both individuals and populations search: adding simple learning to the genetic algorithm. In J. D. Schaffer, editor, *Proceedings of the Third International Conference on Genetic Algorithms*, pages 34–41. Morgan Kaufmann, San Francisco, 1989.
9. F. J. Burkowski. Shuffle crossover and mutual information. In *Congress on Evolutionary Computation*, pages 1574–1580. IEEE Press, New York, 1999.
10. L. Chambers, editor. *Practical Handbook of Genetic Algorithms, Applications (Vol. I) & New Frontiers (Vol. II)*. CRC Press, New York, 1995.
11. T. M. Cover and J. A. Thomas. *Elements of Information Theory*. John Wiley, New York, 1991.

12. L. Davis. Adaptating operator probabilities in genetic algorithms. In J. D. Schaffer, editor, *Proceedings of the International Conference on Genetic Algorithms*, pages 61–69, June 1989.
13. K. DeJong. Genetic algorithms are NOT function optimizers. In D. Whitley, editor, *Foundations of Genetic Algorithms 2*, pages 5–17. Morgan Kaufmann, San Francisco, 1992.
14. L. J. Eshelman, R. A. Caruana, and J. D. Schaffer. Biases in the crossover landscape. In *Proceedings of the Third International Conference on Genetic Algorithms*, pages 10–19. Morgan Kaufmann, San Francisco, 1989.
15. M. L. Furst, J. C. Jackson, and S. W. Smith. Improved learning of AC0 functions. In *The 4th Workshop on Computational Learning Theory*, pages 317–325. Morgan Kaufmann, San Francisco, 1991.
16. G. R. Harik. *Learning Gene Linkage to Efficiently Solve Problems of Bounded Difficulty Using Genetic Algorithms*. PhD dissertation, Computer Science and Engineering, The University of Michigan, 1997.
17. J. C. Jackson. *The Harmonic Sieve: A Novel Application of Fourier Analysis to Machine Learning Theory and Practice*. PhD dissertation, Carnegie Mellon University, CMU-CS-95-183, 1995.
18. H. Kargupta and D. E. Goldberg. SEARCH, blackbox optimization, and sample complexity. In R. K. Belew and M. D. Vose, editors, *Foundations of Genetic Algorithms 4*, pages 291–324. Morgan Kaufmann, San Francisco, 1997.
19. E. Kushilevitz and Y. Mansour. Learning decision trees using the Fourier spectrum. *SIAM Comput.*, 22(6)1331–1348, 1993.
20. R. J. Lechner. Harmonic analysis of switching functions. In Amar Mukhopadhyay, editor, *Recent Developments in Switching Theory*, pages 121–228. Academic Press, New York, 1971.
21. G. E. Liepins and M. D. Vose. Representational issues in genetic optimization. *J. Expt. Theor. Artif. Intell.*, 2:101–115, 1990.
22. M. Manela and J. A. Campbell. Harmonic analysis, epistasis and genetic algorithms. In R. Männer and B. Manderick, editors, *Parallel Problem Solving from Nature 2*, pages 57–64. Elsevier, Amsterdam, 1992.
23. T. M. Mitchell. *Machine Learning*. McGraw–Hall, New York, 1997.
24. N. J. Radcliffe. Non-linear genetic representations. In R. Männer and B. Manderick, editors, *Parallel Problem Solving from Nature 2*, pages 259–268. Elsevier, 1992.
25. C. Reeves and C. Wright. An experimental design perspective on genetic algorithms. In L. D. Whitley and M. D. Vose, editors, *Foundations of Genetic Algorithms 3*, pages 7–22. Morgan Kaufmann, San Francisco, 1995.
26. M. Sebag and M. Schoenauer. Controlling crossover through inductive learning. In *Parallel Problem Solving from Nature 3*, pages 209–218. Springer-Verlag, Berlin Heidelberg New York, 1994.
27. J. E. Smith and T. C. Fogarty. Recombination strategy adaptation via evolution of gene linkage. In *Proceedings of IEEE International Conference on Evolutionary Computing*, pages 826–831, 1996.
28. J. E. Smith and T. C. Fogarty. Operator and parameter adaptation in genetic algorithms. *Soft Computing*, 1(2):81–87. Springer-Verlag, Berlin Heidelberg New York, 1997.
29. L. G. Valiant. A theory of the learnable. *Communications of the ACM*, 27(11):1134–1142, 1984.
30. M. D. Vose and G. E. Liepins. Schema disruption. In *Proceedings of the Fourth International Conference on Genetic Algorithms*, pages 237–242. Morgan Kaufmann, San Francisco, 1991.

How to Detect all Maxima of a Function

J. Garnier and L. Kallel

Centre de Mathématiques Appliquées
UMR CNRS 7641, Ecole Polytechnique
91128 Palaiseau Cedex, France
E-mail: *{garnier,kallel}@cmapx.polytechnique.fr*

Abstract. The first contribution of this paper is a theoretical investigation of a family of landscapes characterized by the number of their local optima N and the distribution of the sizes (α_j) of their attraction basins. For each landscape case, we give precise estimates of the size of the random sample that ensures that at least one point lies in each basin of attraction. A practical methodology is then proposed for identifying these quantities (N and (α_j) distribution) for an unknown landscape, given a random sample on that landscape and a local steepest ascent search. This methodology can be applied to any landscape specified with a modification operator and provides bounds on search complexity to detect all local optima. Experiments demonstrate the efficiency of this methodology for guiding the choice of modification operators, eventually leading to the design of problem-dependent optimization heuristics.

Keywords
Combinatorial complexity, local search, neighborhood graph, randomized starting solution

1 Introduction

In the field of stochastic optimization, two search techniques have been widely investigated during the last decade; simulated annealing [25] and evolutionary algorithms (EAs) [5,6]. These algorithms are now widely recognized as methods of order zero for function optimization as they impose no condition on function regularity. However, the efficiency of these search algorithms, in terms of the time they require to reach the solution, is strongly dependent on the choice of the modification operators used to explore the landscape. These operators in turn determine the neighborhood relation of the landscape under optimization.

This paper provides a new methodology allowing one to estimate the number and the sizes of the attraction basins of a landscape specified in relation to some modification operator. This allows one to derive bounds on the probability that one samples a point in the basin of the global optimum, for example. Further, this method could be used for guiding the choice of efficient problem-dependent modification operators or representations.

Formally, a landscape can be denoted by $\mathcal{L} = (f, \mu)$ where f is the function to optimize and μ a stochastic modification operator that determines the neighborhood relation of the landscape (neighbors are obtained by one application of μ). The structure of the landscape heavily depends on the choice of the modification operators, which in turn may depend on the choice of the representation (the coding of the candidate solutions into binary or Gray strings, for example). Hence, before the optimization process can be started, there are a number of practical choices (representation and operators) that determine the landscape structure. Consequently, these choices are often crucial for the success of stochastic search algorithms.

Some research has studied how the fitness landscape structure impacts the potential search difficulties [13,12,24,26]. It is shown that every complex fitness landscape can be represented as an expansion of elementary landscapes – one term in the Fourier expansion – which are easier to search in most cases. This result has been applied to solve a difficult NP-complete problem [23] (the identification of minimal finite k-state automaton for a given input-output behavior) using evolutionary algorithms. Other theoretical studies of search feasibility consider the whole landscape as a tree of local optima, with a label describing the depth of the attraction basin at each node [18,21]. Such a construction naturally describes the inclusion of the local attraction basins present in the landscape. These studies investigate tree structures that ensure a minimal correlation between the strength of the local optima and their proximity to the global optimum, with respect to an ultra-metric distance on the tree. However, from a practical point of view, the tree describing the distribution of local optima is unknown and too expensive in terms of computational cost to determine for a given landscape.

The lack of an efficient method at reasonable cost that allows one to characterize a given landscape, motivates the construction of heuristics for extracting a priori statistical information about landscape difficulty, for example, based on random sampling of the search space. We cite two examples from the field of EAs: fitness distance relations, first proposed in [7] and successfully used to choose problem dependent random initialization procedures [10,14], and fitness improvement of evolution operators, first proposed in [4], then extended and successfully used to choose binary crossover operators [11] and representations [8]. However, even if such heuristics can guide the a priori choice of some EA parameters, they do not give significant information about landscape structure; for instance, recent work suggests that very different landscapes (leading to different EA behaviors) can share the same fitness distance relation [20,9]. Further, the efficiency of such summary statistics is limited to the sampled regions of the space, and therefore does not necessarily help the long-term convergence results as implicitly illustrated in [11] for example. This gives strong motivation for developing tools that allow one to derive a more global (beyond the sampled regions) information on the landscape at hand, relying on an implicit assumption of stationarity of the landscape. Along that line, this paper proposes a new method to identify the number and the distribution of local optima with respect to a given neighborhood relation of a given landscape. The proposed method applies to any neighborhood relation specified with a modification operator,

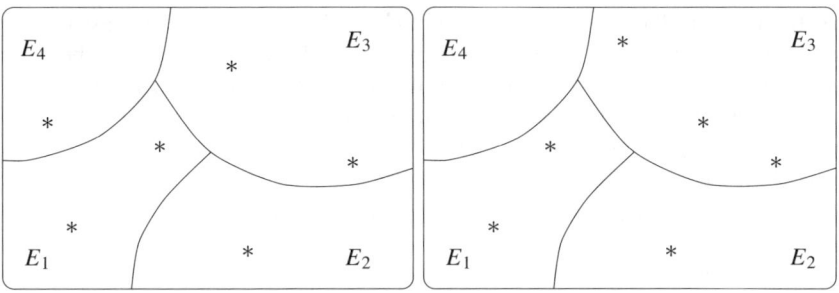

Figure 1. Schematic representations of the search space E with $N = 4$ attraction basins. $M = 6$ points have been randomly placed on both pictures. As a result there is at least one point in each attraction basin in the left picture, but not in the right picture, where E_4 is empty.

and hence provides a practical tool to compare landscapes obtained with different operators and representations.

The framework is the following. We assume that the search space E can be split into the partition $E_1, \ldots, E_N$ of subspaces which are attraction basins of local maxima $m_1, \ldots, m_N$ of the fitness function. We also assume that there exists a local search algorithm (for example a steepest ascent) which is able to find from any point of the search space the closest local maximum

$$\Theta : E \to \{m_1, \ldots, m_N\} : x \mapsto m_j \text{ if } x \in E_j.$$

The basic problem consists in detecting all local maxima m_j. This is equivalent to finding a way to put a point in all attraction basins E_j, because the local search algorithm will complete the job. We shall develop the following strategy. First, we shall study the direct problem, which consists in studying the covering of the search space by a collection of points randomly distributed when the partition (E_j) is known. Second, we shall deal with the inverse problem which consists in estimating the number of local maxima from information deduced from the covering.

Direct Problem (section 4). One puts M points randomly in the search space. The question is, given the statistical distribution of the relative sizes of the attraction basins and their number N, what is the probability $p_{N,M}$ that at least one point lies in every attraction basin? This probability is very important. Indeed, using the local search algorithm, it is exactly equal to the probability to detect all local maxima of the function.

Inverse Problem (section 5). The statistical distribution of the relative sizes of the attraction basins and their number are assumed to be known for computing $p_{N,M}$ in section 4. Unfortunately, this is rarely the case in practical situations, and one wants to estimate both. The strategy is to put randomly M initial points in the search space and to detect the closest local maxima by the local search algorithm. The data we collect is the set $(\beta_j)_{j\geq 1}$ of the number of maxima detected with j initial points. Of course β_0 is unknown (number of local maxima of the landscape that have not been detected). The question is the following: how can the total number of local maxima $N = \sum_{j=0}^{\infty} \beta_j$ be efficiently estimated from the set $(\beta_j)_{j\geq 1}$? A lower bound is $\bar{N} = \sum_{j=1}^{\infty} \beta_j$, but we aim at constructing a better estimator.

Overview. The paper is divided into three parts. First, section 4 addresses the direct problem of sample sizing in the case of basins of random sizes, then in the case of basins of equal sizes. Second, section 5 is devoted to the estimation of the distribution of the relative basins sizes for an unknown landscape, using a random sample from the search space. This is achieved by a two-step methodology. Section 5.2 starts by considering a parametrized family of laws for the relative sizes of basins, for which it derives the corresponding covering of the search space (law of (β_j)). Then section 5.3 comments on how these results can be practically used for characterizing the sizes of basins of an unknown landscape. For instance, it proposes to compare the covering of an unknown landscape (given by the empirically observed (β_j) values) to the coverings studied in section 5.2. Finally, the last part of the paper (section 6) is devoted to some experiments that validate (section 6.1) and illustrate (section 6.2) the methodology: first, a landscape is purposely designed to test the reliability of the method according to the size of the random sample, and to the number of local optima (recall the theoretical results are asymptotic with respect to N and M). Second, the method is used to investigate some problems, known to be difficult to optimize for EAs. For each problem, we also compare the landscapes related to different mutation operators.

2 Notations and Definitions

Consider a fitness function $f : E \to \mathbb{R}$, and a neighborhood relation induced by a modification operator μ, such that the number of different μ-neighbors (neighbors that can be obtained by one application of μ to x) of $x \in E$ is 'bounded'. In the following, we denote by N the number of local optima of $\mathcal{L}$, and by (α_j) the random variables describing the normalized sizes of the attraction basins of $\mathcal{L}$. As shown in [16,17], a local improvement algorithm is efficient to find quickly a local optimum starting from some given point. Among the possible algorithms, we present, in figure 2, the steepest ascent algorithm, also called optimal adjacency algorithm elsewhere [16].

The steepest ascent algorithm thus consists in selecting the best neighbors after the entire neighborhood is examined. An alternative algorithm, the so-called first improvement algorithm (FI), consists in accepting the first favorable neighbor as soon as it is found, without further searching. Notice that in the FI case there are extra free parameters which are the order in which the neighborhood is searched. As pointed

> *Input:* A fitness function $f : E \to \mathbb{R}$, an operator μ and a point $X_0 \in E$.
> *Algorithm:* Modify X by repeatedly performing the following steps:
>
> - Record, for all μ-neighbors of X denoted by $\mu^i(X)$: $(i, f(\mu^i(X)))$
> - Assign $X = \mu^i(X)$ where i is chosen such that $f(\mu^i(X))$ reaches the highest possible value (this is the steepest ascent).
> - Stop when no strictly positive improvement in μ-neighbors fitnesses has been found.
>
> *Output:* The point X, denoted by $\Theta(X_0) = \mu^*(X_0)$.

Figure 2. The steepest ascent algorithm

out in [15, p. 470], the steepest ascent is often not worth the extra computation time, although it is sometimes much quicker. Nevertheless, our focus in this paper is not a complete optimization of the computational time, so we leave this problem as an open question.

Attraction Basin. The attraction basin of a local optimum m_j is the set of points $\{X_1, \ldots, X_k\}$ of the search space such that a steepest ascent algorithm starting from X_i ($1 \le i \le k$) ends at the local optimum m_j ($\Theta(X_i) = m_j$). The normalized size of the attraction basin of the local optimum m_j is then equal to $k/|E|$.

Remark 1. This definition of the attraction basins yields a partition of the search space into different attraction basins, as illustrated in figure 1. The approach proposed in this paper is based on this representation of the search space into a partition of attraction basins, and could be generalized to partitions defined with alternative definitions of attraction basins.

Remark 2. In the presence of local constancy in the landscape, the above definition of the steepest ascent (and hence also the related definition of the attraction basins) is not rigorous. For instance, if the fittest neighbors of point p have the same fitness value, then the steepest ascent algorithm at point p has to make a random or user-defined choice. Nevertheless, even in the presence of local constancy, the comparison of the results (distribution of (α_j)) obtained with different steepest ascent choices, may give useful information about the landscape and guide the best elitism strategy – move to fitter points, or move to strictly fitter points only.

3 Summary of the Results

Given a distribution of (α_j), we determine $M_{\min}$, the minimal size of a random sample of the search space, in order to sample at least one point in each attraction basin of the landscape. Two particular cases are investigated.

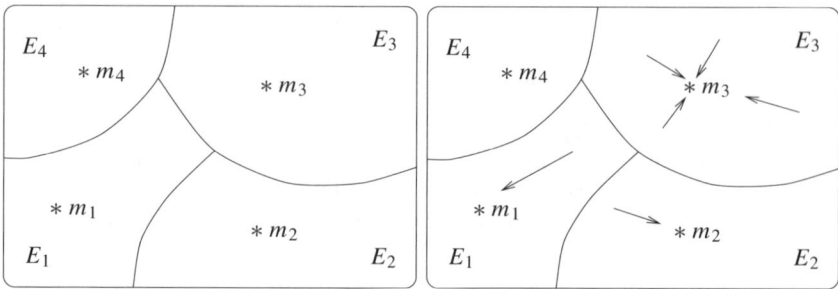

Figure 3. Schematic representation of the search space E with its $N = 4$ attraction basins and the 4 corresponding local maxima $m_1, \ldots, m_4$. In the left picture we have put $M = 6$ points randomly chosen. We apply the search algorithm and detect 3 maxima according to the right picture, so that we have $\beta_1 = 2$, $\beta_2 = 0$, $\beta_3 = 0$, $\beta_4 = 1$, and $\beta_j = 0$ for $j \geq 5$.

1. *Deterministic configuration*: all the attraction basins have the same size ((α_j) are deterministic).
2. *Random configuration*: the sizes of the attraction basins are completely random ((α_j) are uniformly distributed).

In both configurations, we give the value of $M_{\min}$ as a function of the number of local optima N. For instance, a random sample of size $M_{\min} = N(\ln N + \ln a)$ for the deterministic configuration (resp. $M_{\min} = aN^2$ for the random configuration), ensures that a point is sampled in each attraction basin with probability $\exp(-1/a)$. Further, for each configuration, the variance of M is given.

We then address the inverse problem of identifying the distribution of the normalized sizes $(\alpha_j)_{j=1,\ldots,N}$ of the attraction basins, for an unknown landscape. Some direct analysis is first required as discussed below.

3.1 Direct Analysis

Consider a random sample $(X_i)_{i=1,\ldots,M}$ uniformly chosen in the search space. For each $i \in \{1, \ldots, M\}$, a steepest ascent starting from X_i (with the modification operator(s) at hand μ) ends at the local optimum $\mu^*(X_i)$. Define β_j as the number of local optima $(m_.)$ that are reached by exactly j points from (X_i) (see an example in figure 3)

$$\beta_j = \sum_{k=1}^{N} \mathbf{1}_{\#\{i;\mu^*(X_i)=m_k\}=j}.$$

The key result of proposition 2 is the distribution of (β_j) for a class of parametrized distributions Law_γ for (α_j) (asymptotically with respect to N and M). More precisely, if $(Z_j)_{j=1,\ldots,N}$ denotes a family of positive real-valued independent random variables with density

$$p_\gamma(z) = \frac{\gamma^\gamma}{\Gamma(\gamma)} z^{\gamma-1} e^{-\gamma z}$$

and $\alpha_j = \frac{Z_j}{\sum_{i=1}^N Z_i}$, then the expected number $\beta_{j,\gamma} := \mathbb{E}_\gamma[\beta_j]$ is

$$\beta_{j,\gamma} = N \frac{\Gamma(j+\gamma)}{j!\Gamma(\gamma)} \frac{a^j \gamma^\gamma}{(a+\gamma)^{j+\gamma}} \bigg|_{a=M/N}.$$

Moreover, the ratio $r = M/N$ is the unique solution of

$$\frac{\sum_{j=1}^\infty \beta_{j,\gamma}}{M} = \frac{1 - (1+\frac{r}{\gamma})^{-\gamma}}{r}. \tag{1}$$

The latter equation is then used to find a good estimator of N, with observed values of the variables β_j, as explained below.

3.2 Inverse Problem

Given an unknown landscape, we then propose to characterize the distribution of (α_j) through the empirical estimation of the distribution of the random family (β_j). In fact, by construction, the distribution of (α_j) and that of (β_j) are tightly related. We experimentally determine observed values taken by (β_j) (random sampling and steepest ascent search). Then, for each γ value, we use a χ^2 test to compare the observed law for (β_j) to the law β should (theoretically) obey if the law of (α_j) were Law_γ. Naturally, we find a (possible) law for (α_j) if and only if one of the latter tests is positive. Otherwise, we only gain the knowledge that (α_j) does not obey the law Law_γ. Note also that the method can be used to determine sub-parts of the search space with a given distribution for (α_j). In case the law of (α_j) obeys Law_γ, (1) is used to find a good estimator of N.

Finally, section 6 validates the methodology of section 5, by considering known landscapes with random and deterministic sizes of basins, showing that the estimations of the number of local optima N are accurate, even if M is much smaller than N. Further, we apply the methodology on unknown landscapes, and show that the Hamming binary and Gray F1 landscapes contain many more local optima than the 3-bit-flip landscapes.

4 Direct Problem

We assume that the search space E can be split into the partition $E_1, \ldots, E_N$ of subspaces which are attraction basins of local maxima $m_1, \ldots, m_N$ of the fitness function. Let us put a sample of M points randomly in the search space. We aim at computing the probabilities $p_{N,M}$ that at least one point of the random sample lies in each attraction basin. The basic result which will be applied in this section is the following:

Proposition 1. *If we denote $\alpha_j := |E_j|/|E|$, then*

$$p_{N,M} = \sum_{k=0}^{N} (-1)^{N-k} \sum_{1 \leq j_1 < \cdots < j_k \leq N} (\alpha_{j_1} + \cdots + \alpha_{j_k})^M. \tag{2}$$

Proof. Let us denote by $(X_j)_{j=1,\ldots,M}$ the collection of initial points. Since they are randomly chosen uniformly over E, we have for any $(j_1, \ldots, j_M) \in \{1, \ldots, N\}^M$

$$\mathbb{P}\left(X_1 \in E_{j_1}, \ldots, X_M \in E_{j_M}\right) = \alpha_{j_1} \ldots \alpha_{j_M}.$$

By a straightforward combinatorial argument, this implies that the probability that k_1 points be in E_1, k_2 points be in $E_2, \ldots,$ and k_N points be in E_N is given by

$$\mathbb{P}\left(k_1 \text{ points in } E_1, \ldots, k_N \text{ points in } E_N\right) = \frac{M!}{k_1! \ldots k_N!} \alpha_1^{k_1} \ldots \alpha_N^{k_N}$$

where $\sum_{j=1}^{N} k_j = M$ and consequently,

$$p_{N,M}(\alpha) = \sum_{\substack{k_1 \geq 1, \ldots, k_N \geq 1 \\ \sum k_j = M}} \frac{M!}{k_1! \ldots k_N!} \alpha_1^{k_1} \ldots \alpha_N^{k_N}.$$

We introduce some new notation. Let N' be an integer and α' a N'-vector. If $1 \leq k_1 \neq \cdots \neq k_p \leq N'$, then we denote by $\check{\alpha}'^{k_1,\ldots,k_p}$ the $(N' - p)$-vector constructed from α' by deleting the elements corresponding to the coordinates $k_1, \ldots, k_p$. $\overline{\alpha'}$ is shorthand for the sum of the coefficients $\sum_{j=1}^{N'} \alpha'_j$. Furthermore, for any N'-vector α' we introduce

$$p_{N',M}(\alpha') = \sum_{\substack{k_1 \geq 1, \ldots, k_{N'} \geq 1 \\ \sum k_j = M}} \frac{M!}{k_1! \ldots k_{N'}!} \alpha'^{k_1}_1 \ldots \alpha'^{k_{N'}}_{N'}.$$

Since we have

$$\sum_{\substack{k_1 \geq 0, \ldots, k_{N'} \geq 0 \\ \sum k_j = M}} \frac{M!}{k_1! \ldots k_{N'}!} \alpha'^{k_1}_1 \ldots \alpha'^{k_{N'}}_{N'} = \overline{\alpha'}^M$$

we can express $p_{N',M}(\alpha')$ as

$$p_{N',M}(\alpha') = \overline{\alpha'}^M - \sum_{\substack{\exists j \text{ s.t. } k_j=0 \\ \sum k_j = M}} \frac{M!}{k_1! \ldots k_{N'}!} \alpha'^{k_1}_1 \ldots \alpha'^{k_{N'}}_{N'}.$$

We decompose the sum in the right-hand side over the possible subsets $\{k_1, \ldots, k_p\}$ of $\{1, \ldots, N'\}$ which correspond to the k_j's which are equal to 0:

$$p_{N',M}(\alpha') = \overline{\alpha'}^M - \sum_{p=1}^{N'} \sum_{1 \leq k_1 < \cdots < k_p \leq N'} p_{N'-p,M}(\alpha'^{\check{}\,k_1,\ldots,k_p}). \tag{3}$$

This expression is a recursive relation which allows us to compute $p_{N,M}(\alpha)$ from $p_{N',M}(\alpha')$ for $N' \leq N-1$. First check that, for any scalar α we have $p_{1,M}(\alpha) = \alpha^M$ as soon as $M \geq 1$ and 0 otherwise. Now assume that $p_{N',M}(\alpha')$ is given by

$$p_{N',M}(\alpha') = \sum_{k=0}^{N'} (-1)^k \sum_{1 \leq j_1 < \cdots < j_k \leq N'} (\overline{\alpha'} - \alpha'_{j_1} - \cdots - \alpha'_{j_k})^M \tag{4}$$

for any $N' \leq N-1$ and for any N'-vector α'. In this sum the first term corresponding to $k = 0$ is $\overline{\alpha'}^M$. Then (3) implies that $p_{N,M}$ is given by

$$p_{N,M}(\alpha) = \overline{\alpha}^M - \sum_{p=1}^{N} q_{p,N}(\alpha)$$

$$q_{p,N}(\alpha) := \sum_{1 \leq k_1 < \cdots < k_p \leq N} \sum_{p'=0}^{N-p} (-1)^{p'}$$
$$\sum_{1 \leq j_1 < \cdots < j_{p'} \leq N-p} (\overline{\check{\alpha}^{k_1,\ldots,k_p}} - \check{\alpha}^{k_1,\ldots,k_p}_{j_1} - \cdots - \check{\alpha}^{k_1,\ldots,k_p}_{j_{p'}})^M$$

which can also be rewritten as

$$p_{N,M}(\alpha) = \overline{\alpha}^M + q_{0,N}(\alpha) - \sum_{p=0}^{N} q_{p,N}(\alpha) \tag{5}$$

$$q_{p,N}(\alpha) = \frac{1}{p!} \sum_{k_1 \neq \cdots \neq k_p = 1}^{N} \sum_{p'=0}^{N-p} \frac{(-1)^{p'}}{p'!}$$
$$\sum_{j_1 \neq \cdots \neq j_{p'} = 1}^{N-p} (\overline{\check{\alpha}^{k_1,\ldots,k_p}} - \check{\alpha}^{k_1,\ldots,k_p}_{j_1} - \cdots - \check{\alpha}^{k_1,\ldots,k_p}_{j_{p'}})^M. \tag{6}$$

Since $\overline{\check{\alpha}^{k_1,\ldots,k_p}} = \overline{\alpha} - \alpha_{k_1} - \cdots - \alpha_{k_p}$, we get by grouping the terms $p + p' = p''$ together that

$$\sum_{p=0}^{N} q_{p,N}(\alpha)$$

$$= \sum_{p''=0}^{N} \sum_{k_1 \neq \cdots \neq k_{p''} = 1}^{N} (\overline{\alpha} - \alpha_{k_1} - \cdots - \alpha_{k_{p''}})^M \left\{ \sum_{p=0}^{p''} \frac{1}{p!} \frac{1}{(p''-p)!} (-1)^p \right\}.$$

The sum within the brackets is the expansion of $(1-1)^{p''}/(p'')!$ which is equal to zero for $p'' \geq 1$, so that

$$\sum_{p=0}^{N} q_{p,N}(\alpha) = \overline{\alpha}^M. \tag{7}$$

Furthermore, from (6)

$$q_{0,N}(\alpha) = \sum_{p'=0}^{N} \frac{(-1)^{p'}}{p'!} \sum_{j_1 \neq \cdots \neq j_{p'}=1}^{N} (\overline{\alpha} - \alpha_{j_1} - \cdots - \alpha_{j_{p'}})^M$$

which also reads as

$$q_{0,N}(\alpha) = \sum_{p'=0}^{N} (-1)^{p'} \sum_{j_1 < \cdots < j_{p'}=1}^{N} (\overline{\alpha} - \alpha_{j_1} - \cdots - \alpha_{j_{p'}})^M. \tag{8}$$

Substituting (7) and (8) into (5) establishes that (4) holds true for any N-vector α, which completes the proof of the proposition. □

Proposition 1 gives an exact expression for $p_{N,M}$ which holds true for all values of N, M and (α_j), but is quite complicated. The following corollaries show that the expression of $p_{N,M}$ is much simpler in some particular configurations.

Corollary 1. *1. If the attraction basins all have the same size $\alpha_j \equiv 1/N$ (the so-called D-configuration), then*

$$p_{N,M} = \sum_{k=0}^{N} C_N^k (-1)^k (1-k/N)^M.$$

2. If, moreover, the numbers of attractors and initial points are large ($N \gg 1$) and $M = N(\ln N + \ln a)$, $a > 0$, then

$$p_{N,M} = \exp\left(-a^{-1}\right).$$

3. Let us denote by M_D the number of points which are necessary to detect all local maxima. Then in the asymptotic framework $N \gg 1$, M_D obeys the distribution of

$$M_D = N \ln N - N \ln Z$$

where Z is an exponential variable with mean 1.

Proof. The first point is a straightforward application of proposition 1. Let us assume that $N \gg 1$ and $M = N(\ln N + \ln a)$. Then $C_N^k \simeq \frac{N^k}{k!}$ and $(1-k/N)^M \simeq e^{-k(\ln N + \ln a)} = (Na)^{-k}$, which yields the second point of the corollary. The third point then follows readily. □

Corollary 2. *1. If the sizes of the attraction basins are random (the so-called R-configuration), in the sense that their joint distribution is uniform over the simplex of $\mathbb{R}^N$*

$$S_N := \{\alpha_i \geq 0, \sum_{i=1}^{N} \alpha_i = 1\}$$

and the numbers of attractors and initial points are large $N \gg 1$ and $M = N^2 a$, $a > 0$, then

$$p_{N,M} = \exp\left(-a^{-1}\right).$$

2. Let us denote by M_R the number of points which are necessary to detect all local maxima. Then in the asymptotic framework $N \gg 1$, M_R obeys the distribution of

$$M_R = N^2 Z^{-1}$$

where Z is an exponential variable with mean 1.

Remark 3. A construction of the R-configuration is the following. Assume that the search space E is the interval $[0, 1]$. Choose $N-1$ points $(a_i)_{i=1,\ldots,N-1}$ uniformly over $[0, 1]$ and independently. Re-index these points so that $a_0 := 0 \leq a_1 \leq \cdots \leq a_{N-1} \leq a_N := 1$. Denote the spacings by $\alpha_j = a_j - a_{j-1}$ for $j = 1, \ldots, N$. If the j-th attraction basin E_j is the interval $[a_{j-1}, a_j)$, then the sizes $(\alpha_j)_{j=1,\ldots,N}$ of the attraction basins $(E_j)_{j=1,\ldots,N}$ obey a uniform distribution over the simplex S_N.

Proof (of corollary 2). In these conditions, we get from (2) and the relation

$$\sum_{j=1}^{N} \alpha_j = 1$$

that

$$p_{N,M} = \sum_{k=0}^{N} (-1)^k \sum_{1 \leq j_1 < \cdots < j_k \leq N} \mathbb{E}\left[(1 - \alpha_{j_1} - \cdots - \alpha_{j_k})^M\right] \quad (9)$$

where $\mathbb{E}$ stands for the expectation with respect to $(\alpha_j)_{j=1,\ldots,N}$ whose distribution is uniform over S_N.

Let us introduce the order statistics associated with the sequence α, i.e., the unique N-vector $(\tilde{\alpha}_i)_{i=1,\ldots,N}$ which satisfies

$$\{\tilde{\alpha}_1, \ldots, \tilde{\alpha}_N\} = \{\alpha_1, \ldots, \alpha_N\}, \qquad \tilde{\alpha}_1 \leq \cdots \leq \tilde{\alpha}_N.$$

For any fixed k, the k-vector $(\tilde{\alpha}_i)_{i=1,\ldots,k}$ admits the following distribution as N goes to infinity [28, section 9.6]:

$$\tilde{\alpha}_j = \frac{1}{N^2} \sum_{i=1}^{j} Z_i, \text{ for } j = 1, \ldots, k$$

where the family $(Z_i)_{i=1,...,k}$ consists of independent random variables with exponential density and mean 1.

Obviously, we can substitute $\tilde{\alpha}$ for α in (9). From the fact that $\tilde{\alpha}_1$ is of order N^{-2} we guess that the number of points which are necessary to detect all attraction basins is of order N^2. Thus, with $M = aN^2$

$$(1-\tilde{\alpha}_j)^M \simeq \exp-\left(a\sum_{i=1}^{j} Z_i\right).$$

Since the Z_i's are independent, the expectation reads as (for $N \gg 1$)

$$\mathbb{E}\left[(1-\tilde{\alpha}_j)^M\right] \simeq \mathbb{E}\left[\exp(-aZ_1)\right]^j = \left(\frac{1}{1+a}\right)^j$$

and the second term ($k=1$) in the sum of the right-hand member in (9) reads as (we have $\sum_{j\geq 1} 1/(1+r)^j = 1/r$)

$$\sum_{j=1}^{N} \mathbb{E}\left[(1-\alpha_j)^M\right] = \frac{1}{a}.$$

More generally, if $j_1 < \cdots < j_k$

$$(1-\tilde{\alpha}_{j_1} - \cdots - \tilde{\alpha}_{j_k})$$

$$\simeq \exp -a\left(k\sum_{i=1}^{j_1} Z_i + (k-1)\sum_{i=j_1+1}^{j_2} Z_i + \cdots + \sum_{i=j_{k-1}+1}^{j_k} Z_i\right).$$

Taking the expectation

$$\mathbb{E}\left[(1-\tilde{\alpha}_{j_1} - \cdots - \tilde{\alpha}_{j_k})^M\right]$$

$$\simeq \left(\frac{1}{1+ka}\right)^{j_1} \left(\frac{1}{1+(k-1)a}\right)^{j_2-j_1} \cdots \left(\frac{1}{1+a}\right)^{j_k-j_{k-1}}$$

and summing

$$\sum_{j_1<\cdots<j_k} \mathbb{E}\left[(1-\alpha_{j_1} - \cdots - \alpha_{j_k})^M\right]$$

$$= \sum_{j'_1,...,j'_k \geq 1} \mathbb{E}\left[(1-\alpha_{j'_1} - \cdots - \alpha_{j'_1+\cdots+j'_k})^M\right]$$

$$= \underbrace{\sum_{j'_1\geq 1} \left(\frac{1}{1+ka}\right)^{j'_1}}_{=\frac{1}{ka}} \times \cdots \times \underbrace{\sum_{j'_k\geq 1} \left(\frac{1}{1+a}\right)^{j'_k}}_{=\frac{1}{a}}$$

$$= \frac{a^{-k}}{k!}.$$

Substituting into (9) completes the proof of the corollary. □

It follows from the corollaries that about $N \ln N$ points in the D-configuration are needed to detect all maxima, while one needs about N^2 points to expect the same result in the R-configuration. This is due to the fact that there exist very small attraction basins in the R-configuration. In the proof we state the result that the smallest attraction basin in the R-configuration has a relative size which obeys an exponential distribution with mean N^{-2}. That is why one needs of the order of N^2 points to detect this very small basin.

4.1 Mean Values

The expected value of M_D is exactly

$$\mathbb{E}[M_D] = N \ln N + NC$$

where C is Euler's constant whose value is $C \simeq 0.58$. The expected value of M_R/N^2 is equal to infinity. This is due to the fact that the tail corresponding to exceptional large values of M_R is very important

$$\mathbb{P}(M_R \geq N^2 a) = 1 - \exp(-a^{-1}) \underset{a \gg 1}{\simeq} a^{-1}.$$

4.2 Standard Deviations

The normalized standard deviation, which is equal to the standard deviation divided by the mean, of the number of points necessary to detect all local maxima in the D-configuration is equal to

$$\sigma_D := \frac{\sqrt{\mathbb{E}[M_D^2] - \mathbb{E}[M_D]^2}}{\mathbb{E}[M_D]} = \frac{\pi}{\sqrt{6}(\ln N + C)}$$

which goes to 0 as $N \to \infty$, which proves, in particular, that $M_D/(N \ln N)$ converges to 1 in probability. This is, of course, not surprising. The D-configuration has a deterministic environment, since all basins have a fixed size, so that we can expect an asymptotic deterministic behavior. The situation is very different in the R-configuration which has a random environment, and it may happen that the smallest attraction basin is much smaller than its expected size N^{-2}. That is why the fluctuations of M_D, and especially the tail corresponding to exceptional large values, are very important.

5 Inverse Problem

5.1 Formulation of the problem

We now focus on the inverse problem. We look for the number N of local maxima of the fitness function and also some pieces of information on the distribution of the sizes of the corresponding attraction basins. We assume that we can use an algorithm that is able to associate to any point of the search space the closest local maximum. In order to detect all local maxima, we should apply the algorithm to every point of the search space. Nevertheless, this procedure is far too long since the search space has a large cardinality. Practically, we shall apply the algorithm to M points that will be chosen randomly in the search space E. The result of the search process can consequently be summed up by the following set of observed values ($j \geq 1$)

$$\beta_j := \text{number of maxima detected with } j \text{ points.} \tag{10}$$

Our arguments are based upon the following observations. First notice that $\bar{N} := \sum_{j=1}^{\infty} \beta_j$ is the number of detected maxima. It is consequently a lower bound of the total number of local maxima N, but a very rough estimate in the sense that it may happen that many maxima are not detected, especially those whose attraction basins are small. Besides $\bar{N}$ represents less information than the complete set $(\beta_j)_{j \geq 1}$. By a clever treatment of this information, one should be able to find a better estimate of N than $\bar{N}$.

5.2 Analysis

The key point is that the distribution of the set β_j is closely related to the distribution of the sizes of attraction basins. Let us assume that the relative sizes $(\alpha_j)_{j=1,\ldots,N}$ of the attraction basins can be described by a distribution parametrized by some positive number γ as follows. Let $(Z_j)_{j=1,\ldots,N}$ be a sequence of independent random variables whose common distribution has density p_γ with respect to the Lebesgue measure over $(0, \infty)$[1]

$$p_\gamma(z) = \frac{\gamma^\gamma}{\Gamma(\gamma)} z^{\gamma-1} e^{-\gamma z} \tag{11}$$

where Γ is the so-called Euler's Gamma function $\Gamma(s) := \int_0^\infty e^{-t} t^{s-1} dt$. Under p_γ, the expected value of Z_1 is 1 and its standard deviation is $1/\sqrt{\gamma}$. In the following we shall say that we are under H^γ if the relative sizes of the attraction basins $(\alpha_j)_{j=1,\ldots,N}$ can be described as $(Z_1/T_N, \ldots, Z_N/T_N)$ where $T_N := \sum_{j=1}^{N} Z_j$ and the distribution of Z_j has density p_γ. Notice that the large deviations principle (Cramer's theorem [1, Chapter 1]) applied to the sequence (Z_j) yields that for any

---

[1] If γ is a positive integer then p_γ is a negative-binomial distribution.

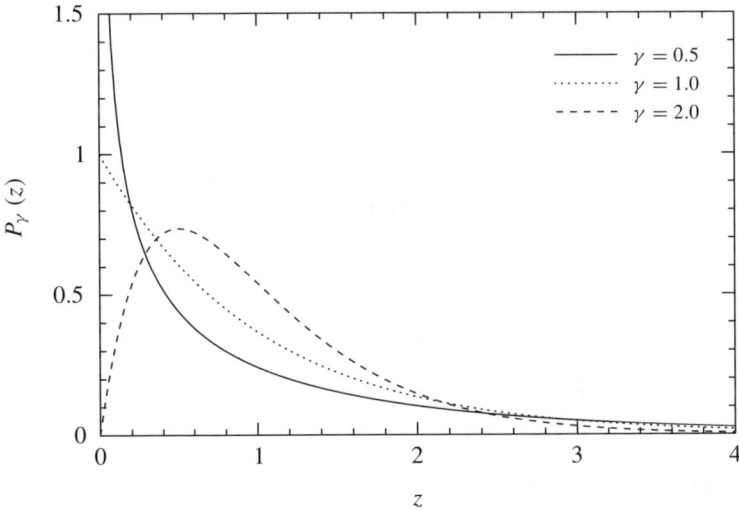

Figure 4. Probability density of the sizes of the attraction basins under H^γ for different γ

$x > 0$ there exists $c_{\gamma,x} > 0^2$ such that

$$\mathbb{P}_\gamma \left(\left| \frac{T_N}{N} - 1 \right| \geq x \right) \leq \exp(-N c_{\gamma,x}) \qquad (12)$$

which shows that, in the asymptotic framework $N \gg 1$, the ratio Z_j/N stands for the relative size α_j up to a negligible correction.

The so-called D- and R-configurations described in section 4 are particular cases of this general framework:

- For $\gamma = \infty$, $Z_j \equiv 1$ and $T_N = N$, so that we get back the deterministic D-configuration.
- For $\gamma = 1$, the Z_j's obey independent exponential distributions with mean 1, and the family $\alpha_j = Z_j/T_N$ obeys the uniform distribution over S_N [19].

The important statement is the following one:

Proposition 2. *Under H^γ, the expected values $\beta_{j,\gamma} := \mathbb{E}_\gamma[\beta_j]$ of the β_j's can be computed and are given by*

$$\beta_{j,\gamma} = N \frac{\Gamma(j+\gamma)}{j!\Gamma(\gamma)} \frac{a^j \gamma^\gamma}{(a+\gamma)^{j+\gamma}} \bigg|_{a=M/N}. \qquad (13)$$

---

[2] Applying the procedure described in [1] establishes that $c_{\gamma,x} = \gamma(x - 1 - \ln x)$.

Proof. Under H^γ, the probability that j of the M points lie in the k-th attraction basin can be computed explicitly

$$\mathbb{P}_\gamma(j \text{ points in } E_k) = C_M^j \mathbb{E}_\gamma \left[\alpha_k^j (1 - \alpha_k)^{M-j} \right]$$

where $\alpha_k = Z_k / \sum_i Z_i$ and $\mathbb{E}_\gamma$ stands for the expectation of Z_j with distribution p_γ. From (12), if $N \gg 1$ we can substitute N for $\sum_j Z_j$ so that

$$\mathbb{P}_\gamma(j \text{ points in } E_k) = C_M^j N^{-j} \mathbb{E}_\gamma \left[Z_k^j (1 - Z_k/N)^{M-j} \right].$$

If $N \gg 1$ and $M = aN$, then we have

$$\mathbb{P}_\gamma(j \text{ points in } E_k) = \frac{a^j}{j!} \mathbb{E}_\gamma \left[Z_k^j e^{-a Z_k} \right].$$

By computing the expectation and summing over $k = 1, \ldots, N$, we finally get (13). $\square$

In particular, the distribution of the β_j's under the D-configuration is Poisson

$$\beta_{j,\infty} = N e^{-M/N} \frac{1}{j!} \left(\frac{M}{N} \right)^j$$

while it is geometric under the R-configuration

$$\beta_{j,1} = N \frac{1}{1 + \frac{M}{N}} \left(\frac{\frac{M}{N}}{1 + \frac{M}{N}} \right)^j.$$

From (13) one can deduce that the following relation is satisfied by the ratio $r = M/N$:

$$\frac{\sum_{j=1}^\infty \beta_{j,\gamma}}{M} = \frac{1 - (1 + \frac{r}{\gamma})^{-\gamma}}{r}. \tag{14}$$

5.3 Estimator of the Number of Local Maxima

We have now sufficient tools to exhibit a good estimator of the number of local maxima. We remind the reader of the problem at hand. We assume that some algorithm is available to determine from any given point the closest local maximum. We choose randomly M points in the search space and detect the corresponding closest local maxima. We thus obtain a set of values $(\beta_j)_{j \geq 1}$ as defined by (10). We can then determine from the set of values $(\beta_j)_{j \geq 1}$ which configuration H^{γ_0} is the most probable, or at least which H^{γ_0} is the closest configuration of the real underlying distribution of the relative sizes of the attraction basins. The statistic used to compare observed and expected results is the so-called χ^2 goodness of fit test [27, section 8.10], which consists first in calculating for each γ

$$T_\gamma := \sum_{j \in \Omega} \frac{(\beta_j - \beta_{j,\gamma})^2}{\beta_{j,\gamma}}$$

where $\Omega \subset \mathbb{N}$ is the set of the indices j for which $\beta_j \geq 1$

$$\Omega := \{j, \beta_j \geq 1\}.$$

Obviously, a large value for T_γ indicates that the corresponding $\beta_{j,\gamma}$ are far from the observed ones, that is to say H^γ is unlikely to hold. Conversely, the smaller T_γ, the more likely H^γ holds true. In order to determine the significance of various values of T, we need the distribution of the statistic. A general result says that if the hypothesis H^{γ_0} does hold true, then the distribution of T_{γ_0} is approximatively the so-called χ^2-distribution with degrees of freedom equal to the cardinality of the set Ω minus 1. Consequently, we can say that the closest configuration of the real underlying distribution of the relative sizes of the attraction basins is H^{γ_0} where γ_0 is given by

$$\gamma_0 = \arg\min \{T_\gamma, \gamma > 0\}. \tag{15}$$

Furthermore, one can estimate the accuracy of the configuration H^{γ_0} by referring T_{γ_0} to tables of the χ^2-distribution for $\text{card}(\Omega) - 1$ degrees of freedom. A value of T_{γ_0} much larger than the one indicated in the tables means that none of the configurations H^γ hold true. Nevertheless, H^{γ_0} is the closest distribution of the real one.

Remark 4. The distribution theory of χ^2 goodness of fit statistic can be found in [3, Chapter 30]. The result is in any case approximate, and all the poorer as there are many expected $\beta_{j,\gamma}$ less than five. These cases must be avoided by combining cells. But one then loses power in the tail regions, where differences are more likely to show up.

Defining γ_0 as (15), we denote by $\bar{\beta}$ the quantity

$$\bar{\beta} = \frac{\sum_{j=1}^\infty \beta_j}{M}.$$

From (14), under H^{γ_0} the ratio $\alpha = M/N$ is the unique solution of

$$\bar{\beta} = \frac{1 - (1 + \frac{r}{\gamma_0})^{-\gamma_0}}{r}. \tag{16}$$

Consequently, once we have determined γ_0, formula (16) is a good estimator of the ratio $\alpha = M/N$, hence N.

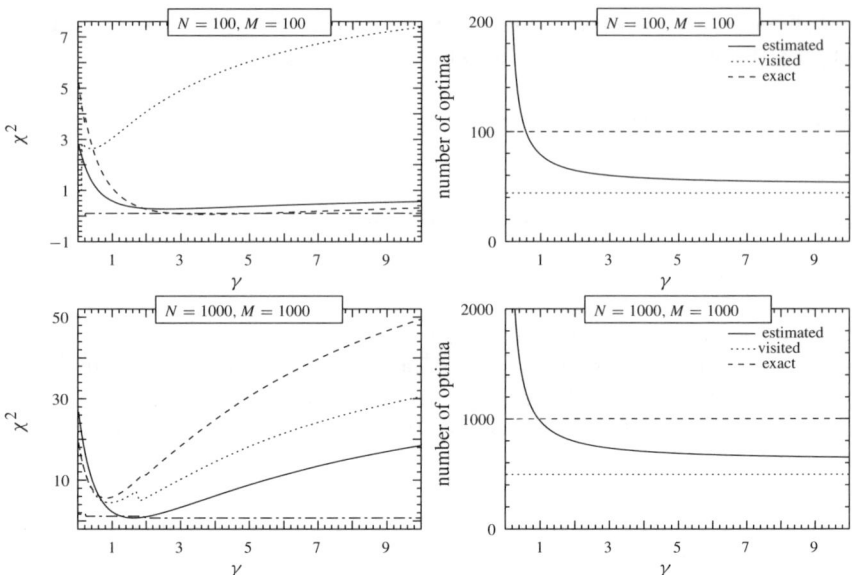

Figure 5. Basins with random uniform sizes. The left figures plot the χ^2 test comparing the empirical β distribution (with three independent simulations) with γ-parametrized distributions. The horizontal line in the left plots show the 95% confidence level beyond which the χ^2-test is positive (accepts the hypothesis that the compared distributions are equal with confidence 0.95). The right figures plot the estimation of the number of local optima computed by (16), and the number of optima actually visited by the steepest ascent. The χ^2 results are unstable for $M = N$ when N is small.

6 Experiments

Given a landscape $\mathcal{L}$, the following steps are performed in order to identify a possible law for the number and size of the attraction basins of $\mathcal{L}$, among the family of laws Law_γ studied above.

1. Choose a random sample $(X_i)_{i=1,...,M}$ uniformly in E.
2. Perform a steepest ascent starting from each X_i up to $\mu^*(X_i)$.
3. Compute β_j defined as the number of local optima reached by exactly j initial points X_i.
4. Compare the observed law of β to the laws of $\beta(\gamma)$ for different γ values, using the χ^2 test.

To visualize the comparison of the last item, we propose to plot the obtained χ^2 value for different γ values. We also plot the corresponding χ^2 value below which the test is positive with a confidence of 95 %.

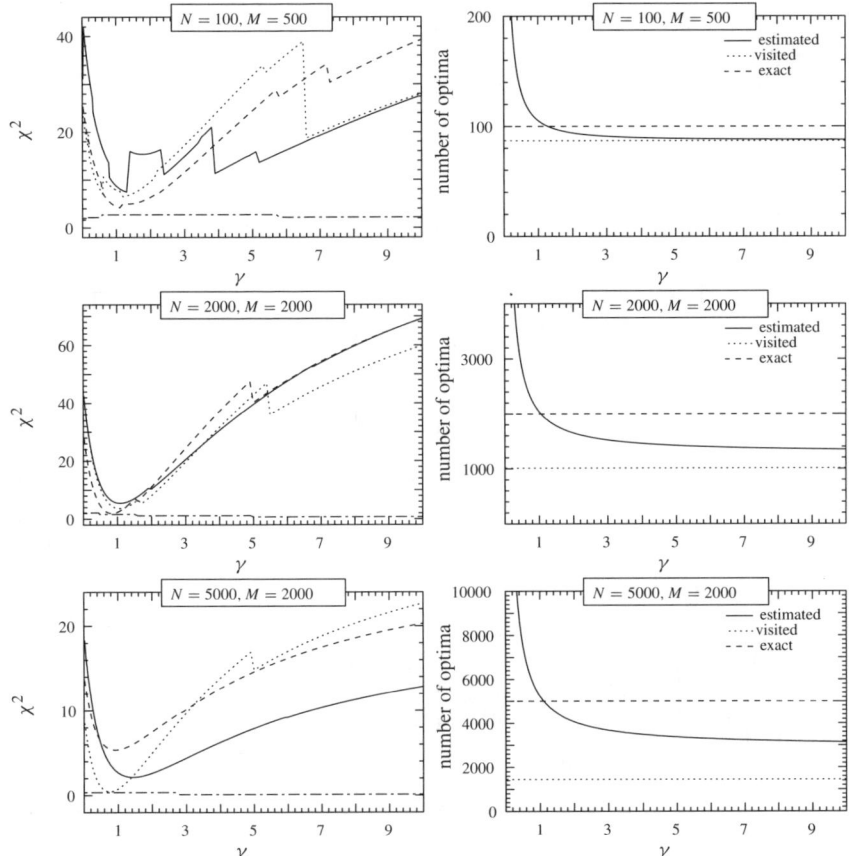

Figure 6. The same as in figure 5 with different values for N and M. Stable results are obtained when N increases, and M is bounded ($M \leq \min(2000, 3N)$ here). The estimation of N corresponding to the smallest χ^2 value ($\gamma = 1$) is very accurate.

6.1 Experimental Validation

The results obtained in section 5 are asymptotic with respect to the number of local optima N and the size of the random sample M. Hence, before the methodology can be applied, some experimental validation is required in order to determine practical values for M and N for which the method is reliable. This is achieved by applying the methodology to determine the distribution of (α_j) (normalized size of the attraction basins) in two known purposely constructed landscapes: the first contains basins with random sizes, the second contains basins with equal sizes.

Results are plotted in figures 5, 6 and 7. Samples with smaller size than those shown in these figures, yield β_j values which are not rich enough to allow a significant χ^2 test comparison. For instance, the χ^2 test requires that observed β_j are non-null

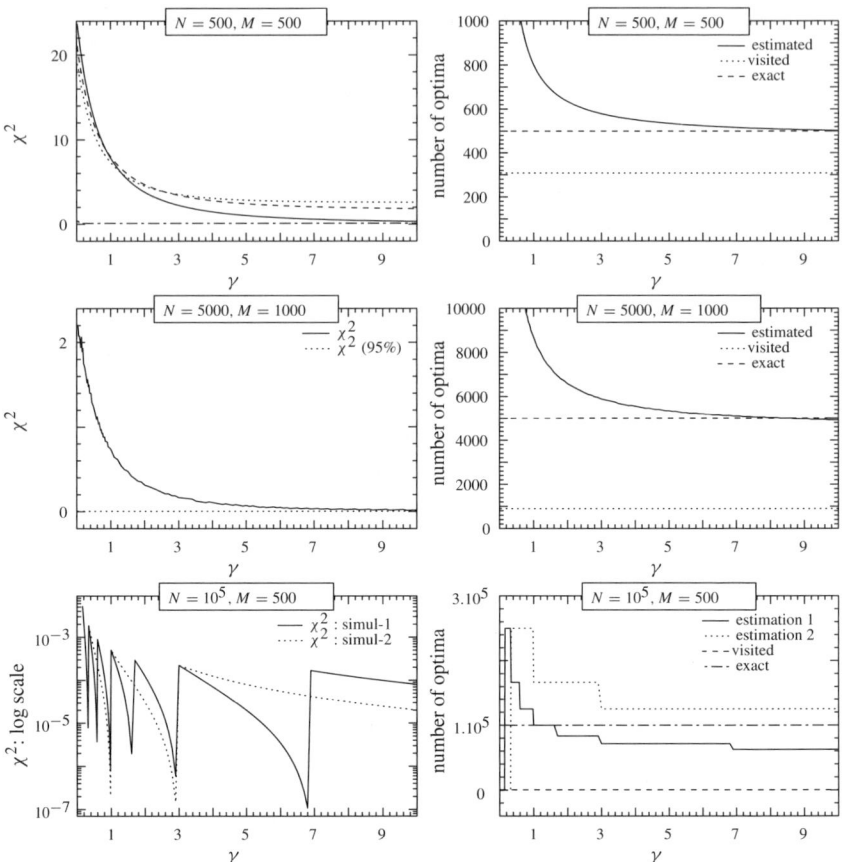

Figure 7. Basins with deterministic equal sizes. The χ^2 results are stable for smaller sample sizes than those of the random configuration. The bottom figures correspond to the case $N = 10^5$ and $M = 500$, where the χ^2 test is not significant, yet the predicted number of local optima is very accurate! With 500 initial points, 497 local optima have been visited, while there are actually 10^5 optima. Yet, formula (16) is able to estimate the true number with an error of less than 30% when the adequate γ value is used.

for some $j > 1$ at least (some initial points are sampled in the same attraction basin). In case all initial points are sampled in different attraction basins the χ^2 test comparison is not significant.

These experiments give practical bounds on the sample sizes (in relation to the number of local optima) for which the methodology is reliable. The numerical simulations exhibit unstable results for the χ^2 test for $M = N$ and small N values (figures 5). When N increases and M is bounded ($M \leq \min(2000, 3N)$ in the experiments), results become stable and accurate (figures 6). Further, we demonstrate that the estimation of number of local optima is accurate, even when initial points visit a small

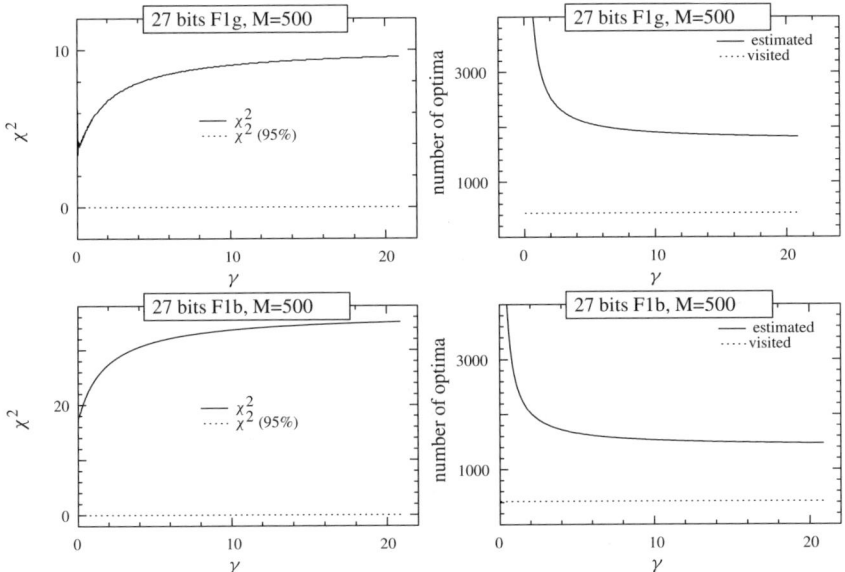

Figure 8. The difficult Baluja 27-bits F1 Gray (F1g) and binary (F1b) landscapes with a single-bit-flip mutation. Experiments with samples of size $M = 2000$ and $M = 5000$ show the same results for the χ^2 test, and the corresponding estimations of the number of local maxima converge to a stable value around 4000.

number of attraction basins of the landscape (figure 7). This situation is even more striking in the experiments of the following section on Baluja F1 problems.

6.2 The Methodology at Work

Having seen that the methodology is a powerful tool, provided that the information obtained for β is rich enough, we apply it to investigate the landscape structure of the difficult Gray and binary coded F1 Baluja problems [2], for single-bit-flip and 3-bit-flip neighborhood relations.

Gray-Coded Baluja F1 Functions. Consider the following function [2] of k variables $(x_0, \ldots x_{k-1})$, with $x_i \in [-2.56, 2.56]$:

$$F1(\vec{x}) = \frac{100}{10^{-5} + \sum_{i=0}^{k-1} |y_i|}$$

with $y_0 = x_0$ and $y_i = x_i + y_{i-1}$ for $i = 1, \ldots, k-1$. It reaches its maximum value of 10^7 at point $(0, \ldots, 0)$. The Gray-encoded F1g and binary F1b versions, with respectively 2, 3, and 4 variables encoded on 9 bits, are considered. This encoding consequently corresponds to the binary search space with $\ell = 9k$.

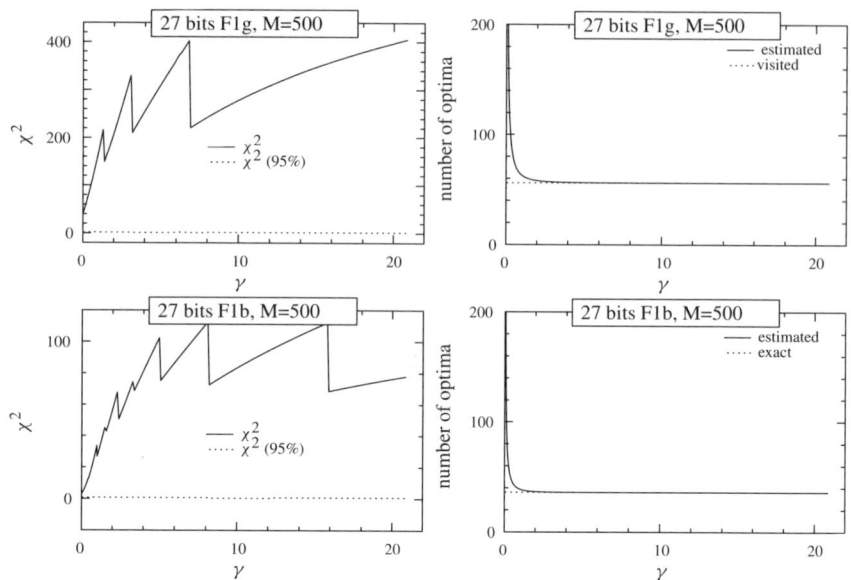

Figure 9. The difficult Baluja 27-bits F1 Gray (F1g) and binary (F1b) landscapes with a 3-bit-flip mutation: the number of local optima drops significantly compared to the Hamming single-bit-flip landscape. These results are confirmed by experiments using samples of sizes $M = 2000$ and $M = 5000$ which give the same estimation for the number of local optima.

Considering the single-bit-flip mutation (Hamming landscape), figure 8 shows that the distribution of the sizes of the basins is closer to the random configuration than to the deterministic one, and that the estimated number of local optima is similar for the binary and Gray codings. On the other hand, considering the 3-bit-flip mutation (figure 9), the estimated number of local optima drops significantly for both problems: less than 250 for both binary and Gray landscapes, whereas the Hamming landscape contains thousands of local optima (figure 8).

Experiments at problem sizes $\ell = 18$ and $\ell = 36$ have been carried out in addition to the plotted ones ($\ell = 27$), leading to similar results for both F1g and F1b problems. The number of local optima of the 3-bit-flip landscape is significantly smaller than that of the Hamming landscape. For example, when $\ell = 18$, there are less than 10 local optima in the 3-bit-flip landscape versus hundreds in the Hamming landscape. When $\ell = 36$, the estimations for the Hamming landscape show about two times more local optima for the Gray than for the binary encoding (45000 and 25000, respectively). Still, for $\ell = 36$, but for the 3-bit-flips landscape, the estimated number of local optima drops to 1400 and 1000, respectively.

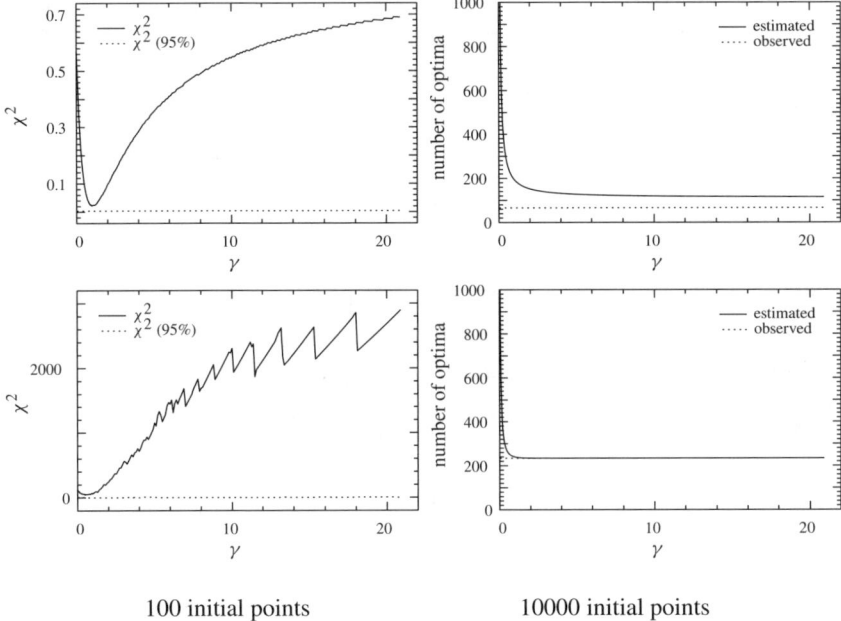

| 100 initial points | 10000 initial points |

Figure 10. Permuted one-max landscape in $E = \{0, 1\}^{15}$. The left figures plot the χ^2 test results comparing the empirically observed β distribution to the family of γ-parametrized distributions. The right figures plot, for different γ values, the estimation of the number of local optima computed by (16). These estimations are rather stable when the size of the initial random sample increases. The same figures also show the visited numbers of optima actually visited by the steepest ascent ($\bar{N} = \sum_{j=1}^{\infty} \beta_j$).

Permuted One-Max Landscapes. This section applies the methodology to explore new functions, namely those obtained when a permutation is applied to the points of the search space.

Define a permuted binary one-max function $g(X) = \mathcal{O}(\sigma^{-1}(X))$, where σ is a random permutation in $E = \{0, 1\}^{\ell}$ and $\mathcal{O}(X)$ is the number of ones in string X (see [22] for more details about the permutation scheme).

Typical landscapes obtained by simple one-max permutations show to be intractable for the single-bit-flip mutation hill climber, and mostly easy to optimize by a generational GA with ranking selection (population size equal to string size ℓ, uniform crossover and mutation at rate $1/\ell$, each applied with probability 0.5): one run over five is trapped in a local optimum. All successful runs converge within 3ℓ generations.

These experiments give examples of landscapes with random sizes of the basins of attraction, yet that are basically easy for GAs.

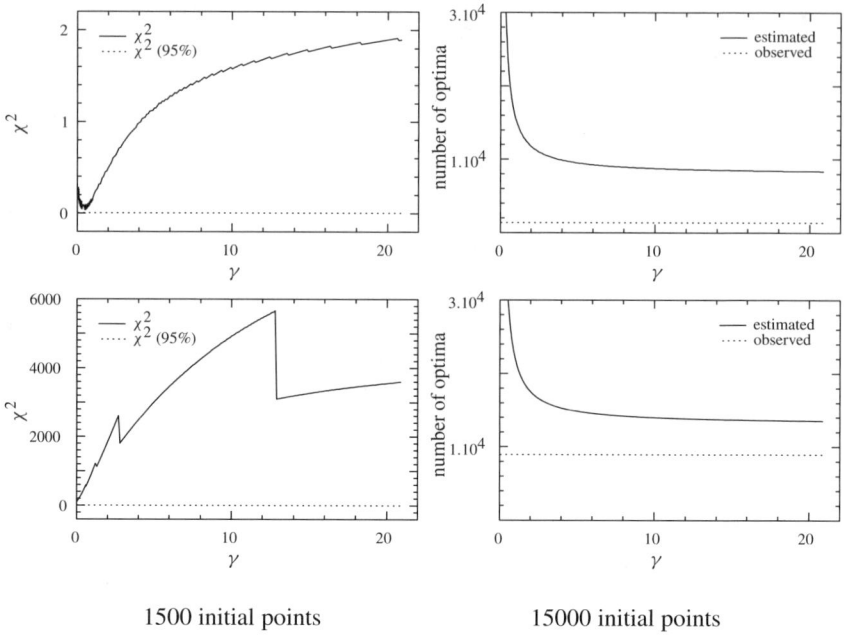

Figure 11. Permuted one-max landscape. The same as in figure 6, but in $E = \{0, 1\}^{30}$. Note that the estimated number of local optima is significantly larger for $\ell = 30$ than for $\ell = 15$.

6.3 A New Optimization Heuristic

A simple search strategy for solving difficult problems naturally follows from the methodology presented in this paper. Once the number N and distribution of the attraction basins is estimated following the guidelines summarized in the beginning of section 6, one generates a random sample whose size is $N(\ln N + \ln a)$ if the sizes of the basins are close to the deterministic configuration (aN^2 if the sizes of the basins are close to random, respectively). Then, a simple steepest ascent starting from each point of the sample ensures that the global optimum is found with probability $\exp(-1/a)$.

In the 27-bits F1 problem, this heuristic demonstrates to be very robust and efficient in solving the problem with the 3-bit-flip operator. Using a 3-bit-flip mutation steepest ascent, an initial random sample of 5 points (versus 100 with single-bit-flip mutation) is enough to ensure that one point at least lies in the global attraction basin (experiments based on 50 runs). This is due to the fact that the basin of the global optimum is larger than the basins of the other local optima. In order to detect all attraction basins, we can estimate the required sample size to 62500 (250 × 250 using corollary 2 and the estimation of $N = 250$ in the experiments of figure 9).

6.4 Discussion

This paper provides a new methodology allowing to estimate the number and the sizes of the attraction basins of a landscape specified in relation to some modification operator. This allows us to derive bounds on the probability to sample a point in the basin of the global optimum, for example. Furthermore, it allows us to compare landscapes related to different modification operators or representations, as illustrated with the Baluja problem.

The efficiency of the proposed method is certainly dependent on the class of laws of (α_j) (the sizes of the attraction basins) for which the distribution of β is known. We have chosen a very particular family of distributions p_γ to represent all possible distributions for the relative sizes of attraction basins. The constraints for this choice are twofold, and contradictory. On the one hand, a large family of distributions is required to be sure that at least one of them is sufficiently close of the observed distribution (β_j). On the other hand, if we choose too large a family, then we need a lot of parameters to characterize the distributions. It is then very difficult to estimate all parameters and consequently to decide which distribution is the closest to the observed one. That is why the choice of the family is very delicate and crucial. We feel that the particular family p_γ (11) fulfills important conditions. First, it contains two very natural distributions, the so-called D and R configurations that we have studied with great detail. Second, it is characterized by a single parameter which is easy to estimate. Third, it contains distributions with a complete range of variances, from 0 (the D-configuration) to infinity, by going through 1 (the R-configuration).

However, the experiments with the Baluja problem appeal for refining the class of laws of (α_j) around basins with random sizes. We may propose $\alpha_j = Z_j / \sum_{i=1}^{N} Z_i$ where Z_j are independent and identically distributed with one of the distributions of the bidimensional family $p_{\gamma,\delta}(.), \gamma > 0, \delta > 0$

$$p_{\gamma,\delta}(z) = \frac{\delta^\gamma}{\Gamma(\gamma)} z^{\delta-1} e^{-\gamma z}.$$

The parameter δ characterizes the distribution of the sizes of the small basins, since $p_{\gamma,\delta}(z) \sim z^{\delta-1}$ as $z \to 0$, while γ characterizes the distribution of the sizes of the large basins, since the decay of $p_{\gamma,\delta}(z)$ as $z \to \infty$ is essentially governed by $e^{-\gamma z}$. This family presents more diversity than the family $p_\gamma(.)$ which we considered in section 5.2. The expected value of β_j is under $p_{\gamma,\delta}$

$$\beta_{j,\gamma,\delta} = N \frac{\Gamma(j+\gamma)}{j!\Gamma(\gamma)} \frac{a^j \delta^\gamma}{(a+\delta)^{j+\gamma}} \bigg|_{a=M/N}.$$

The method of estimating the number of local minima described in section 5.3 could then be applied with this family.

To apply our method, we have also made a crucial choice: we executw the local search algorithm from randomly distributed points. We do so because we have no a priori information on the landscape at hand. However, assume that one has some a priori information about the fitness function, for instance, its average value. Consequently, one could hope that starting with points whose fitnesses are better than average would improve the detection of the local maxima. Nevertheless, extensive computer investigations of some particular problems have shown that this is not the case [15, page 456], possibly because a completely random sample of starting points allows one to get a wider sample of local optima.

A first application of the methodology presented in this paper is to compare landscapes obtained when different operators are used (k-bit-flip binary mutations for different k values, for example). However, the complexity of this method is directly related to size of the neighborhood of a given point. Hence, its practical usefulness to study k-bit-flip landscapes is limited when k value increases. Hence, it seems most suited to investigate different representations. Its extension to non-binary representations is straightforward, provided that a search algorithm that leads to the closest local optimum can be provided for each representation. Furthermore, this methodology can be used to determine sub-parts of the search space where the (α_j) obey a particular law, hence guiding a hierarchical search in different subparts of the space.

A possible limitation of this methodology is that we have no guarantee on the time to convergence of the steepest ascent search. Even if in average, time to convergence of local search is linear in problem size [16,17], there exists many functions were the steepest ascent takes a very long time to find the closest local optimum.

Note finally that the distributions of the sizes of basins do not fully characterize landscape difficulty. Depending on the relative position of the attraction basins, the search still may range from easy to difficult. The permuted one-max landscape illustrates an example of a GA-easy landscape with a random distribution of basin sizes.

Additional information is necessary to compare landscape difficulty. Further work may address such issues as extracting additional significant information in order to guide the choice, or designing problem dependent operators and representations.

References

1. R. Azencott. Grandes déviations et applications. In P. L. Hennequin, editor, *Proceedings of Ecole d'Eté de Probabilités de Saint-Flour*. Springer-Verlag, Berlin Heidelberg New York, 1978.
2. S. Baluja. An empirical comparison of seven iterative and evolutionary function optimization heuristics. Technical Report CMU-CS-95-193, Carnegie Mellon University, 1995.
3. H. Cramer. *Mathematical methods of statistics*. Princeton University Press, Princeton, NJ, 1946.
4. D. B. Fogel and A. Ghozeil. Using fitness distributions to design more efficient evolutionary computations. In T. Fukuda, editor, *Proceedings of the Third IEEE International Conference on Evolutionary Computation*, pages 11–19. IEEE, New York, 1996.

5. D. E. Goldberg. *Genetic algorithms in search, optimization and machine learning.* Addison–Wesley, Englewood Cliffs, NJ, 1989.
6. J. H. Holland. *Adaptation in natural and artificial systems.* University of Michigan Press, Ann Arbor, MI, 1975.
7. T. Jones and S. Forrest. Fitness distance correlation as a measure of problem difficulty for genetic algorithms. In L. J. Eshelman, editor, *Proceedings of the 6th International Conference on Genetic Algorithms*, pages 184–192. Morgan Kaufmann, San Francisco, 1995.
8. L. Kallel. *Convergence des algorithmes génétiques: aspects spatiaux et temporels.* PhD thesis, Ecole Polytechnique, Palaiseau, France, 1998.
9. L. Kallel, B. Naudts, and M. Schoenauer. On functions with a fixed fitness distance relation. Technical Report 399, CMAP, Ecole Polytechnique, Palaiseau, France, 1998.
10. L. Kallel and M. Schoenauer. Alternative random initialization in genetic algorithms. In T. Bäck, editor, *Proceedings of the 7th International Conference on Genetic Algorithms*, pages 268–275. Morgan Kaufmann, San Francisco, 1997.
11. L. Kallel and M. Schoenauer. A priori predictions of operator efficiency. In J.-K. Hao, E. Lutton, E. Ronald, M. Schoenauer, and D. Snyers, editors, *Artificial Evolution'97*, LNCS. Springer-Verlag, Berlin Heidelberg New York, 1997.
12. P. Stadler. Towards a theory of landscapes. In R. Lopèz-Pena, R. Capovilla, R. Garcia-Pelayo, H. Waelbroeck, and F. Zertouche, editors, *Complex systems and binary networks*, pages 77–163. Springer-Verlag, Berlin Heidelberg New York, 1995.
13. B. Manderick, M. de Weger, and P. Spiessens. The genetic algorithm and the structure of the fitness landscape. In R. K. Belew and L. B. Booker, editors, *Proceedings of the 4th International Conference on Genetic Algorithms*, pages 143–150. Morgan Kaufmann, San Francisco, 1991.
14. P. Merz and B. Freisleben. Memetic algorithms and the fitness landscape of the graph bi-partitioning problem. In A. E. Eiben, T. Bäck, M. Schoenauer, and H.-P. Schwefel, editors, *Proceedings of the 4th Conference on Parallel Problems Solving from Nature*, volume 1498 of *LNCS*, pages 765–774. Springer-Verlag, Berlin Heidelberg New York, 1998.
15. C. H. Papadimitriou and K. Steiglitz. *Combinatorial optimization: Algorithms and complexity.* Prentice–Hall, Englewoods Cliffs, NJ, 1982.
16. C. A. Tovey. Local improvement on discrete structures. In E. Aarts and J. K. Lenstra, editors, *Local search and combinatorial optimization.* John Wiley, New York, 1997.
17. C. A. Tovey. Hill climbing with multiple local optima. *SIAM J. Alg. Disc. Math.*, 6:384–393, 1985.
18. G. Parisi, M. Mezard, and M. A. Virasoro. *Spin glass theory and beyond.* World Scientific, 1987.
19. R. Pyke. Spacings. *J. Roy. Statis. Soc.*, 27:395–436, 1965.
20. R. J. Quick, V. J. Rayward-Smith, and G. D. Smith. Fitness distance correlation and ridge functions. In A. E. Eiben, T. Bäck, M. Schoenauer, and H.-P. Schwefel, editors, *Proceedings of the 4th Conference on Parallel Problems Solving from Nature*, volume 1498 of *LNCS*, pages 77–86. Springer-Verlag, Berlin Heidelberg New York, 1998.
21. R. Rammal, G. Toulouse and M. A. Virasoro. *Ultrametricity for Physicists.* Reviews of Modern Physics 58:765–788, 1986.
22. S. Rochet, G. Venturini, M. Slimane, and E. M. El Kharoubi. A critical and empirical study of epistasis measures for predicting GA performances: A summary. In J.-K. Hao, E. Lutton, E. Ronald, M. Schoenauer, and D. Snyers, editors, *Artificial Evolution'97*, LNCS, pages 287–299. Springer-Verlag, Berlin Heidelberg New York, 1997.

23. V. Slavov and N. I. Nikolaev. Genetic algorithms, fitness sublandscapes and subpopulations. In W. Banzhaf and C. R. Reeves, editors, *Foundations of Genetic Algorithms 5*, pages 199–218. Morgan Kaufmann, San Francisco, 1999.
24. P. Stadler. Landscapes and their correlation functions. *J. Math. Chem.*, 20:1–45, 1996.
25. P. J. van Laarhoven and E. H. L. Aarts. *Simulated Annealing: Theory and Applications*. Kluwer Academic, Dordrecht, The Netherlands, 1987.
26. E. D. Weinberger. Correlated and uncorrelated fitness landscapes and how to tell the difference. *Biological Cybernetics*, 63:325–336, 1990.
27. G. B. Wetherill. *Elementary statistical methods*. Chapman and Hall, London, 1972.
28. S. S. Wilks. *Mathematical statistics*. John Wiley, New York, 1962.

On Classifications of Fitness Functions

T. Jansen

FB 4, LS 2,
Universität Dortmund
44221 Dortmund, Germany
E-mail: *jansen@ls2.cs.uni-dortmund.de*

Abstract. It is well-known that evolutionary algorithms succeed in optimizing some functions efficiently and fail for others. Therefore, one would like to classify fitness functions as more or less hard to optimize for evolutionary algorithms. The aim of this paper is to clarify limitations and possibilities for classifications of fitness functions from a theoretical point of view. We distinguish two different types of classifications, descriptive and analytical ones. We shortly discuss three widely known approaches, namely the *NK*-model, epistasis variance, and fitness distance correlation. Furthermore, we consider another recent measure, bit-wise epistasis. We discuss shortcomings and counter examples for all four measures and use this to motivate a discussion about possibilities and limitations of classifications of fitness functions in a broader context.

Keywords
Predictive measures of problem difficulty, epistasis, fitness distance correlation

1 Introduction

Evolutionary algorithms are general search heuristics that can be used for global optimization. They are typically applied when there is not much knowledge about the objective function and there is no time for intense investigation of the problem. This is due to the circumstance that evolutionary algorithms are relatively simple to implement and that they are thought to be robust. Robustness means that evolutionary algorithms show above-average performance on almost any objective function, whereas more specialized optimization tools are superior on the class of objective functions they are designed for, but inferior on all of the other functions. This popular point of view is visualized in a well-known figure in Goldberg's famous book on genetic algorithms (GAs) [12] and can be found implicitly and explicitly in many papers.

On the other hand, it is quite well-known that, averaged over *all* different fitness functions, all optimization algorithms (including a pure random walk) perform equally if one uses the number of different function evaluations as a performance measure [28]. It is not useful to argue that taking only the different function evaluations into account is not appropriate for evolutionary algorithms. Using a dictionary, it can easily be achieved that no point in the search space is actually sampled twice. Furthermore, sampling any point more often than once obviously cannot improve the performance of an algorithm.

In spite of the correctness of this 'no-free-lunch theorem' [28] the result is not too interesting. It is easy to see that averaging over all different fitness functions does not match the situation of black-box optimization in practice. It can even be shown that in more realistic optimization scenarios there can be no such thing as a no-free-lunch theorem [8].

We conclude that the classification of objective functions, in order to determine whether they can be successfully optimized by (certain) evolutionary algorithms, is both practically relevant and theoretically justified. Before objective functions are optimized by evolutionary algorithms there usually is some mapping to a fitness function. We omit this step, that is subject to research in itself, and directly consider the optimization of fitness functions. In order to simplify the task we restrict ourselves to fitness functions $f : \{0, 1\}^n \to \mathbb{R}$. We know that the optimization of functions $g : \mathbb{R}^n \to \mathbb{R}$ may lead to different techniques. Already the analysis of evolutionary algorithms heavily depends on the chosen representation, whether it is discrete or continuous, see, for example, Rudolph's analyses of evolution strategies [25]. But in concrete implementations on digital computers real numbers are (in most cases implicitly) mapped to bit strings. Thus, we do not consider our simplification to be a fundamental restriction.

We distinguish two different types of classifications, namely descriptive and analytical ones. A descriptive classification defines a class of fitness functions with some common property. It can be considered to be helpful in our context if this property can be related to the difficulty of optimizing these functions. Note, that it may be a hard problem to decide whether a given fitness function belongs to the defined class.

An analytical classification is a kind of algorithm that takes a fitness function as input and yields some kind of classifying attribute as output. Typically, the output is some number, but more complex attributes may be, and are also used.

Maybe the best kind of analytical classification one would like to have is the following. Assume we have a set of optimization algorithms $\{A_1, \ldots, A_n\}$. Then we look for a classification algorithm that takes a fitness function as input and computes the index i of the optimal algorithm A_i for f. By applying first the classification algorithm and then the optimization algorithm A_i we want to reduce the time needed to optimize f. The no-free-lunch theorem rules out that such a classification is possible for all fitness functions. In fact, the best that can be achieved is an improvement on a proper subclass of all functions. Thus, it is a necessary element of such an analytical classification that the class of functions it works well for is defined. We see the need for a combination of analytical and descriptive classifications here.

The problem of classification of fitness functions has already been subject to intense research, probably most often in the context of GA-hardness. Among the most prominent hardness measures are epistasis variance [4] and fitness distance correlation [17]. An overview can be found in [20]. Another approach can be seen in the *NK*-model [18], though introduced in another context and for other reasons. We cannot hope to solve the multitude of open problems that are still unsolved. We thus intend to clarify and enrich the field by providing another perspective of the problem.

In section 2 we describe three well-known classifications and one more recent one. In section 2.1 we discuss the *NK*-model, a descriptive classification, which is of special interest to us. We give a very brief overview of epistasis variance and fitness distance correlation in section 2.2 and section 2.3, respectively. Both are analytical classifications. In section 2.4 we consider another quite recent approach to the classification of fitness functions, namely bit-wise epistasis [9], which is an analytical classification, too. We present arguments why bit-wise epistasis is not an appropriate measurement for the hardness of functions with respect to function optimization by means of evolutionary algorithms. Then we try to gain a broader perspective and look for fundamental restrictions and possible directions of research in the field of fitness function classification in section 3. Finally, section 4 summarizes and gives perspectives.

2 Four Different Classifications

In this section we discuss four different classifications, one descriptive and three analytical ones. We give serious reasons why the classifications are of limited use only, and may be considered as not too helpful in practice. This section is intended to clarify the common problems of classifications and shows the way to a more constructive discussion in the following section.

2.1 The *NK*-model

As first example of a classification of fitness functions we consider the *NK*-model introduced by Kauffman [18] (for an overview see [2]). Originally, it defines a kind of probabilistic class of fitness functions from $\{0, 1\}^N$ (where N is n in our notation) to $[0, 1]$, with tunable 'ruggedness' or 'epistasis', i.e., interaction between different bits. We present a version of the *NK*-model (and turn to the notation *nk*-model), that fits better within the context of classification.

Definition 1 ([18]). *A fitness function $f: \{0, 1\}^n \to \mathbb{R}$ belongs to the class of nk functions, where $k \in \{0, \ldots, n-1\}$, if there are n functions*

$$g_1: \{0, 1\}^{k+1} \to \mathbb{R}, \ldots, g_n: \{0, 1\}^{k+1} \to \mathbb{R}$$

and n sets

$$I_1 = \{i_{1,1}, \ldots, i_{1,k}\}, \ldots, I_n = \{i_{n,1}, \ldots, i_{n,k}\}$$

such that

$$f(x) = g_1\left(x_1, x_{i_{1,1}}, \ldots, x_{i_{1,k}}\right) + \cdots + g_n\left(x_n, x_{i_{n,1}}, \ldots, x_{i_{n,k}}\right)$$

for all $x = x_1 x_2 \ldots x_n \in \{0, 1\}^n$. The fitness function f is called *adjacent nk function*, if additionally $I_1 = \{2, 3, \ldots, 1 + k \bmod n\}, \ldots, I_n = \{1, 2, \ldots, k\}$ holds.

Obviously, the nk-model defines a partition of the class of all fitness functions into sets S_k of functions that are nk functions but not $n(k-1)$ functions. The idea is that the parameter k defines the degree of interaction between different bits that is allowed. For $k = 0$ no interaction is permitted. Such functions are called bitwise separable (or linear) and can be optimized easily by many algorithms (see, for example, [7] for the expected running time of a simple evolutionary algorithm on the class of linear functions). Furthermore, for $k = n - 1$, difficult functions like the class of needle-in-a-haystack functions (that is, $\{f_a \mid a \in \{0, 1\}^n\}$ with $f_a(a) = 1$ and $f_a(x) = 0$ for $x \neq a$) are included, that require exponential computational effort for all optimization algorithms.

But the classes with intermediate k cause some problems. For the moment, we leave the field of evolutionary algorithms and take a look at the nk-model from a complexity theoretical point of view. Then we look for upper resp. lower bounds on the computational effort that is sufficient resp. needed to optimize a nk function. It is easy to see that even for $k = 2$ optimizing such a function is NP-hard [26]. In fact, it can be reduced to the NP-hard MAX-2-SAT problem (for an introduction see [10]). For $k = 1$ a polynomial algorithm can be found [26]. Things change dramatically, if we consider adjacent nk functions. In this case, using a dynamic programming approach (see [3] for an introduction) leads to an optimization algorithm with running time $\mathcal{O}\left(2^k n\right)$, which is polynomial for $k = \mathcal{O}(\log n)$ [27].

For evolutionary algorithms no results are proven for general k. Experiments show little difference for the two cases of the adjacent nk functions as well as the ones with arbitrary neighborhood [18]. Weinberger [27] computes the correlation between pairs of fitness in the two different versions of the nk-model and finds only little difference, too.

As general nk functions are NP-hard for $k \geq 2$, there is no hope that they are appropriate for partitioning the class of all fitness functions according to their difficulty for evolutionary algorithms. For adjacent nk functions things may be different. But as empirical experiences suggest that for evolutionary algorithms there is little difference between the two models, we may speculate that adjacent nk-models do not deliver a useful partition, either. Furthermore, a recent paper of Naudts and Kallel [21] demonstrates in an empirical way that already the class of adjacent $n1$ functions contains functions that can be of extremely different difficulty to a (standard) GA.

2.2 Epistasis Variance

Epistasis variance is introduced by Davidor [4] as a simple statistic for measuring the hardness of a function. It gives a measure of the amount of nonlinearity that can be found. It is formally given by

$$\varepsilon = \sqrt{2^{-n} \sum_{x \in \{0,1\}^n} \left(f(x) - 2^{1-n} \sum_{i=1}^{n} \sum_{\substack{y \in \{0,1\}^n \\ y_i = x_i}} f(y) + \frac{n-1}{2^n} \sum_{y \in \{0,1\}^n} f(y) \right)^2 }$$

[22] and is obviously an analytical classification. Note, that the computation of the exact value takes $\Omega(2^n)$ steps in general. Naudts, Suys, and Verschoren define a normalized epistasis value and remark that the maximal value is obtained by so called 'camel functions'. These functions are constant for all points in the search space except for one point s and its bitwise complement $\bar{s}$, which are the global maxima. Obviously, these functions are almost as hard to optimize as the needle-in-a-haystack functions discussed above. Let as consider the instance of *Camel* with $s = \mathbf{0}$, the all zero bit string (so the complement is $\mathbf{1}$, the all one string), and the two different function values 0 and 1. Then we have

$$\sum_{y \in \{0,1\}^n} Camel(y) = 2$$

and for all $x \in \{0, 1\}^n$ and all $i \in \{1, 2, \ldots, n\}$

$$\sum_{\substack{y \in \{0,1\}^n \\ y_i = x_i}} Camel(y) = 1.$$

This leads to

$$\varepsilon_{Camel} = \left(2^{-n} \left(2 \cdot \left(1 - \frac{n}{2^{n-1}} + \frac{n-1}{2^{n-1}} \right)^2 \right.\right.$$
$$\left.\left. + (2^n - 2) \cdot \left(0 - \frac{n}{2^{n-1}} + \frac{n-1}{2^{n-1}} \right)^2 \right) \right)^{1/2}$$
$$= \sqrt{ 2^{-n} \left(2 \cdot \left(1 - \frac{1}{2^{n-1}} \right)^2 + (2^n - 2) \cdot \frac{1}{2^{2n-2}} \right) }$$

for the epistasis variance of *Camel*. We do not care about the size of this value here, but compare it with the epistasis variance of another function denoted as $\overline{Camel}$. In particular consider the instance of $\overline{Camel}$ with $\overline{Camel}(\mathbf{0}) = \overline{Camel}(\mathbf{1}) = 0$ and $\overline{Camel}(x) = 1$ for $x \notin \{\mathbf{0}, \mathbf{1}\}$. Obviously, $\overline{Camel}$ is extremely simple to maximize for any reasonable algorithm. But we have

$$\sum_{y \in \{0,1\}^n} \overline{Camel}(y) = 2^n - 2$$

and for all $x \in \{0, 1\}^n$ and all $i \in \{1, 2, \ldots, n\}$

$$\sum_{\substack{y \in \{0,1\}^n \\ y_i = x_i}} \overline{Camel}(y) 2^{n-1} - 1$$

leading to

$$\varepsilon_{\overline{Camel}} = \left(2^{-n}\left(2 \cdot \left(0 - \frac{n\left(2^{n-1}-1\right)}{2^{n-1}} + \frac{(n-1)\left(2^{n-1}-1\right)}{2^{n-1}}\right)^2\right.\right.$$
$$\left.\left. + (2^n - 2) \cdot \left(1 - \frac{n\left(2^{n-1}-1\right)}{2^{n-1}} + \frac{(n-1)\left(2^{n-1}-1\right)}{2^{n-1}}\right)^2\right)\right)^{1/2}$$

$$= \sqrt{2^{-n}\left(2 \cdot \left(\frac{1}{2^{n-1}} - 1\right)^2 + (2^n - 2) \cdot \frac{1}{2^{2n-2}}\right)}.$$

We conclude that epistasis variance may associate exactly the same value to extremely simple and extremely difficult functions.

2.3 Fitness Distance Correlation

Fitness distance correlation is introduced by Jones and Forrest [17] as an alternative analytical measure of search difficulty. It is based on the idea of a 'fitness landscape' where the points in the search space together with their fitness form a kind of landscape in the following sense: there is a weighted neighborhood relationship defined in the search space and we interpret each point's fitness as its height [16]. We assume that we are maximizing, so one is looking for a point with maximal height. To establish a meaningful relationship between the optimization problem at hand and the fitness landscape, it is necessary that the neighborhood relationship depends on the algorithm used. However, often simply Hamming distance is employed for the definition of the fitness landscape.

Fitness distance correlation measures the extent to which high fitness values correlate with small distance to a global optimum. If $F = \{f_1, \ldots, f_k\}$ is the set of fitness values and $D = \{d_1, \ldots, d_k\}$ the corresponding distances to the nearest global optimum, let $\bar{f}$ denote the mean of F, $\bar{d}$ the mean of D over all $x \in \{0, 1\}^n$. Let s_F and s_D denote the standard deviation of F and D. Let c_{FD} with

$$c_{FD} = k^{-1} \sum_{i=1}^{k} (f_i - \bar{f})(d_i - \bar{d})$$

be the covariance of F and D. Then, the fitness distance correlation r is formally defined as

$$r = \frac{c_{FD}}{s_F \cdot s_D}.$$

Again, the computational effort for the computation of the exact value is $\Omega(2^n)$. Though it is reported [17] that fitness distance correlation correctly classifies the hardness of many functions, it is well-known that there are counterexamples. Altenberg [1] constructs a function that is easy to optimize but shows no relationship between Hamming distance from the global optimum and fitness. Quick, Rayward-Smith and Smith [24] present a class of functions with the same property and the advantage to be well-structured and thus understandable. Formally, the ridge function is defined by

$$Ridge(x) = \begin{cases} n + 1 + \|x\|_1 & \text{if } \exists i \in \{0, 1, \ldots, n\} : x = 1^i 0^{n-i} \\ n - \|x\|_1 & \text{otherwise} \end{cases}$$

where $\|x\|_1$ denotes the number of ones in x and $1^i 0^{n-i}$ is the bit string of length n with first i consecutive ones followed by $n - i$ consecutive zeros.

The global optimum of *Ridge* is **1**, the all one string, where we have $Ridge(\mathbf{1}) = 2n + 1$. But for all 2^n strings except for the $n + 1$ strings of the form $1^i 0^{n-i}$ higher fitness values are correlated with fewer ones in the string. Thus, as fitness distance correlation averages over all 2^n strings, the correlation is very small indicating a very difficult function. Nevertheless, *Ridge* is fairly easy to optimize. We consider a simple hill climber to see that. Obviously, the all zero string **0** is easy to find. Then, a path of length n that can be easily followed leads to the global optimum: there is always exactly one string with Hamming distance 1 to the current string x that has better function value than x. Any reasonable hill climber finds this string in $\mathcal{O}(n)$ steps. So, after $\mathcal{O}(n^2)$ steps the global optimum is reached.

2.4 Bit-Wise Epistasis

Another quite recent attempt of a classification of fitness functions is the introduction of bit-wise epistasis as measurement of the problem difficulty by Fonlupt, Robilliard, and Preux [9]. The basis of this measurement is the epistasis, i.e., the interaction between different bits, like for nk-models. Like epistasis it is an analytical classification.

Definition 2 ([9]). The bit-wise epistasis is defined for a bit i, where $1 \leq i \leq n$, and a fitness function $f : \{0, 1\}^n \to \mathbb{R}$. Let $\Sigma_i \subseteq \{0, 1, *\}^n$ be the set of all schemata, such that for all $x = b_1 b_2 \ldots b_n \in \Sigma_i$ we have $b_i = *$ and $b_j \in \{0, 1\}$ for $j \neq i$. Obviously, we have $|\Sigma_i| = 2^{n-1}$. For $x \in \Sigma_i$ we define $x_0 \in \{0, 1\}^n$ as the bit string that is obtained by replacing the '*' at the i-th position in x by a 0. Analogously, x_1 is defined as x with a 1 at the i-th position. Then, *bit-wise epistasis* σ_i^2 for bit i is given by

$$\sigma_i^2 := 2^{1-n} \sum_{x \in \Sigma_i} \left(2^{1-n} \left(\sum_{y \in \Sigma_i} f(y_0) - f(y_1) \right) - (f(x_0) - f(x_1)) \right)^2.$$

We present two classes of functions and one special function that we use to exemplify the inherent problems of this new difficulty measure. The first class of functions $\{f_a : \{0,1\}^n \to \mathbb{R} \mid a \in \{0,1\}^n\}$ consists of the already mentioned needle-in-a-haystack functions with $f_a(a) = 1$ and $f_a(x) = 0$ for $x \neq a$. The second class of functions $\{g_k : \{0,1\}^n \to \mathbb{R} \mid k \in \mathbb{N}\}$ is defined by $g_k(x) := k \cdot \prod_{i=1}^{n} x_i$ and is simply a variant of $f_{11...1}$. Finally, the function LeadingOnes : $\{0,1\}^n \to \mathbb{R}$ is defined by

$$LeadingOnes(x) := \sum_{i=1}^{n} \prod_{j=1}^{i} x_j$$

and simply counts the number of leading ones in x.

The computation of the precise bit-wise epistasis for an objective function f is extremely computationally expensive, in general. All 2^n function values of f have to be known for that. It follows that, in practice, one has to use an estimation of the bit-wise epistasis that is based on a polynomially bounded number of function evaluations. Before we discuss this difficulty in more detail, we consider the bit-wise epistasis of f_a to demonstrate that even the knowledge of the exact bit-wise epistasis is not necessarily too helpful when one wants to determine a global optimum.

Due to the definition of f_a the exact values of the bit-wise epistasis can be easily determined. For all i with $1 \leq i \leq n$ we have

$$\sigma_i^2 = \frac{1}{2^{n-1}} \sum_{x \in \Sigma_i} \left(f_a(x_1) - \frac{1}{2^{n-1}} \right)^2 = \frac{1}{2^{n-1}} - \frac{1}{2^{2n-2}}.$$

We recognize that the values σ_i^2 are equal for all i and do not depend on a, so that no information about the global optimum a can be won from the exact values of the bit-wise epistasis.

For g_k the bit-wise epistasis σ_i^2 can be determined analogously. We have

$$\sigma_i^2 = \frac{1}{2^{n-1}} \sum_{x \in \Sigma_i} \left(g_k(x_1) - \frac{1}{2^{n-1}} \right)^2 = \frac{(2^{n-1}-1) \cdot k^2}{2^{3n-3}} + \frac{1 - 2^n + 2^{2n-2}}{2^{2n-2}}.$$

Obviously, all σ_i^2 grow with k. Since k may take any value in $\mathbb{N}$, the bit-wise epistasis may grow arbitrarily large.

For practical purposes the bit-wise epistasis is computed by choosing for each position i a polynomial number $p(n)$ of schemata from Σ_i. There is only one schema $x \in \Sigma_i$ that yields the all one string for x_1. Therefore, with probability $\rho = (1 - 2^{-(n-1)})^{n \cdot p(n)}$ the estimated bit-wise epistasis is 0 for each i. Obviously, we have $\lim_{n \to \infty} \rho = 1$, so with g_k we have functions for which almost surely the difference between the estimated bit-wise epistasis and the exact values can be arbitrarily large.

The functions f_a and g_k have in common that they are all very difficult to optimize. The expected number of function evaluations before the optimum is found is exponential for f_a; the same holds for a generalized version of g_k where the bit string with optimal function value k is not the same for all functions. So, one may want to argue that for more practical, i.e., less hard functions, the bit-wise epistasis may well be a valuable measure of problem difficulty and – what is more important – may be a valuable tool to speed up optimization. Fonlupt, Robilliard, and Preux [9] suggest that bits with higher bit-wise epistasis are mutated with a higher probability than bits with lower bit-wise epistasis.

We consider the bit-wise epistasis of *LeadingOnes*, which is not hard to compute, to investigate this idea. We consider σ_i^2 and distinguish two cases concerning the bit strings in $x \in \Sigma_i$. If there is at least one zero in x, let j be the position of the leftmost zero in x. If $j < i$ holds, we have

$$LeadingOnes(x_0) - LeadingOnes(x_1) = 0.$$

If $j > i$ holds, we have

$$LeadingOnes(x_0) - LeadingOnes(x_1) = (i-1) - (j-1) = i - j.$$

In the second case, x does not contain any zero at all and we have

$$LeadingOnes(x_0) - LeadingOnes(x_1) = (i-1) - n.$$

This leads directly to

$$\sigma_i^2 = \frac{1}{2^{n-1}} \sum_{x \in \Sigma_i} \left(\frac{1}{2^{n-1}} \left(i - 1 - n + \sum_{j=i+1}^{n} 2^{n-j}(i-j) \right) \right.$$
$$\left. - (LeadingOnes(x_0) - LeadingOnes(x_1)) \right)^2$$

$$= 2^{1-n} \left(\left(2^{1-n} \left(i - 1 - n + \sum_{j=i+1}^{n} 2^{n-j}(i-j) \right) - (i - 1 - n) \right)^2 \right.$$
$$\left. + \sum_{k=i+1}^{n} 2^{n-k} \left(2^{1-n} \left(i - 1 - n + \sum_{j=i+1}^{n} 2^{n-j}(i-j) \right) - i + k \right)^2 \right).$$

For $i = n$ we get

$$\sigma_n^2 = 2^{-n+1} \left(2^{-n+1} - 1 \right)^2 = \mathcal{O}(2^{-n}).$$

For $i = 1$ a lengthy but straightforward calculation shows that

$$\sigma_1^2 = 2 + 2^{-n+1} - n 2^{-n+2} - 2^{-2n+2} = \Omega(1)$$

holds. Fonlupt, Robilliard, and Preux [9] suggest that the mutation probabilities of the single bits are chosen proportionally to the bit-wise epistasis. For *LeadingOnes* this implies that the last bit gets an exponentially small mutation probability, immediately implying exponentially large expected running times. We would like to mention that *LeadingOnes* is in fact quite easy to optimize. Rudolph [25] proves that the well-known (1 + 1) evolutionary algorithm optimizes it in expected time $\mathcal{O}(n^2)$; indeed the expected running time of this simple EA is $\Theta(n^2)$ [6]. But even increasing the mutation probability for the leading bits in a less dramatic way is apparently an idea that is not very clever. It implies that the probability for a successful mutation is decreased so that overall the expected running time is increased.

We see that bit-wise epistasis is a measurement for the difficulty of objective functions that is at least problematic. Namely, we proved:

1. that knowing the exact values of the bit-wise epistasis may be of no help at all for finding an optimum,
2. that the estimates that have to be used in practice may be arbitrarily bad with high probability, and
3. that even an heuristic use of bit-wise epistasis for adjusting mutation probabilities may be of no help.

It is not excluded that in practice bit-wise epistasis may turn out to be a helpful tool in some cases. But it is definitely not the ultimate answer to the question of how the difficulty of objective functions can be measured, and more sound arguments in favor of its usefulness have still to be found.

3 Limits and Perspectives of Fitness Function Classification

The previous section, in particular the work of Fonlupt, Robilliard, and Preux [9] can be seen as a motivation for this section. They present a new way to measure the difficulty of a function with respect to optimization by evolutionary algorithms. As we have shown in the previous section, there are serious reasons to doubt the usefulness of this measure. The same holds for the other measures considered. They all may deliver wrong predictions. Obviously, it is not sensible to continue presenting new ways of function classifications while leaving the task of proving that the presented classifications are inappropriate in some way to other researchers. Therefore, we try to clarify the task of classifying functions and gain a hopefully helpful perspective from a theoretical point of view.

In the context of optimization with evolutionary algorithms usually the underlying optimization scenario is that of black-box optimization. It therefore seems reasonable to perform the task of classification of fitness functions in a black-box scenario, too. In this case there is no hope to find a measure that allows us to distinguish between very simple and very difficult functions with polynomially bounded computational effort and acceptable probability in all cases. In order to distinguish between a constant function and a needle-in-a-haystack function f_a, one has to sample the point a. But, if, for a polynomial p, at most $p(n)$ points in the search space are sampled, the probability of finding a is

$$1 - \left(1 - 2^{-n}\right)^{p(n)} \leq \frac{2p(n)}{2^n} = 2^{-\Omega(n)}$$

so with probability $1 - 2^{-\Omega(n)}$ we fail to distinguish f_a, which is hard to optimize, from a trivially optimized constant function.

As Heckendorn and Whitley [14] point out, things may change, if we assume that the fitness function is given as a mathematical expression, e.g., a polynomial. But it is obvious that the optimization of polynomials of degree 2 is NP-hard (as MAX-2-SAT can be coded that way). Furthermore, for a simple EA, Droste, Jansen, and Wegener [5] prove in a more direct way that already polynomials of degree 2 can be very hard to optimize. Therefore, the question remains unanswered how a useful classification can be extracted from the representation of a fitness function as mathematical expression.

One may ask whether concentrating on concrete, maybe simplified, algorithms enables us to find a helpful classification of the set of all fitness functions with respect to this chosen algorithm. Though this cannot in general be ruled out, there are well-known examples that make a success of this approach for interesting algorithms unlikely. Local search is a well-known, simple general search heuristic, that is, for example, employed in simulated annealing [19]. We consider only such local search heuristics and optimization problems, that it is easy to decide whether a local optimum is reached. It is a natural question to ask *which* local optimum will be reached by the considered local search heuristic if started in some given point in the search space. It is known that answering this question is NP-hard [15], in fact, it is PSPACE-complete [23]. We see that, even for local search, deciding whether the global optimum is reached at all can be PSPACE-complete. Doing this for some evolutionary algorithm is probably too ambitious.

We recognize that solving the problem of classifying fitness functions in a black-box scenario is not possible when restricted to measurements with polynomially bounded computational effort. Furthermore, there are good reasons to believe that restricting the interest to some specific evolutionary algorithm does not simplify this task significantly. So, we are left with the question of what can be achieved at all. We try to characterize three different ways, where fruitful research seems possible.

We know that in practice it is not possible to distinguish between a constant and a needle-in-a-haystack function. This is due to their extreme 'closeness', i.e., the two functions differ at only one point in the search space. If two functions differ at substantially more points, then there is a not too small probability to distinguish between them by random sampling. This approach has similarities to learning theory; we refer to an article on property testing by Goldreich, Goldwasser, and Ron for an introduction and interesting examples [13]. It is an important open question how property testing can be related to the classification of fitness functions. What is needed are sets of fitness functions, such that two functions of different sets are, in a well-defined sense, far from each other, and that all functions of one set are of similar difficulty with respect to optimization by evolutionary algorithms. An interesting approach that can be seen as a kind of classification in this sense is presented by Garnier and Kallel in this volume, page 343. There, estimates are given for the number of random starting points for local search that are needed in order to find all local maxima of a given function. For certain classes of functions tight estimates are derived. But furthermore, the practically more relevant inverse situation is investigated, too. For an unknown function the number of local optima is estimated by the application of random sampling, local search, and the confidence level is estimated using a statistical test. We remark that the problem of guaranteeing a reasonable running time is left open as the running time of the local search is not bounded.

Another approach is the identification of more or less natural classes of fitness functions that have similar difficulty. This means constructing a purely descriptive classification. If functions from such a class can be expected to occur in practice, this approach leads to valuable knowledge. We mention linear functions as a successful example [7]. As a negative example we mention unimodal functions, that can be of highly differing difficulty [6].

The third possible way in which one may hope to gain a deeper understanding of evolutionary algorithms is the perhaps least ambitious one. It is the analysis of concrete, maybe simplified, evolutionary algorithms on concrete fitness functions. There are many examples for this kind of research; we name [7,11,25] as examples. Though it may seem that such an approach is quite far away from the classification of fitness functions, this is not necessarily the case. An in-depth analysis of a concrete evolutionary algorithm provides us with insights into the principles governing the different search operators employed by evolutionary algorithms. Finding good examples, functions that paradigmatically show successes and failures of an EA, can enable us to formulate more general circumstances under which such algorithms will succeed or fail. Proven results may serve as a more solid ground for an understanding of when and why evolutionary algorithms should be applied. And this is the aim of the classification of fitness functions, after all.

4 Conclusions

It is clear by now that there are serious limitations to what can be achieved by a classification of fitness functions. An efficient, overall improving, and perfect classification scheme cannot exist. Nevertheless, there are partial successes in the field and we speculate that more are possible.

We summarize what we have learned by reviewing four existing classifications. There are two different ways to approach the problem of classification of fitness functions, a descriptive and an analytical one. Though there are fundamental differences between the two approaches, there are strong connections between them. A descriptive classification is a valuable piece of knowledge, since it tells us what functions share common properties that are relevant to the task of optimization. In order to become practically relevant, a method is needed that allows us to find out whether a given function belongs to the defined class. We see that a kind of analytical classification is needed, that allows us to efficiently determine whether a given function belongs to a specific class.

On the other hand, an analytical classification alone cannot be a useful tool in all situations. Either it needs exponential running time to classify some functions or wrong predictions will be drawn. In section 2 we discussed examples that prove that all the analytical classifications we considered can lead to wrong predictions even if they are allowed to use an exponential number of calculation steps. Of course, all example functions used are purely artificial and may be seen as 'pathological cases'. But as long as an analytical classification is not accompanied by a definition of the classes of functions, i.e., a descriptive classification, such 'pathological' examples are allowed.

Acknowledgments

This work was supported by the Deutsche Forschungsgemeinschaft (DFG) as part of the Collaborative Research Center "Computational Intelligence" (531). The author wishes to thank Leila Kallel, Bart Naudts, Karsten Tinnefeld, and Ingo Wegener for helpful comments and discussions.

References

1. L. Altenberg. Fitness distance correlation analysis: an instructive counter-example. In T. Bäck, editor, *Proceedings of the 7th International Conference on Genetic Algorithms (ICGA '97)*, pages 57–64. Morgan Kaufmann, San Francisco, 1997.
2. L. Altenberg. NK fitness landscapes. In *Handbook of Evolutionary Computation*, page B2.7.2. Oxford University Press, Oxford, 1997.
3. T. H. Cormen, C. E. Leiserson, and R. L. Rivest. *Introduction to Algorithms*. McGraw-Hill, New York, 1990.
4. Y. Davidor. Epistasis variance: a viewpoint on GA-hardness. In G. J. E. Rawlins, editor, *Proceedings of the 1st Workshop on Foundations of Genetic Algorithms (FOGA)*, pages 23–35. Morgan Kaufmann, San Francisco, 1991.
5. S. Droste, Th. Jansen, and I. Wegener. On the analysis of the (1 + 1) evolutionary algorithm. Technical Report CI-21/98, University of Dortmund, SFB 531, Dortmund, Germany, 1998.
6. S. Droste, T. Jansen, and I. Wegener. On the optimization of unimodal functions with the (1+1) evolutionary algorithm. In A. E. Eiben, T. Bäck, M. Schoenauer, and H.-P. Schwefel, editors, *Parallel Problem Solving from Nature – PPSN V*, volume 1498 of LNCS, pages 47–56. Springer-Verlag, Berlin Heidelberg New York, 1998.
7. S. Droste, T. Jansen, and I. Wegener. A rigorous complexity analysis of the (1 + 1) evolutionary algorithm for separable functions with Boolean inputs. *Evolutionary Computation*, 6(2):185–196, 1998.

8. S. Droste, T. Jansen, and I. Wegener. Perhaps not a free lunch but at least a free appetizer. In W. Banzhaf et al., editors, *Proceedings of the Genetic and Evolutionary Computation Conference (GECCO '99)*, pages 833–839. Morgan Kaufmann, San Francisco, 1999.
9. C. Fonlupt, D. Robilliard, and P. Preux. A bit-wise epistasis measure for binary search spaces. In A. E. Eiben, T. Bäck, M. Schoenauer, and H.-P. Schwefel, editors, *Parallel Problem Solving from Nature – PPSN V*, volume 1498 of LNCS, pages 47–56. Springer-Verlag, Berlin Heidelberg New York, 1998.
10. M. R. Garey and D. S. Johnson. *Computers and Intractability. A Guide to the Theory of NP-Completeness*. W. H. Freeman, New York, 1979.
11. J. Garnier, L. Kallel, and M. Schoenauer. Rigorous hitting times for binary mutations. *Evolutionary Computation*, 7(2):173–203, 1999.
12. D. E. Goldberg. *Genetic Algorithms in Search, Optimization, and Machine Learning*. Addison-Wesley, Reading, MA, 1989.
13. O. Goldreich, S. Goldwasser, and D. Ron. Property testing and its connections to learning and approximation. *Journal of the ACM*, 45(4):563–750, 1998.
14. R. B. Heckendorn and D. Whitley. Predicting epistasis from mathematical models. *Evolutionary Computation*, 7(1):69–101, 1999.
15. D. S. Johnson, C. H. Papadimitriou, and M. Yannakakis. How easy is local search? *Journal of Computer and System Sciences*, 37:79–100, 1988.
16. T. Jones. *Evolutionary Algorithms, Fitness Landscapes and Search*. PhD thesis, University of New Mexico, Albuquerque, NM, 1995.
17. T. Jones and S. Forrest. Fitness distance correlation as a measure of problem difficulty for genetic algorithms. In L. J. Eshelman, editor, *Proceedings of the 6th International Conference on Genetic Algorithms (ICGA '95)*, pages 184–192. Morgan Kaufmann, San Francisco, 1995.
18. S. A. Kauffman. *The Origins of Order: Self-Organization and Selection in Evolution*. Oxford University Press, Oxford, 1993.
19. S. Kirkpatrick, C. D. Gelatt, and M. P. Vecchi. Optimization by simulated annealing. *Science*, 220:671–680, 1983.
20. B. Naudts. *Measuring GA-hardness*. PhD thesis, University of Antwerp, Department of Mathematics and Computer Science, Antwerpen, Belgium, 1998.
21. B. Naudts and L. Kallel. A Comparison of Predictive Measures of Problem Difficulty in Evolutionary Algorithms. *IEEE Transactions on Evolutionary Computing*, 4(1):1–15, 2000.
22. B. Naudts, D. Suys, and A. Verschoren. Epistasis as a basic concept in formal landscape analysis. In T. Bäck, editor, *Proceedings of the 7th International Conference on Genetic Algorithms (ICGA '97)*, pages 65–72. Morgan Kaufmann, San Francisco, 1997.
23. C. H. Papadimitriou, A. A. Schäffer, and M. Yannakakis. On the complexity of local search (extended abstract). In *ACM Symposium on Theory of Computing*, pages 438–445, 1990.
24. R. J. Quick, V. J. Rayward-Smith, and G. D. Smith. Fitness distance correlation and ridge functions. In A. E. Eiben, T. Bäck, M. Schoenauer, and H.-P. Schwefel, editors, *Parallel Problem Solving From Nature – PPSN V*, volume 1498 of LNCS, pages 77–86. Springer-Verlag, Berlin Heidelberg New York, 1998.
25. G. Rudolph. *Convergence Properties of Evolutionary Algorithms*. Verlag Dr. Kovač, Hamburg, Germany, 1997.
26. R. K. Thompson and A. H. Wright. Additively decomposable fitness functions. Technical Report, University of Montana, Computer Science Department, 1996.

27. E. D. Weinberger. NP completeness of Kauffman's *n-k* model, a tuneably rugged fitness landscape. Technical Report SFI-TR-96-02-003, The Sante Fe Institute, Santa Fe, NM, 1996.
28. D. H. Wolpert and W. G. Macready. No free lunch theorem for optimization. *IEEE Transactions on Evolutionary Computation*, 1(1):67–82, 1997.

Genetic Search on Highly Symmetric Solution Spaces: Preliminary Results

A. Marino

Department of Electronics and Computer Science
University of Southampton
Southampton SO17 1BJ, UK
E-mail: *am97r@ecs.soton.ac.uk*

Abstract. Several optimisation problems have a highly symmetric solution space. In this case, there exist multiple equivalent solutions with respect to their structure and cost. Genetic algorithms do not generally perform well in this domain due to the lack of an effective search. In fact, the algorithm is very likely to explore a relatively small part of the space without being able to detect that a solution has been previously evaluated. As a consequence, the algorithm either converges to poor solutions or does not converge at all. This paper introduces two *ad hoc* genetic operators to deal directly with highly symmetric solution spaces. Both operators replace the standard crossover procedure of genetic algorithms. We present results of an investigation on combinatorial optimisation problems and in particular the graph colouring problem.

Keywords
Highly symmetrical spaces, greedy mutations, permutation crossover, graph colouring problem

1 Introduction

Optimisation can be defined as the process of finding the *best* value for some given variables (called *independent variables*) with respect to some evaluation criteria. We are, of course, implicitly assuming that the adopted criteria actually allow the existence of such a good value. In many cases, the search of the right answer for the independent variable is subject to restrictions (the *constraints*). This added feature may increase the difficulty of the search.

There exist a variety of techniques to generate good approximate solutions for optimisation problems. This paper addresses the use of a stochastic search mechanism known as evolutionary algorithms [7]. In particular the investigation refers to genetic algorithms [6,13], which have proven successful for many applications [23,2,22]. Unfortunately, the standard algorithm (i.e., with flip/swap mutations and one/two points crossover) does not perform very well on problems with large symmetries. In fact, the genetic search fails to explore efficiently the solution space due to the presence of multiple equivalent solutions. The algorithm is not able to identify those genotypes that differ only by permutations of the elements used to encode solutions. Therefore, genetic algorithms are likely to evaluate the *same* solution repeatedly during the evolution.

A major drawback of the lack of the efficiency is that the performance of a standard genetic algorithm (on symmetric solution spaces) is very poor. In most of the cases, there is very little improvement of the solution cost even after a large number of evaluations.

This paper introduces two novel genetic recombination schemes that improve the search mechanism of simple genetic algorithms on this specific domain. After a brief introduction on the concept of spaces with large symmetries (section 2) and the definition of the graph colouring problem (section 3), an outline of the proposed solution is given in section 4. Experimental results of the simulations are reported in section 5 followed by our future research plan.

2 Solution Spaces With a High Degree of Symmetry

Combinatorial optimisation problems have discrete solution spaces. For this investigation, structural symmetries of the problem at hand (i.e., rotations, topological invariances, etc.) have not been taken into account. The only symmetries which are considered refer to the fact that, for some optimisation problems, there exist a large number of possible representations for any given solution. These representations are *equivalent* since they generate the same structure of the solution and the same cost[1]. This feature can be explained in a more formal way using the notion of *group actions on sets* which we briefly recall[2].

Let X be a non-empty set with a certain structure and let G be a group. We say that G *acts on* X (or *permutes* X) if there exists a homomorphism from G to the symmetry group S_X of X (i.e the set of all structure-preserving permutations of X). Then, to each $g \in G$ there corresponds a map $\pi_g : X \to X$ (i.e., $\pi_g : x \to xg$) which identifies a permutation of X.

In other words, if G acts on X, we can define a relation $\mathfrak{R}$ on X. Then, for any x' and $x'' \in X$ we set $x'\mathfrak{R}x'' \iff$ there is an element $g \in G$ such that $x'g = x''$. It can be proved that $\mathfrak{R}$ is an equivalence relation. Consequently, the set X is partitioned into disjoint equivalence classes (with respect to the relation $\mathfrak{R}$) called *orbits* of the action. For a given $x \in X$ its orbit is the set $\{xg : g \in G\}$ of all elements $x' \in X$ that have a relation with x.

Given an optimisation problem, the set X corresponds to the set of solutions while G identifies the set of transformations which induce the symmetries. Assume X is the set of all possible combinations of m letters (not necessarily distinct) drawn from an

---
[1] The converse is not true. In fact, there exist solutions at the same cost with different structures.
[2] For more information about this topic, please refer to [1].

alphabet of size k. A typical element $x = (x_1, x_2, \ldots, x_m)$ is a sequence of letters. An element $x' = (x'_1, x'_2, \ldots, x'_m)$ is equivalent to x (indicated as $x' \approx x$) if there exists a permutation that sends each letter x_i to the corresponding x'_i

$$x' \approx x \Leftrightarrow \exists g \in G \text{ with } x'_i = g(x_i) \ \forall i = 1, \ldots, m. \qquad (1)$$

This means that we are interested in analyzing optimisation problems whose symmetries can be identified as permutations of the alphabet that encodes their solutions. For the purpose of the investigation we selected the graph colouring problem which is introduced in the next section.

3 The Graph Colouring Problem

Let $\mathcal{G}_{n,p}$ be a graph with a non-empty set of n vertices and probability p of joining any two given ones. Graphs are often represented by their incidence matrix $W = \{w_{ij}\}$, where each entry w_{ij} is a binary variable representing the existence of an edge between vertices i and j of the graph. For weighted graphs $w_{ij} \in \mathcal{R}_0^+$ and indicates the strength of the connection.

The problem of colouring graphs became relevant already in 1840 [4] with the colouration of geographical maps using four colours. Since then, many other variations and instances have been proposed and nowadays the colouring of graphs has several different formulations. This paper investigated the *vertex-graph colouring* problem which can be stated in two possible ways:

- Given a graph $\mathcal{G}_{n,p}$, what is its chromatic number $\chi(\mathcal{G})$? In other words, what is the minimum number of colours required to assign adjacent vertices of the graph different colours?
- Given a graph $\mathcal{G}_{n,p}$ and a number of colours k, what is the minimum number of conflicts for the graph (i.e., how can we allocate the k colours in such a way that the number of adjacent vertices being equally coloured is minimized)?

For this investigation the second formulation has been selected. Despite the presence of a vast literature on this problem [3,14,17,18,21,25,26],

there does not exist a standard approach to the presentation and comparison of results since investigations have different objectives. Measures of performance include the *success ratio* (SR) and the *average number of function evaluations to solution* (AES) which are used in [9,10] or the chromatic number mentioned in [5,15]. Unfortunately, all these different approaches also make comparisons more difficult because in many cases the graph instances tested are not the same. In fact, in some cases the graphs are k-colourable, i.e., their chromatic number is k (often 3-colourable), in order to evaluate the precision of the algorithm being tested. In other cases, papers report results on specific benchmark problems. Section 5 will specify the graph instances used for this investigation and the measures of performance adopted.

The main techniques used to produce approximate solutions for the graph colouring problem are heuristics like simulated annealing [16] or tabu search [15] and probabilistic algorithms [11,19]. More recently, Culberson [5] presented some results obtained applying to the same graph several heuristic algorithms in sequence. His idea was to improve each approximation using a different technique.

4 Our Approach

This paper addresses the features of spaces with a high degree of symmetry by means of *ad hoc* operators for genetic algorithms. In particular, a major issue comes from the recombination operator since it destroys important information for the solution of the graph colouring problem. As pointed out also by Falkanauer [12], whose work focuses on suitable representation schemes for grouping problems, the most relevant feature which needs to be inherited by offspring refers (in the case of the graph colouring problem) to the partition of the set of vertices (or nodes). Each subset of the partition corresponds to a specific colour, however this colour is named. The recombination of two randomly chosen solutions in the population destroys the original partition of the graph in all cases, even when the two solutions are equivalent (i.e., they generate the same partition of the graph). As a consequence, the quality of the offspring is on average worse than the one of its parents. Hence, the search process is not effective!

We observed that, in many cases, the recombination can be more advantageous if one could only *match* in a better way the parents' genomes. In fact, without changing the partition being inherited, the cost of the child solution might be improved by a simple re-allocation of the names of the colours in one of the parents. The solutions of the graph colouring problem identify the set X on which acts the group Q of all possible colour permutations. The two operators we are about to describe incorporate permutations to speed up the search. In one case the use of an operational research technique determines the *best* choice of colour labels for one part of the inherited genome, while in the other we try to identify the element $q \in Q$ under which the two parents are equivalent (this is of course not always possible since not all individuals of the population represent equivalent solutions).

The purpose of this investigation is mainly to evaluate and improve the search mechanism of a simple genetic algorithm on spaces like the ones introduced in section 2. In other words, we are looking for genetic operators which search the solution space without getting trapped in orbits that correspond to highly sub-optimal solutions.

4.1 Representation of Solutions

There are different ways of representing a colouring of a graph in a genotype. In a preliminary study we tested a naive representation and a grouping representation schemes. In the first case, each gene of the genotype encodes the colour of the vertex of the graph it represents. In other words, each solution is encoded with a genome of length n, if n is the order of the graph to colour. The i-th entry of the genome is a positive integer[3] representing the colour allocated to the i-th vertex of the graph. The enumeration of the vertices of the graph (i.e., which vertex corresponds to each gene) is implicitly deduced from its incidence matrix and has only the purpose of identifying them. In the other case (i.e., the grouping representation), the genome is divided in *groups* of dynamic size. Each group represents a set of nodes which have been assigned the same colour. An extra set of k genes (k being the number of colours used) is added at the end of the genotype to determine the size of each group. Hence, while the genotype of a naive representation encodes a sequence of colours allocated to the vertices of the graph, the grouping one encodes a permutation of the labels of the nodes. The nodes are dynamically grouped according to the information of the extra set of genes.

Our simulations did not show the superiority of any of these schemes. The grouping scheme was computationally more expensive though, mainly due to additional routines to keep feasibility of the solutions after the application of the genetic operators. Consequently, the naive representation was preferred.

4.2 The Fitness Function

A possible way of measuring the performance of a given individual is to count the number of conflicts generated by the encoded colouring of the graph. Therefore, for a given individual x, its fitness $f(x)$ can be expressed as

$$f(x) = \sum_{i=1}^{n-1} \sum_{j=i+1}^{n} w_{ij} \delta_{\sigma_x(i), \sigma_x(j)} \tag{2}$$

where $\delta_{\rho,\tau}$ equals 1 if and only if both nodes i and j are equally coloured. The objective of the search is to minimize $f(x)$ during the evolution.

4.3 The Mutation Operator

Random mutations are not very effective for the graph colouring problem since they have beneficial effects only seldomly. Consequently, a genetic algorithm with random mutations will converge very slowly to a solution (if it converges at all!). We investigated the possibility of replacing this operator with a more sophisticated one which could speed up the local search of the solution space. Due to the nature of the graph colouring problem, a suitable mutation scheme should be able to change the partition of the graph. A mere swap mutation may, in fact, not satisfy this requirement when the random vertices selected for the swap belong to the same partition. Therefore, we decided to test a mutation operator that applies partly swaps

---
[3] Colours have been assigned numerical labels only to allow easy re-allocation routines.

and partly flips with a given probability. A different approach has also been tested. Rather than selecting the new colour of a given vertex of the graph at random, a greedy selection rule can be applied instead. In this case, the colour can be determined on the basis of the colouring of its adjacent nodes. The selection reduces the conflicts between the given node and its neighbors.

Figure 1 (upper plot) compares swap mutation and a mixture of swap and flip mutations to the greedy heuristic mentioned above. Results refer to a graph of 250 vertices with probability $p = 0.5$ and 28 colours. Data have been averaged on 5 different seeds and show the average number of conflicts for the best individual of the population against the number of applied mutations. A simple genetic algorithm with elitism and mutation operator only has been used. Since the greedy mutation procedure searches all k colours each time, its complexity equals that of k different mutations. Hence, in order to present a fair comparison of the efficiency of each operator, results have been scaled to be consistent with the number of mutations. A measure of the CPU and kernel time has also been taken and it is reported in the same figure (lower plot). It compares the three mutation schemes against a fixed number of applied mutations.

It is clear that the greedy approach outperformed the other two on the selected graph. The average number of conflicts is significantly reduced at the beginning of the search, where the curve reports a steep descend from the initial value but the convergence is highly reduced soon after. It is worth mentioning, though, that the same algorithms would have different running times if they were compared on the basis of the number of generations. Section 5 reports results based on the use of these mutation schemes with different crossover operators.

4.4 The Linear Assignment Crossover

Crossing two individuals that encode colourings of a graph does not often improve the quality of the offspring solution. In fact, let p_1 and p_2 be two parent solutions and let p_{i_a} and p_{i_b} ($i = \{1, 2\}$) be the resulting sub-genotypes after a single cut has been applied. Let also $o = p_{1_a} \cup p_{2_b}$ be the offspring obtained by the combination of the first sub-genotype from parent p_1 and the second one from parent p_2[4]. Let us assume that both p_{1_a} and p_{2_b} are good colourings of the subgraphs induced by the vertices represented in each sub-genotype. The *merging* process is very likely to produce conflicts within the subgraphs as shown in figure 2. Conflicts are often generated by the labelling of the colours within each subgraph. This *mismatch* could in many cases be removed by an appropriate re-labelling of the colours in either p_{1_a} or p_{2_b}. In our case a good re-labelling has to minimize the number of conflicts between the two subgraphs while leaving the colouring of them unchanged (i.e., we do not want to change the partition of the vertices in the induced subgraphs).

The operator we are about to describe incorporates the above idea in a more intelligent way in order to maximize the improvement of the offspring solution. Each

---

[4] A similar reasoning can be applied for the other offspring (i.e., $o_2 = p_{2_a} \cup p_{1_b}$), when generated.

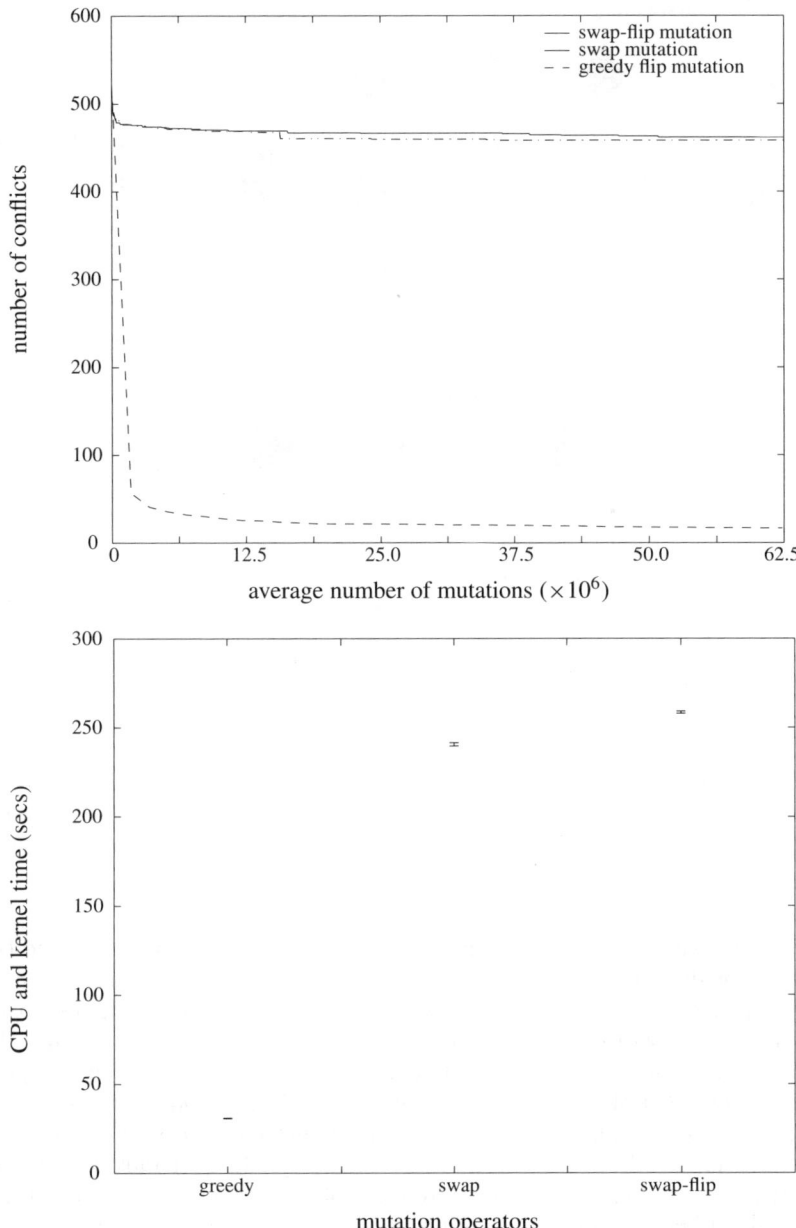

Figure 1. Average number of conflicts for the best individual of a genetic algorithm with swap-flip mutation, swap mutation, and greedy flip mutation only. Results refer to the colouration of a graph with 250 nodes and edge probability $p = 0.5$ using 28 colours, a population of 50 individuals, and 50% mutation rate. The lower plot shows the average CPU and kernel time in seconds for the mentioned simulations.

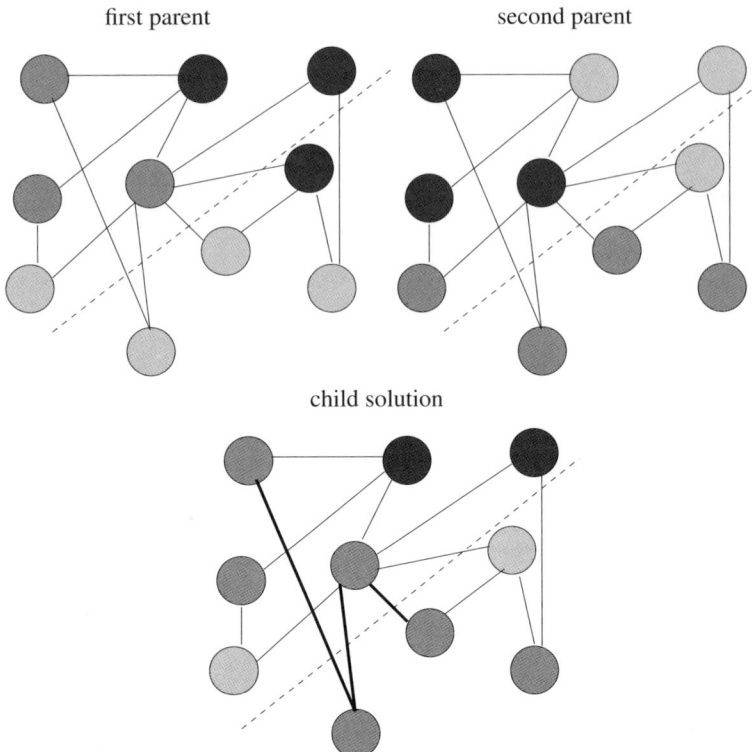

Figure 2. Schematic representation of the effects of recombination of two good colourings of a graph. The conflicts generated in the child solution have been highlighted using thicker lines.

time the crossover is applied, the graph $\mathcal{G}$ is split into two subgraphs: $\mathcal{G}_1$ and $\mathcal{G}_2$. This operation does not guarantee a substantial improvement when the re-labelling is done. In fact, even when the two recombined sub-genotypes do not originate many conflicts, the quality of the solution may still be very poor due to the fact that the original colouring of each subgraph was not a good one. Therefore, rather than generating a random cut of $\mathcal{G}$, we identify a bipartition of the graph where one of the subgraphs, say $\mathcal{G}_1$, has been properly coloured (i.e., the assignment of colours does not produce any conflict). This subgraph is created by adding iteratively nodes to it with the property that they do not introduce colour conflicts. Hence, the procedure just described maximizes the cardinality of $\mathcal{G}_1$. As a consequence, there will certainly be some conflicts across the cut of the graph (if this were not the case, we would add other nodes to $\mathcal{G}_1$) and possibly some conflicts in subgraph $\mathcal{G}_2 = \mathcal{G} \setminus \mathcal{G}_1$.

The next step is to generate the new labels for the colours in subgraph $\mathcal{G}_2$. This is done using the linear assignment algorithm[5] which searches all possible allocations

---

[5] More information about the algorithm can be found in [24].

in polynomial time. Its application requires a *cost matrix* to measures the cost of renaming a given colour to another one. The cost indicates, in our case, the number of conflicts (across the two subgraphs) that the re-labelling would introduce.

Let $\sigma_x(i)$ denote the colour of node i for the x-th individual of the population and $C = \{c_{ab}\}$ the cost matrix, where

$$c_{ab} = \sum_{\substack{i \in \mathcal{G}_2 : \sigma_x(i)=a \\ j \in \mathcal{G}_1 : \sigma_x(j)=b}} w_{ij} \qquad a, b = 1, \ldots, k \qquad (3)$$

is the number of conflicts that would be generated (across the cut) if colour a is replaced with colour b in $\mathcal{G}_2$. It is worth observing that

$$c_{aa} \neq 0 \qquad \forall a = 1, \ldots, k \qquad (4)$$

since if any colour is not re-labelled it might still create colour conflicts. We need to select some of the c_{ab} in such a way that we minimize the total number of conflicts for the graph. Let $K = \{1, 2, \ldots, k\}$ denote the set of usable colours. This problem can be formulated as

$$\text{minimize } z = \sum_{a \in K} \sum_{b \in K} c_{ab} p_{ab}$$

$$\text{subject to: } \sum_{a \in K} p_{ab} = 1 \qquad \forall b \in K \qquad (5)$$
$$\sum_{b \in K} p_{ab} = 1 \qquad \forall a \in K$$
$$p_{ab} \in \{0, 1\} \qquad \forall a, b \in K$$

where p_{ab} is a binary variable representing that colour a is changed to colour b[6]. Once the solution is found, each vertex of $\mathcal{G}_2$ is allocated its new colour. This crossover will be referred as *LAP crossover* to distinguish it from the one introduced in the next section. This operator has been implemented for both sexual and asexual mating. In the former case, we generate two offspring, each of which takes subgraph $\mathcal{G}_1$ from one parent and the *permuted* version of subgraph $\mathcal{G}_2$ from the other[7]. Despite the fact that the asexual version performs better than the sexual one (when applied without mutations) as observed in [20], a genetic algorithm with greedy mutations and sexual LAP crossover outperforms a similar one with the asexual implementation. Unless differently stated, this paper refers to the sexual LAP crossover only.

4.5 The Cyclic Permutation Crossover

The idea behind this operator is to imitate the inheritance of good building blocks. In our specific application a building block is the colouration of a subgraph of the original graph. Given the naive representation adopted, the subgraph is not guaranteed

---
[6] It is interesting to observe that the set $\{p_{ab}|p_{ab} = 1\}$ that satisfies the assignment formulation is an element of the group Q acting on the set X.

[7] Subgraph $\mathcal{G}_1$ is not guaranteed to have zero conflicts for both parents due to a different partition of the nodes.

to be represented in contiguous genes. In order to avoid the *mismatch* disruption, genotypes are re-mapped so that the first node of both parents are always assigned the same colour[8]. For this reason, this crossover has only been implemented for sexual recombination. Given the purpose of the mapping (i.e., to identify if there exists a $q \in Q$ which transforms one parent into the other), we do not need to re-allocate the colours of both parents. Therefore, each gene of one of the parents' genotypes, say p_2, is transformed according to the rule

$$\sigma_{p'_2}(i) = (\sigma_{p_2}(i) + (\sigma_{p_1}(1) - \sigma_{p_2}(1))) \bmod k \tag{6}$$

where $s = \sigma_{p_1}(1) - \sigma_{p_2}(1)$ identifies a subset of elements of Q, i.e., a set of cyclic permutations where each colour is *shifted* s positions. Hence, we refer to this crossover as the *cyclic crossover*.

Once the re-mapping procedure has been done, the actual recombination takes place between parents p_1 and p'_2. If $\sigma_{p_1}(i) = \sigma_{p_2}(i)$ for all nodes of the graph, the two genotypes encode equivalent solutions and the recombination will simply produce a copy of it. In all other cases, the offspring inherits genes from both parents according to the greedy rule

$$\sigma_o(i) = \begin{cases} \sigma_{p_1}(i) & \text{if } LF(\sigma_{p_1}(i)) \leq LF(\sigma_{p'_2}(i)) \\ \sigma_{p'_2}(i) & \text{otherwise} \end{cases} \quad \forall i = 1, \ldots, n. \tag{7}$$

LF is a local fitness measure which counts the number of conflicts among adjacent vertices of node i with respect to the specified parent.

It is worth observing that the offspring o is not guaranteed to improve its fitness but has probability non-zero of being worse than its parents. Why? Because, since *LF* takes into account colour conflicts with respect to the colouring of the parents' genotypes, we cannot guarantee that the combination of colours selected for the offspring will improve or keep the same fitness value. A more detailed investigation revealed that this crossover outperforms approaches based on pure correlation or decorrelation. In fact, when both parents differ only on the first gene, the re-mapping procedure decorrelates the two parents while in the case of parents which are totally different, (included the first node), the mapping introduces correlation between the two. Unfortunately, when both parents happen to have the same colour for the first node, the re-mapping has no effect.

From the point of view of the search space, the cyclic operator *merges* together partial good colouration of subgraphs. If one verifies the source of the inherited genes in the offspring solutions, it can be observed that they are transmitted in *blocks* from one of the parents. It is very likely, in fact, that if a subgraph has a good colouration (however the colours are named), the local fitness *LF* will be very good for all nodes in the subgraph. Therefore, the genes corresponding to the nodes of that subgraph will be inherited by the offspring when crossover is applied.

---

[8] This does not imply that $\sigma_{p_1}(1)$ is always the same value, but it simply means that $\sigma_{p_1}(1) = \sigma_{p_2}(1)$ any time the crossover is applied.

5 Experimental Settings and Results

The results presented in this paper refer to graph instances from the class $\mathcal{G}_{n,\frac{1}{2}}$. This class includes graphs with n vertices and probability $p = 1/2$ of connecting any two given nodes. Graph instances from this class are hard to colour[9] and generally use about twice the number of colours needed [14]. Some instances have been selected among benchmark problems[10] while others have been generated using a self-made software and their chromatic number is not known. This lack of information does not affect our possibility to compare results since this class has been investigated already in other works [5,15]. Despite the fact that the proposed approach is suitable for weighted and unweighted graphs, the results presented in this paper will refer to unweighted graphs only to provide a clear outline of the achievements.

Simulations have been run on a shared machine running Linux (Redhat 6.0) using C++ software. The software includes some classes from a genetic algorithms library called GAlib2.4.4[11]. A measure of the CPU and kernel time[12] gives an estimation of the computational cost of the algorithm against a fixed number of generations. The results presented in this section refer to a simple genetic algorithm with elitism[13]. The algorithm has a population of 50 individuals which are uniformly selected[14]. Mutation and crossover operators may vary depending on the results being plotted. Simulations terminate either when there exists a solution with no conflict or after 10,000 generations. For this investigation mutations are applied with probability $p_m = 0.5$ while individuals are recombined with probability $p_c = 0.1$.

Figures 3, 4 and 5 depict the average number of conflicts for a specific graph instance (DSJC250.5.col) with 250 nodes. Plots compare a genetic algorithm with different mutation and crossover operators. In particular, we selected a uniform and a one point crossover together with the operators mentioned in sections 4.4 and 4.5. Each of them has been applied together with swap mutations, a mixture of swap and flip mutations, and the greedy heuristic presented in section 4.3. Results have been averaged on 5 different seeds. Figures plot the average number of conflicts against the number of generations.

It can be observed that the different settings generate quite a variety of performance levels. Moreover, the simulations with the greedy mutation operator greatly outperformed the others. This is mainly due to the search mechanism of this operator rather than to the particular crossover being used. In fact, as we observed earlier in section 4.3, the greedy heuristic provides a good method to improve substantially

---

[9] See [8] for more information about the graph colouring transition phase.
[10] Available via ftp from http://mat.gsia.cmu.edu/COLOR/instances.html.
[11] Available via ftp from http://lancet.mit.edu/ga.
[12] The time refers to the initialization and actual evolutionary computation, but excludes the time required to write statistics to files at the end of the simulations.
[13] Earlier investigations on Steady State and Deme instances did not perform as good as the simple one. Elitism has been applied to avoid large fluctuations for the best solution.
[14] Tournament and rank-based selection schemes led to premature convergence of the population in all simulations.

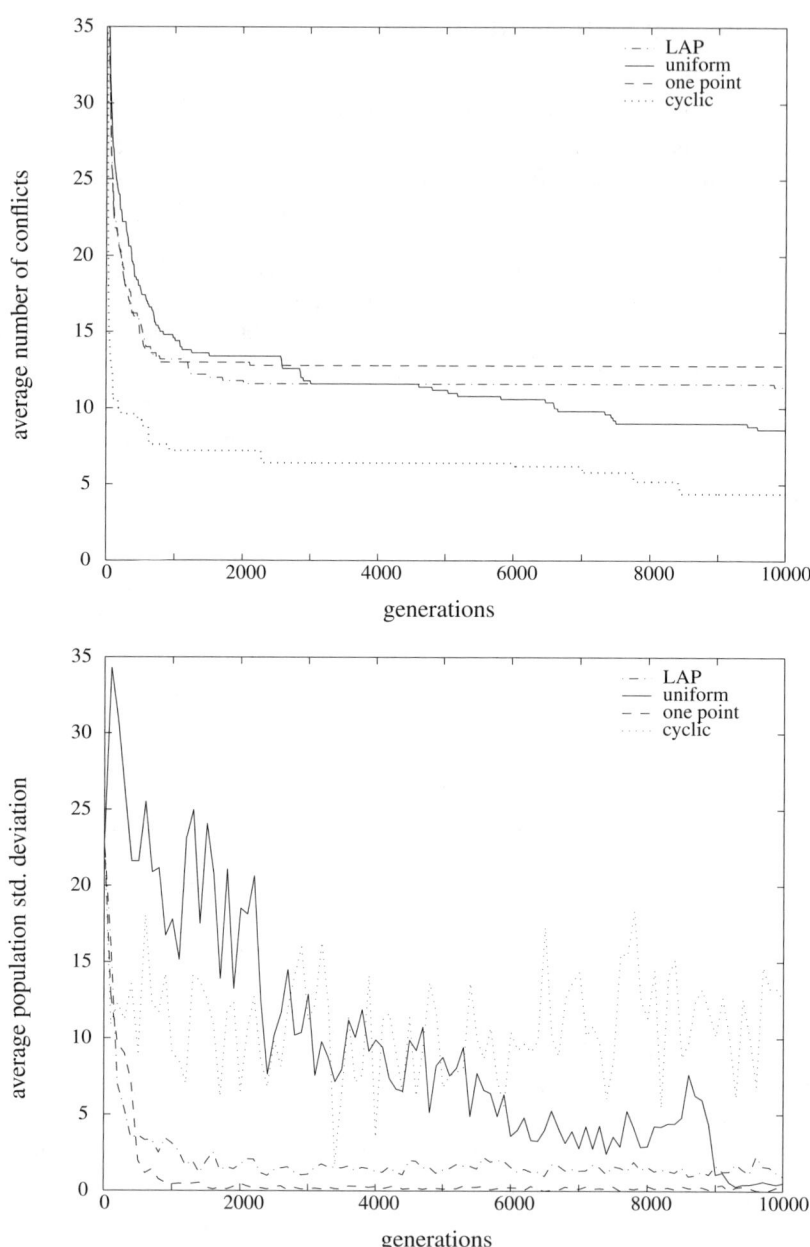

Figure 3. Average number of conflicts for the best individual *(upper plot)* and average standard deviation of the population *(lower plot)* on DSJC250.5.col with 28 colours. Performance refers to the use of a simple genetic algorithm with greedy mutations and four different types of crossover (i.e., uniform, one point, LAP, and cyclic). Results are the average of 5 different seeds.

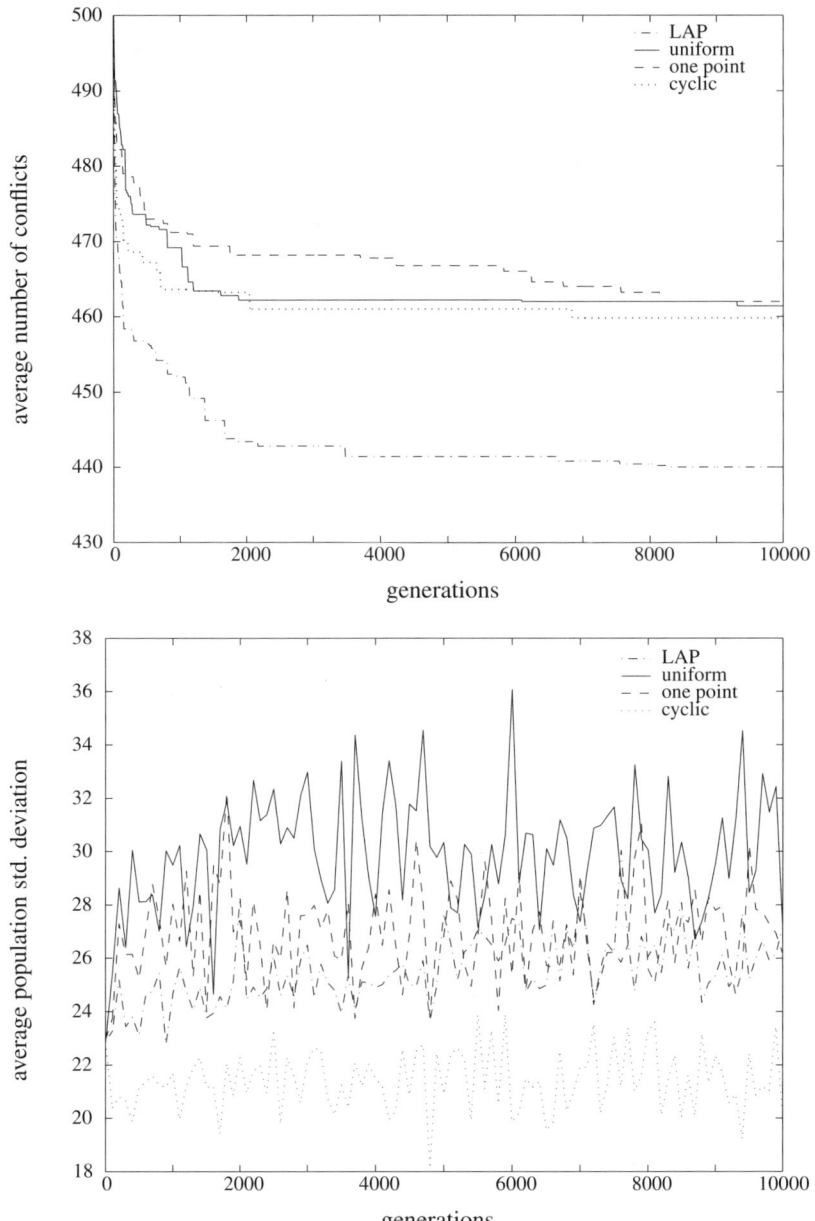

Figure 4. Average number of conflicts for the best individual *(upper plot)* and average standard deviation of the population *(lower plot)* on DSJC250.5.col with 28 colours. Performance refers to the use of a simple genetic algorithm with swap mutations and four different types of crossover (i.e., uniform, one point, LAP, and cyclic). Results are the average of 5 different seeds.

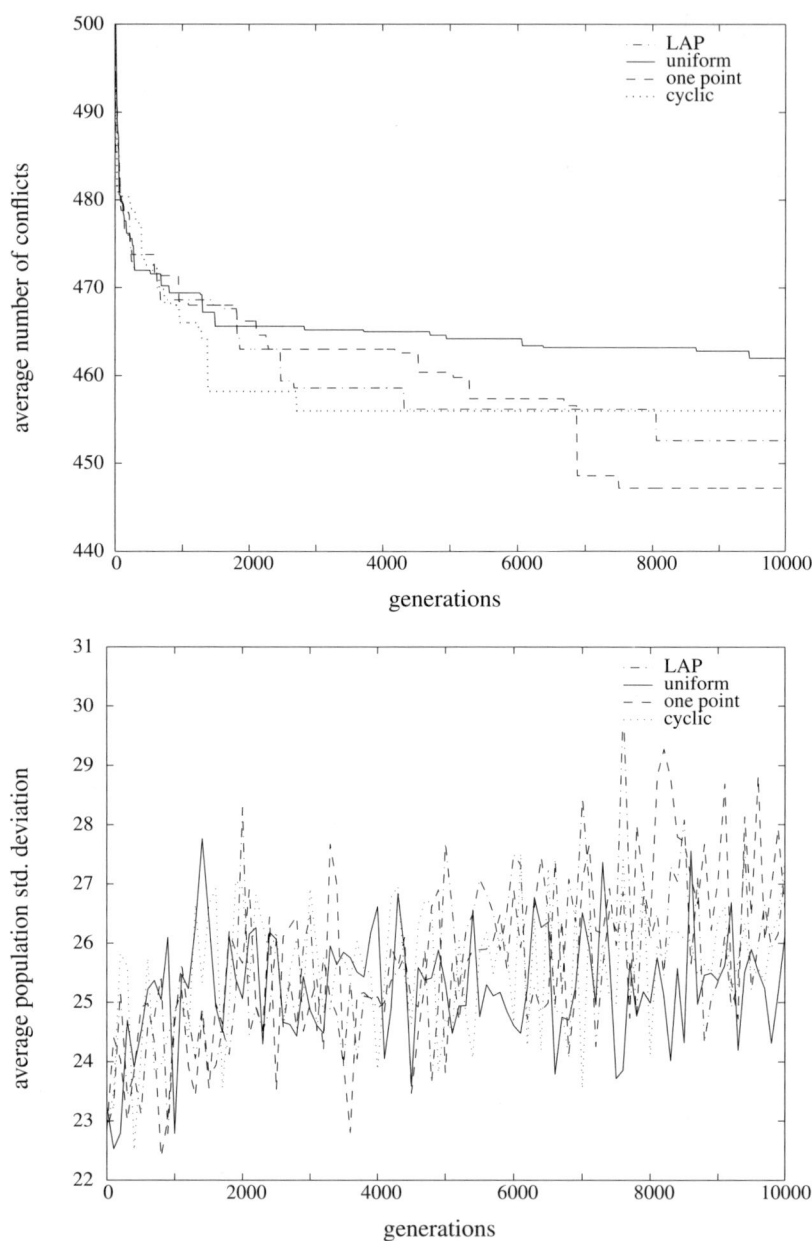

Figure 5. Average number of conflicts for the best individual *(upper plot)* and average standard deviation of the population on *(lower plot)* DSJC250.5.col with 28 colours. Performance refers to the use of a simple genetic algorithm with a mixture of swap and flip mutations and four different types of crossover (i.e., uniform, one point, LAP, and cyclic). Results are the average of 5 different seeds.

the colouration of the graph. The LAP crossover, on the contrary, places an individual (i.e., the offspring solution) in the best possible orbit given that part of the colouration (i.e., subgraph $\mathcal{G}_1$) is fixed. In other words, the linear assignment algorithm searches among the orbits of the space, the best *match* between the colouration of subgraph $\mathcal{G}_1$ and the one of subgraph $\mathcal{G}_2$. Unfortunately, this does not guarantee that improvements can lead to no conflicts at all since subgraph $\mathcal{G}_2$ might be badly coloured. Hence, despite the fact that many orbits of the solution space are quickly eliminated from the search, the algorithm still fails to escape orbits which correspond to few conflicts.

According to the results of figure 3, the performance of the LAP crossover is not very different to that of the one point crossover operator, while the cyclic crossover behaves in a similar way to the uniform crossover operator. This result is not surprising given the implementation of the two operators. In fact, the cyclic crossover performs a uniform recombination where genes are selected on the basis of a greedy rule (see equation (7)). This mechanism increases the possibility of generating better offspring. The LAP one generates multiple cuts[15] on the genotypes and alternate blocks are re-coloured using the linear assignment algorithm.

The CPU and kernel time of each algorithm variant is displayed in figure 6, where genetic algorithms with LAP crossover require on average a longer time to complete the search. This is mainly due to the fact that the LAP crossover incorporates the linear assignment algorithm and creates each time two offspring. The values reported in the figure do not represent *absolute* measurements due to the shared CPU time with other users. Therefore, they can only give a trend of the running time required by each instance. It can be observed that genetic algorithms with greedy mutations require more than double the CPU time compared with similar instances with swap (and swap-flip) mutations, when compared against a fixed number of generations. The extra running time allows, though, a significant improvement of performance (which in many cases can be achieved with a smaller number of generations). In fact, neither of the two other variants tested produced comparable results. Therefore, both the LAP and the cyclic crossover together with the greedy mutation operator search the space more efficiently and produce good approximations within a reasonable amount of computation.

To evaluate how performance scales up with the size of the graph, the investigation just outlined has been extended to graphs with larger orders. In this case, different probabilities of mutations and recombinations were tested but the paper reports only results for the best combinations.

In particular, figures 7 and 8 refer to an investigation on 5 random·graphs from the class $\mathcal{G}_{1000, \frac{1}{2}}$ using some of the algorithms described above. Figure 7 depicts results of a simple genetic algorithms with swap mutations and one point crossover for this class of graphs. Simulations applied mutations and crossover with probabilities $p_m = 0.5$ and $p_c = 0.1$, respectively.

---

[15] Even if the LAP crossover splits the graph into two subgraphs, the genome representation does not reflect the division as a single cut.

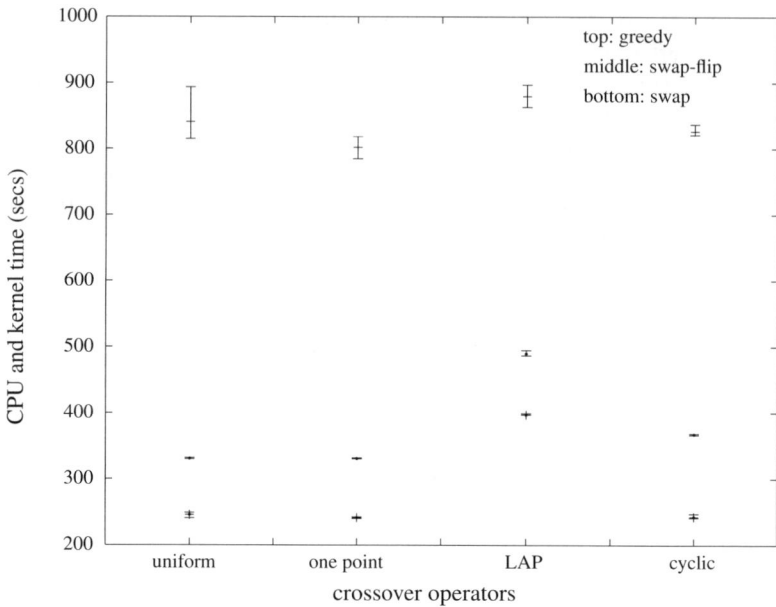

Figure 6. CPU and kernel time in seconds for the algorithms in figures 3, 4 and 5. The plot refers to the use of greedy mutations, swap mutations, and swap/flip mutations with different types of crossover operators. Values represent the average of 5 simulations.

The algorithm shows a similar behaviour to the one depicted in figure 3 for the graph DSJC250.5.col. In particular, due to the slow convergence, the algorithm does not perform very well even if the number of generations is doubled or tripled. A better result is obtained when applying mutations with probability $p_m = 1/n = 0.001$ as depicted in the lower plot (figure 7).

When greedy mutations and LAP crossover are applied instead, very different behaviour can be observed (see figure 8). The evolutionary process is able to eliminate most of the conflicts in the first 100 generations (the initial average value of 2600 drops to just above 200) converging to a fairly good solution.

The bottom plot of figure 8 refers to the performance of a simple genetic algorithm with greedy mutations and cyclic crossover on the random graphs mentioned earlier. Simulations apply recombination with probability $p_c = 0.2$. It is interesting to observe that the cyclic crossover keeps a higher diversity in the population. A better comparison between the two algorithms is reported in figure 9, where the genetic algorithm with the cyclic crossover show slightly better solutions. It seems that the more diverse the population, the better the results that can be achieved. Although this hypothesis can be confirmed for the genetic algorithms depicted in figures 3 and 8, it can be rejected for the simple genetic algorithm reported in the upper plot of figure 7 where, despite the high diversity of the population, performance is very

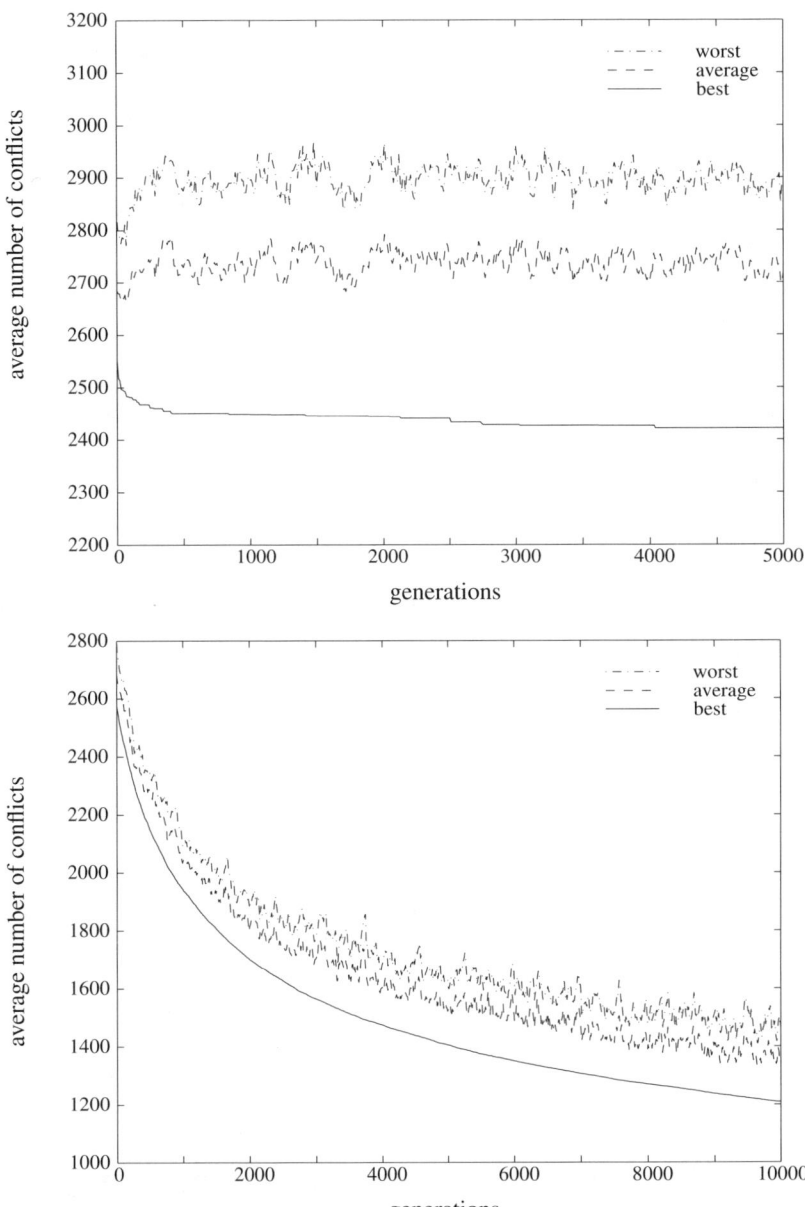

Figure 7. Performance of a simple GA with swap mutations and one point crossover on random graphs from the class $\mathcal{G}_{1000,\frac{1}{2}}$ using 93 colours. The plot reports the average number of conflicts for the best, the average, and the worst individual of the population. Mutations are applied with probabilities $p_m = 0.5$ *(upper plot)* and $p_m = 0.001$ *(lower plot)*, respectively.

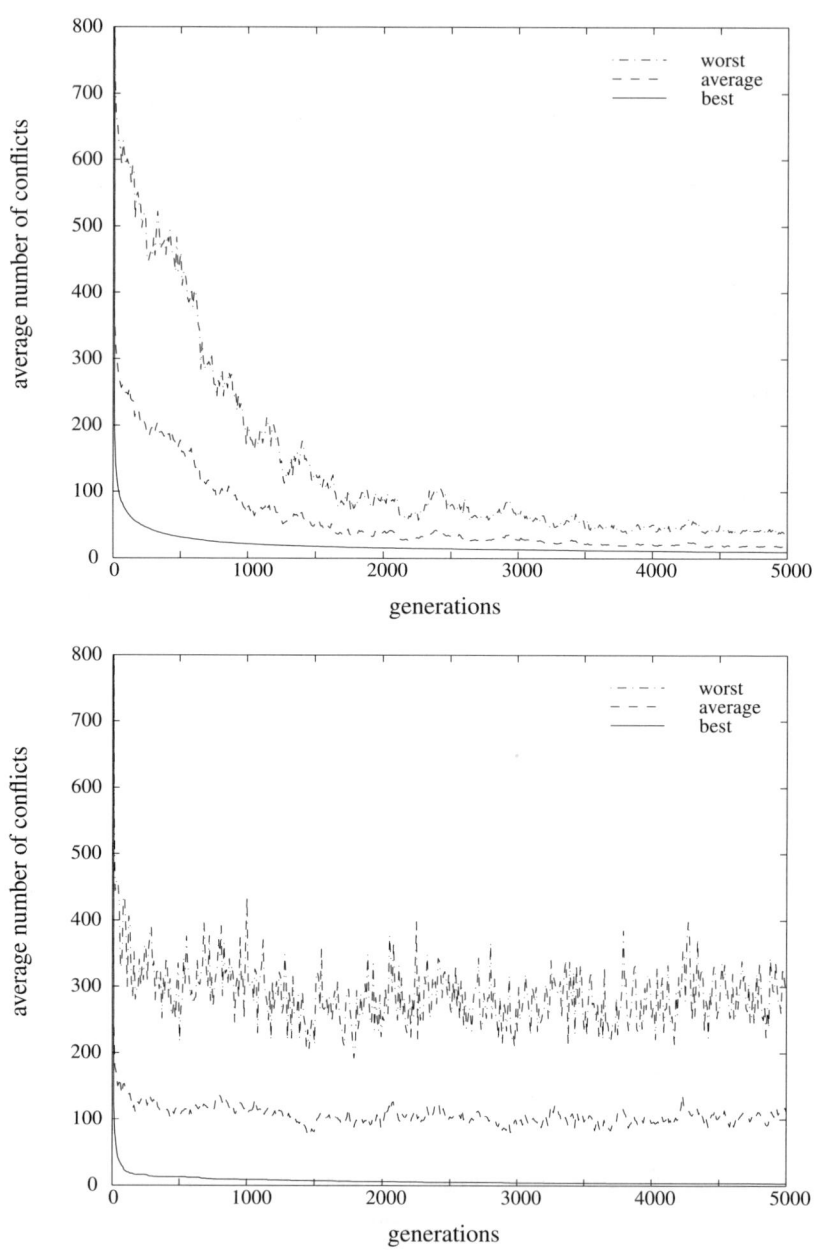

Figure 8. Average number of conflicts for a simple GA with greedy mutations and LAP crossover *(upper plot)* and a simple GA with greedy mutations and cyclic crossover *(lower plot)* on random graphs from the class $\mathcal{G}_{1000, \frac{1}{2}}$ using 93 colours. The plot reports the average number of conflicts for the best, the average, and the worst individual of the population.

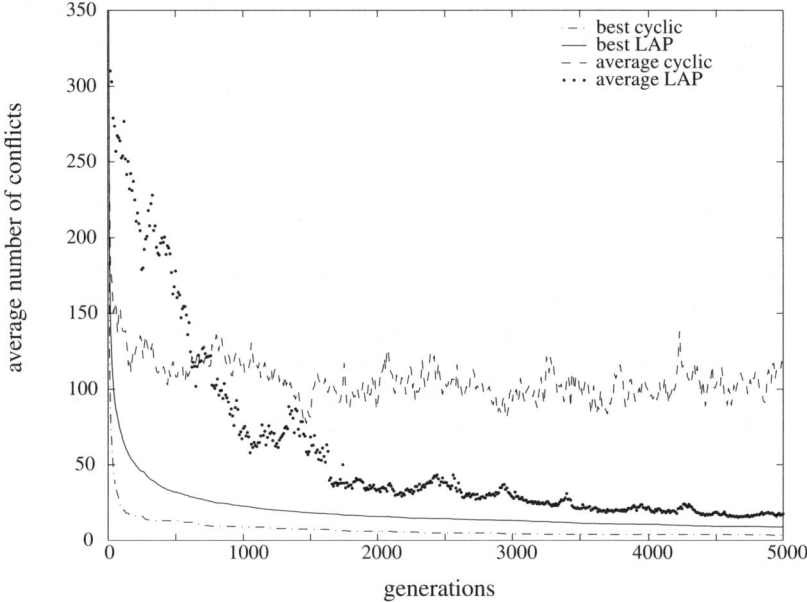

Figure 9. Comparison between a simple GA with LAP crossover and a simple GA with cyclic crossover. In both cases elitism and greedy mutations have been applied. Results refer to the colouration of random graphs form the class $\mathcal{G}_{1000,\frac{1}{2}}$ using 93 colours.

poor. The CPU time for this simulation is not reported but reflects the same proportions shown in figure 6. A comparison based on the average number of mutations and recombination would still depict a better performance for the algorithms applying the new crossover operators. Hence, the LAP and cyclic crossover (applied with greedy mutations) improve the use of the genetic algorithms for the colouration of random graphs from the class $\mathcal{G}_{1000,\frac{1}{2}}$.

6 Conclusions

This work represents a first approach to the evaluation of genetic algorithms search mechanism on spaces with a high degree of symmetry under a group action. The graph colouring problem, like many other combinatorial optimisation ones, is a good representative for these spaces. In this investigation we did not take into account *at all* any symmetries in the topology of the graph which may be considered in further investigations. Our major goal is, in fact, to develop an efficient search mechanism for problems with highly symmetrical solution spaces. In order to accomplish that, a better understanding of the search strategies of genetic algorithms on this particular domain is required. This will be investigated by the analysis of artificial spaces with controlled symmetries. A first investigation is going to evaluate how the population of a genetic algorithm moves among the orbits of the solution space. Also, we are currently testing the hypothesis of generating unique representatives for each orbit of the space in order to reduce the search significantly. Initial tests seem to reveal that this approach still produces sub-optimal solutions.

Another issue which we would like to address relates to the slow convergence of the genetic algorithm after the initial steep improvement of the solution. One possible explanation of this phenomena is that it becomes harder to find better solutions, but at the same time it might be the case that the search is *trapped* in a large orbit of the space. In several simulations, in fact, the evolution could have been stopped much earlier since there was no further improvement of the solutions. For this purpose, we planned to improve the operators presented in this paper. For instance, the LAP crossover might perform better if the subgraph $\mathcal{G}_2$ had a proper colouration. We are currently investigating the possibility of implementing this feature at the least running cost.

Acknowledgments

The author wishes to thank Dr. C. A. Glass, who suggested the use of the linear assignment algorithm to re-allocate colours to a subgraph at the least cost and Dr. J. J. Wood for the useful discussions on permutation groups and his valuable comments to improve the manuscript.

References

1. M. A. Armstrong. *Groups and Symmetry*. Springer-Verlag, Berlin Heidelberg New York, 1980.
2. S. Bhattacharyya. Direct marketing performance modelling using genetic algorithms. *Informs Journal on Computing*, 11(3):248–257, 1999.
3. D. Brélaz. New methods to color the vertices of a graph. *Communications of the ACM*, 22(4):251–2566, 1979.
4. A. Cayley. On the 4-colouring problem (original title not known). *Proc. London Math. Society*, 9:148, 1878.
5. J. Culberson and F. Luo. *Cliques, Coloring and Satisfiability: Second DIMACS Implementation Challenge*, chapter: Exploring the k-Colorable Landscape with Iterated Greedy, pages 245–284. American Mathematical Society, 1996.
6. L. Davis. *Handbook of Genetic Algorithms*. Van Nostrand Reinhold, New York, 1991.
7. A. E. Eiben. *Handbook of Evolutionary Computation*. IOP Press and Oxford University Press, New York, 1997.
8. A. E. Eiben and J. K. van der Hauw. Graph colouring with adaptive genetic algorithms. Technical report, Dept. of Comp. Science, University of Leiden, 1996.
9. A. E. Eiben and J. K. van der Hauw. Adaptive penalties for evolutionary graph-coloring. In *Artificial Evolution'97*, pages 95–106. Springer-Verlag, Berlin Heidelberg New York, 1997.

10. A. E. Eiben, J. K. van der Hauw, and J. I. van Hemert. Graph coloring with adaptive evolutionary algorithms. *Journal of Heuristics*, 4:25–46, 1998.
11. J. A. Ellis and P. M. Lepolesa. A Las Vegas graph coloring algorithm. *The Computer Journal*, 32(5):474–476, 1989.
12. E. Falkanauer. A new representation and operators for genetic algorithms applied to grouping problems. *Evolutionary Computation*, 2(2):123–144, 1994.
13. D. E. Goldberg, K. Deb, and J. H. Clark. Genetic algorithms, noise and the sizing of populations. *Complex Systems*, 6:333–362, 1992.
14. G. R. Grimmett and C. J. H. McDiarmid. On colouring random graphs. In *Mathematical Proceedings of the Cambridge Philosophical Society*, volume 77, pages 313–324, 1975.
15. A. Hertz and D. de Werra. Using tabu search techniques for graph coloring. *Computing*, 39(4):345–351, 1987.
16. D. S. Johnson, C. R. Aragon, L. A. McGeoch, and C. Schevon. Optimization by simulated annealing: an experimental evaluation; part II, graph coloring and number partitioning. *Operational Research*, 39(3):378–406, 1991.
17. A. Johri and D. W. Matula. Probabilistic bounds and heuristic algorithms for coloring large random graphs. Technical Report TR 82-CSE-06, Dept. Computer Science, Southern Methodist University, Dallas, TX, June 1982.
18. L. Kučera. Graphs with small chromatic numbers are easy to color. *Information Processing Letters*, 30(5):233–236, 1989.
19. B. Manvel. Extremely greedy coloring algorithms. In F. Harary and J. S. Maybee, editors, *Graphs and Applications*, pages 257–270, 1985.
20. A. Marino, A. Prügel-Bennett, and C. A. Glass. Evolutionary graph colouring: a new perspective. *IEEE Trans. on Evolutionary Computation*, 1999. To be submitted for publication.
21. C. McDiarmid. Colouring random graphs badly. In *Graph Theory and Combinatorics*, volume 34, 1979.
22. A. C. Nearchou. A genetic navigation algorithm for autonomous mobile robots. *Cybernetics and systems*, 30(7):629–661, 1999.
23. C. A. PenaReyes and M. Sipper. A fuzzy-genetic approach to breast cancer diagnosis. *Artificial Intelligence in Medicine*, 17(2):131–155, 1999.
24. A. Schrijver. *Theory of linear and integer programming*. John Wiley, New York, 1998.
25. J. P. Spinrad and G. Vijayan. Worst case analysis of a graph coloring algorithm. *Discrete Applied Mathematics*, 12(1):89–92, 1985.
26. J. S. Turner. Almost all k-colourable graphs are easy to color. *Journal of Algorithms*, 9:63–82, 1988.

Structure Optimization and Isomorphisms

P. Stagge and C. Igel

Institut für Neuroinformatik
Ruhr-Universität Bochum
Bochum, Germany
E-mail: {peter.stagge,christian.igel}@neuroinformatik.ruhr-uni-bochum.de

Abstract. In this article we deal with a quite general topic in evolutionary structure optimization, namely, redundancy in the encoding due to isomorphic structures. This problem is well known in topology optimization of neural networks (NNs) and we study it in this framework. Choosing a good NN structure for a given problem is still a difficult task for which we do not know any successful analytical means. But problem specific structures can lead to significantly improved results; evidence for this is given in this contribution by an NN that was evolutionarily adapted for a benchmark problem. The degree to which isomorphic structures, i.e., classes of equivalent NN topologies, enlarge the search space depends on the restrictions of the allowed structures and on the representation of the search space. In the context of structure optimization of NNs we observe similar phenomena of rare and frequent structures as are known from molecular biology. To cope with isomorphisms we make use of the relation between NN topologies and graphs. Exploiting methods from graph theory we demonstrate a general way to deal with isomorphic structures. The presented approach applied to NN optimization can be regarded as a solution to the so-called *competing conventions* problem [18]. Further, it can be used to save fitness evaluations when used in combination with a graph-database.

Keywords
Structure and topology optimization, recurrent neural networks, isomorphism, graph representation, competing conventions problem, rare and frequent structures

1 Introduction

In this paper we present a way to deal with isomorphic structures in evolutionary structure optimization, see also the chapters of Marino and Van Hoyweghen. Topology adaptation of artificial neural networks (NNs) [1,8] serves as an example.

When searching for a good NN with an evolutionary algorithm (EA) one has to code for the network structure; it can, for example, be described by its connection matrix. In general, there are several connection matrices leading to functionally equivalent network topologies, i.e., isomorphic structures, see figure 1.

Such a redundant representation can be harmful or beneficial for evolutionary search processes. For example, the mapping from RNA sequences to RNA secondary structures in biology is very well studied and proves to be extremely redundant. This

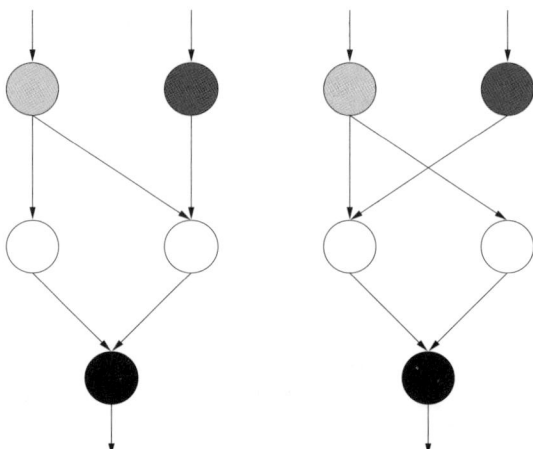

Figure 1. Two isomorphic network structures. Each input neuron and the output neuron have different colours.

redundancy is assumed to be helpful to avoid local optima by constituting *neutral networks* [19]. However, it seems to be a waste to evaluate isomorphic structures several times according to different connection matrices. Here, we present a way to circumvent this problem. We propose to construct graph representations that are free of isomorphic topologies and preserve neighborhood relations with respect to "canonical" mutations (addition or deletion of a node or connection).

The article begins with an introduction into structure optimization of NNs with EAs, then the concepts of isomorphic graphs, canonically labeled graphs, and rare and frequent structures are given. These are applied to artificial NNs. The investigated benchmark example is described concisely. Finally, the results are given and a conclusion ends this article.

2 Evolutionary Structure Optimization of Neural Networks

The motivation for optimizing the structure of NNs is the fact that we know about the approximation capability of NNs [9], but we do not know what a beneficial structure looks like for a given problem. Often we evaluate an NN according to its generalization error and the time it takes to train the network, but there are only very general hints that tell us how the structure of an NN influences these values. The fact that standard structures are widely used (multi-layer perceptrons with 1, 2, or 3 hidden layers, fully connected) does not reflect deeper insight into the theory of NNs but demonstrates that there is no systematic and reliable way to generate problem specific network topologies. Results from structure optimization of NNs show that evolutionary algorithms are very suitable to optimize the topology of NNs [2,18,20,25].

In order to tackle the structure optimization problem in the framework of EAs one has to define a genotype space which constitutes the search space, and search operators, e.g., mutation or crossover. Each genotype in the search space can be mapped to an NN (simple example: connection matrix as genotype being mapped to an NN). Representation and operators strongly depend on one another.

In this work we consider only evolutionary optimization of network topologies. Neither the weights nor weight initialisations are included in the genotype. Each NN structure has to be trained from a random weight initialisation and the resulting weights are not coded back into the genotype. There are a lot of different approaches using different codings and operators which can be crudely divided into two categories:

In *direct encoding* each connection in the NN is explicitly represented in the genotype, e.g., in a connection matrix [13]. Thus, one can get a good idea of the effects the search operators on the genotypes will have in the structure space, e.g., connection deletion.

In *indirect encoding* schemes the genotype does not represent the NN topology directly, but it consists of – for example, grammar based – rules that describe the construction process for the NN. The mapping from a genotype to an NN structure might even be non-deterministic; in some cases connections are given according to a coded connection density. There are a lot of different indirect encoding schemes some of which are biologically inspired as models for the natural neural growth processes, for an overview see [20].

Any approach has to cope with several problems. There might be structures which are represented by numerous genotypes, others only by few or one genotype, and in indirect encoding schemes some might not be coded for at all. Search operators can have the tendency to produce NNs with specific characteristics (bias in structure space), e.g., NNs with many layers or sparsely connected NNs. The influence of the operators on the structure and on the fitness is not obvious. With regard to EAs one asks for operators obeying the principle of "strong causality" [15], i.e., a small change in the genotype causes a small change in the fitness value (the distance in genotype space should be defined by the search operators [21]). Up to now there is neither a proven best representation–operator setting nor a constructive way to generate a good setting for a given problem except some general design heuristics.

3 NNs and Graph Isomorphisms

3.1 Graph Isomorphism

In this section, we summarize some classical graph theory needed to identify isomorphic NNs. The notation and results used in this section are based on McKay [10,11].

Let $G = (V, E)$ be a directed graph, V and E denote the set of vertices and edges, respectively. A *colouring* π denotes an ordered partition of V into disjoint, non-empty colour classes, i.e., for k different colours $\pi = (V_1, V_2, \ldots, V_k)$. The order of the colour classes is significant, but the order of the vertices within each class is not.

Let γ be a permutation of the vertex set V of a graph G and let v^γ denote the image of the vertex $v \in V$ under γ. Similarly, we define $W^\gamma = \{w^\gamma | w \in W\}$ with $W \subseteq V$ and G^γ as the graph in which vertices v^γ and w^γ are adjacent iff v and w are adjacent vertices in G. The image of a colouring π under γ is defined as $\pi^\gamma = (V_1^\gamma, V_2^\gamma, \ldots, V_k^\gamma)$. A permutation γ of a vertex set with colouring π is called *colour-preserving* iff $\pi^\gamma = \pi$, i.e., each $v \in V$ has the same colour as v^γ. Two coloured graphs G_1 and G_2 are isomorph iff there exists a colour-preserving permutation γ that transforms one into the other, say $G_1 = G_2^\gamma$.

Let $\pi = (V_1, V_2, \ldots, V_k)$ be a partition of $\{0, 1, \ldots, n-1\}$, then we define $c(\pi)$ as the partition $(\{0, 1, \ldots, |V_1|-1\}, \{|V_1|, \ldots, |V_1|+|V_2|-1\}, \ldots, \{n-|V_k|, \ldots, n-1\})$, i.e., $c(\pi)$ is independent of π except it has the same number of classes with the same sizes in the same order.

A *canonical labelling map* is a function $\mathcal{C}$ such that for any graph G, partition π of V, and permutation γ of V we have

1. $\mathcal{C}(G, \pi) = G^\delta$ for some permutation δ such that $\pi^\delta = c(\pi)$, and
2. $\mathcal{C}(G^\gamma, \pi^\gamma) = \mathcal{C}(G, \pi)$.

The usefulness of canonical labelling maps is shown by the following theorem [11].

Theorem 1 (McKay 1990). *Suppose the graphs G_1 and G_2 are coloured using the same number of vertices of each colour. Then $\mathcal{C}(G_1, \pi_1) = \mathcal{C}(G_2, \pi_2)$ iff $G_1^\gamma = G_2$ for some colour-preserving permutation γ. (Here, π_1 and π_2 are the colourings, with the colours in the same order in each.)*

The general graph-isomorphism problem, and therefore the canonical labelling, is a potential member of the problem class NPI which contains the so-called NP-incomplete problems. NPI is defined as NP excluding P and all NP-complete problems [6]. So we cannot expect to have an algorithm which solves the general graph-isomorphism problem in polynomial time. However, efficient heuristic algorithms exist that can handle the canonical labelling for graphs generated from NNs. The graphs representing NNs which can be regarded as tractable with today's learning algorithms have moderate size, so that the costs of computing their canonical representation can almost be neglected compared to the fitness evaluation (learning) time.

Fast[1], well-documented software that implements a canonical labeling map is freely available: the *nauty* package by Brendan D. McKay is used in this investigation.

---
[1] Computing the canonically labeled graph representing an NN consisting of 100 nodes takes in the order of 10^{-2}s on a 250 MHz UltraSparc.

3.2 Neural Networks as Graphs

Feed-Forward NNs. We want to use graph theory to determine whether two NN topologies are isomorphic. Therefore, we need a way to map a network to a coloured graph. Assume a feed-forward network $\mathcal{N}$ with n neurons $x_0, \ldots, x_{n-1}$. Without loss of generality, we assume that the neurons $x_0, \ldots, x_{d-1}$ are the d input neurons, $x_{n-m}, \ldots, x_{n-1}$ are the m output neurons, and that each neuron x_i gets only input from neurons x_j with $j < i$. The output y_i of each neuron i is then given by

$$y_i = g\left(\sum_{j=0}^{i-1} w_{ij}^0 y_j + \vartheta_i\right), \qquad (1)$$

where g is a so-called activation function, ϑ_i is the threshold (bias), and w_{ij}^0 is the strength (weight) of the connection between neuron j and i.

The vertex set of the corresponding directed graph $G_\mathcal{N} = (V_\mathcal{N}, E_\mathcal{N}^0)$ is $\{0, \ldots, n-1\}$ and there is an edge between two vertices u and v iff there is a connection from neuron x_u to neuron x_v. As $\mathcal{N}$ is a feed-forward network, $G_\mathcal{N}$ is acyclic. We use a coloured graph to distinguish between different classes of neurons, e.g., neurons that are connected to certain inputs or outputs. We define the colouring

$$\pi_\mathcal{N} = (\{0\}, \{1\}, \ldots, \{d-1\}, \{d, d+1, \ldots, n-m-1\},$$
$$\{n-m\}, \{n-m+1\}, \ldots, \{n-1\}) \qquad (2)$$

to fix the input and output units, see figure 1 for an example.

Recurrent NNs. The recurrent networks considered in this investigation are generalized Elman networks [3,23,24]. Let $y_i(t)$ denote the output of neuron i at time t. A network with S memory layers obeys the following dynamics

$$y_i(t) = g\left(\sum_{j=0}^{i-1} w_{ij}^0 y_j(t) + \sum_{s=1}^{S}\sum_{j=0}^{n-1} w_{ij}^s y_j(t-\tau_s) + \vartheta_i\right). \qquad (3)$$

The w_{ij}^s for $s \geq 1$ are the weights for recurrent connections from memory layer s, and τ_s is the corresponding time-delay, see [23,24] for details. The structure of such a network $\mathcal{N}$ can be described by a set of vertices $V_\mathcal{N} = \{0, \ldots, n-1\}$ and $S+1$ sets of edges, $E^s, s \in \{0, \ldots, S\}$. $E_\mathcal{N}^0$ is the same as for the feed-forward network. Additionally, there is an edge (u, v) in $E_\mathcal{N}^s$ iff there is a connection from x_u to x_v in memory layer s. For $s > 0$ there may be edges $(u, v) \in E_\mathcal{N}^s$ with $u \geq v$.

3.3 Detecting Isomorphic NNs

Feed-Forward NNs. We want to use theorem 1 to determine whether the topologies of two NNs as defined in section 3.2 are isomorphic.

The structure of two NNs $\mathcal{N}_1$ and $\mathcal{N}_2$ is said to be isomorphic iff $V_{\mathcal{N}_1} = V_{\mathcal{N}_2}$ and there is a bijection $\gamma : V_{\mathcal{N}_1} \to V_{\mathcal{N}_1}$ so that for each s, $0 \leq s \leq S$

$$(u, v) \in E^s_{\mathcal{N}_1} \leftrightarrow (\gamma(u), \gamma(v)) \in E^s_{\mathcal{N}_2} \; . \tag{4}$$

Further, we require that input and output units of one network are mapped to the corresponding units of the other.

We can prove the following statement.

Proposition 1. *Two feed-forward NN structures $\mathcal{N}_1$ and $\mathcal{N}_2$ are isomorphic iff the coloured graphs $G_{\mathcal{N}_1}$ and $G_{\mathcal{N}_2}$ with colouring $\pi_{\mathcal{N}_1}$ and $\pi_{\mathcal{N}_2}$, respectively, are isomorphic.*

In particular, the networks are isomorphic iff

$$\mathcal{C}(G_{\mathcal{N}_1}, \pi_{\mathcal{N}_1}) = \mathcal{C}(G_{\mathcal{N}_2}, \pi_{\mathcal{N}_2}) \tag{5}$$

for some canonical labelling map $\mathcal{C}$. For a fixed function $\mathcal{C}$, for each feed-forward NN $\mathcal{N}$ there exists exactly one isomorphic *canonical network* which is uniquely defined by the graph $\mathcal{C}(G_{\mathcal{N}}, \pi_{\mathcal{N}})$.

The colouring ensures that input and output units of one network are mapped to the corresponding units of the other network. Further, if different types of neurons are encoded, the colouring can be used to group these neurons. For example, when a *sigma-pi* network [17] is described, colouring can prevent from mapping a sigma unit to a pi unit and vice versa.

Recurrent NNs. In order to use theorem 1 for recurrent NNs, we have to represent the network by a single directed coloured graph.

Let $\phi(.)$ be an injective mapping from the space of Elman network structures into the space of directed coloured graphs.

Proposition 2. *Two Elman topologies $\mathcal{N}_1$ and $\mathcal{N}_2$ are isomorphic iff there is a colour-preserving isomorphism from $\phi(\mathcal{N}_1)$ onto $\phi(\mathcal{N}_2)$.*

Let $(V_{\phi,\mathcal{N}}, E_{\phi,\mathcal{N}}, \pi_{\phi,\mathcal{N}})$ be the coloured directed graph generated by $\phi(.)$ given an Elman network $\mathcal{N}$. If δ is a colour-preserving isomorphism from $\phi(\mathcal{N}_1)$ onto $\phi(\mathcal{N}_2)$, then $\gamma = \phi^{-1} \circ \delta \circ \phi$ is a colour-preserving isomorphism from $\mathcal{N}_1$ onto $\mathcal{N}_2$.

$$\begin{array}{ccc} \mathcal{N}_1 & \xrightarrow{\phi} & (V_{\phi,\mathcal{N}_1}, E_{\phi,\mathcal{N}_1}, \pi_{\phi,\mathcal{N}_1}) \\ \gamma \downarrow & & \downarrow \delta \\ \mathcal{N}_2 & \xleftarrow{\phi^{-1}} & (V_{\phi,\mathcal{N}_2}, E_{\phi,\mathcal{N}_2}, \pi_{\phi,\mathcal{N}_2}) \end{array}$$

In the following, we give one possible choice for ϕ. Assume an Elman topology $\mathcal{N}$ with n neurons and S memory layers. The injective mapping $\phi(.)$ generates a coloured directed graph with $(S+1)n$ vertices. The set of edges is constructed the following way. Starting from $E_{\phi,\mathcal{N}} = \emptyset$

$$\forall s = 0,\ldots,S \ \forall [x,y] \in E_{\mathcal{N}}^s : \ E_{\phi,\mathcal{N}} = E_{\phi,\mathcal{N}} \cup \{[ns+x, ns+y]\}$$

$$\forall x \in V_{\mathcal{N}} : \ E_{\phi,\mathcal{N}} = E_{\phi,\mathcal{N}} \cup \bigcup_{s=1}^{n} \{[x, ns+x]\} \ .$$

The colouring $\pi_{\phi,\mathcal{N}}$ is used to distinguish between vertices that "represent" different memory layers. As in (2), the input and output nodes are assigned different colours

$$\begin{aligned}\pi_{\phi,\mathcal{N}} = (&\{0\}, \{1\}, \ldots, \{d-1\}, \{d, d+1, \ldots, n-m-1\},\\ &\{n-m\}, \{n-m+1\}, \ldots, \{n-1\},\\ &\{n\}, \{n+1\}, \ldots, \{n+d-1\}, \{n+d, n+d+1, \ldots, 2n-m-1\},\\ &\{2n-m\}, \{2n-m+1\}, \ldots, \{2n-1\},\\ &\{2n\}, \{2n+1\}, \ldots, \{2n+d-1\}, \{2n+d, d+1, \ldots, 3n-m-1\},\\ &\{3n-m\}, \{3n-m+1\}, \ldots, \{3n-1\}, \ldots,\\ &\{Sn\}, \{Sn+1\}, \ldots, \{Sn+d-1\}, \{Sn+d, d+1, \ldots, (S+1)n-m-1\},\\ &\{(S+1)n-m\}, \{(S+1)n-m+1\}, \ldots, \{(S+1)n-1\}).\end{aligned} \quad (6)$$

4 Frequent and Rare Structures

Having defined isomorphic networks, a simple example shall give an idea about the number of equivalent NNs. Consider a feed-forward NN with d input nodes, l hidden nodes in one hidden layer, i.e., they are only connected to the inputs and the output, and 1 output node – a d-l-1 NN. The connections between the hidden nodes and the output node are fixed. For the sake of simplicity, we allow hidden neurons to have no connection to the input. In an encoding with a connection matrix, there is a field of $d \cdot l$ bits specifying the connections between input and hidden neurons. Thus, there are $G = 2^{d \cdot l}$ different matrices in the genotype space. The following reasoning gives the number of different structures. For each hidden neuron the connection to the input neurons is given by a bitstring of size d, yielding $T = 2^d$ different possible types for a hidden node. For a d-l-1 NN the multi-set of the types of the hidden nodes, $\{t_1, \cdots, t_l\}$, with $t_i \in \{0, \cdots, T-1\}$ specifies a class of equivalent NNs. Combinatorics gives an expression for the number of these different sets, in our case for the number of not-equivalent NNs

$$S = \binom{T+l-1}{l} \ .$$

We can define the average number of matrices per NN class as: $\bar{R} = G/S$. For example, $\bar{R} = 67.6326$ for a 4-5-1 NN.

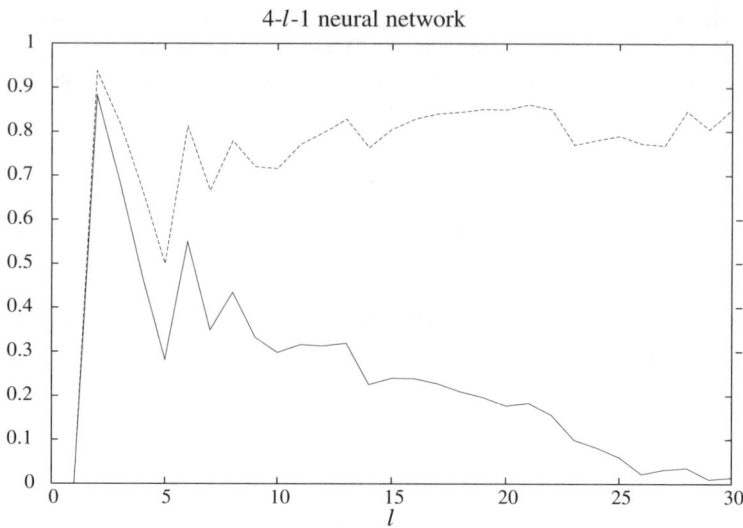

Figure 2. Fraction of *frequent* structures *(solid line)* and of matrices coding for these structures *(dashed line)* as the number of the hidden neurons increases

There are equivalent NNs for which a large number of matrices code and others for which only few matrices or even just one matrix codes. To get an impression of how often equivalent NNs are represented in the genotype space, let us define *frequent* and *rare structures* [7].

Definition 1. An equivalence class (a structure) is frequent iff the number of matrices generating a network of this class is above average $\bar{R}$. Otherwise it is rare.

For an increasing number of hidden nodes l, the following observation is made. The fraction of frequent structures decreases, whereas the fraction of matrices that code for a frequent structure increases, see figure 2. That is, the majority of genotypes – matrices – codes for a minority of structures – equivalent NNs. Similar considerations may hold for other structure optimization problems, too. An encoding that is based on equivalent classes can circumvent this artefact.

5 Experiments

5.1 Evolutionary Algorithm

For the simulations presented here we employ a direct encoding and a small number of intuitive operators.

In genotype space an individual consists of a connection matrix; additionally, each neuron belongs to a layer. The NNs are not allowed to have more than a maximum number of layers. All feed-forward connections including all short-cuts are allowed as long as the layer constraint is met.

One of the used operators is to split a layer into two layers; this does not change the topology of the network but it allows new connections that were not allowed before. There is also an operator that merges two successive layers. This may delete connections between neurons in the originally separated layers. We also use an operator that adds or removes a connection. Adding a new connection increases the capabilities of the NN but might make learning more difficult and can lead to over-fitting. Deleting a connection decreases the capabilities of the NN but can benefit the learning process. Thus, the effect of these operators is not easily assessed. Furthermore, there is a node insertion operator. The new node will get 2 inputs and 1 output. The output is directed into the output layer, the input is received from two randomly chosen preceding nodes. Finally, there is a node deletion operator. As this causes a great structural change, those nodes which received input from the deleted node will get new connections from those nodes that provided the input to the deleted node. So, this operator possibly introduces a lot of new connections by deleting a node. We ensure that the operators only produce feasible networks.

One effect of our coding obviously is the restriction in the number of layers. This is not only a topological constraint, which relies on the idea that layers are meaningful to the adaptation process of NNs, but it also reduces the search space. As there are many more NNs with many layers we also reduce in this way a possible bias towards deep structures: assume that a connection is set with a probability p, then the probability that n subsequent neurons belong to one layer is

$$P(n) = (1 - p)^{\frac{n(n-1)}{2}} . \tag{7}$$

This shows that large layers become very rare which in turn leads to deep structures.

The used EA consists of a parent population with 15 individuals which produce an offspring population with 90 individuals, by randomly selecting one parental genotype and applying one of the introduced operators drawn according to different probabilities until a topological change occurs. Thereafter, each offspring is evaluated, i.e., the NN is trained. If the NN has been trained before during the course of running the EA, the formerly computed fitness will be used. The parent population for the next generation is determined in the following way: out of the old parents and the offspring the very best individual is placed in the new parental population, i.e., we keep one elitist. The other 14 individuals are deterministically selected from the offspring population according to their rank.

5.2 Test Problem

The example data set called *diabetes1* is taken from the PROBEN1 [14] benchmark collection. The goal of this real-world classification task is to decide whether a Pima Indian individual is diabetes positive or not. There are eight inputs and a 1-of-2 encoding is used for the output. The size of the training, validation, and test set is

384, 192, and 192, respectively. Some of the patterns are disturbed by noise; they contain senseless zero entries.

Following [14], the quality of the evolved networks was determined by the error percentage[2]. The resilient backpropagation (RPROP) algorithm [16] was used to train the nets. The RPROP parameters were $\eta^+ = 1.1$, $\eta^- = 0.5$, $\Delta_0 = 0.01$, $\Delta_{\min} = 0$ and $\Delta_{\max} = 50.0$. The initial weights were chosen randomly between -0.5 and 0.5. The hidden units used a sigmoidal activation function given by $g(x) = x/(1 + |x|)$ and the output units were linear. All nets were trained for 300 iterations. The weight configuration with the lowest error on the validation set was stored. In this experiment, we wanted to force adaptation to the test set, so the error of the network on the test set determined the fitness[3]. Of course, normally the test set must not be used in the model adaptation process.

In [14], 12 different standard network architectures and different initialisations were tested. The networks had one or two hidden layers and up to 32 hidden units. Each architecture was tested with sigmoidal and linear output units. The best network (using the *diabetes1* data set) found had a test error of 16.47 and a classification error of 25%. The same problem was used as a benchmark in [12] for 23 different classification techniques, where the best algorithm achieved a classification error of 22.3%.

5.3 Graph Database

During the evolutionary process each generated network was canonically labeled and this canonically labeled graph was stored in a database. In advance, it was tested whether this network or an isomorphic one was evaluated before by looking up the canonically labeled graph in the database. Such a database can be used to reduce the number of computational expensive fitness evaluations. This idea was first mentioned in the work of Christoph M. Friedrich [4,5] but no experiments were performed.

5.4 Results

We conducted two runs of the algorithm described in section 5.1 on the diabetes problem. A maximum of 4 hidden layers was allowed.

The first run found a network with 49 connections and a classification error on the test set of 19.27% in generation 43. The error on the test set dropped to 15.40 in generation 200. The second run evolved a network with 49 connections and a

---

[2] The error percentage is defined as $E = \frac{100}{N \cdot d \cdot (t_{\max} - t_{\min})^2} \sum_{n=1}^{N} (o_n - t_n)^2$, t: target, o: net output, N: size of training data set, $t_{\min}$ ($t_{\max}$): minimum (maximum) target value.

[3] This was done in order to show how different statistics of test and validation data influence the topology optimization process, see also section 5.4.

minimum test error of 15.22 in generation 95. The classification error of this network on the test set was 19.79%. We performed a single additional run with a slightly changed selection scheme and found a net with a classification error of 18.75% and 51 connections. The results cannot be directly compared to those in the literature [12,14] as we used the test data set to calculate the fitness. So, we do not get a true generalization error. Moreover, training a standard architecture on the training data and monitoring validation and test error reveals that these data sets contain quite different statistics. Thus, including the test error in the fitness value may lead to an overfitting of the network structure to that data set. Nevertheless, the experiments demonstrate that the proposed method is an efficient and robust method to determine NN topologies.

In the runs of the algorithm outlined in section 5.1 the average number of isomorphic networks detected per generation was 2.22 and 2.75, respectively, i.e., more than 2.75% of the evaluations could be saved.

6 The Competing Conventions Problem

One point that has been addressed by several researchers in the field of structure optimization of NNs is the *competing conventions* problem [18]. That is, many genotypes (conventions) map to equivalent structures, i.e., the corresponding network topologies are isomorphic. It was argued that this has a harmful effect on recombination, "Crossover among parents utilizing the same convention are likely to be successful while crossovers between dissimilar conventions are unlikely to be effective" [18]. Our results show that there are not so many isomorphic nets present in one generation as might be expected (for the population size and selection mechanism used in our investigation). So, the probability of selecting two competing conventions of one net for recombination is rather small. Further, if only canonically labeled networks are stored in the population, there is exactly one representation of each network and the competing conventions problem is solved (at least if the weights are not encoded into the genotype). This requires that each genotype which is created by the search operators is replaced by the genotype belonging to the canonical representative of the network structure.

However, we do not believe that crossover on the genotype level is an important operator for the evolution of NNs using a direct encoding, therefore it has been omitted in our algorithm. The isomorphism-free representation ensures a single "convention", but does not necessarily lead to something like a "homologous" crossover between NNs.

7 Discussion

In structure optimization problems the space of structures is given by the problem, but for an EA the genotype space and the search operators are chosen according to design criteria for successful evolutionary search processes. This can lead to a genotype space that is much larger than the number of different structures. In these cases it seems very reasonable to organize the fitness evaluation process in an isomorphism-free space. We exemplified this procedure in a simple but realistic problem. The chosen setup did not try to enforce isomorphic network structures. The number of isomorphic structures increases if one reduces the number of allowed hidden layers and if one forbids short-cut connections. Additionally, different choices of the selection procedure can possibly lead to a more localized search yielding a higher rate of already evaluated network structures. Furthermore, our results show that NN structures which are adapted to the problem can lead to much better results than those standard structures achieve.

Keeping an elitist with its fitness in an EA with a noisy fitness function – in our case the learning process is noisy due to the random initialization – might lead to a good individual with an exceptionally good learning result blocking further developments. In this case, one can train the NN several times, possibly in the next generations, until there is a reliable statistic. This, and further refinements concerning the selection process on noisy fitness functions [22], work very well together with the proposed method of using canonical labeled graphs.

Summarizing our investigation, we have shown that:

- The phenomena of rare and frequent structures exist in technical real-world applications.
- Smart techniques to deal with isomorphic structures can be beneficial in evolutionary structure optimization.
- The labelled, coloured graph isomorphism question can be reduced to a coloured graph isomorphism problem.
- The competing conventions problem can be circumvented in the field of NN topology adaptation.

Acknowledgments

C. Igel thanks C. M. Friedrich for fruitful discussions and acknowledges support from the BMBF under grant LEONET 01IB802A9.

References

1. C. M. Bishop. *Neural Networks for Pattern Recognition*. Oxford University Press, Oxford, 1995.
2. H. Braun. *Neuronale Netze: Optimierung durch Lernen und Evolution*. Springer-Verlag, Berlin Heidelberg New York, 1997.
3. J. Elman. Finding structure in time. *Cognitive Science*, 14:179–211, 1990.

4. C. M. Friedrich. Entwicklung und Analyse Evolutionärer Algorithmen zur Optimierung der Struktur und Parameter von künstlichen Neuronalen Netzen. Master's thesis, Universität Dortmund, Fachbereich Informatik, 1995.
5. C. M. Friedrich and C. Moraga. An evolutionary method to find good building-blocks for architectures of artificial neural networks. In *Sixth International Conference on Information Processing and Management of Uncertainty in Knowledge-Based Systems (IPMU'96)*, pages 951–956, Granada, Spain, 1996.
6. M. R. Garey and D. S. Johnson. *Computers and Intractability, A Guide to the Theory of NP-Completeness*. W. H. Freeman, New York, 1979.
7. W. Grüner, R. Giegerich, D. Strothmann, C. Reidys, J. Weber, I. Hofacker, P. Stadler, and P. Schuster. Analysis of RNA sequence structure maps by exhaustive enumeration II. Structures of neutral networks and shape space covering. *Monatshefte für Chemie/Chemical Monthly*, 127:375–389, 1996.
8. S. Haykin. *Neural Networks: A Comprehensive Foundation*, 2nd edition. Prentice Hall, Englewood Cliffs, NJ, 1998.
9. K. Hornik, M. Stinchcombe, and H. White. Multilayer feedforward networks are universal approximators. *Neural Networks*, 2:359–366, 1989.
10. B. D. McKay. Practical graph isomorphism. *Congressus Numerantium*, 30(30):47–87, 1981.
11. B. D. McKay. nauty user's guide (version 1.5). Technical Report TR-CS-90-02, Australian National University, Computer Science Department, 1990.
12. D. Michie, D. J. Spiegelhalter, and C. C. Taylor, editors. *Machine Learning, Neural and Statistical Classification*. Ellis Horwood, Chichester, 1994.
13. G. Miller and P. Todd. Designing neural networks using genetic algorithms. In J. Schaffer, editor, *Proceedings of the 3rd International Conference on Genetic Algorithms and their Applications*, pages 379–384. Morgan Kaufmann, San Francisco, 1989.
14. L. Prechelt. PROBEN1 — A set of benchmarks and benchmarking rules for neural network training algorithms. Technical Report 21/94, Fakultät für Informatik, Universität Karlsruhe, 1994.
15. I. Rechenberg. *Evolutionsstrategie '94*. Friedrich Frommann Holzboog, Stuttgart, 1994.
16. M. Riedmiller and H. Braun. A direct adaptive method for faster backpropagation learning: The RPROP algorithm. In *Proceedings of the IEEE International Conference on Neural Networks*. IEEE Press, New York, 1993.
17. D. E. Rummelhart, G. E. Hinton, and R. J. Williams. Learning internal representations by error backpropagation. In D. E. Rummelhart, J. L. McClelland, and the PDP Research Group, editors, *Parallel Distributed Processing: Explorations in the Microstructure of Cognition*, volume 1, pages 318–362. MIT Press, Cambridge, MA, 1986.
18. J. D. Schaffer, D. Whitley, and L. J. Eshelman. Combinations of genetic algorithms and neural networks: a survey of the state of the art. In L. D. Whitly and J. D. Schaffer, editors, *International Workshop on Combinations of Genetic Algorithms and Neural Networks – COGANN'92*, pages 1–3. IEEE Press, New York, 1992.
19. P. Schuster and W. Fontana. Continuity in evolution: On the nature of transitions. *Science*, 280:1451-1455, 1998.
20. B. Sendhoff. *Evolution of Structures*. PhD thesis, Institut für Neuroinformatik, Ruhr–Universität Bochum, 1998.
21. B. Sendhoff, M. Kreutz, and W. von Seelen. A condition for the genotype–phenotype mapping: Causality. In T. Bäck, editor, *Proceedings of the 7th International Conference on Genetic Algorithms*, pages 73–80. Morgan Kaufmann, San Francisco, 1997.

22. P. Stagge. Averaging efficiently in the presence of noise. In A. Eiben et al., editors, *Parallel Problem Solving from Nature – PPSN V*, volume 1498 of Lecture Notes in Computer Science, pages 188–197. Springer-Verlag, Berlin Heidelberg New York, 1998.
23. P. Stagge and B. Sendhoff. An extended Elman net for modeling time series. In W. Gerstner et al., editors, *Proceedings of the 7th International Conference on Artificial Neural Networks – ICANN'97*, volume 1327 of Lecture Notes in Computer Science, pages 427–432. Springer-Verlag, Berlin Heidelberg New York, 1997.
24. P. Stagge and B. Sendhoff. Organisation of past states in recurrent neural networks: implicit embedding. In M. Mohammadian, editor, *Computational Intelligence for Modelling, Control and Automation*, pages 21–27. IOS Press, Amsterdam, 1999.
25. X. Yao. Evolving artificial neural networks. In *Proceedings of the IEEE*, 87(9):1423–1447, 1999.

Detecting Spin-Flip Symmetry in Optimization Problems

C. Van Hoyweghen

Departement Wiskunde–Informatica
Universiteit Antwerpen (RUCA)
Groenenborgerlaan 171
B-2020 Antwerpen, Belgium
E-mail: *hoyweghe@ruca.ua.ac.be*

Abstract. The presence of symmetry in the representation of an optimization problem can cause the search algorithm to get stuck before reaching the global optimum. The traveling salesman problem and the graph coloring problem are some well-known NP-complete optimization problems containing symmetry in their usual representation. This paper describes a class of symmetry, called spin-flip symmetry, which one finds in functions like the one-dimensional nearest neighbor interaction functions. Spin-flip symmetry indicates that bit-complementary strings have the same function value. This notion can be generalized to substrings, called spin-flip blocks, in a canonical way. We distinguish two specific cases of spin-flip symmetry and introduce a spin-flip detection algorithm. The performance of the algorithm strongly depends on the initial sample the algorithm uses to detect the spin-flip blocks. The algorithm is designed to detect spin-flip symmetry in the whole search space, as well as in its hyperplanes. The difficulty with detection in hyperplanes is related to the detection of the correct hyperplane, which can be time consuming. Once the desired hyperplane is found, the spin-flip block can be detected using the normal detection procedure. The one-max problem shows that spin-flip symmetry in other types of subspaces, like the subspace in which the number of 1s equals the number of 0s, is not detected. For these kinds of subspaces the method needs to be extended.

Keywords
Symmetry, spin-flip block, nearest neighbor interactions

1 Introduction

In the context of black-box optimization using heuristic techniques, the representation of the search space is a determining factor of the performance of the search technique used, be it a hill-climber, simulated annealing, a sophisticated genetic algorithm or an evolution strategy. Bearing the *no free lunch* theorem [8] in mind, an important question can be asked; which heuristic technique is the best for a certain class of optimization problems? One way to solve this question is to look for characteristics of the representation and determine the algorithms which perform well on functions exhibiting these characteristics. The presence of symmetry in the representation is a factor which can cause the search algorithm to get stuck before reaching the global optimum. However, it can also be beneficial [1]. Therefore, symmetry is worth studying.

Next to symmetry, the interactions between several string positions in the representation, also called epistasis [2], is an important characteristic determining the behavior of an iterative black-box optimization algorithm. The *one-dimensional nearest neighbor interaction functions* (*NNIs*) contain a minimal amount of second-order interaction. We can easily write them in the form

$$f : \{0, 1\}^\ell \to \mathbb{R} : s \to \sum_{i=0}^{\ell-1} g_i(1 - s_i)s_{i+1} + \sum_{i=0}^{\ell-1} (k_i s_i + l_i(1 - s_i))$$

where ℓ is the string length, $s \in \{0, 1\}^\ell$, $s_0 \equiv s_\ell$, and g_i, k_i and l_i are non-negative integers. *NNIs* have their origin in statistical physics. They can easily be transformed to instances of the *generalized Ising model* [7]

$$H = - \sum_{0 \leq i < j < \ell} J_{ij} s_i s_j - \sum_{i=0}^{\ell-1} h_i s_i \qquad (1)$$

with $s \in \{-1, 1\}^\ell$, $s_0 \equiv s_\ell$, J_{ij} and h_i integers, and for *NNIs* $J_{ij} = 0$ when $i+1 \neq j$. The optima of an arbitrary *NNI* can be found by a deterministic algorithm in linear time, but several optimization heuristics stay a long time in suboptimal parts of the search space.

The presence of symmetry in a fitness function influences the dynamics of an optimization algorithm. For example, the existence of spin-flip symmetry (described in section 3) in a problem allows for strings with an objective value close to that of the two optima to be in Hamming distance far from the optima. This prevents simulated annealing, and typically also a genetic algorithm (GA) with uniform crossover, to get to the optima quickly. We shall see that some *NNIs* contain spin-flip symmetry.

This paper describes an algorithm which tries to detect spin-flip symmetry in the representation of an arbitrary black-box optimization problem. Later on we will construct a method which can break the detected symmetry in case it is harmful. In this context *NNIs* are useful test functions to get started.

2 Problems Containing Symmetry in Their Representation

In this section we consider some common NP-complete problems [4] which contain symmetry in their representation. For now, we distinguish two types of problems. One type are problems containing symmetry on their alphabet, for which there exists a permutation on the alphabet which leaves the objective value of a string unchanged, and the other, problems containing symmetry on their string positions, for which there exists a permutation on the string positions, e.g., a rotation, which leaves the objective value of a string unchanged.

2.1 Symmetry on the Alphabet

The Graph Coloring Problem. Given a graph G and k colors, the graph coloring problem is to assign all vertices on the graph a color from the set of k colors under the restriction that no pair of connected vertices have the same color. An alternative problem is to find the least number of colors as possible for which a coloring without clashes exists. A possible representation for this problem is to number the vertices of the graph and assign to the i-th string position the color j, indicating that vertex i has color j.

The order of the k colors we use in the representation is not relevant. Without loss of generality, assume we have a solution represented by the string $(c_1, c_2, c_3, c_2, c_2)$ for some graph G, with c_1, c_2, c_3 three different colors. This solution is equivalent to the solutions

$(c_2, c_3, c_1, c_3, c_3)$
$(c_3, c_1, c_2, c_1, c_1)$
$(c_1, c_3, c_2, c_3, c_3)$
$(c_2, c_1, c_3, c_1, c_1)$
$(c_3, c_2, c_1, c_2, c_2)$.

In general, any possible permutation on the k colors leaves the objective value of a string unchanged, which makes them equivalent solutions. The symmetry on the alphabet of k colors can cause a search algorithm to get trapped in a local area of the search space without improving its solutions anymore. A new GA crossover operator to cope with this symmetry has been previously published [6] and is discussed by Marino in this volume (page 387).

The Graph Partitioning Problem. Given a set of an even number of vertices $V = \{v_0, v_1, \ldots, v_{N-1}\}$, and a set of edges E, the graph partitioning problem is to partition the N vertices into two sets V_1 and V_2 of equal size such that the number of edges joining V_1 and V_2 is minimized. This problem can be easily transformed into one of statistical mechanics [3]. We define the Hamiltonian for this problem as follows. Associate with each vertex v_i an Ising spin s_i and set $s_i = -1$ when v_i belongs to V_1 and $s_i = 1$ if v_i belongs to V_2. Since the two sets must have equal size, $\sum_{i=0}^{N-1} s_i$ must be zero. For each pair of vertices (v_i, v_j) we introduce a coupling J_{ij} which is a positive integer J if the pair is connected and 0 if it is not. The problem is now translated into minimizing the Hamiltonian

$$H = -\sum_{i<j} J_{ij} s_i s_j \qquad (2)$$

under the constraint of $\sum_{i=0}^{N-1} s_i = 0$. It is easy to see that this representation of the problem contains symmetry on the alphabet $\{-1, 1\}$. Swapping the two sets V_1 and V_2 does not change the objective value.

The Random Number Partitioning Problem. Given a set of N positive integers $a_0, a_1, \ldots, a_{N-1}$, with N an even number, the problem is to partition the integers into two sets G_1 and G_2 of equal size, such that the cost $C = |\sum_{i \in G_1} a_i - \sum_{i \in G_2} a_i|$ is minimized. Again, this problem can be transformed in a Hamiltonian to be minimized [3]. With each number a_i we associate an Ising spin s_i with value -1 when a_i belongs to G_1, and value 1 if a_i belongs to G_2. To keep the size of G_1 and G_2 equal, we require $\sum_{i=0}^{N-1} s_i = 0$. The problem can then be translated into minimizing

$$H = \frac{1}{2} \sum_{i,j=0}^{N-1} a_i a_j s_i s_j$$

under the constraint of $\sum_{i=0}^{N-1} s_i = 0$. Just like the problem above, this problem contains symmetry on the alphabet $\{-1, 1\}$ because of the exchangeability of G_1 and G_2.

2.2 Symmetry on the String Positions

The Traveling Salesman Problem. Given N cities and a non-negative integer distance d_{ij} between any two cities i and j ($d_{ij} = d_{ji}$), the traveling salesman problem is to find the shortest tour visiting all cities exactly once. A possible representation of a candidate solution is a string $s_0 s_1 \ldots s_{N-1}$ where s_0 has as value the city we visit first, s_1 has as value the city we visit next, etc. Of course, the city in which we start our tour is not important. Only the sequence of our tour influences the objective value. Therefore, the traveling salesman problem is a simple example of a problem containing symmetry on the string positions. The symmetry can easily been broken by always starting from the same city.

In the rest of the paper we will only concentrate on problems containing symmetry on the alphabet, in particular alphabets of size 2, which we will call spin-flip symmetry.

3 Spin-Flip Symmetry

The symmetry we try to detect in this paper is called spin-flip symmetry. A function f contains spin-flip symmetry on spin-flip block $(p_0, \ldots, p_n)$, if each string in the search space has an objective value that equals the objective value of the string constructed by taking the bit-complementary value for the string positions in $(p_0, \ldots, p_n)$ and keeping the value for the other string positions.

Formally, let P be a binary optimization problem with fitness function f and string length ℓ, $f : \{0, 1\}^\ell \to \mathbb{R}$. A spin-flip block $(p_0, \ldots, p_n)$ is defined by the function

$$(p_0, \ldots, p_n) : \{0, 1\}^\ell \to \{0, 1\}^\ell : s \to \bar{s}$$

with $\bar{s} = \bar{s}_0 \ldots \bar{s}_{\ell-1}$ and

$$\bar{s}_i = \begin{cases} 1 - s_i & \text{for all } i \in \{p_0, \ldots, p_n\} \\ s_i & \text{otherwise.} \end{cases}$$

We now say that problem P (or its fitness function f) contains spin-flip symmetry on spin-flip block $(p_0, \ldots, p_n)$, with $n \leq \ell - 1$ iff

$$\forall s \in \{0, 1\}^\ell : f(s) = f((p_0, \ldots, p_n)(s)).$$

Having defined spin-flip symmetry, we distinguish two different types of spin-flip blocks. A function contains a type A spin-flip block if it also contains every sub-spin-flip block. A function contains a type B spin-flip block if it does not contain sub-spin-flip blocks whose union is the spin-flip block itself. Let us now define these two concepts more formally.

Given a spin-flip block $(p_0, \ldots, p_n)$, we define the spin-flip block $(p_{i_0}, \ldots, p_{i_k})$ as a sub-spin-flip block of $(p_0, \ldots, p_n)$ iff $\{p_{i_0}, \ldots, p_{i_k}\} \subset \{p_0, \ldots, p_n\}$. A problem P with fitness function f contains a spin-flip block $(p_0, \ldots, p_n)$ of type A iff

$$\forall \{p_{i_0}, \ldots, p_{i_k}\} \subseteq \{p_0, \ldots, p_n\} : \forall s \in \{0, 1\}^\ell : f(s) = f((p_{i_0}, \ldots, p_{i_k})(s)).$$

A problem P with fitness function f contains a spin-flip block $(p_0, \ldots, p_n)$ of type B iff

$$\forall s \in \{0, 1\}^\ell : f(s) = f((p_0, \ldots, p_n)(s))$$

and

$$\nexists (p_{i_{0_0}}, \ldots, p_{i_{k_0}}), \ldots, (p_{i_{0_n}}, \ldots, p_{i_{k_n}}) \text{ sub-spin-flip blocks of } (p_0, \ldots, p_n)$$

such that

$$\bigcup_{j=0}^{n} \{p_{i_{0_j}}, \ldots, p_{i_{k_j}}\} = \{p_0, \ldots, p_n\}$$

and

$$\forall 0 \leq j \leq n : f(s) = f((p_{i_{0_j}}, \ldots, p_{i_{k_j}})(s)).$$

Obviously, a spin-flip block of type A cannot be a spin-flip block of type B within the same problem, and vice versa.

The distinction between these two types of spin-flip symmetry is only necessary to evaluate the performance of the spin-flip detection algorithm later on. We will see that the detection algorithm easily finds a spin-flip block of type A, since it can construct the spin-flip block by combining some of its detected sub-spin-flip blocks. On the other hand, a spin-flip block of type B is hard to detect using random samples.

3.1 Subspaces

The aim of this research is to detect symmetry in some interesting industrial problems. We assume that most of the symmetry in those problems is situated in subspaces of the search space, instead of in the whole search space. Therefore, we have to extend the definitions of spin-flip symmetry to subspaces. A function f contains a spin-flip block $(p_0, \ldots, p_n)$ in a subspace S, if each string in S has an objective value that equals the objective value of the string constructed by taking the bit-complementary value for the string positions in $(p_0, \ldots, p_n)$ and keeping the value for the other string positions. Of course, all the newly constructed strings have to belong to the subspace S.

The spin-flip detection algorithm we introduce in section 5 is designed to detect spin-flip symmetry in the whole search space as well as in its hyperplanes. An example of a hyperplane is 1#0#...#. This hyperplane characterizes the subspace containing all strings having a 1 on their first string position and a 0 on their third. We will see that for other types of subspaces the algorithm needs to be extended.

4 Test Functions Containing Spin-Flip Symmetry

4.1 One-Dimensional Nearest Neighbor Interaction Functions

In section 1 we introduced the one-dimensional nearest neighbor interaction functions (*NNIs*) as interesting functions for detecting spin-flip symmetry. Now we know the exact definition of this type of symmetry, we will show the presence of spin-flip symmetry in an *NNI* in more detail.

Example 1. If we ignore the first-order component and set all g_i to the same positive value g, we get *NNIs* of the form

$$f : s \rightarrow \sum_{i=0}^{\ell-1} g(1 - s_i)s_{i+1}. \tag{3}$$

Note the similarity between (2) and (3). The *NNIs* in (3) have the maxima 1010... and 0101... and it is clear that they contain the type B spin-flip block $(0, \ldots, \ell-1)$.

Example 2. We can also construct *NNIs* containing a spin-flip block of type A. Consider, for example, arbitrary *NNIs* satisfying following constraints:

$$\forall j \leq i \leq j + n : l_i = k_i,$$
$$\forall j - 1 \leq i \leq j + n : g_i = 0,$$

with $0 \leq j, j + 1, \ldots, j + n \leq \ell - 1$. They contain the type A spin-flip block $(j, j+1, \ldots, j+n)$.

4.2 Spin-Flip Symmetry in Subspaces

We also need some test functions containing spin-flip symmetry in a subspace. Therefore, we construct a new type of function which is composed of two sub-functions, f_1 and f_2, and where the value of the last string position determines which sub-function will be used. If the value is 1 function f_1 will be used, if the value is 0 function f_2 will be used.

Example 3. NNIorSum1 function:

- NNI on the first $\ell - 1$ sites: $f_1 : \{0, 1\}^{\ell-1} \to \mathbb{R} : s \to \sum_{i=0}^{\ell-2}(1 - s_i)s_{i+1}$
- sum function on the first $\ell - 1$ sites: $f_2 : \{0, 1\}^{\ell-1} \to \mathbb{R} : s \to \sum_{i=0}^{\ell-2} \frac{s_i}{2}$

NNIorSum function:

$$f : \{0, 1\}^\ell \to \mathbb{R} : s \to \begin{cases} f_1(s_0, \ldots, s_{l-2}), & \text{if } s_{\ell-1} = 1 \\ f_2(s_0, \ldots, s_{l-2}), & \text{if } s_{\ell-1} = 0. \end{cases}$$

The value of the last site determines whether the *NNI* or the sum function will be used. The *NNIorSum1* function has the optima 10100, 01010 and 11111 for string length $\ell = 5$ and contains type B spin-flip block $(0, \ldots, 3)$ in hyperplane ####1.

Example 4. If we exchange f_1 by the function described in example 2 we create a new function, *NNIorSum2*, which contains the type A spin-flip block $(j, j + 1, \ldots, j + n)$ in hyperplane #...#1.

5 A Spin-Flip Detection Algorithm

The spin-flip detection algorithm presented below is designed to detect spin-flip blocks in an arbitrary unknown function. The algorithm looks for spin-flip blocks in the whole search space as well as in its hyperplanes. Figure 1 shows an overview of the algorithm. The algorithm first detects spin-flip blocks in the whole search space. Afterwards, it constructs some interesting hyperplanes and looks for spin-flip blocks in these hyperplanes.

The most important part of the algorithm is the detector, depicted in figure 2. The detector detects spin-flip blocks in a specific space. The detected spin-flip blocks of a space can be combined to form larger spin-flip blocks of that space. This helps to find spin-flip blocks of type A.

5.1 The Detector

The detector's input is a sample of the space in which it looks for spin-flip blocks. The strings of this sample are evaluated and classified by their fitness value. In

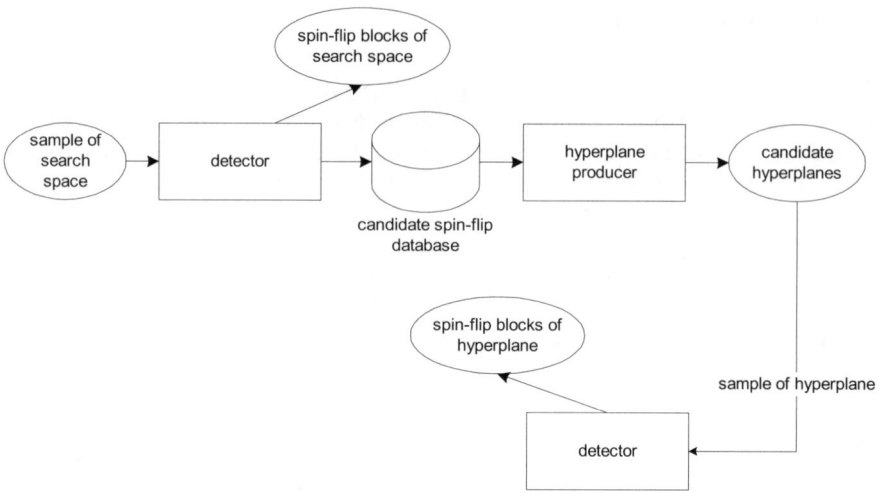

Figure 1. Overview of the spin-flip detection algorithm

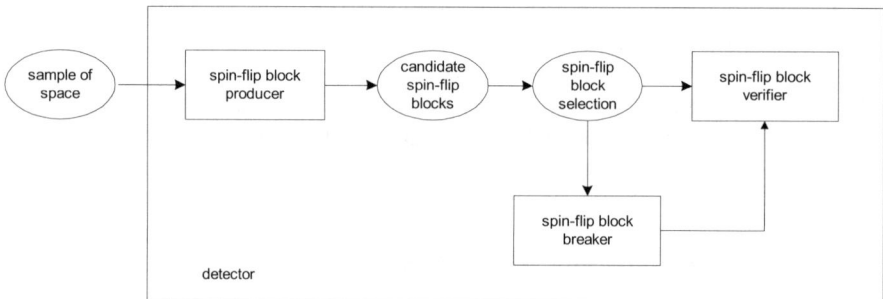

Figure 2. Schematic representation of the detector

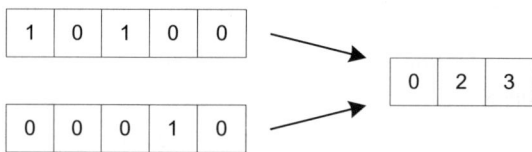

Figure 3. Example of the production of a candidate spin-flip block. Two strings of length 5 produce a spin-flip of size 3.

each of these classes, every pair of strings is compared to produce candidate spin-flip blocks. A pair produces a candidate spin-flip block $(p_0, \ldots, p_n)$ if they have bit-complementary values on the string positions in this block. Figure 3 gives an example of the production of a spin-flip block.

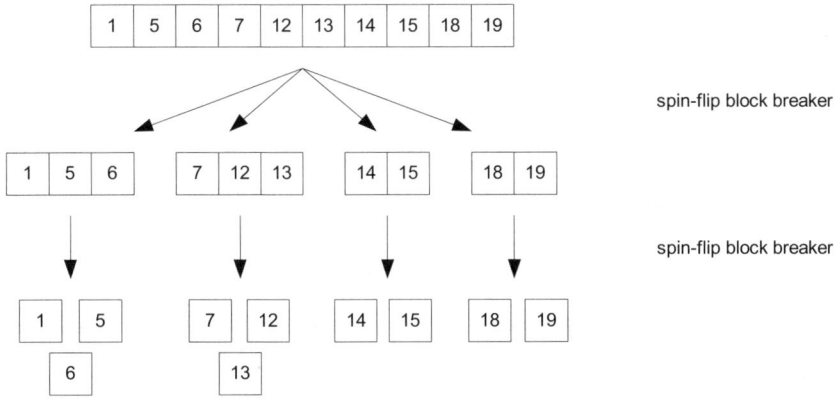

Figure 4. Example of a spin-flip block breaker which breaks a spin-flip block in $\frac{\ell}{10} + 2$ pieces, if possible. We assume the string length $\ell = 20$, so a block is broken into a maximum of $\frac{20}{10} + 2 = 4$ pieces.

Having produced all the candidate spin-flip blocks, we assign to each block the number of times it was produced and select 200 of them. Priority is given to candidate blocks with a higher occurrence and a larger size. The selected candidate blocks are fed to the spin-flip block verifier which checks if the candidate spin-flip blocks are real spin-flip blocks of the function or not. To do this, the verifier takes 20 random strings of the space and checks for every selected candidate spin-flip block if it is valid on these randomly generated strings. A spin-flip block is valid on a string if the string constructed by taking the bit-complementary values for the string positions in the spin-flip block and taking the same values as the original string on the other string positions, has the same fitness value as the original string. Each time the spin-flip block is valid we increase its number with one. If the candidate spin-flip block is valid for all the 20 random strings it is declared to be a real spin-flip block of the function. The number 200 in the selection procedure and the number 20 in the verifier are chosen arbitrary for the first set of experiments.

After checking the selected candidate spin-flip blocks, the selected blocks are broken in pieces to detect spin-flip blocks of type A more easily. The pieces of spin-flip blocks are treated as normal candidate spin-flip blocks. They are tested for real in the spin-flip block verifier and afterwards, if possible, broken in pieces again. Figure 4 gives an example of how the spin-flip breaker works. The spin-flip breaker breaks every spin-flip block into $\frac{\ell}{10} + 2$ pieces of (almost) the same length.

5.2 Detection in Hyperplanes

The difficulty with symmetry detection in hyperplanes is related to the detection of the correct hyperplane. Once the desired hyperplane is found, the spin-flip block can be detected by using the normal detection procedure described above. We cannot say

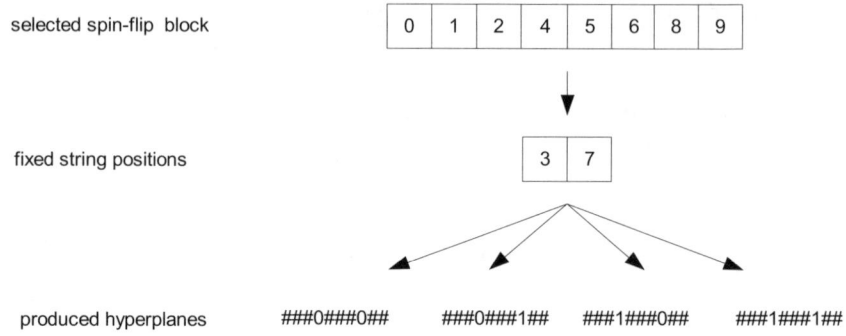

Figure 5. Details of the hyperplane producer for strings of length 10

in advance whether a hyperplane is interesting enough to be investigated for spin-flip blocks or not. Investigating all hyperplanes is also impossible. Therefore, the hyperplane producer has to make a selection.

The hyperplane producer selects a number of candidate spin-flip blocks produced while detecting spin-flip blocks in the whole search space. Hereby it uses the following selection rule. First, multiply the number assigned to a candidate spin-flip block by the length of the block to emphasize the attention to large blocks. Then, priority is given to candidate blocks with a high product and if two blocks have the same product the spin-flip block with the largest size is selected first. Having selected the candidate spin-flip blocks, the hyperplane producer takes for each candidate block the string positions which are not in the block and constructs every possible hyperplanes which fixes these positions to a value. Figure 5 shows an example of the construction of 4 hyperplanes starting from a selected spin-flip block.

Because we cannot investigate all hyperplanes, we are interested in hyperplanes with small sizes. Therefore, it is possible to define a maximum size and forbid the hyperplane producer to construct hyperplanes having a larger size than this maximum size. Once the hyperplanes are created we construct a sample of each of these hyperplanes. These samples are given one at a time as input to the detector which starts looking for spin-flip blocks in these hyperplanes.

6 Experimental Results

All experiments are done using the bIOS optimization framework [5]. To investigate the quality of the spin-flip detection algorithm we separate the test functions into 4 groups:

- **G1:** Functions with type B spin-flip symmetry in the whole search space.
- **G2:** Functions with type A spin-flip symmetry in the whole search space.

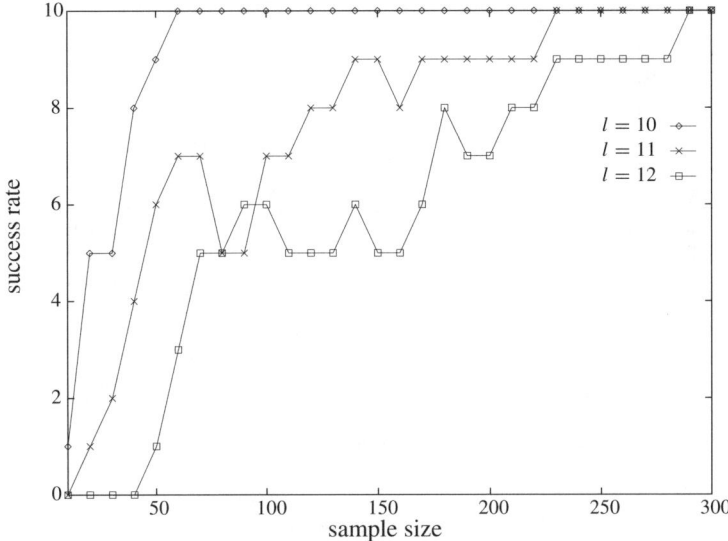

Figure 6. The number of successful runs out of 10 of the spin-flip detection algorithm using a random sample on functions containing spin-flip symmetry of type B in the whole search space

- **G3:** Functions with type B spin-flip symmetry in a subspace.
- **G4:** Functions with type A spin-flip symmetry in a subspace.

In our experiments we use the following test functions. For functions of group G1 we use the *NNI* functions of example 1 with $g = 1$ and vary the string length ℓ. For functions of group G2 we use the functions of example 2 and let the functions for $\ell = 10, 15$ and 20, respectively, contain a spin-flip block of length 3, 5, and 5. For functions of group G3 we use the *NNIorSum1* functions of example 3. For functions of group G4 we use the *NNIorSum2* functions of example 4, in which we use for f_1 the functions of G2.

In our first experiments the detection algorithm uses a random sample. For the functions containing spin-flip symmetry of type B in the whole search space (G1), the sample size is increased with 10 strings at a time. Every experiment is done for 10 different seeds. In figure 6 we see that for string length $\ell = 10$ an optimal result is obtained if the detector gets a random sample of size 60 as input. For $\ell = 11$ we already need a random sample of size 230 to detect every type B spin-flip block. For $\ell = 12$ the smallest random sample which detects every type B spin flip block has size 290. These results show that it is very difficult to find type B spin-flip blocks in a search space with a large string length using a random sample. This is not the case with type A spin-flip blocks (G2). Type A spin-flip blocks can be found more easily by taking the union of a set of detected sub-spin-flip blocks. In figure 7 we see that a random sample of size 10 is sufficient to detect type A spin-flip block in search spaces with string length $\ell = 10$, $\ell = 15$ and even $\ell = 20$.

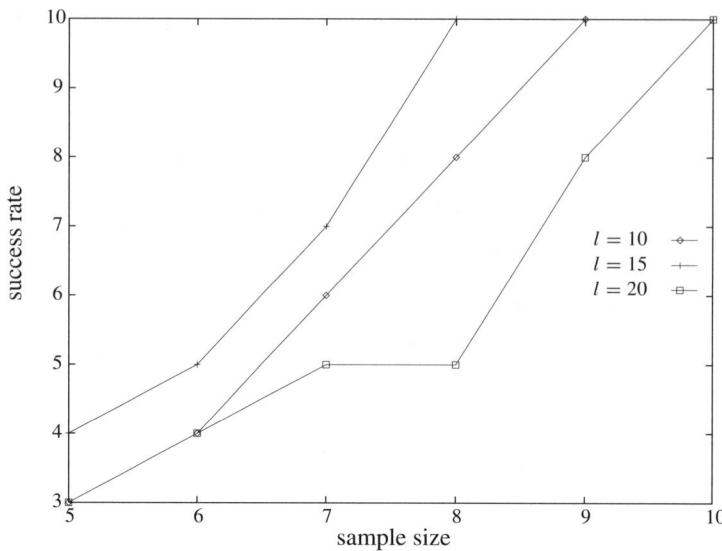

Figure 7. The number of successful runs out of 10 of the spin-flip detection algorithm using a random sample on functions containing spin-flip symmetry of type A in the whole search space

- Add string $00\ldots 0$ to S.
- Add string $11\ldots 1$ to S to detect sequential spin-flip blocks of size ℓ.
- For $1 \leq j \leq \lfloor\frac{\ell}{2}\rfloor$: add ℓ strings to S to detect sequential spin-flip blocks of size j and size $\ell - j$:

$$\underbrace{11\ldots 1}_{j}00\ldots 0$$
$$0\underbrace{11\ldots 1}_{j}0\ldots 0$$
$$\ldots$$

Figure 8. Construction of block sample S of length ℓ

6.1 Block Sample

The very bad result for type B spin-flip blocks makes it necessary to evaluate our detection algorithm with other samples. The best result is found with a prepared sample which we will call the block sample, see figure 8. The block sample is constructed to detect all sequential spin-flip blocks $(i, i+1, \ldots, i+k)$ of type B. Every block sample contains exactly $\lfloor\frac{\ell}{2}\rfloor * \ell + 2$ strings.

To evaluate the detection algorithm using the prepared block sample, we extend the test functions of G1 with the *NNIandSum* functions.

Example 5. Take two positive natural values a and b. a is the begin position of the *NNI* and b the length of the *NNI*. So, $I = \{a, \ldots, a+b-1\}$ is the set of positions used by the *NNI*. The other string positions are used by the sum function:

- NNI: $f_1 : \{0,1\}^\ell \to \mathbb{R} : s \to \sum_{i \in I}(1 - s_i)s_{i+1}$ with $s_{a+b} \equiv s_a$
- sum: $f_2 : \{0,1\}^\ell \to \mathbb{R} : s \to \sum_{i \notin I} s_i$

The *NNIandSum* function is now the composition of the two functions f_1 and f_2. This function contains type B spin-flip block $(a, \ldots, a+b-1)$. By varying the a and b parameters, we manipulate the position and the length of the type B spin-flip block.

Using the block sample, the detection algorithm performed well for all functions of group G1, *NNIandSum* functions included. We also tried the block sample on functions containing type A spin-flip blocks in the whole search space (G1), and every spin-flip block was detected.

6.2 Subspaces

If we use the detection algorithm to find spin-flip blocks in subspaces, we notice again that the results depend strongly on the initial sample the algorithm uses. If it uses a random sample, the algorithm only works well for strings with a small length, for example $\ell = 10$ or less. For larger string lengths we need a tremendously large random sample and a portion of good luck to find the spin-flip block in the search space. The difficulty here is the creation of the right hyperplane, which depends strongly on the the initial sample. The *NNIorSum* functions, for example, contain spin-flip symmetry in hyperplane #...#1. To construct this hyperplane with the hyperplane producer spin-flip block $(0, \ldots, \ell-2)$ has to be a candidate spin-flip block. If we extend the block sample and let j vary between 1 and ℓ instead of 1 and $\lfloor \frac{\ell}{2} \rfloor$ (see figure 8) the spin-flip block $(0, \ldots, \ell-2)$ will be generated and the hyperplane #...#1 will be constructed. The experiments showed that using the extended block sample, the algorithm detects all spin-flip blocks in subspaces of functions in group G3 and G4, with string lengths varying between 10 and 40. Unfortunately, spin-flip blocks in more exotic hyperplanes, like #0###1#1001#1, will not been found.

7 Conclusions and Future Work

This paper stressed the fact that some common NP-complete problems contain symmetry in their representation. We chose one particular class of symmetry, namely spin-flip symmetry, which we tried to detect using a spin-flip detection algorithm. When we know that a function contains symmetry, we can try to break this symmetry in case it is harmful or use this knowledge to find the optima more quickly. The hybrid GA in [6] and discussed by Marino in this volume is a good example of this. It uses a linear assignment algorithm in the crossover operator to cope with the symmetry in the representation of the graph coloring problem. By writing this paper we hope that more research will be done towards symmetry in optimization problems and the consequences it has on optimization techniques. We also would like to find some interesting industrial problems containing some kind of symmetry in them.

The spin-flip detection algorithm we described in this paper is very limited. We only find spin-flip blocks in the whole search space and in some subspaces which can be described by a hyperplane. The one-max problem, for instance, tries to optimize the function $f : s \rightarrow \sum_{i=0}^{\ell-1} s_i$. This function has optimum $1111\ldots 1$ and contains the type B spin-flip block $(0, \ldots \ell-1)$ in the subspace containing all strings in which the number of 1s equals the number of 0s. The hyperplane producer will never construct this subspace and so the spin-flip block in this subspace will never be found. For this and other kinds of subspaces, the algorithm needs to be extended.

Acknowledgments

This research is supported by the Flemish Institute for the Encouragement of Scientific and Technological Research in the Industry–(IWT) (Flanders) (Belgium). The author would like to thank B. Naudts for his support and useful comments on the manuscript.

References

1. Philippe Collard and Jean-Philippe Aurand. DGA: an efficient genetic algorithm. In A. G. Cohn, editor, *ECAI'94: European Conference on Artificial Intelligence*, pages 487–492. John Wiley, New York, 1994.
2. Y. Davidor. Epistasis variance: a viewpoint on GA-hardness. In G. J. E. Rawlins, editor, *Foundations of Genetic Algorithms*, pages 23–35. Morgan Kaufmann, San Francisco, 1991.
3. Yaotian Fu. The use and abuse of statistical mechanics in computational complexity. In D.L. Stein, editor, *Lectures in the Sciences of Complexity, The Proceedings of the 1988 Complex Systems Summer School*, pages 815–826. Santa Fe, NM, 1989.
4. M. R. Garey and D. S. Johnson. *Computers and Intractability, A Guide to the Theory of NP-completeness*. W. H. Freeman, New York, 1979.
5. I. Landrieu and B. Naudts. An object model for search spaces and their transformations. In: *Proceedings of the 2000 Congress on Evolutionary Computation*, pages 811–816. IEEE press, 2000.
6. A. Marino, A. Prügel-Bennett, and C. A. Glass. Improving graph colouring with linear programming and genetic algorithms. In *Proceedings of Eurogen'99*, pages 113–118.

Dept. of Mathematical Information Technology Report, Series A, No. A2/1999, University of Jyväskylä, Finland.
7. B. Naudts and J. Naudts. The effect of spin-flip symmetry on the performance of the simple GA. In A.E. Eiben, T. Bäck, M. Schoenauer, and H.-P. Schwefel, editors, *Proceedings of the 5th Conference on Parallel Problem Solving from Nature*, LNCS 1498, pages 67–76. Springer-Verlag, Berlin Heidelberg New York, 1998.
8. D. H. Wolpert and W. G. Macready. No Free Lunch Theorems For Search. Technical Report, Santa Fe Institute, 1995.

Asymptotic Results for Genetic Algorithms with Applications to Nonlinear Estimation

P. Del Moral and L. Miclo

Laboratoire de Statistiques et Probabilités
CNRS-UMR C5583
Université Paul Sabatier
118, route de Narbonne
31062 Toulouse cedex, France
E-mail: {delmoral,miclo}@cict.fr

Abstract. Genetic algorithms (GAs) are stochastic search methods based on natural evolution processes. They are defined as a system of particles (or individuals) evolving randomly and undergoing adaptation in a time non-necessarily homogeneous environment represented by a collection of fitness functions. The purpose of this work is to study the long-time behavior as well as large population asymptotic of GAs. Another side topic is to discuss the applications of GAs in numerical function analysis, Feynman–Kac formulae approximations, and in nonlinear filtering problems. Several variations and refinements will also be presented including continuous-time and branching particle models with random population size.

Keywords
Branching and interacting particle systems, Feynman–Kac formula, nonlinear filtering, numerical function optimization

1 Introduction

J. H. Holland [37] introduced GAs as a kind of universal and global search method based on natural evolution processes. During the last two decades they have been used as an optimization tool in a variety of research areas; to name a few: machine learning [34], control systems [33], electromagnetics [39,53], economics and finance [42,49], aircraft landing [1,16], topological optimum design [40], and identification of mechanical inclusions [51,52].

More recently, GAs have appeared naturally in the study of Feynman–Kac formulas and nonlinear filtering problems (the reader is recommended to consult the survey paper [20] and references therein). These particle interpretations of Feynman–Kac models have had numerous applications in many nonlinear filtering problems: to name a few; radar signal processing ([28,30]), Global Positioning System ([6,7]), as well as in tracking problems ([41,45,46,35]). Other numerical experiments are also given in [11] and [17].

In contrast to the applications in numerical function analysis, GAs are not used here to approximate the extrema of a given numerical function but a flow of conditional distributions. In addition the genetic structure of the algorithm (such as the mutation and the selection transitions) is not only designed as an instrumental tool to mimic natural evolution, but is, in fact, dictated by the structure of the dynamics of the so-called nonlinear filtering equations.

The main purpose of this article is to introduce the reader to the asymptotic theory of GAs. We also explain the use of these stochastic methods for the numerical solving of nonlinear filtering problems and in numerical function optimization problems. We also give a detailed discussion on several variations and refinements for algorithms recently proposed in the literature on nonlinear estimation problems.

This work is essentially divided into two main parts devoted, respectively, to the applications of GAs for the numerical solving of the so-called nonlinear filtering equations and the convergence of GAs towards the global minima of a given numerical function. Our presentation of this material has relied heavily on the two papers [20] and [21].

In the opening section 2 we introduce the two-step mutation-selection procedure and the time-inhomogeneous Markov model of GAs treated in this work. As mentioned above, this model will then be regarded as a global stochastic search method for studying the set of the global minima of a given numerical function or as a stochastic adaptive grid approximation of a flow of conditional distributions in nonlinear filtering settings.

To each of these applications correspond a specific asymptotic analysis. In section 2 we lay the foundations of the work that follows by explaining the general methodologies needed to study the large population asymptotic and the long-time behavior of the algorithm. In section 2.1 we give an alternative description of the genetic model presented in section 2 in terms of an N-interacting particle system approximating model associated with a measure-valued dynamical system. This formulation enables us to identify the limit of the empirical measures associated with the GA in terms of a Feynman–Kac formula. The modeling impact of this approach will be illustrated in nonlinear filtering in section 3.

Section 2.2 is devoted to the study of the long-time behavior of the genetic model presented in section 2. The idea here is to connect GAs with the so-called generalized simulated annealing. We describe a general methodology to conclude that a GA converges in probability, as time tends to infinity, to the set of global minima of a virtual energy function. We will combine this general convergence result with a natural test set approach in section 4 to prove that the resulting stochastic algorithm converges towards the set of global minima of the desired fitness function as the time parameter tends to infinity and when the population size is sufficiently large.

The GA presented in section 2 and further developed in section 3 and section 4 is the crudest of the evolutionary particle methods. In section 5 we discuss several variations and refinements arising in the literature about nonlinear filtering and generalized simulated annealing.

The final section discusses continuous-time GAs. In contrast to the classical Moran-type genetic model commonly used in GA literature, our interacting particle system model converges to a deterministic distribution flow. This model has been proposed in [23] for solving continuous-time nonlinear filtering problems. Several variants based on an auxiliary time discretization procedure can be found in [11] and [13]. The fundamental difference between the Moran-type particle scheme and the algorithms presented in [11,13] lies in the fact that, in the former, competitive interactions occur at random times. The resulting scheme is therefore a genuine continuous-time particle approximating model.

The interested reader is referred to [20] for a detailed description of the robust and pathwise filter and for a complete proof of the convergence results in a context more general than we have given. Here, we have chosen to restrict our attention to continuous-time Feynman–Kac formulae. We will also discuss the connections between this scheme and the generalized and spatially homogeneous Boltzmann models presented in [36,44].

We end this paper with a novel branching GA in which the size of the population is not necessarily fixed, but random.

Only a selection of existing results is presented here. Deeper information is available in [19–21].

2 Description of the Models and Statement of Some Results

The simplest GA is a two-stage and time-inhomogeneous Markov chain given for each $n \geq 0$ by setting

$$\xi_n \stackrel{\text{def.}}{=} \left(\xi_n^1, \ldots, \xi_n^N\right) \xrightarrow{\text{Selection}} \widehat{\xi}_n \stackrel{\text{def.}}{=} \left(\widehat{\xi}_n^1, \ldots, \widehat{\xi}_n^N\right) \xrightarrow{\text{Mutation}} \xi_{n+1}$$

and taking values in a product space E^N where $N \geq 1$ and E is an abstract topological space. The coordinates of points of E^N are seen as positions of N particles and the integer parameter N represents the size of the population.

- The initial system $\xi_0 = \left(\xi_0^1, \ldots, \xi_0^N\right)$ consists of N independent random particles with a common law η_0 on E.
- In the selection transition the particles $\widehat{\xi}_n = \left(\widehat{\xi}_n^1, \ldots, \widehat{\xi}_n^N\right)$ are chosen randomly and independently in the previous configuration $\xi_n = \left(\xi_n^1, \ldots, \xi_n^N\right)$ according to a given non-necessarily homogeneous fitness function

$$g_n : E \to \mathbb{R}_+$$

namely

$$\mathbb{P}\left(\widehat{\xi}_n \in dx \mid \xi_n = y\right) = \prod_{p=1}^N \sum_{i=1}^N \frac{g_n(y^i)}{\sum_{j=1}^N g_n(y^j)} \delta_{y^i}(dx^p) \tag{1}$$

where $dx \stackrel{\text{def}}{=} dx^1 \times \cdots \times dx^N$ is an infinitesimal neighborhood of the point $x = (x^1, \ldots, x^N) \in E^N$, $y = (y^1, \ldots, y^N) \in E^N$ and δ_a stands for the Dirac measure at $a \in E$.

- The mutation transition is modelled by independent motions of each particle that is

$$\mathbb{P}\left(\xi_{n+1} \in dx \,\big|\, \widehat{\xi}_n = y\right) = \prod_{p=1}^{N} K_{n+1}\left(y^p, dx^p\right) \quad (2)$$

where $\{K_n \,;\, n \geq 1\}$ is a collection on Markov transition kernels from E into itself.

The study of the convergence as $n \to \infty$ or as $N \to \infty$ of this algorithm requires specific developments.

To explain and motivate the organization of our work in the next two sections we describe the main ideas involved in the study of these different asymptotics as well as some of their consequences in the study of nonlinear estimation problems.

Before turning to further details it is convenient at this point to make a couple of remarks. As we said above, the previous selection-mutation Markov chain is the crudest of the genetic-type methods. There are, in fact, a number of ways to construct variations on this model (see for instance [12,20,21] and section 5). In particular, the definition of the initial system as N i.i.d. particles is not really essential. In numerical function analysis the asymptotic results as $n \to \infty$ (and fixed N) presented here (see also [8] and [21]) are valid for any choice of N starting points. In nonlinear filtering settings we will be interested in the asymptotic behavior of the empirical measures of the system as $N \to \infty$. In this framework the initial distribution η_0 is not arbitrarily chosen, but represents the initial law of the state signal. Therefore, the initial configuration of the particle systems will be chosen so that the associated empirical measure is an N-approximating measure of η_0.

Another more general remark is that in filtering problems the choice of quantities (η_0, g_n, K_n) is dictated by the problem at hand. In some situations the initial law η_0, the transitions of the state signal K_n and/or the corresponding fitness functions g_n are not explicitly known and/or we cannot simulate random variables exactly according to η_0 and/or K_n. Therefore, we need to introduce additional approximating quantities $(\eta_0^{(M)}, g_n^{(M)}, K_n^{(M)})$, where the parameter $M \geq 1$ is a measure of the quality of the approximation so that in some sense $(\eta_0^{(M)}, g_n^{(M)}, K_n^{(M)}) \to (\eta_0, g_n, K_n)$ as $M \to \infty$. The way the two asymptotics $N \to \infty$ and $M \to \infty$ combine are studied in full detail in [18].

2.1 Large Population Asymptotic

To show one of the central roles played by the selection/mutation transitions (1) and (2), we start with the study of the asymptotic behavior of the empirical measures associated with the systems of particles ξ_n and $\widehat{\xi}_n$

$$m(\xi_n) \stackrel{\text{def.}}{=} \frac{1}{N} \sum_{i=1}^{N} \delta_{\xi_n^i} \quad \text{and} \quad m(\widehat{\xi}_n) \stackrel{\text{def.}}{=} \frac{1}{N} \sum_{i=1}^{N} \delta_{\widehat{\xi}_n^i}, \quad n \geq 0$$

as the number of particles N tends to infinity. It is transparent from the previous construction that the pair selection/mutation transition can be summarized, for each each $n \geq 0$, as follows

$$\mathbb{P}(\xi_{n+1} \in dx \mid \xi_n = y) = \prod_{p=1}^{N} \sum_{i=1}^{N} \frac{g_n(y^i)}{\sum_{j=1}^{N} g_n(y^j)} K_{n+1}\left(y^i, dx^p\right). \tag{3}$$

In order to obtain a more tractable description of (3) in terms of a transition which only depends on the empirical measure of the system $m(\xi_n)$, it is convenient to introduce some additional notations. We recall that any transition probability kernel $K(x, dy)$ on E generates two integral operators. One is acting on the set $\mathcal{B}_b(E)$ of bounded Borel test functions $f : E \to \mathbb{R}$ endowed with the supremum norm, defined by

$$\|f\| = \sup_{x \in E} |f(x)|$$

and the other on the set $\boldsymbol{M}_1(E)$ of probability measures μ on E

$$K(f)(x) \stackrel{\text{def.}}{=} \int K(x, dz) f(z)$$

and

$$(\mu K)(f) \stackrel{\text{def.}}{=} \mu(Kf) = \int \mu(dx) K(x, dz) f(z).$$

If K_1 and K_2 are two integral operators on $\mathcal{B}_b(E)$ we denote by $K_1 K_2$ the composite operator on $\mathcal{B}_b(E)$ defined for any $f \in \mathcal{B}_b(E)$ by

$$K_1 K_2 f(x) = \int_E K_1(x, dy) K_2(y, dz) f(z).$$

Using these notations (3) can be rewritten as

$$\mathbb{P}(\xi_{n+1} \in dx \mid \xi_n = y) = \prod_{p=1}^{N} \Phi_{n+1}\left(\frac{1}{N} \sum_{i=1}^{N} \delta_{y^i}\right)(dx^p), \quad n \geq 0 \tag{4}$$

where for all $n \geq 0$, $\Phi_{n+1} : \boldsymbol{M}_1(E) \to \boldsymbol{M}_1(E)$ is the mapping defined by

$$\Phi_{n+1}(\eta) \stackrel{\text{def.}}{=} \Psi_n(\eta) K_{n+1} \quad \text{with} \quad \Psi_n(\eta)(f) \stackrel{\text{def.}}{=} \frac{\eta(g_n f)}{\eta(g_n)} \quad \forall f \in \mathcal{B}_b(E)$$

$$\tag{5}$$

We note that the one-step mapping Φ_{n+1} involves two separate transitions: the first one $\eta \mapsto \Psi_n(\eta)$ is nonlinear and will be called the updating step, and the second one $\eta \mapsto \eta K_{n+1}$ will be called the prediction transition with reference to filtering theory.

With this formulation it also becomes quite clear that the flow of empirical measures $\{m(\xi_n) \,;\, n \geq 0\}$ converge in some sense as $N \to$ to the solution $\{\eta_n \,;\, n \geq 0\}$ of the following measure-valued process

$$\eta_n = \Phi_n(\eta_{n-1}), \qquad n \geq 1. \tag{6}$$

Intuitively speaking, if $m(\xi_{n-1})$ is close to the desired distribution η_{n-1} then one expects that $\Phi_n(m(\xi_{n-1}))$ is a nice approximating measure for η_n. Therefore, at the next step the particle system $\xi_n = (\xi_n^1, \ldots, \xi_n^N)$ looks like a sequence of independent random variables with common law η_n and therefore $m(\xi_n)$ is close to the desired distribution η_n …

As a parenthesis, and along the same idea, we can associate with any abstract measure-valued process (6) an N-interacting particle approximating model as in (4). In other words, the previous algorithm is a particular example of particle approximating model and the mutation/selection transitions are dictated by the form of the limiting measure-valued dynamical system (6).

In our situation, the preceding scheme is clearly a system of interacting particles undergoing adaptation in a time-nonhomogeneous environment represented by the fitness functions $\{g_n; \, n \geq 0\}$ and the selection/mutation transitions are dictated by the nature of the two-step mappings $\{\Phi_n \,;\, n \geq 1\}$. Roughly speaking, the natural idea is to approximate the two-step transitions

$$\eta_n \xrightarrow{\text{Updating}} \widehat{\eta}_n \stackrel{\text{def}}{=} \psi_n(\eta_n) \xrightarrow{\text{Prediction}} \eta_n = \widehat{\eta}_n K_{n+1}, \qquad n \geq 0$$

of the system (6) by a two-step Markov chain taking values in the set of finitely discrete probability measures with atoms of size some integer multiple of $1/N$. Namely, for each $n \geq 0$

$$\eta_n^N \stackrel{\text{def.}}{=} \frac{1}{N}\sum_{i=1}^{N} \delta_{\xi_n^i} \xrightarrow{\text{Selection}} \widehat{\eta}_n^N \stackrel{\text{def.}}{=} \frac{1}{N}\sum_{i=1}^{N} \delta_{\widehat{\xi}_n^i} \xrightarrow{\text{Mutation}} \eta_{n+1}^N = \frac{1}{N}\sum_{i=1}^{N} \delta_{\xi_{n+1}^i}. \tag{7}$$

These constructions first appeared in [26] and [27] and they were developed in [24]. In [20] the authors present an exposé of the mathematical theory that is useful in analyzing the convergence of such particle-approximating models including the law of large numbers, large deviations principles, fluctuations, and empirical process theory, as well as semigroup techniques and limit theorems for processes. In section 3 we briefly indicate some of the main directions explored in this recent research and we will introduce the reader to some mathematical tools upon which the theory dwells.

Anticipating section 3.1, we also mention that the measure-valued dynamical system (6) can be explicitly solved. More precisely, if

$$X = \{X_n \,;\, n \geq 0\}$$

denotes a time-inhomogeneous Markov chain with transition probability kernels $\{K_n\ ;\ n \geq 1\}$ and initial distribution η_0 and if $\gamma_n(f)$, $f \in \mathcal{B}_b(E)$, represents the Feynman–Kac formula

$$\gamma_n(f) = \mathbb{E}\left(f(X_n) \prod_{k=0}^{n-1} g_k(X_k)\right)$$

(with the convention $\prod_\emptyset = 1$), then the distribution flow $\{\eta_n;\ n \geq 0\}$ defined for any $n \geq 0$ and for any test function $f \in \mathcal{B}_b(E)$ as the ratio

$$\eta_n(f) = \frac{\gamma_n(f)}{\gamma_n(1)} \tag{8}$$

is solution of the measure-valued dynamical system (6). In fact, as we shall see in the further development of section 3, the classical nonlinear filtering problem can be summarized as to find distributions of the form (8). In this framework, the probability kernels $\{K_n\ ;\ n \geq 1\}$ represent the transitions of the signal process and the fitness functions $\{g_n\ ;\ n \geq 0\}$ depend on the observation data and on the density of the noise source.

2.2 Long-Time Behavior

Our next objective is to initiate the study of the long-time behavior of the genetic-type algorithms. In contrast to the situation presented in section 2.1, the size N of the particle systems is fixed, the genetic model is thought of as a global search procedure for studying the set $U^\star$ of global minima of a given numerical function $U : E \to \mathbb{R}_+$, and the state space E is assumed to be finite, namely

$$U^\star \stackrel{\text{def.}}{=} \left\{x \in E\ ;\ U(x) = \min_E U\right\}.$$

To clarify the notations we shall use the following notations

$$Q_n^{(1)}(x, \mathrm{d}y) = \prod_{p=1}^{N} K_n(x^p, \mathrm{d}y^p)$$

and

$$Q_n^{(2)}(x, \mathrm{d}y) = \prod_{p=1}^{N} \sum_{i=1}^{N} \frac{g_n(y^i)}{\sum_{j=1}^{N} g_n(y^j)}\, \delta_{y^i}(\mathrm{d}x^p).$$

Thus, $\xi = \{\xi_n\ ;\ n \geq 0\}$ is a time-inhomogeneous Markov chain with transition probability kernel

$$Q_n = Q_{n-1}^{(2)} Q_n^{(1)}$$

and $\widehat{\xi} = \{\widehat{\xi}_n \; ; \; n \geq 0\}$ is a time-inhomogeneous Markov chain with transition probability kernel

$$\widehat{Q}_n = Q_n^{(1)} Q_n^{(2)}. \tag{9}$$

In time-homogeneous settings (that is, if $K_n = K$ and $g_n = g$) the general theory of time-homogeneous Markov chains can be used to study the long-time behavior of these two chains, but to our knowledge the stochastic stability results which can be stated are not really useful to calibrate the convergence of GAs to the desired extrema of a given numerical function.

One of the apparent difficulties in establishing a useful convergence result as $n \to \infty$ is finding a candidate invariant measure which enables us to describe some interesting aspects of the limiting behavior of the algorithm.

The key idea is to introduce an inverse cooling schedule parameter $\beta : \mathbb{R}_+ \to \mathbb{R}_+$ with $\lim_{t \to \infty} \beta(t) = \infty$ to reduce the analysis to the study of a generalized simulated annealing. This idea has been initiated in [8–10] and has been simplified and further extended in [21]. As the time parameter is growing, the arbitrary exploration of the path space by the particles during the mutation step will progressively disappear. The precise choice of the mutation transitions K_n in terms of the parameter $\beta(n)$ will be given in section 4.1. We have already mentioned that in the selection transitions the fitness functions g_n will take the form

$$g_n(x) = e^{-\beta(n) U(x)}, \qquad n \geq 1$$

and, as the time is growing, the randomness in the selection will also tend to disappear so that the particles with below peak fitness will progressively not be selected.

The purpose of this paper is to present some theoretical background needed to analyze the convergence of the algorithm. The results presented here will be restricted to the transition probability kernel (9) and can be found with complete proof in [21]. In this opening section we describe the basic but general idea which is, in fact, quite simple. This methodology will be used in several parts of this paper. It is also quite general and can be used in other contexts.

We have tried to present easily verifiable conditions and results at a relevant level of generality. Our claim that this description of the mutation and selection transitions is the natural framework for formulating and studying the long-time behavior of GAs in numerical function analysis will be amply justified by the results that follow.

We provide no examples in this short section; this choice is deliberate. In section 4 we will show how to obtain the transitions $Q_\beta^{(1)}$ and $Q_\beta^{(2)}$ in terms of the mutation kernels and the fitness functions. We will also use this framework in section 5 for studying a related GA in which the mutation and the selection stage take place randomly at each time step. We also believe that it is possible to use this formulation to analyze the convergence of the branching genetic-type variants presented in section 5.

To commence to formalize this we first chose the mutation/selection transitions $\mathcal{Q}_n^{(1)}$ and $\mathcal{Q}_n^{(2)}$ as governed by $\beta(n)$, that is

$$\mathcal{Q}_n^{(1)} = \mathcal{Q}_{\beta(n)}^{(1)} \quad \text{and} \quad \mathcal{Q}_n^{(2)} = \mathcal{Q}_{\beta(n)}^{(2)} \tag{10}$$

so that for any $\beta > 0$, $Q_\beta^{(1)}$ and $Q_\beta^{(1)}$ take the form

$$Q_\beta^{(1)}(x,y) = q_\beta^{(1)}(x,y)\, e^{-\beta V^{(1)}(x,y)}$$

and

$$Q_\beta^{(2)}(x,y) = q_\beta^{(2)}(x,y)\, e^{-\beta V^{(2)}(x,y)}$$

for some numerical functions $q_\beta^{(1)}, q_\beta^{(2)} : E^N \times E^N \longrightarrow \mathbb{R}_+$ and

$$V^{(1)}, V^{(2)} : E^N \times E^N \longrightarrow \overline{\mathbb{R}}_+$$

($\overline{\mathbb{R}}_+ \stackrel{\text{def.}}{=} \mathbb{R}_+ \cup \{+\infty\}$).

It is then straightforward to check that the transition probability kernels

$$\widehat{Q}_\beta(x,y) \stackrel{\text{def.}}{=} Q_\beta^{(1)} Q_\beta^{(2)}(x,y)$$

take the form

$$\widehat{Q}_\beta(x,y) = \sum_{v \in \mathcal{V}} \widehat{q}_\beta(x,v,y)\, e^{-\beta \widehat{V}(x,v,y)} \tag{11}$$

with $\mathcal{V} = E^N$ and

$$\widehat{q}_\beta(x,v,y) = q_\beta^{(1)}(x,v) q_\beta^{(2)}(v,y)$$

and

$$\widehat{V}(x,v,y) = V^{(1)}(x,v) + V^{(2)}(v,y).$$

These kinds of mathematical models naturally arise when studying the long-time behavior of stochastic algorithms such as the generalized simulated annealing. The parameter β in (10) will be regarded as the inverse freezing schedule in classical simulated annealing and will be used to control the random perturbations of the stochastic algorithm. When $\beta \to \infty$ the random perturbations will progressively disappear and the two different cost functions $V^{(1)}$ and $V^{(2)}$ will be regarded, respectively, as the mutation and selection costs to communicate from one population to another.

The objective is to prove that the law of a well-chosen time-inhomogeneous genetic particle scheme concentrates as times tends to infinity to the set $U^\star$ of global minima of a desired numerical function $U : E \to \mathbb{R}_+$. In order to prove this asymptotic result we need to characterize more explicitly the long-time behavior of the algorithm in terms of the communication cost functions $V^{(1)}$ and $V^{(2)}$.

As traditional, under some nice conditions, the first step consists in proving that the algorithm converges to the set of the global minima $W^\star$ of a virtual energy function $W : E^N \to \mathbb{R}_+$ defined explicitly in terms of the communication cost functions $V^{(1)}$ and $V^{(2)}$. The second subtle step will be to find conditions on the population size which ensure that $W^\star$ is contained in the subset $U^\star \times \ldots \times U^\star (\subset E^N)$. We will settle this question in section 4 and 5 by using a natural test set approach.

Under appropriate continuity and irreductibility conditions the first step can be solved using quite general results on the generalized simulated annealing. Anticipating section 5.1, we also notice that the transition probability kernel $\tilde{Q}_\beta$ defined by

$$\tilde{Q}_\beta = \alpha_1 \, Q_\beta^{(1)} + \alpha_2 \, Q_\beta^{(2)} \qquad \alpha_1 + \alpha_2 = 1 \quad (\alpha_1, \alpha_2 \in (0,1)) \tag{12}$$

can be written as in (11) with $\mathcal{V} = \{1, 2\}$.

The precise continuity and irreductibility conditions needed to handle the first step are summarized in the following assumption:

H *The transition probability kernels $\widehat{Q}_\beta$ take the form*

$$\widehat{Q}_\beta(x, y) = \sum_{v \in \mathcal{V}} \widehat{q}_\beta(x, v, y) \, e^{-\beta \widehat{V}(x, v, y)}$$

where $\mathcal{V}$ is a finite set and there exists a nonnegative function

$$\widehat{q} : E^N \times \mathcal{V} \times E^N \longrightarrow \mathbb{R}_+$$

so that

- *For any $x, y \in E^N$ and $v \in \mathcal{V}$ and $\beta > 0$ we have*

$$\lim_{\beta \to +\infty} \widehat{q}_\beta(x, v, z) = \widehat{q}(x, v, z)$$

and

$$\widehat{q}_\beta(x, v, z) > 0 \iff \widehat{q}(x, v, z) > 0.$$

- *For every $\widehat{q}(x, v, z) > 0$ and for some $\beta_0 \geq 0$*

$$\sup_{\beta \geq \beta_0} |\frac{d \log \widehat{q}_\beta}{d\beta}(x, v, z)| < +\infty.$$

- *For any $x, y \in E^N$ there exists an integer $r \geq 1$ and sequence of elements $(p_k, v_k)_{0 \leq k \leq r}$ in $E^N \times \mathcal{V}$ such that*

$$p_0 = x \qquad \text{and} \qquad \widehat{q}(p_k, v_k, p_{k+1}) > 0 \quad \forall 0 \leq k < r \quad \text{and} \quad p_r = y.$$

It is transparent from these conditions that we have some suitable function $\epsilon(\beta) \to 0$, as $\beta \to +\infty$, such that

$$(1 - \epsilon(\beta)) Q_\beta(x, y) \leq \widehat{Q}_\beta(x, y) \leq (1 + \epsilon(\beta)) Q_\beta(x, y) \tag{13}$$

where

$$Q_\beta(x, y) = \sum_{v \in \mathcal{V}(x,y)} \widehat{q}(x, v, y)\, e^{-\beta \widehat{V}(x,v,y)}$$

and

$$\mathcal{V}(x, y) = \{v \in \mathcal{V} : \widehat{q}(x, v, y) > 0\}.$$

But, if we write

$$V(x, y) = \min_{v \in \mathcal{V}(x,y)} \widehat{V}(x, v, y)$$
$$q(x, y) = \sum_{v \in \mathcal{V}^*(x,y)} \widehat{q}(x, v, y)$$
$$\mathcal{V}^*(x, y) = \{v \in \mathcal{V}(x, y) : \widehat{V}(x, v, y) = V(x, y)\}$$

then we also have that

$$Q_\beta(x, y)$$
$$= \sum_{v \in \mathcal{V}^*(x,y)} \widehat{q}(x, v, y)\, e^{-\beta V(x,y)}$$
$$\quad + e^{-\beta V(x,y)} \sum_{v \in \mathcal{V}(x,y) - \mathcal{V}^*(x,y)} \widehat{q}(x, v, y)\, e^{-\beta(\widehat{V}(x,v,y) - V(x,y))}$$
$$= q(x, y)\, e^{-\beta V(x,y)}$$
$$\quad + e^{-\beta V(x,y)} \sum_{v \in \mathcal{V}(x,y) - \mathcal{V}^*(x,y)} \widehat{q}(x, v, y)\, e^{-\beta(\widehat{V}(x,u,y) - V(x,y))}.$$

Note that condition (**H**) implies that q is irreducible. Furthermore, if we write

$$I = \{(x, y) \in E^2 : \mathcal{V}(x, y) \neq \emptyset\}$$
$$J = \{(x, v, y) \in E^N \times \mathcal{V} \times E^N : (x, y) \in I \ v \in \mathcal{V}(x, y)\}$$

and

$$h_1 = \min_{(x,y) \in I} \sum_{v \in \mathcal{V}(x,y) - \mathcal{V}^*(x,y)} \widehat{q}(x, v, y) / q(x, y)$$
$$h_2 = \min_{(x,v,y)\, :\, v \notin \mathcal{V}^*(x,y)} (\widehat{V}(x, v, y) - V(x, y))$$

using (13) we get the system of inequalities

$$(1 - \epsilon(\beta)) \, q(x, y) \, e^{-\beta V(x,y)} \leq \widehat{Q}_\beta(x, y)$$

and

$$\widehat{Q}_\beta(x, y) \leq (1 + \epsilon(\beta)) \, (1 + h_1 \, e^{-\beta h_2}) \, q(x, y) \, e^{-\beta V(x,y)}. \tag{14}$$

As a parenthesis, if we choose $q(x, y) > 0$, after some elementary computations, then we find that

$$\left| \frac{d \log \widehat{Q}_\beta(x, y)}{d\beta} \right| \leq \sup_{v \in \mathcal{V}(x,y)} \left| \frac{d \log \widehat{q}_\beta}{d\beta}(x, v, y) \right| + \sup_{v \in \mathcal{V}(x,y)} \widehat{V}(x, v, y).$$

The inequality (14) shows that the transition probability kernels

$$\{\widehat{Q}_\beta \, ; \, \beta > 0\}$$

are of the general form of generalized simulated annealing models studied in [54] and [21].

In [54] the author studies the asymptotic behavior of such chains using large deviation techniques and in [21] the authors propose an alternative approach based on semi-group techniques. Both approaches give a precise study of the convergence of the time-inhomogeneous Markov process controlled by a suitably chosen cooling schedule and associated with the family of Markov transitions $\widehat{Q}_\beta$ of the form (11) when $\mathcal{V}$ is an auxiliary finite set.

The first method in [54] is developed for discrete-time models whereas the convergence analysis in [21] is centered around continuous-time models. There is a vast literature on discrete-time simulated annealing (see, for instance [54], and references therein). For this reason we have chosen to give a more detailed description of the second approach.

It is now convenient to introduce some additional notations. In discrete-time or continuous-time settings the asymptotic behavior of the desired time-inhomogeneous Markov processes will be strongly related to the virtual energy function $W : E^N \to \overline{\mathbb{R}}_+$ defined as follows:

$$W(x) = \min_{g \in G(x)} \sum_{(y \to z) \in g} V(y, z) - \min_{x' \in E^N} \min_{g \in G(x')} \sum_{(y \to z) \in g} V(y, z) \tag{15}$$

where $G(x)$ is the set of x-graphs over E^N (we recall that an x-graph is an oriented tree over the vertex set E^N such that for any $x \neq y$ there exists a unique path in the x-graph leading from x to y. See also [5] or [32] for more details), and $V : E^N \times E^N \to \widehat{\mathbb{R}}_+$ is the virtual communication cost function given by

$$V(x, y) = \min \{\widehat{V}(x, v, y) \, ; \, v \in \mathcal{V} \, \, \widehat{q}(x, v, y) > 0\}.$$

We will also use the notation

$$W^\star = \{x \in E^N \ : \ W(x) = \min_E W\}.$$

As mentioned, the first approach presented in [54] gives a complete answer for the convergence in discrete-time settings and in the time-inhomogeneous case when the parameter $\beta(n)$ is an increasing function of the time parameter n. With some obvious abusive notations let us denote by $\{\widehat{\xi}_n \ ; \ n \geq 0\}$ the discrete-time and time-inhomogeneous Markov chain starting at some point $x \in E^N$ and associated with the collection of time-inhomogeneous transitions $\{\widehat{Q}_{\beta(n)} \ ; \ n \geq 1\}$.

Theorem 1 ([54]). *There exists a constant C_0 (which can be explicitly described in terms of V) such that if $\beta(n)$ takes the parametric form $\beta(n) = \frac{1}{C}\log n$ for sufficiently large n and $C > C_0$ then*

$$\lim_{n \to \infty} \mathbb{P}\left(\widehat{\xi}_n \in W^\star\right) = 1$$

The semi-group approach presented in [21] is based on log-Sobolev inequalities and on the notion of relative entropy. We recall that the relative entropy $\mathrm{Ent}_\pi(\mu)$ of a measure μ with respect to a measure π (charging all the points) is defined by

$$\mathrm{Ent}_\pi(\mu) = \sum_{x \in E} \mu(x) \, \log\left(\mu(x)/\pi(x)\right).$$

In contrast to the latter, the former approach is based entirely on considerations of the time-continuous semi-group associated with the Markov kernels $\widehat{Q}_\beta$, $\beta > 0$. Namely, define, for $f : E^N \to \mathbb{R}$

$$L_\beta(f)(x) = \sum_{y \in E^N} (f(y) - f(x)) \, \widehat{Q}_\beta(x, y).$$

Instead of the discrete-time model introduced above we are now concerned with the continuous-time Markov process defined as follows. For a probability measure μ on E, and an inverse-freezing schedule $\beta \in C^1(\mathbb{R}_+, \mathbb{R}_+)$, we slightly abuse notation and write $\{\widehat{\xi}_t \ ; \ t \in \mathbb{R}_+\}$ the canonical process associated with the family of generators

$$(L_{\beta(t)})_{t \geq 0} = (\widehat{Q}_{\beta(t)} - I)_{t \geq 0}$$

and whose initial condition is $\mu_0 = \mu$. We also write $\mu(t)$ the distribution of $\widehat{\xi}_t$.

Before we turn to the long-time behavior of ξ_t we first give a more tractable description of this process. Let $\Delta = \{\Delta_k \ ; \ k \geq 0\}$ be independent and exponentially distributed random variables with parameter 1 and, given Δ, let $\widehat{\zeta} = \{\widehat{\zeta}_n \ ; \ n \geq 0\}$ be a time-inhomogeneous Markov chain on E^N with initial distribution μ and time-inhomogeneous transition probability kernels

$$\widehat{K}_n \stackrel{\mathrm{def.}}{=} \widehat{Q}_{\beta(T_n)}, \qquad n \geq 1$$

where

$$T_n \stackrel{\text{def.}}{=} \sum_{k=0}^{n} \Delta_k, \quad n \geq 0.$$

Then

$$\widehat{\xi}_t = \begin{cases} \widehat{\zeta}_0 & 0 \leq t < T_0 \\ \widehat{\zeta}_n & T_{n-1} \leq t < T_n \end{cases}$$

defines a time-inhomogeneous Markov process $\widehat{\xi} = \{\widehat{\xi}_t \; ; \; t \in \mathbb{R}_+\}$ with initial law μ and infinitesimal generators $\{L_{\beta(t)} \; ; \; t \in \mathbb{R}_+\}$.

Whenever $\widehat{\xi}$ is time-homogeneous (i.e., $\beta(t) = \beta$) it is well-known that L_β has a unique invariant probability measure π_β so that

$$\forall f \in B_b(E^N) \quad \pi_\beta(L_\beta(f)) = 0$$

and π_β charges all the points. Asymptotically, the behavior of the invariant measure π_β as $\beta \to \infty$ depends principally on the virtual energy function W defined in (15). To be more precise, we recall that π_β can be written as follows:

$$\pi_\beta(x) = \frac{\widehat{R}_\beta(x)}{\sum_{z \in E^N} \widehat{R}_\beta(z)} \quad \text{where} \quad \widehat{R}_\beta(x) = \sum_{g \in G(x)} \prod_{(y \to z) \in g} \widehat{Q}_\beta(y, z).$$

Now, from the inequality (14) one concludes that

$$\epsilon_1(\beta) \, R_\beta(x) \leq \widehat{R}_\beta(x) \leq \epsilon_2(\beta) \, R_\beta(x)$$

where $\epsilon_i(\beta)$, $i = 1, 2$, are some functions such that

$$\lim_{\beta \to \infty} \epsilon_i(\beta) = 1, \quad i = 1, 2$$

and

$$R_\beta(x) = \sum_{g \in G(x)} \prod_{(y \to z) \in g} q(y, z) \, e^{-\beta V(y, z)}.$$

This can also be rewritten in the form

$$R_\beta(x) = \sum_{g \in G(x)} q(g) \, e^{-\beta V(g)}$$

with

$$q(g) = \prod_{(y \to z) \in g} q(y, z) \quad \text{and} \quad V(g) = \sum_{(y \to z) \in g} V(y, z).$$

Therefore, we clearly have the estimate

$$\lim_{\beta\to\infty} -\frac{1}{\beta} \log \pi_\beta(x) = \lim_{\beta\to\infty} -\frac{1}{\beta} \log R_\beta(x) - \lim_{\beta\to\infty} -\frac{1}{\beta} \log \sum_{z\in E^N} R_\beta(z)$$

$$= \min_{g\in G(x)} V(g) - \min_{z\in E^N} \min_{g\in G(z)} V(g)$$

$$= W(x).$$

Due to this estimate, for each $\beta > 0$, if $\{\widehat{\xi}_{\beta,t} \; ; \; t \geq 0\}$ denotes the time-homogeneous Markov process associated with L_β then we have that

$$\lim_{\beta\to\infty} \lim_{t\to\infty} \mathbb{P}\left(\widehat{\xi}_{\beta,t} \in W^\star\right) = 1.$$

In the time-inhomogeneous situation the convergence of the algorithm to $W^\star$ is guaranteed by the following result:

Theorem 2 ([21]). *Let $\{\widehat{Q}_\beta \; ; \; \beta > 0\}$ be a collection of general Markov kernels of the form*

$$\widehat{Q}_\beta(x, y) = \sum_{v \in \mathcal{V}} \widehat{q}_\beta(x, v, y) \, e^{-\beta \widehat{V}(x,v,y)}$$

*where $\mathcal{V}$ is a given finite set, $\widehat{V} : E^N \times \mathcal{V} \times E^N \to \mathbb{R}_+$ and $\widehat{q}_\beta : E^N \times U \times E^N \to \mathbb{R}_+$, $\beta \in \mathbb{R}_+$, is a family of functions satisfying condition (**H**). There exist a constant C_0 (which can be explicitly described in terms of V) such that if $\beta(t)$ takes the parametric form $\beta(t) = \frac{1}{C} \log t$ for sufficiently large t and $C > C_0$ then*

$$\lim_{t\to\infty} \mathrm{Ent}_{\pi_{\beta(t)}}(\mu(t)) = 0 \quad \text{and} \quad \lim_{t\to\infty} \mathbb{P}\left(\widehat{\xi}_t \in W^\star\right) = 1.$$

This theorem is quite general and it will be used to study the convergence of GAs when the corresponding transitions have the form (11) or (12). It is also powerful enough to allow one to treat the classical simulated annealing algorithm. In this situation $N = 1$ and $\widehat{Q}_\beta$ takes the form

$$\widehat{Q}_\beta(x, y) = q(x, y) \, e^{-\beta V(x,y)}$$

with

$$V(x, y) = \max\left(U(y) - U(x), 0\right)$$

for $x \neq y$, where q is an irreducible transition probability kernel on E and $U : E \to \mathbb{R}_+$. In the special case where q is symmetric (that is, $q(x, y) = q(y, x)$), it is also well-known that the corresponding virtual energy function $W = U$ and the previous theorem implies convergence to the desired subset of the global minima $U^\star$. For genetic-type algorithms the virtual energy function depends on the function

U in a more subtle way and we need to work harder to check that $W^\star$ is contained in the desired subset of global minima.

The results developed here are, in fact, a particular form of those in [21] which also apply to the study of the convergence of generalized simulated annealing with random and time-inhomogeneous communication cost functions. Although this subject is tangential to the main object of this article; let us discuss how these results may be useful in solving mean cost optimization problems.

In some practical problems the object is to find the global minima of a function $U : E \to \mathbb{R}_+$ defined by

$$U(x) = \mathbb{E}\left(\mathcal{U}(x, Z)\right) = \int_F \mathcal{U}(x, z)\, \nu(\mathrm{d}z)$$

where Z is a random variable taking values in a finite set F with distribution ν and $\mathcal{U} : E \times F \to \mathbb{R}_+$. The essential problem is to compute at each time step the mean cost function U, and the huge size of the set F often precludes the use of the previous stochastic algorithms.

To solve this problem an additional level of approximation is needed. The natural idea proposed in [21] consists in replacing at each moment of time in the description of the stochastic algorithm the function U by the time-inhomogeneous and random function

$$U_t(x) \stackrel{\text{def.}}{=} \frac{1}{t^A} \int_0^{t^A} \mathcal{U}(x, Z_s)\, \mathrm{d}s$$

where $A > 0$ and $\{Z_t \,;\, t \geq 0\}$ is a given time-homogeneous Markov process associated with the generator $\mathcal{G} = \mathcal{K} - \mathrm{Id}$ where $\mathcal{K}$ is an irreducible transition probability kernel on E with invariant measure ν. A full discussion of the convergence of the resulting stochastic algorithm to the desired subset $U^\star$ is outside the scope of this work; the interested reader is referred to [21].

3 Feynman–Kac and Nonlinear Filtering Models

3.1 Description of the Models

The nonlinear filtering problem consists in computing the conditional distribution of internal states in dynamical systems, when partial observations are made and random perturbations are present in the dynamics as well as in the sensors. In discrete-time settings the state signal $X = \{X_n \,;\, n \geq 0\}$ is a discrete-time Markov chain taking values in a Polish space E (i.e., a complete separable metric space) with transition probabilities $\{K_n \,;\, n \geq 1\}$ and initial distribution η_0. The observation sequence $Y = \{Y_n \,;\, n \geq 0\}$ are $\mathbb{R}^q$-valued random variables and take the form

$$Y_n = H_n(X_n, V_n)$$

where the V_n are independent and q-dimensional variables, independent of X, and with a law having a known density, and H_n is a measurable function from $E \times \mathbb{R}^q$ into $\mathbb{R}^q$. For any $x \in E$ we assume that the variable $Y_n = H_n(x, V_n)$ admits a positive density $y \mapsto \varphi_n(x, y)$ and the function φ_n is bounded. To clarify the notation we fix the observations $Y_n = y_n$, $n \geq 0$ and we write

$$g_n(x) \stackrel{\text{def.}}{=} \varphi_n(x, y_n).$$

For somewhat technical reasons we will assume that H_n and φ_n and the observation sequence $\{y_n \,;\, n \geq 0\}$ are chosen so that g_n is a positive and bounded function on E. These assumptions can be relaxed considerably; a more complete and general set of assumptions is formulated in [17] and [18].

Given the stochastic nature of the pair signal/observation process, the nonlinear filtering problem consists in computing recursively in time the one-step predictor conditional probabilities η_n and the filter conditional distributions $\widehat{\eta}_n$ given for any bounded Borel test function f by

$$\eta_n(f) = \mathbb{E}(f(X_n)|Y_0 = y_0, \ldots, Y_{n-1} = y_{n-1})$$
$$\widehat{\eta}_n(f) = \mathbb{E}(f(X_n)|Y_0 = y_0, \ldots, Y_{n-1} = y_{n-1}, Y_n = y_n).$$

As usual, the n-step filter $\widehat{\eta}_n$ is written in terms of η_n as

$$\widehat{\eta}_n(f) = \Psi_n(\eta_n)(f) = \frac{\eta_n(f\, g_n)}{\eta_n(g_n)} \tag{16}$$

and the n-step predictor is defined in terms of the Feynman–Kac type formula

$$\eta_n(f) = \frac{\gamma_n(f)}{\gamma_n(1)} \quad \text{with} \quad \gamma_n(f) = \mathbb{E}\left(f(X_n) \prod_{k=0}^{n-1} g_k(X_k) \right) \tag{17}$$

with the convention $\prod_\emptyset = 1$. By (16), the n-step filter $\widehat{\eta}_n$ may also be expressed as the ratio

$$\widehat{\eta}_n(f) = \frac{\widehat{\gamma}_n(f)}{\widehat{\gamma}_n(1)} \tag{18}$$

with

$$\widehat{\gamma}_n(f) = \gamma_n(g_n f) = \mathbb{E}\left(f(X_n) \prod_{k=0}^{n} g_k(X_k) \right).$$

It is also not difficult to check that γ_n and $\widehat{\gamma}_n$ are connected by

$$\gamma_n = \widehat{\gamma}_{n-1} K_n, \qquad n \geq 1. \tag{19}$$

It then follows from the relations (16) and (19) that for any $n \geq 1$

$$\eta_n = \Phi_n(\eta_{n-1}) \tag{20}$$

with

$$\Phi_n(\eta) = \Psi_{n-1}(\eta) K_n.$$

Another interesting feature of the GA defined by (3) is that it can be used to approximate the Feynman–Kac formulas $\gamma_n(f)$ and $\widehat{\gamma}_n(f)$ defined in (17) and (18). One of the best ways for introducing the corresponding particle approximating models is through the following observation. By definition it is easy to establish that for any $n \geq 0$

$$\eta_n(g_n) = \frac{\gamma_n(g_n)}{\gamma_n(1)} = \frac{\gamma_{n+1}(1)}{\gamma_n(1)}.$$

This yields that

$$\gamma_n(1) = \prod_{p=0}^{n-1} \eta_p(g_p) \quad \text{and} \quad , \gamma_n(f) = \eta_n(f) \prod_{p=0}^{n-1} \eta_p(g_p)$$

with the usual convention $\prod_\emptyset = 1$. Taking these relations into consideration, we define a natural N-approximating measure γ_n^N for γ_n by setting

$$\gamma_n^N(1) = \prod_{p=0}^{n-1} \eta_p^N(g_p) \quad \text{and} \quad \gamma_n^N(f) = \eta_n^N(f)\, \gamma_n^N(1). \tag{21}$$

In view of (7) and using the same line of ideas we can define the corresponding N-approximating measures of $\widehat{\gamma}_n$ and $\widehat{\eta}_n$. We have chosen here to restrict our attention to the distributions γ_n and η_n.

3.2 Asymptotic Behavior

One of the simplest ways for studying the asymptotic behavior as $N \to \infty$ of the GA presented in section 2 is through the analysis of the un-normalized distributions $\{\gamma_n \,;\, n \geq 0\}$. This approach has been initiated in [26] and further developed in [19] and [20]. Here we follow line-by-line the synthetic presentation given in [19]. This approach is based on the observation that the structure of the dynamics of the latter is linear and one might expect that the analysis of the corresponding approximating measures will be simplified. In view of (17) and (18) we have that

$$\forall 0 \leq p \leq n, \qquad \gamma_n = \gamma_p L_{p,n}, \qquad (\gamma_0 = \eta_0) \tag{22}$$

where $\{L_{p,n} \,;\, 0 \leq p \leq n\}$ is the time-inhomogeneous semi-group defined by the relations

$$L_{p,n} = L_{p+1} L_{p+2} \ldots L_n \quad \text{with} \quad L_n(f) = g_{n-1}.K_n(f)$$

and the convention $L_{n,n} = \text{Id}$. Using these notations one can also check that the one-step mappings Φ_n can be rewritten as

$$\Phi_n(\eta)(f) = \frac{\eta(L_n(f))}{\eta(L_n(1))}$$

for any $\eta \in M_1(E)$ and $f \in \mathcal{B}_b(E)$. Using these notations, we notice that for any $n \geq 0$ and $f \in \mathcal{B}_b(E)$ the stochastic process

$$\{M_q^N(f)\,;\ 0 \leq q \leq n\}$$

defined as

$$\begin{aligned}
M_q^N(f) &\stackrel{\text{def.}}{=} \gamma_q^N(L_{q,n}f) - \gamma_q(L_{q,n}f) \\
&= \sum_{p=0}^{q} \left(\gamma_p^N(L_{p,n}f) - \gamma_{p-1}^N(L_p L_{p,n}f)\right) \\
&= \sum_{p=0}^{q} \gamma_p^N(1) \left(\eta_p^N(L_{p,n}f) - \Phi_p\left(\eta_{p-1}^N\right)(L_{p,n}f)\right)
\end{aligned} \tag{23}$$

with the convention $\Phi_0\left(\eta_{-1}^N\right) = \eta_0$, is a martingale with respect to the natural filtration $F^N = \{F_n^N\,;\ n \geq 0\}$ associated with the N-particle system $\{\xi_n\,;\ n \geq 0\}$, and its angle bracket is given by

$$\langle M^N(f)\rangle_q = \frac{1}{N}\sum_{p=0}^{q} \left(\gamma_p^N(1)\right)^2 \Phi_p(\eta_{p-1}^N)\left(\left(L_{p,n}f - \Phi_p(\eta_{p-1}^N)L_{p,n}f\right)^2\right). \tag{24}$$

One concludes easily that γ_n^N is an approximating measure of γ_n without any bias, that is, for any bounded Borel test function f

$$\mathbb{E}\left(\gamma_n^N(f)\right) = \gamma_n(f) \tag{25}$$

and

$$\mathbb{E}\left((\gamma_n^N(f) - \gamma_n(f))^2\right)$$

$$= \frac{1}{N}\sum_{p=0}^{n} \mathbb{E}\left(\left(\gamma_p^N(1)\right)^2 \Phi_p(\eta_{p-1}^N)\left(\left(L_{p,n}f - \Phi_p(\eta_{p-1}^N)L_{p,n}f\right)^2\right)\right). \tag{26}$$

Under our assumptions it is also clear that there exist some finite constants $C(n) < \infty$ such that

$$\mathbb{E}\left((\gamma_n^N(f) - \gamma_n(f))^2\right)^{1/2} \leq \frac{C(n)}{\sqrt{N}} \|f\|.$$

Exponential bounds can also be obtained using the decomposition (23). For instance, by definition of η_t^N, Hoeffding's inequality implies that for each $0 \leq p \leq n$ and for any $\epsilon > 0$

$$\mathbb{P}\left(\left|\eta_p^N(L_{p,n}f) - \Phi_p\left(\eta_{p-1}^N\right)(L_{p,n}f)\right| > \epsilon \,\Big|\, \eta_{n-1}^N\right) \leq 2\, e^{-\frac{N}{8}\frac{\epsilon^2}{\|L_{p,n}f\|^2}}.$$

From which one concludes that

$$\mathbb{P}\left(\sup_{0\leq p\leq n}\left|\eta_p^N(L_{p,n}f) - \Phi_p\left(\eta_{p-1}^N\right)(L_{p,n}f)\right| > \epsilon\right) \leq 2\sum_{p=0}^{n} e^{-\frac{N}{8}\frac{\epsilon^2}{\|L_{p,n}f\|^2}}.$$

Since the fitness functions are assumed to be bounded this exponential bound implies that there exists some finite constants $C_1(n)$ and $C_2(n)$ such that for any bounded Borel function f, $\|f\| \leq 1$ and for every $\epsilon > 0$ we have that

$$\mathbb{P}\left(\sup_{0\leq p\leq n}\left|\gamma_p^N(L_{p,n}f) - \gamma_p(L_{p,n}f)\right| > \epsilon\right) \leq C_1(n)\,\exp{-\frac{N\epsilon^2}{C_2(n)}}.$$

We now give a brief indication of how these results can be used to obtain useful estimates for the N-approximating measures η_n^N and $\widehat{\eta}_n^N$. From the previous displayed exponential rate one can also prove that there exists some finite constants $C_1(n)$, $C_2(n)$ such that for any $\epsilon > 0$ and for any bounded Borel test function f, $\|f\| \leq 1$

$$\mathbb{P}\left(\left|\eta_n^N(f) - \eta_n(f)\right| > \epsilon\right) \leq C_1(n)\,\exp{-\frac{N\epsilon^2}{C_2(n)}}. \tag{27}$$

Precise estimates of these exponential rates are studied in [24] using large deviations techniques. The previous exponential rates also imply $\mathbb{L}^p$ mean errors

$$\forall p \geq 1 \quad \mathbb{E}\left(\left|\eta_n^N(f) - \eta_n(f)\right|^p\right)^{1/p} \leq \frac{C(p,n)}{\sqrt{N}}\|f\|$$

for some constant $C(p,n) < \infty$ which only depends on the parameters p and n. With little work one can use (25) and (26) to prove that there exist some finite constants $C(n)$ such that for any bounded Borel function f, such that $\|f\| \leq 1$,

$$\left|\mathbb{E}\left(\eta_n^N(f)\right) - \eta_n(f)\right| \leq \frac{C(n)}{N}. \tag{28}$$

Taking this inequality into consideration, by the exchangeability of the particles and the definition of the total variation distance of probability measures one can check that for each $1 \leq i \leq N$

$$\|\text{Law}(\xi_n^i) - \eta_n\|_{\text{tv}} \leq \frac{C(n)}{N}. \tag{29}$$

The precise magnitude of variability of these mean errors is given by central limit theorems. A full discussion on these fluctuations would be too great a digression here, but as the form of the angle bracket (24) indicates, one can prove that the sequence of random fields

$$U_n^N(f) \stackrel{\text{def.}}{=} \sqrt{N} \left(\gamma_n^N(f) - \gamma_n(f) \right), \qquad f \in \mathcal{B}_b(E)$$

converges in law as $N \to \infty$ to a centered Gaussian field

$$\{U_n(f) \; ; \; f \in \mathcal{B}_b(E)\}$$

satisfying

$$\mathbb{E}\left(U_n(f)^2\right) = \sum_{p=0}^{n} \left(\gamma_p(1)\right)^2 \eta_p\left(\left(L_{p,n}f - \eta_p L_{p,n}f\right)^2 \right)$$

for any $f \in \mathcal{B}_b(E)$ (in the sense of convergence of finite dimensional distributions). The previous fluctuations imply that the sequence of random fields

$$W_n^N(f) \stackrel{\text{def.}}{=} \sqrt{N} \left(\eta_n^N(f) - \eta_n(f) \right), \qquad f \in \mathcal{B}_b(E)$$

converges in law as $N \to \infty$ to the centered Gaussian field

$$W_n(f) \stackrel{\text{def.}}{=} U_n\left(\frac{1}{\gamma_n(1)} (f - \eta_n(f)) \right), \qquad f \in \mathcal{B}_b(E).$$

We conclude this section with some comments on the long-time behavior of the N-interacting particle system approximating models. If the measure-valued dynamical system (20) is sufficiently stable in the sense that it forgets any erroneous initial condition, then one can prove uniform convergence results with respect to the time parameter (see for instance [20,22] and [25] and references therein). For instance, with some suitable stability properties for the Markov kernels $\{K_n \; ; \; n \geq 1\}$ one can find some coefficient $\alpha \in (0, 1/2)$ such that for any bounded Borel test function f, $\|f\| \leq 1$,

$$\forall p \geq 1 \qquad \sup_{n \geq 0} \mathbb{E}\left(\left| \eta_n^N(f) - \eta_n(f) \right|^p \right)^{1/p} \leq \frac{c(p)}{N^\alpha}$$

for some constant $c(p) < \infty$ which only depends on the parameter p. This uniform convergence result with respect to the time parameter leads us to hope that maybe we can construct an asymptotic method to study the convergence of GAs in numerical function optimization in more general settings than the one treated in section 2.2 and in the next section.

4 Numerical Function Analysis

4.1 Description of the Models

The objective of this section is to formulate more precisely the mutation and selection transitions (10) so that the resulting empirical measures of the GA presented in section 2.2 will concentrate in probability, as the time parameter tends to infinity, on the set U^* of the global minima of a given numerical function $U : E \to \mathbb{R}_+$.

As in section 2.2 we assume that E is a finite state space and

$$\beta : \mathbb{N} \to \mathbb{R}_+$$

is an inverse cooling schedule. Let $a : E \times E \to \mathbb{R}_+$ be a numerical function which induces an equivalence relation on E defined by

$$x \sim y \iff a(x, y) = 0.$$

This leads us naturally to consider the partition

$$S_1, \ldots, S_{n(a)}, \qquad n(a) \geq 1$$

induced by $\sim$.

If x is a typical element of E then the equivalence class of x will be denoted by $S(x)$

$$S(x) = \{y \in E \; : \; x \sim y\}.$$

We further require that

$$a(x, y) = 0 \implies U(x) = U(y).$$

A trivial example of the equivalence relation satisfying this condition is given by the following function a

$$a(x, y) = a_0 \, (1 - 1_x(y)), \qquad a_0 > 0.$$

In this case we clearly have $a(x, y) = 0 \iff x = y$.

The mutation kernels K_n and the fitness functions g_n are related to $\beta(n)$ as

$$g_n(x) = e^{-\beta(n) \, U(x)} \qquad \text{and} \qquad K_n(x, y) = k_{\beta(n)}(x, y)$$

with for any $\beta > 0$

$$k_\beta(x, y) = \begin{cases} k(x, y) \, e^{-\beta a(x,y)} & \text{if } a(x, y) > 0 \\ \frac{1}{|S(x)|} \left(1 - \sum_{z \notin S(x)} k(x, z) \, e^{-\beta a(x,z)}\right) & \text{otherwise} \end{cases}$$

where $k : E \times E \to \mathbb{R}_+$ is an irreducible Markov kernel, that is, for any $x \in E$

$$\sum_{y \in E} k(x, y) = 1$$

and for any $(x, y) \in E \times E$ there exists a sequence $x_0, x_1, \ldots, x_r \in E, r \geq 1$ such that

$$x_0 = x, \qquad k(x_k, x_{k+1}) > 0 \qquad (\forall 0 \leq k < r), \qquad x_r = y.$$

We now describe a general construction which allows us to find the asymptotics of the desired transition kernels

$$\widehat{Q}_\beta \stackrel{\text{def.}}{=} Q_\beta^{(1)} Q_\beta^{(2)} \tag{30}$$

where

$$Q_\beta^{(1)}(x, y) = \prod_{p=1}^N k_\beta(x^p, y^p)$$

and

$$Q_\beta^{(2)}(x, y) = \prod_{p=1}^N \sum_{i=1}^N \frac{e^{-\beta U(x^i)}}{\sum_{j=1}^N e^{-\beta U(x^j)}} 1_{x^i}(y^p).$$

It can be directly checked that

$$Q_\beta^{(1)}(x, y) = \left(\prod_{p: a(x^p, y^p)=0} k_\beta(x^p, y^p) \right)$$

$$\times \left(\prod_{p: a(x^p, y^p)>0} k(x^p, y^p) \right) e^{-\beta \sum_{p=1}^N a(x^p, y^p)}$$

$$= \theta_\beta^{(1)}(x, y) \; q^{(1)}(x, y) \; e^{-\beta V^{(1)}(x, y)} \tag{31}$$

with

$$\theta_\beta^{(1)}(x, y) = \prod_{p: a(x^p, y^p)=0} k_\beta(x^p, y^p) |S(x^p)|$$

$$q^{(1)}(x, y) = \left(\prod_{p: a(x^p, y^p)>0} k(x^p, y^p) \right) \left(\prod_{p: a(x^p, y^p)=0} |S(x^p)|^{-1} \right)$$

and

$$V^{(1)}(x, y) = \sum_{p=1}^N a(x^p, y^p).$$

We also notice that

$$\theta_\beta^{(1)}(x, y) \to 1 \text{ as } \beta \to \infty.$$

To describe the asymptotic of $Q_\beta^{(2)}$ as $\beta \to \infty$ we need to recall some terminology introduced in [8]. We will use the superscript $f^\star$ to denote the set of global minima of a given numerical function $f : \mathcal{E} \to \overline{\mathbb{R}}$ on a given finite state space $\mathcal{E}$ so that

$$f^\star \stackrel{\text{def.}}{=} \left\{ x \in \mathcal{E} \; ; \; f(x) = \min_\mathcal{E} f \right\}.$$

The cardinality of a finite set $\mathcal{E}$ will be denoted by $|\mathcal{E}|$ and if x and y belong to E^N and $z \in E$ we write

$$x(z) = |\{p : 1 \leq p \leq N, x_p = z\}|$$

and

$$\widehat{x} = \{p : 1 \leq p \leq N, U(x_p) = \widehat{U}(x)\} \quad \text{and} \quad \widehat{U}(x) = \min_{1 \leq p \leq N} U(x_p).$$

A similar discussion to that above leads to the decomposition

$$\begin{aligned} Q_\beta^{(2)}(x, y) &= \prod_{p=1}^N \sum_{i: x_i = y_p} \frac{e^{-\beta U(x^i)}}{\sum_{j=1}^N e^{-\beta U(x^j)}} \\ &= \prod_{p=1}^N \frac{x(y_p)}{|\widehat{x}|} \frac{e^{-\beta (U(y^p) - \widehat{U}(x))}}{1 + |\widehat{x}|^{-1} \sum_{j \notin \widehat{x}} e^{-\beta (U(x^j) - \widehat{U}(x))}} \\ &= \theta_\beta^{(2)}(x, y) \; q^{(2)}(x, y) \; e^{-\beta V^{(2)}(x, y)}, \end{aligned} \tag{32}$$

with

$$\theta_\beta^{(2)}(x, y) = [1 + |\widehat{x}|^{-1} \sum_{j \notin \widehat{x}} e^{-\beta (U(x^j) - \widehat{U}(x))}]^{-N}$$

$$q^{(2)}(x, y) = \prod_{p=1}^N \frac{x(y_p)}{|\widehat{x}|} \quad \text{and} \quad V^{(2)}(x, y) = \sum_{p=1}^N (U(y^p) - \widehat{U}(x)).$$

As before we also notice that

$$\theta_\beta^{(2)}(x, y) \to 1 \text{ as } \beta \to \infty.$$

If we combine (31) and (32), we conclude that the transition (30) has the same form as in (11), namely

$$\widehat{Q}_\beta(x, z) = \sum_{y \in E^N} \widehat{q}_\beta(x, y, z) \, e^{-\beta \widehat{V}(x, y, z)}$$

with

$$\widehat{q}_\beta(x, y, z) = \widehat{q}(x, y, z) \, \theta_\beta(x, y, z), \quad \widehat{V}(x, y, z) = V^{(1)}(x, y) + V^{(2)}(y, z)$$
$$\theta_\beta(x, y, z) = \theta_\beta^{(1)}(x, y) \, \theta_\beta^{(2)}(y, z), \quad \widehat{q}(x, y, z) = q^{(1)}(x, y) \, q^{(2)}(y, z).$$

Using the fact that $q^{(1)}$ is irreducible, $q^{(2)}(x, x) > 0$ and using the form of $\theta_\beta^{(1)}$, $\theta_\beta^{(2)}$, one can also check that the assumption (H) introduced on page 448 is satisfied and therefore theorem 2 applies to our situation with

$$W(x) = \min_{g \in G(x)} \sum_{(y \to z) \in g} V(y, z) - \min_{x' \in E^N} \min_{g \in G(x')} \sum_{(y \to z) \in g} V(y, z) \tag{33}$$

and

$$V(x, z) = \min\left\{V^{(1)}(x, y) + V^{(2)}(y, z) \; ; \; q^{(1)}(x, y)q^{(2)}(y, z) > 0\right\}. \tag{34}$$

Furthermore, we proved in [21] that there exists a critical population size $N(a, U)$ depending on the function U and on the equivalence relation a such that

$$N \geq N(a, U) \implies W^\star \subset U^\star \cap A$$

where

$$A \stackrel{\text{def.}}{=} \{x \in E^N \; : \; x_i \sim x_j \quad \forall 1 \leq i, j \leq N\}$$

and

$$U^\star \stackrel{\text{def.}}{=} \{x \in E^N \; : \; \widehat{U}(x) = \min_E U\}.$$

4.2 A Test Set Method

To be more precise about this critical population size we need to investigate more closely the properties of the virtual energy function W. We now describe a natural test set approach to study its global minima. This approach is based on the following concept of λ-stability:

Definition 1. Let λ be a nonnegative real number. A subset $H \subset E^N$ is called λ-stable with respect to a communication cost function V when the following conditions are satisfied:

1. $\forall x \in H \; \forall y \notin H \quad V(x, y) > \lambda$,
2. $\forall x \notin H \; \exists y \in H \quad V(x, y) \leq \lambda$.

The importance of the notion of λ-stability resides in the following result which extends lemma 4.1 of Freidlin–Wentzell [32].

Proposition 1 ([21]). *Let λ be a nonnegative real number and $H \subset E^N$. Any λ-stable subset H with respect to V contains $W^\star$.*

One remark is that the subset A is 0-stable with respect to the communication cost function V defined in (34). From this observation one concludes that the canonical process $\{\widehat{\xi}_t \; ; \; t \geq 0\}$ associated with the family of generators $\{L_{\beta_t} = \widehat{Q}_{\beta_t} - \text{Id} \; ; \; t \geq 0\}$, converges as $t \to \infty$ in probability to the set A and

$$\min_{x \in A} W(x) = \min_{x \in E^N} W(x) = 0. \tag{35}$$

Using classical arguments (35) implies that for any $x \in A$

$$W(x) = W_A(x) \stackrel{\text{def.}}{=} \min_{g \in G_A(x)} \sum_{(y \to z) \in g} V_A(y,z) - \min_{x' \in A} \min_{g \in G_A(x')} \sum_{(y \to z) \in g} V_A(y,z)$$

where $G_A(x)$ is the set of x-graphs over A (here the starting and end points of the x-graphs are in A), and $V_A : A \times A \to \overline{\mathbb{R}}_+$ is the *taboo* communication cost function defined by setting for any $x, y \in A$

$$V_A(x,y) = \min \left\{ \sum_{k=0}^{|p|-1} V(p_k, p_{k+1}) \; ; \; p \in C_{x,y} \text{ with } \forall 0 < k < |p| \; p_k \notin A \right\},$$

where $C_{x,y}$ is the set of all paths $p = (p_0, \ldots, p_{|p|})$, with some length $|p|$, admissible for q (that is $q(p_k, p_{k+1}) > 0$ for each $0 \le k < |p|$) leading from x to y (that is $p_0 = x$ and $p_{|p|} = y$).

Let $\mathcal{A} = \{A_1, \ldots, A_{n(a)}\}$ be the partition of A induced by the partition $\mathcal{S} = \{S_1, \ldots, S_{n(a)}\}$ of E associated with the relation $\sim$

$$\forall 1 \le i \le n(a) \qquad A_i \stackrel{\text{def.}}{=} A \cap S_i^N = S_i^N$$

with

$$S_i^N \stackrel{\text{def.}}{=} \underbrace{S_i \times \ldots \times S_i}_{N \text{ times}}.$$

We observe that for any $1 \le i \le n(a)$ and $x, y \in A_i$, $V(x, y) = 0$. Using this observation one can prove that for any $x \in A$

$$W_{\mathcal{A}}(x) = W_A(x) \stackrel{\text{def.}}{=} \min_{g \in G_A(x)} \sum_{(y \to z) \in g} V_{\mathcal{A}}(y,z) - \min_{x' \in A} \min_{g \in G_A(x')} \sum_{(y \to z) \in g} V_{\mathcal{A}}(y,z)$$

where $V_{\mathcal{A}}$ is the communication cost function defined by setting for any $x \in A_i$ and $y \in A_j$ and $1 \le i, j \le n(a)$

$$V_{\mathcal{A}}(x,y) = \min \Bigg\{ \sum_{k=0}^{|p|-1} V(p_k, p_{k+1}) \; : \; p \in C_{x,y}, \quad \exists 0 \le n_1 < n_2 \le |p| \cdot$$
$$\forall 0 \le k \le n_1 \cdot p_k \in A_i, \quad \forall n_1 < k < n_2 \cdot p_k \notin A,$$
$$\forall n_2 \le k \le |p| \cdot p_k \in A_j \Bigg\}. \tag{36}$$

As is easily seen, $V_{\mathcal{A}}(x,y)$ does not depend on the choice of $x \in A_i$ and $y \in A_j$. Another remark is that

$$W^\star = W_A^\star = W_{\mathcal{A}}^\star$$

and therefore the following implication holds for any subset $H \subset A$

$$\exists \lambda \geq 0 \quad : \quad H \ \lambda - \text{stable w.r.t.} \ V_A \Longrightarrow W^* \subset H. \tag{37}$$

In other words, $\widehat{\xi}_t$ converges in probability as $t \to \infty$ to any λ-stable subset $H \subset A$ with respect to V_A. The technical trick now is to find a critical size $N(a, U)$ and a nonnegative constant $\lambda(a, U)$ such that the subset $U^* \cap A$ is $\lambda(a, U)$-stable with respect to V_A.

To describe $N(a, U)$ and $\lambda(a, U)$ precisely, we need to introduce some additional notation. By $\Gamma_{x,y}$, $x, y \in E$, we denote the paths q in E joining x and y, that is,

$$\forall 0 \leq l < |q| \quad k(x_l, x_{l+1}) > 0 \quad q_0 = x \quad q_{|q|} = y.$$

We will also denote as $R(a)$ the smallest integer such that for every $x, y \in E$ in two different classes there exists a path joining x and y with length $|q| \leq R(a)$, namely

$$R(a) = \max_{1 \leq i, j \leq n(a)} \min_{(x_i, x_j) \in S_i \times S_j} \min_{q \in \Gamma_{x_i, x_j}} |q|.$$

It will be also convenient to use the following definitions

$$\Delta a = \min\{a(x, y) \ : \ x, y \in E \ \ a(x, y) \neq 0\},$$
$$\delta(a) = \sup\{a(x, y) \ : \ x, y \in E\}$$

and

$$\Delta U = \min\{|U(x) - U(y)| \ : \ x, y \in E \ \ U(x) \neq U(y)\},$$
$$\delta(U) = \sup\{|U(x) - U(y)| \ : \ x, y \in E\}.$$

To formulate our convergence result precisely we need the following lemma.

Lemma 1 ([21]). *For every $x \in A$ there exists a state $y \in U^* \cap A$ such that*

$$V_A(x, y) \leq (\delta(a) + \delta(U)) \ R(a).$$

For every $x, y \in A$ such that $\widehat{U}(x) < \widehat{U}(y)$ we have

$$V_A(x, y) \geq \min(\Delta a, \Delta U) \ N.$$

Let us write

$$\lambda(a, U) \stackrel{\text{def.}}{=} (\delta(a) + \delta(U)) \ R(a)$$

and

$$N(a, f) \stackrel{\text{def.}}{=} \lambda(a, U) / \min(\Delta a, \Delta U).$$

Using the above lemma one concludes that

$$N > N(a, f) \Longrightarrow U^* \cap A \ \text{is} \ \lambda(a, U) - \text{stable with respect to} \ V_A. \tag{38}$$

If we combine (37) and (38), with theorem 2 one concludes that:

Theorem 3 ([21]). *There exists a constant C_0 (which can be explicitly described in terms of V), such that if $N \geq N(a, U)$ and if $\beta(t)$ takes the parametric form $\beta(t) = \frac{1}{C} \log t$ for sufficiently large t and $C > C_0$ then*

$$\lim_{t \to \infty} \mathbb{P}\left(\widehat{\xi}_t \in U^* \cap A\right) = 1.$$

5 Refinements and Variants

The research literature abounds with variations of the GA described in section 2. Each of these variants is intended to make the selection and/or the mutation more efficient in some sense. The convergence analysis of all these alternative schemes is far from being complete. We also emphasize that these variations come from different sources of inspiration. Some of them are strongly related to traditional weighted re-sampling plans in weighted bootstrap theory (see [4] and references therein). Another source of inspiration was provided by branching and interacting particle system theory. The aim of this section is to introduce the reader to these recently established connections between branching and interacting particle systems, GAs, simulated annealing, and bootstrap theory.

We begin our program with an alternative GA whose transitions are obtained through choosing randomly at each step the selection or the mutation transition. This variation has been presented for the first time in [21] to improve the convergence results of the classical GA studied in section 4.

We will use the general methodology presented in section 2.2 and the test set approach of section 4 to prove that the corresponding genetic-type algorithm converges towards the set of the global minima of a desired numerical function. These results can be found with a complete proof in [21]. We will make some comments on how these results improve the one of section 4.

The second variation has been presented in [27] for solving non linear filtering problems. The main difference with the classical GA of section 2 lies in the fact that in the former the mutation kernels also depend on the fitness function. The corresponding mutation transition has, in fact, a natural interpretation in nonlinear filtering and can be regarded as a conditional transition probability. In reference to nonlinear filtering we will call this kind of mutation a conditional mutation.

We end this section with a brief presentation of several branching genetic-type algorithms. These branching strategies are strongly related to weighted bootstrap techniques [4].

There are many open problems concerning these variations such as finding a way to study the convergence in global optimization problems.

5.1 Random Selection/Mutation Transitions

The setting here is exactly as in section 2.2 and section 4 but the genetic-type algorithm is now described by the transition probability kernels

$$\tilde{Q}_\beta = \alpha_1 Q_\beta^{(1)} + \alpha_2 Q_\beta^{(2)} \qquad \alpha_1 + \alpha_2 = 1 \quad (\alpha_1, \alpha_2 \in (0, 1)).$$

Returning to the definition of $Q_\beta^{(1)}$ and $Q_\beta^{(2)}$ given in (31) and (32) and using the same notation, one concludes that $\tilde{Q}_\beta$ has the same form as in (11)

$$\tilde{Q}_\beta(x, z) = \sum_{v \in \mathcal{V}} \widehat{q}_\beta(x, v, z)\, e^{-\beta \widehat{V}(x,v,z)}$$

with $\mathcal{V} = \{1, 2\}$ and for any $v \in \mathcal{V}$

$$\widehat{q}_\beta(x, v, y) = \theta_\beta(x, v, y)\, \widehat{q}(x, v, y) \qquad \theta_\beta(x, v, y) = \theta_\beta^{(v)}(x, y)$$
$$\widehat{q}(x, v, y) = \alpha_v\, q^{(v)}(x, y) \qquad \widehat{V}(x, v, y) = V^{(v)}(x, y).$$

To clarify the presentation we use the superscript $(\tilde{\cdot})$ to denote the communication cost function $\tilde{V}$, the critical height constant $\tilde{C}_0$ arising in theorem 2 and the virtual energy function $\tilde{W}$ associated with the transition probability kernels $\tilde{Q}_\beta$. From the above observations and theorem 2, choosing β of the form

$$\beta(t) = \frac{1}{C} \log t \quad \text{where} \quad C > \tilde{C}_0$$

for sufficiently large t, yields that the canonical process

$$(\Omega, P, (F_t)_{t \geq 0}, (\tilde{\xi}_t)_{t \geq 0})$$

associated with the family of generators

$$L_{\beta(t)} = \tilde{Q}_{\beta(t)} - \mathrm{Id}$$

converges in probability to the set of the global minima $\tilde{W}^*$ of the virtual energy $\tilde{V}$ associated with $\tilde{Q}_\beta$ and defined as in (33) by replacing the communication cost functions V by $\tilde{V}$ where

$$\tilde{V}(x, y) = \min\left\{ V^{(v)}(x, y) \,;\, v \in \mathcal{V},\; q^{(v)}(x, y) > 0 \right\}.$$

By the same test set approach we used in section 4 the technical trick here is to find a critical size $\tilde{N}(a, U)$ and a nonnegative constant $\tilde{\lambda}(a, U)$ such that the subset $U^* \cap A$ is $\tilde{\lambda}(a, U)$-stable with respect to $\tilde{V}_A$, where $\tilde{V}_A$ is defined as in (36) by replacing the communication cost function V by $\tilde{V}$. In this setting the analogue of lemma 1 is the following

Lemma 2 ([21]). *For every $x, y \in A$ such that $\widehat{U}(x) \geq \widehat{U}(y)$ we have*

$$\widetilde{V}_A(x, y) \leq \delta(a) \, R(a).$$

For every $x, y \in A$ such that $\widehat{U}(x) < \widehat{U}(y)$ we have

$$\widetilde{V}_A(x, y) \geq \min(\Delta a, \Delta U) \, N.$$

Now, if we write

$$\widetilde{\lambda}(a, U) = \delta(a) \, R(a) \qquad \text{and} \qquad \widetilde{N}(a, U) = \widetilde{\lambda}(a, U) / \min(\Delta a, \Delta U)$$

one concludes that

$$N > \widetilde{N}(a, U) \implies U^\star \cap A \text{ is } \widetilde{\lambda}(a, U) - \text{stable with respect to } V_A.$$

Using the same line of arguments as in the end of section 4 one finally obtains

Theorem 4 ([21]). *If $N \geq \widetilde{N}(a, U)$ and if $\beta(t)$ takes the parametric form $\beta(t) = \frac{1}{C} \log t$ for sufficiently large t and $C > \widetilde{C}_0$ then*

$$\lim_{t \to \infty} \mathbb{P}\left(\widetilde{\xi}_t \in U^\star \cap A\right) = 1.$$

Several comments are in order. The first remark is that in contrast to $\lambda(a, U)$, the constant $\widetilde{\lambda}(a, U)$ does not depend any more on U. Furthermore, the critical population size $\widetilde{N}(a, U)$ does not depend on $\delta(U)$. In addition, the bound

$$\lambda(a, U) > \widetilde{\lambda}(a, U)$$

seems to indicate that it is more difficult for the algorithm associated with the communication cost function V to move from one configuration to a better one. This observation also implies that for the critical size values we obtained we have that

$$N(a, U) = \left(1 + \frac{\delta(U)}{\delta(a)}\right) \widetilde{N}(a, U) > \widetilde{N}(a, U).$$

Let us see what happens when this alternative genetic-type model specializes to the case where the state is

$$E = \{-1, +1\}^S \qquad S = [-n, n]^p \qquad p \geq 1$$

and the fitness function U is given by

$$U(x) = \sum_{s \in S} \sum_{s' \in V_s} \mathcal{I}_{s,s'} \, x(s) \, x(s') + \sum_{s \in S} h(s) \, x(s)$$

where $\mathcal{I}_{s,s'}, h(s) \in \mathbb{Z}$, and

$$\forall s \in S \qquad V_s = \{s' \in S : |s_k - s'_k| \leq 1, \ 1 \leq k \leq p\}.$$

Let k be the Markovian mutation kernel on S given by

$$k(x, y) = \frac{1}{|\mathcal{V}(x)|} 1_{\mathcal{V}(x)}(y)$$

with

$$\mathcal{V}(x) \stackrel{\text{def}}{=} \{y \in E \;:\; \text{card}\{s \in \mathcal{S} \;:\; x(s) \neq y(s)\} \leq 1\}.$$

Suppose that the function a is given by

$$a(x, y) = (1 - 1_x(y)) \qquad \forall (x, y) \in E^2.$$

Then, one can check that

$$R(a) \leq \max_{x,y} \min_{q \in C_{x,y}} |q| = \text{card}(\mathcal{S}) = (2n+1)^p$$

and

$$\delta(a) = \Delta(a) = 1.$$

Let $\mathcal{I}_{s,s'}$ and $h(s)$ be chosen so that $\Delta U \geq 1$ and let N be an integer that $N > (2n+1)^p$. The above theorem shows that N individuals will solve the optimization problem when using the GA associated with $\tilde{Q}_\beta$.

5.2 Conditional Mutations

We now present some genetic-type variants arising in nonlinear filtering literature (see [12,20] and references therein). For the sake of unity and to highlight issues in both nonlinear filtering and numerical function analysis, we place ourselves in the abstract setting of section 2 and section 3.

The first variation is based on the observation that the distribution flow $\{\widehat{\eta}_n \;;\; n \geq 0\}$ is a solution of a measure-valued dynamical system defined as in (5), by replacing the transitions K_n and the fitness functions g_n by the transitions $\widehat{K}_n$ and the fitness functions $\widehat{g}_n$ defined for any $f \in \mathcal{B}_b(E)$ by setting

$$\widehat{K}_n(f) \stackrel{\text{def.}}{=} \frac{K_n(g_n f)}{K_n(g_n)} \quad \text{and} \quad \widehat{g}_n \stackrel{\text{def.}}{=} K_n(g_n).$$

More precisely, one can check that

$$\widehat{\eta}_n = \widehat{\Phi}_n(\widehat{\eta}_{n-1}), \qquad n \geq 1 \tag{39}$$

with

$$\widehat{\Phi}_n(\eta) \stackrel{\text{def.}}{=} \Psi_n(\eta) \widehat{K}_n$$

and

$$\widehat{\Psi}_n(\eta)(f) \stackrel{\text{def.}}{=} \frac{\eta(\widehat{g}_n f)}{\eta(\widehat{g}_n)} \qquad \forall f \in \mathcal{B}_b(E).$$

As in section 2.1 we can associate with (39) an N-interacting particle system $\{\zeta_n \ ; \ n \geq 0\}$ which is a Markov chain in E^N with transitions

$$\mathbb{P}\left(\zeta_{n+1} \in dx \mid \zeta_n = y\right) = \prod_{p=1}^{N} \widehat{\Phi}_{n+1}\left(\frac{1}{N}\sum_{i=1}^{N} \delta_{y^i}\right)(dx^p), \qquad n \geq 0$$

and initial law $\widehat{\eta}_0 = \Psi(\eta_0)$, where, as usual, $dx \stackrel{\text{def}}{=} dx^1 \times \cdots \times dx^N$ is an infinitesimal neighborhood of the point $x = (x^1, \ldots, x^N) \in E^N$, $y = (y^1, \ldots, y^N) \in E^N$. Arguing as in section 2.1, it is transparent that this transition is decomposed into two separate mechanisms, namely, for each $n \geq 0$

$$\zeta_n \stackrel{\text{def.}}{=} \left(\zeta_n^1, \ldots, \zeta_n^N\right) \xrightarrow{\text{Selection}} \widehat{\zeta}_n \stackrel{\text{def.}}{=} \left(\widehat{\zeta}_n^1, \ldots, \widehat{\zeta}_n^N\right) \xrightarrow{\text{Mutation}} \zeta_{n+1}.$$

The selection transition is now defined by

$$\mathbb{P}\left(\widehat{\zeta}_n \in dx \mid \zeta_n = y\right) = \prod_{p=1}^{N} \sum_{i=1}^{N} \frac{\widehat{g}_n(y^i)}{\sum_{j=1}^{N} \widehat{g}_n(y^j)} \delta_{y^i}(dx^p)$$

and the mutation step

$$\mathbb{P}\left(\zeta_{n+1} \in dx \mid \widehat{\zeta}_n = y\right) = \prod_{p=1}^{N} \widehat{K}_{n+1}\left(y^p, dx^p\right).$$

We emphasize that in contrast to the latter genetic model, this genetic particle scheme involves mutation transitions that depend on the fitness functions. The study of this variant has been initiated in [27], and large population asymptotic are described in [24] and [20].

5.3 Branching Genetic-Type Algorithms

We end this section with a brief description of branching and genetic-type variants presented in [20]. Here again, we place ourselves in the abstract setting of section 2 and section 3.

All these branching strategies are based on the same natural idea. Namely, how to approximate an updated empirical measure of the following form

$$\Psi_n\left(\frac{1}{N}\sum_{i=1}^{N}\delta_{\xi_n^i}\right) = \sum_{i=1}^{N} \frac{g_n(\xi_n^i)}{\sum_{j=1}^{N} g_n(\xi_n^j)} \delta_{\xi_n^i} \qquad (40)$$

by a new probability measure with atoms of size integers multiples of $1/N$. In the GA presented in section 2.1 this approximation is done by sampling N-independent random variables

$$\{\widehat{\xi}_n^i \, ; \, 1 \leq i \leq N\}$$

with common law (40) and the corresponding approximating measure is given by

$$\frac{1}{N} \sum_{i=1}^{N} \delta_{\widehat{\xi}_n^i} = \sum_{i=1}^{N} \frac{M_n^i}{N} \delta_{\xi_n^i}$$

where

$$\left(M_n^1, \ldots, M_n^N\right) \stackrel{\text{def.}}{=} \text{Multinomial}\left(N, W_n^1, \ldots, W_n^N\right)$$

and for any $1 \leq i \leq N$

$$W_n^i \stackrel{\text{def.}}{=} \frac{g_n(\xi_n^i)}{\sum_{j=1}^{N} g_n(\xi_n^j)}.$$

Using these notations the random and $\mathbb{N}$-valued random variables

$$\left(M_n^1, \ldots, M_n^N\right)$$

can be seen as random numbers of offsprings created at the positions $(\xi_n^1, \ldots, \xi_n^N)$. The above question is strongly related to weighted bootstrap and GA theory (see for instance [4] and references therein). In this connection the above multinomial approximating strategy can be viewed as a weighted Efron bootstrap.

Let us present several examples of branching laws. The first one is known as the *remainder stochastic sampling* in GA literature. It has been presented for the first time in [2,3]. From a purely practical point of view this sampling technique seems to be the more efficient since it is extremely time-saving, and if the branching particle model is only based on this branching selection scheme then the size of the system remains constant.

In what follows we denote by $[a]$ (respectively $\{a\} = a - [a]$) the integer part (respectively the fractional part) of $a \in \mathbb{R}$.

1. Remainder Stochastic Sampling

At each time $n \geq 0$, each particle ξ_n^i branches directly into a fixed number of offsprings

$$\overline{M}_n^i \stackrel{\text{def.}}{=} [NW_n^i] \qquad \forall 1 \leq i \leq N$$

so that the intermediate population consists of $\overline{N}_n \stackrel{\text{def.}}{=} \sum_{i=1}^{N} \overline{M}_n^i$ particles. To prevent extinction and to keep the size of the system fixed it is convenient to introduce in this population $\tilde{N}_n$ additional particles with

$$\tilde{N}_n \stackrel{\text{def.}}{=} N - \overline{N}_n = \sum_{i=1}^{N} NW_n^i - \sum_{i=1}^{N} [NW_n^i] = \sum_{i=1}^{N} \{NW_n^i\}$$

One natural way to do this is to introduce the additional sequence of branching numbers

$$\left(\tilde{M}_n^1, \ldots, \tilde{M}_n^N\right)$$

$$\stackrel{\text{def.}}{=} \text{Multinomial}\left(\tilde{N}_n, \frac{\{NW_n^1\}}{\sum_{j=1}^{N}\{NW_n^j\}}, \ldots, \frac{\{NW_n^N\}}{\sum_{j=1}^{N}\{NW_n^j\}}\right). \tag{41}$$

More precisely, if each particle ξ_n^i again produces a number of $\tilde{M}_n^i$ additional offspring, $1 \leq i \leq N$, then the total size of the system is kept constant.
At the end of this stage, the particle system $\widehat{\xi}_n$ again consists of N particles denoted by

$$\widehat{\xi}_n^i = \xi_n^k$$

with

$$1 \leq k \leq N, \quad \sum_{l=1}^{k-1} \overline{M}_n^l + 1 \leq i \leq \sum_{l=1}^{k-1} \overline{M}_n^l + \overline{M}_n^k$$

and for

$$1 \leq k \leq N, \quad \sum_{l=1}^{k-1} \tilde{M}_n^l + 1 \leq i \leq \sum_{l=1}^{k-1} \tilde{M}_n^l + \tilde{M}_n^k$$

$$\widehat{\xi}_n^{\overline{N}_n + i} = \xi_n^k.$$

The multinomial (41) can also be defined as follows

$$\tilde{M}_n^k = \text{card}\left\{1 \leq j \leq \tilde{N}_n \ ; \ \tilde{\xi}_n^j = \xi^k\right\} \qquad 1 \leq k \leq N$$

where $(\tilde{\xi}_n^1, \ldots, \tilde{\xi}_n^{\tilde{N}_n})$ are $\tilde{N}_n$ independent random variables with common law

$$\sum_{i=1}^{N} \frac{\{NW_n^i\}}{\sum_{j=1}^{N} \{NW_n^j\}} \delta_{\xi_n^i}.$$

2. Independent Branching Numbers

In the next examples the branching numbers are, at each time step, independent one of each other (conditionally on the past). As a result, the size of the population at each time n is not fixed but random. The corresponding branching genetic-type algorithms can be regarded as a two-step Markov chain

$$(N_n, \xi_n) \xrightarrow{Branching} (\widehat{N}_n, \widehat{\xi}_n) \xrightarrow{Mutation} (N_{n+1}, \xi_{n+1}) \qquad (42)$$

with product state space $\mathcal{E} = \bigcup_{\alpha \in \mathbb{N}} (\{\alpha\} \times E^\alpha)$ with the convention that $E^\alpha = \{\Delta\}$ is a cemetery if $\alpha = 0$. We will note

$$\mathcal{F} = \{F_n, \widehat{F}_n : n \geq 0\}$$

the canonical filtration associated with (42) so that

$$F_n \subset \widehat{F}_n \subset F_{n+1}.$$

(a) Bernoulli branching numbers

The Bernoulli branching numbers were introduced in [11] and further developed in [12]. They are defined as a sequence $M_n = (M_n^i, 1 \leq i \leq N_n)$ of conditionally independent random numbers with respect to F_n with distribution given for any $1 \leq i \leq N_n$ by

$$P(M_n^i = k | F_n) = \begin{cases} \{N_n W_n^i\} & \text{if } k = [N_n W_n^i] + 1 \\ 1 - \{N_n W_n^i\} & \text{if } k = [N_n W_n^i] \end{cases}$$

In addition, it can be seen from the relation

$$\sum_{i=1}^{N_n} (N_n W_n^i) = N_n$$

that at least one particle has one offspring (see [11] for more details). Therefore, using the above branching correction the particle system never dies. It is also worth observing that the Bernoulli branching numbers are defined as in the *remainder stochastic sampling* by replacing the multinomial remainder branching law (41) by a sequence of N_n independent Bernoulli random variables $\left(\tilde{M}_n^1, \ldots, \tilde{M}_n^{N_n}\right)$ given by

$$P(\tilde{M}_i^{N_n} = 1 | F_n) = 1 - P(\tilde{M}_i^{N_n} = 0 | F_n) = \{N_n W_n^i\}.$$

(b) Poisson branching numbers

The Poisson branching numbers are defined as a sequence $M_n = (M_n^i, 1 \leq i \leq N_n)$ of conditionally independent random numbers with respect to F_n with distribution given for any $1 \leq i \leq N_n$ by

$$\forall k \geq 0 \qquad P(M_n^i = k | F_n) = \exp(-N_n W_n^i) \frac{(N_n W_n^i)^k}{k!}.$$

(c) **Binomial branching numbers**

The binomial branching numbers are defined as a sequence

$$M_n = (M_n^i, \ 1 \le i \le N_n)$$

of conditionally independent random numbers with respect to F_n with distribution given for any $1 \le i \le N_n$ by

$$P(M_n^i = k | F_n) = \binom{N_n}{k} (W_n^i)^k (1 - W_n^i)^{N_n - k}$$

for any $0 \le k \le N_n$

The previous models are described in full detail in [12]. In particular, it is shown that the GA with multinomial branching laws arises by conditioning a GA with Poisson branching laws. The law of large numbers and large deviations for the genetic model with Bernoulli branching laws are studied in [12] and [14]. The convergence analysis of these particle approximating schemes is still in progress.

6 Continuous-Time Genetic Algorithms

We shall now describe the continuous-time version of the GA discussed in section 2. This particle algorithm has been introduced in [23] for solving a flow of distributions defined by the ratio

$$\eta_t(f) = \frac{\gamma_t(f)}{\gamma_t(1)} \qquad \forall f \in \mathcal{B}_b(E) \qquad t \in \mathbb{R}_+ \qquad (43)$$

where $\gamma_t(f)$ is defined through a Feynman–Kac formula of the following form:

$$\gamma_t(f) = \mathbb{E}\left(f(X_t) \ \exp\left(\int_0^t U_s(X_s) \, ds \right) \right)$$

where $\{X_t \ ; \ t \in \mathbb{R}_+\}$ is a càdlàg and time-inhomogeneous Markov process taking values in a Polish space E and $\{U_t \ ; \ t \in \mathbb{R}_+\}$ is a measurable collection of locally bounded (in time) and measurable nonnegative functions. Here, we merely content ourselves in describing the mathematical models of such particle numerical schemes. The detailed convergence analysis as the size of the system tends to infinity can be founded in [20] or [23]. In order to illustrate the idea in a simple form we will also make the sanguine assumption that X is a time-homogeneous Markov process with initial law η_0, its infinitesimal generator is a bounded linear operator on the set on bounded Borel test functions $\mathcal{B}_b(E)$ and $U_t = U$ is a time-homogeneous function. The interested reader is referred to [20] for a more general presentation including Riemannian or Euclidean diffusions X.

To motivate our work we also mention that the Feynman–Kac model (43) has different interpretations coming from quite distinct research areas. First it can be regarded as the distributions of a random particle X killed at a given rate and conditioned by non-extinction (see for instance [48]). Second the previous Feynman–Kac formula may serve to model the robust version of the optimal filter in nonlinear filtering settings (see [20] and [23]). Finally, as pointed out in [20], the ratio distributions (43) can also be regarded as the solution flow of a simple generalized and spatially homogeneous Boltzmann equation as defined in [36,44].

As for the discrete-time models discussed in section 2.1 and section 3, one of the best ways to define the genetic particle approximating models of (43) is through the dynamical structure of (43). By definition, one can easily check that for any bounded Borel test function $f \in \mathcal{B}_b(E)$

$$\frac{d}{dt}\eta_t(f) = \eta_t(L(f)) + \eta_t(fU) - \eta_t(f)\eta_t(U) = \eta_t(L_{\eta_t}(f)) \tag{44}$$

where L_η, for any fixed distribution η on E, is the bounded linear operator on $\mathcal{B}_b(E)$ defined by

$$L_\eta(f)(x) = L(f)(x) + \int (f(z) - f(x))\, U(z)\, \eta(dz). \tag{45}$$

As its discrete-time analogue (20), we want to solve a nonlinear and measure-valued dynamical system (44), and the associate generator $\mathcal{L}_\eta$ is decomposed into two separate generators.

To highlight the quadratic nature of (44) and the connections with spatially-homogeneous Boltzmann equations we also notice that (44) can be rewritten as

$$\frac{d}{dt}\eta_t(f) = \eta_t(L(f)) + \frac{1}{2}\int \eta_t(dx)\,\eta_t(dy)$$
$$\times \left((f(x^\star) - f(x)) + (f(y^\star) - f(y))\right) Q(x, y; dx^\star, dy^\star)$$

with

$$Q(x, y; .) = U(y)\,\delta_{(y,y)} + U(x)\,\delta_{(x,x)}.$$

In section 6.1 we discuss a Moran-type particle approximation of the Feynman–Kac formula (43). We also give an illustration of the semi-group techniques introduced in [20] for proving useful convergence results as the size of the population tends to infinity including the central limit theorem and exponential bounds. In the final section 6.2 we propose a branching and interacting particle approximating scheme. To the best of our knowledge this branching-type particle approximation of the Feynman–Kac formula (43) has not been covered in the literature. We will also give the connections between this particle scheme and the previous Moran particle model.

6.1 A Moran Particle Model

Description of the Model

As traditional, starting from a family $\{\mathcal{L}_\eta \; ; \; \eta \in M_1(E)\}$, we consider an interacting N-particle system

$$(\xi_t)_{t \geq 0} = ((\xi_t^1, \ldots, \xi_t^N))_{t \geq 0}$$

which is Markov process on the product space E^N, $N \geq 1$, whose infinitesimal generator acts on bounded Borel functions $f : E^N \to \mathbb{R}$ by setting for any $x = (x_1, \ldots, x_N) \in E^N$

$$\mathcal{L}(f)(x) = \sum_{i=1}^{N} L_{m(x)}^{(i)}(f)(x) \qquad \text{with} \quad m(x) \stackrel{\text{def.}}{=} \frac{1}{N} \sum_{i=1}^{N} \delta_{x_i}$$

and where the notation $\mathcal{G}^{(i)}$ has been used instead of $\mathcal{G}$ when an operator $\mathcal{G}$ on $\mathcal{B}_b(E)$ acts on the i-th variable of $f(x_1, \ldots, x_N)$. This abstract and general formulation is well-known in mean field interacting particle system literature (the interested reader is for instance referred to [44] and [50] and references therein). Taking into consideration definition (45) we get

$$\mathcal{L} = \tilde{\mathcal{L}} + \widehat{\mathcal{L}} \tag{46}$$

where

$$\tilde{\mathcal{L}}(f)(x) = \sum_{i=1}^{N} L^{(i)}(f)(x)$$

and

$$\widehat{\mathcal{L}}(f)(x) = \sum_{i=1}^{N} \sum_{j=1}^{N} \left(f(x^{(i,j)}) - f(x) \right) \frac{1}{N} U(x_j)$$

and where for $1 \leq i, j \leq N$ and $x = (x_1, \ldots, x_N) \in E^N$, $x^{(i,j)}$ is the element of E^N given by

$$\forall \, 1 \leq k \leq N, \qquad x_k^{(i,j)} = \begin{cases} x_k, & \text{if } k \neq i \\ x_j, & \text{if } k = i \end{cases}$$

In order to describe more explicitly the time evolution of the E^N-valued Markov process $\{\xi_t \; ; \; t \geq 0\}$ with infinitesimal generator $\mathcal{L}$, it is convenient to write (46) as follows

$$\begin{aligned}\mathcal{L}(f)(x) &= \tilde{\mathcal{L}}(f)(x) + \lambda(x) \int_{E^N} (f(y) - f(x)) \, Q(x, dy) \\ &= \tilde{\mathcal{L}}(f)(x) + \widehat{\lambda} \int_{E^N} (f(y) - f(x)) \, \widehat{Q}(x, dy)\end{aligned} \tag{47}$$

with

$$\lambda(x) = \sum_{i=1}^{N} U(x_i) = N\, m(x)(U) \quad \text{and} \quad \widehat{\lambda} = N\, \|U\|$$

and

$$Q(x, dy) = \sum_{i,j=1}^{N} \frac{1}{N} \frac{U(x_i)}{\sum_{k=1}^{N} U(x_k)} \delta_{x^{(i,j)}}(dy)$$

$$\widehat{Q}(x, dy) = \left(1 - m(x)\left(\frac{U}{\|U\|}\right)\right) \delta_x(dy) + m(x)\left(\frac{U}{\|U\|}\right) Q(x, dy).$$

The construction of $\{\xi_t\ ;\ t \geq 0\}$ on an explicit probability space is now classical (see for instance [23] or [31]). For the convenience of the reader we propose a basic construction based on the second decomposition (47).

Let $\{X^{(k,i)}(a)\ ;\ (k, i) \in \mathbb{N}^2,\ a \in E\}$ be a collection of independent copies of $\{X(a)\ ;\ a \in E\}$ where for any $a \in E$, $X(a)$ denotes the process X starting at a. Let $\{T_k\ ;\ k \in \mathbb{N}\}$ ($T_0 = 0$) be a sequence of independent and identically distributed random variables on $\mathbb{R}_+$ with a common exponential law with parameter $N\|U\|$.

The random times $\{T_k\ ;\ k \in \mathbb{N}\}$ ($T_0 = 0$) will be regarded as the random dates at which competitive interaction occurs. The initial particle system $\xi_0 = (\xi_0^1, \ldots, \xi_0^N)$ consists of N independent random variables with common law η_0.

1. **Mutation**

    Between the dates T_{k-1} and T_k the particles evolve randomly and independently according the law of the time-inhomogeneous Markov process X. That is for any $1 \leq i \leq N$

    $$\xi_t^i = X^{(k,i)}_{t-T_{k-1}}\left(\xi_{T_{k-1}}^i\right), \quad \forall t \in [T_{k-1}, T_k[\quad k \geq 1.$$

2. **Competitive Selection**

    At the time $t = T_k$, $\xi_{T_k} = (\xi_{T_k}^1, \ldots, \xi_{T_k}^N)$ is an E^N-valued random variable with law $\widehat{Q}(\xi_{T_k^-}, \cdot)$.

The important difference between this Moran-type particle model and the classical one is that for the former N-particles system, the total rate of selection jumps $\widehat{\lambda}$ is of order N, while for the classical N-particle Moran model it is of order N^2. It is that difference of scaling, with comparatively less frequent selections, which enables us to end up with a deterministic process in the limit.

Furthermore, even if we had multiplied the rate of selection by N, the limit exists (as a right-continuous measure-valued stochastic process) only if the weight of replacing the particle ξ_t^i by the particle ξ_t^j is symmetrical in ξ_t^i and ξ_t^j, a condition which is not satisfied here, since due to the fitness functions, its value is $U_t(\xi_t^j)/N$. In our case, more frequent selections would oblige the limit measure-valued process to jump instantaneously from a probability to another one better suited for the maximization of U. In fact, an asymmetrical weighted sampling needs a selection total rate of order N (this can be deduced from the calculations given in the section 5.7.8 of [15]), if one wants to end up with a bounded selection generator. Then one can add the natural non-weighted sampling selection (see section 2.5 of [15], or more generally, any other symmetrical weighted sampling selection) with a total rate of order N^2, to obtain in the limit a Fleming–Viot process with selection, as it is defined directly in the level of the measure-valued process (and not at the particles system approximation level) in section 10.1.1 of [15] (or, more generally, p. 175 of this review).

Asymptotic Behavior

The interpretation of the distribution flow $\{\eta_t \; ; \; t \geq 0\}$ in terms of the limit of the empirical measures

$$\eta_t^N \stackrel{\text{def.}}{=} \frac{1}{N} \sum_{i=1}^{N} \delta_{\xi_t^i} \tag{48}$$

as $N \to \infty$ is given in [20,23] including the central limit theorem and exponential bounds, see also [36,44] for an alternative approach using coupling techniques. To see that (48) is a reasonable approximation of η_t observe that for any bounded Borel function $\varphi \in \mathcal{B}_b(E)$ if

$$f(x_1, \ldots, x_N) \stackrel{\text{def.}}{=} \frac{1}{N} \sum_{i=1}^{N} \varphi(x_i)$$

then for any $x = (x_1, \ldots, x_N) \in E^N$

$$\mathcal{L}(f)(x) = m(x) \left(L_{m(x)}(\varphi) \right).$$

Our aim is now to make some comments on the semi-group approach presented in [20] to study the asymptotic behavior of η_t^N as the population size N tends to infinity.

Under our assumption, it is well-known (see lemma 3.68, p. 446 in [38]) that for any bounded Borel test function $f \in \mathcal{B}_b(E^N)$ the stochastic process

$$M_t(f) \stackrel{\text{def.}}{=} f(\xi_t) - f(\xi_0) - \int_0^t \mathcal{L}(f)(\xi_s) \, ds$$

is a square integrable martingale, and its angle bracket is given by

$$\langle M(f) \rangle_t = \int_0^t \Gamma(f, f)(\xi_s) \, ds$$

where Γ is the "carré du champ" associated with $\mathcal{L}$

$$\forall f \in \mathcal{B}_b(E^N), \qquad \Gamma(f, f) = \mathcal{L}(f^2) - 2f\,\mathcal{L}(f).$$

Using the decomposition (46) and the definition of $\tilde{\mathcal{L}}$ and $\widehat{\mathcal{L}}$ it is easy to establish that

$$\Gamma(f, f) = \tilde{\Gamma}(f, f) + \widehat{\Gamma}(f, f)$$

with

$$\tilde{\Gamma}(f, f) = \tilde{\mathcal{L}}(f^2) - 2f\,\tilde{\mathcal{L}}(f) \qquad \widehat{\Gamma}(f, f) = \widehat{\mathcal{L}}(f^2) - 2f\,\widehat{\mathcal{L}}(f)$$

and if $f \in \mathcal{B}_b(E^N)$ is chosen so that

$$f(x) = m(x)(\varphi)$$

for some $\varphi \in \mathcal{B}_b(E)$ then

$$\tilde{\Gamma}(f, f)(x) = \frac{1}{N}\, m(x)\,(\Gamma_L(\varphi, \varphi))$$

with

$$\Gamma_L(\varphi, \varphi) = L(\varphi^2) - 2\varphi\, L(\varphi)$$

and

$$\widehat{\Gamma}(f, f)(x) = \frac{1}{N}\, m(x)\left((\varphi - m(x)(\varphi))^2\,(U + m(x)(U))\right).$$

Using these notations one concludes that

$$d\eta_t^N(\varphi) = \eta_t^N(L_{\eta_t^N}(\varphi))\, dt + dM_t(f)$$

with

$$|\langle M(f)\rangle_t| \leq \frac{C_t}{N}\, \|\varphi\|^2, \qquad C_t < \infty \qquad \forall t \geq 0.$$

One can use this result to check that the sequence of distributions $\{\eta_t^N\ ;\ t \geq 0\}$ is weakly compact and any weak limit point is concentrated on the set of solutions of (44). Using the continuity of the angle bracket and the construction of ξ_t, one can check that there exists some finite constant $C'_t < \infty$ such that the jumps $\Delta M_t(f)$ of the previously defined martingale are bounded by $C'_t\|\varphi\|/N$, that is $\mathbb{P}$-a.s.

$$|\Delta M_t(f)| \leq \frac{C'_t}{N}\, \|\varphi\|.$$

Let us recall a classical exponential inequality for martingales M_t starting at 0 and whose jumps are bounded uniformly by $a \in]0, \infty[$: for all $0 < \epsilon \leq \frac{b}{a}$ and $t > 0$

$$\mathbb{P}\left(\sup_{s \in [0,t]} |M_s| > \epsilon, \ \langle M \rangle_t \leq b\right) \leq 2 \exp -\frac{\epsilon^2}{4b}. \tag{49}$$

This inequality may be established using calculations from section 4.13 of [43] (see corollary 3.3 in [47]). Now, if we apply this inequality to the martingale $M_t(f)$ one obtains the following result:

Proposition 2. *For any bounded Borel test function $\varphi \in \mathcal{B}_b(E)$, and $T > 0$, and $0 < \epsilon \leq \|\varphi\|$, we have that*

$$\mathbb{P}\left(\sup_{t \in [0,T]} |\eta_t^N(\varphi) - \eta_0^N(\varphi) - \int_0^t \eta_s^N(L_{\eta_s^N}(\varphi)) \, ds| > \epsilon\right)$$
$$\leq 2 \exp -\frac{N\epsilon^2}{C(t)\|\varphi\|^2}$$

for some finite constant $C(t) < \infty$.

To get some more precise estimates we proceed as in discrete-time settings. We start by noting that

$$\gamma_t(1) = \exp \int_0^t \eta_s(U) \, ds$$

and therefore for any $\varphi \in \mathcal{B}_b(E)$,

$$\gamma_t(\varphi) = \eta_t(\varphi) \, \exp \int_0^t \eta_s(U) \, ds.$$

As in section 3 we introduce the N-approximating measures

$$\gamma_t^N(\varphi) \stackrel{\text{def.}}{=} \eta_t^N(\varphi) \, \exp \int_0^t \eta_s^N(U) \, ds.$$

On the other hand, using the Markovian property of X we observe the simple but essential fact that

$$\gamma_t(\varphi) = \gamma_s(K_{t-s}(\varphi)),$$

where $\{K_\tau \ ; \ \tau \geq 0\}$ is the semi-group defined by

$$\forall \varphi \in \mathcal{B}_b(E), \quad (K_\tau(\varphi))(x) = \mathbb{E}\left(\varphi(X_\tau(x)) \, \exp \int_0^\tau U(X_s(x)) \, ds\right)$$

where $\{X_\tau(x)\ ;\ \tau \geq 0\}$ is the time-homogeneous Markov process with infinitesimal generator L and starting at $x \in E$. From this simple observation one concludes that for any fixed $T > 0$ and for any $t \in [0, T]$, $x \in E$ and $\varphi \in \mathcal{B}_b(E)$

$$\frac{d}{dt}(K_{T-t}(\varphi))(x) = -L(K_{T-t}(\varphi))(x) - U(x)\,(K_{T-t}(\varphi))(x).$$

By definition of $\gamma_t^N(1)$ and using the same kind of arguments as before, one can check that the stochastic process

$$\mathcal{M}_t(\varphi, T) \stackrel{\text{def.}}{=} \sqrt{N}\left(\gamma_t^N(K_{T-t}(\varphi)) - \gamma_0^N(K_T(\varphi))\right), \qquad 0 \leq t \leq T$$

is a martingale, and its angle bracket is given by

$$\langle \mathcal{M}(\varphi, T)\rangle_t = \int_0^t \gamma_s^N(1) \left\{ \eta_s^N\left(\Gamma_L(K_{T-s}(\varphi), K_{T-s}(\varphi))\right) \right. \tag{50}$$

$$\left. + \eta_s^N\left(\left(K_{T-s}(\varphi) - \eta_s^N(K_{T-s}(\varphi))\right)^2 \left(U + \eta_s^N(U)\right)\right)\right\} ds.$$

Recalling that $\gamma_t(K_{T-t}(\varphi)) = \gamma_T(\varphi) = \gamma_0(K_T(\varphi))$ and $\gamma_0^N = \eta_0^N$, $\gamma_0 = \eta_0$ one concludes that

$$\gamma_t^N(K_{T-t}(\varphi)) - \gamma_t(K_{T-t}(\varphi))$$
$$= \eta_0^N(K_T(\varphi)) - \eta_0(K_T(\varphi)) + \frac{1}{\sqrt{N}} \mathcal{M}_t(\varphi, T).$$

From which it becomes clear that:

Proposition 3. *For any $N \geq 1$ and for any bounded Borel test function $\varphi \in \mathcal{B}_b(E)$ we have that*

$$\mathbb{E}\left(\gamma_T^N(\varphi)\right) = \gamma_T(\varphi)$$

and

$$\mathbb{E}\left(\left(\gamma_T^N(\varphi) - \gamma_T(\varphi)\right)^2\right)^{1/2} \leq \frac{C_T}{\sqrt{N}} \|\varphi\| \tag{51}$$

for some finite constant C_T which does not depend on the test function.

Using the same line of argument as the one we used in discrete-time settings (see section 3), it is possible to obtain central limit theorems for the N-approximating measures γ_T^N and η_T^N as well as error bounds for the total variation distance. For instance, using the decomposition

$$\eta_T^N(\varphi) - \eta_T(\varphi) = \frac{\gamma_T^N(\varphi)}{\gamma_T^N(1)} - \frac{\gamma_T(\varphi)}{\gamma_T(1)}$$
$$= \frac{1}{\gamma_T(1)}\left(\left(\gamma_T^N(\varphi) - \gamma_T(\varphi)\right) + \eta_T^N(\varphi)\left(\gamma_T(1) - \gamma_T^N(1)\right)\right) \tag{52}$$

and (51) one gets the following result:

Proposition 4. *For any $N \geq 1$ and for any bounded Borel test function $\varphi \in \mathcal{B}_b(E)$*

$$\mathbb{E}\left(\left(\eta_T^N(\varphi) - \eta_T(\varphi)\right)^2\right)^{1/2} \leq \frac{C_T}{\sqrt{N}} \|\varphi\|$$

for some finite constant C_T which does not depend on the test function.

Using the decomposition (52) and Proposition 3 one obtains that

$$\mathbb{E}\left(\eta_T^N(\varphi)\right) - \eta_T(\varphi) = \mathbb{E}\left(\eta_T^N(\varphi)\left(1 - \frac{\gamma_T^N(1)}{\gamma_T(1)}\right)\right)$$

$$= \mathbb{E}\left(\left(\eta_T^N(\varphi) - \eta_T(\varphi)\right)\left(1 - \frac{\gamma_T^N(1)}{\gamma_T(1)}\right)\right).$$

Thus, a simple application of Cauchy-Schwartz's inequality yields that for any test function φ, $\|\varphi\| \leq 1$

$$\left|\mathbb{E}\left(\eta_T^N(\varphi)\right) - \eta_T(\varphi)\right| \leq \frac{C_T}{N}$$

for some finite constant C_T which only depends on the time parameter T. By exchangeability of the particles and the definition of the total variation distance of probability measures, this implies that

$$\|\mathcal{L}aw(\xi_t^i) - \eta_t\|_{\text{tv}} \leq \frac{C_T}{N}.$$

Finally, as the form of the angle bracket (50) indicates one can prove the following result:

Theorem 5 ([20]). *The sequence of random fields*

$$U_T^N(f) \stackrel{\text{def.}}{=} \sqrt{N}\left(\gamma_T^N(f) - \gamma_T(f)\right), \qquad f \in \mathcal{B}_b(E)$$

converges in law as $N \to \infty$, in the sense of finite distributions, to a centered Gaussian field $\{U_n(f) \; ; \; f \in \mathcal{B}_b(E)\}$ satisfying

$$\mathbb{E}\left(U_T(f)^2\right) = \eta_0\left((K_T(\varphi) - \eta_0(K_T(\varphi)))^2\right)$$
$$+ \int_0^t \gamma_s(1) \Big\{\eta_s\left(\Gamma_L(K_{T-s}(\varphi), K_{T-s}(\varphi))\right)$$
$$+ \eta_s\left((K_{T-s}(\varphi) - \eta_s(K_{T-s}(\varphi)))^2 (U + \eta_s(U))\right)\Big\} ds.$$

Arguing as in discrete-time settings, the previous fluctuation result implies that the sequence of random fields

$$W_T^N(f) \stackrel{\text{def.}}{=} \sqrt{N}\left(\eta_T^N(f) - \eta_T(f)\right), \qquad f \in \mathcal{B}_b(E)$$

converges in law as $N \to \infty$ to the centered Gaussian field

$$W_T(f) \stackrel{\text{def.}}{=} U_T\left(\frac{1}{\gamma_T(1)}(f - \eta_T(f))\right), \qquad f \in \mathcal{B}_b(E).$$

Finally, setting

$$M_t(T, \varphi) \stackrel{\text{def.}}{=} \gamma_t^N(K_{T-t}(\varphi)) - \gamma_0^N(K_T(\varphi)), \qquad 0 \le t \le T$$

and, using the same reasoning as before, one can prove that for any $0 \le t \le T$

$$|<M(T,\varphi)>_t| \le \frac{1}{N} C_T \|\varphi\|^2 \quad \text{and} \quad |\Delta M_t(T,\varphi)| \le \frac{1}{N} C_T \|\varphi\|$$

for some finite constant $C_T < \infty$. Thus, the exponential bound (49) implies that for any $0 < \epsilon \le \|\varphi\|$

$$\mathbb{P}\left(\sup_{t \in [0,T]} |\gamma_t^N(K_{T-t}(\varphi)) - \gamma_0^N(K_T(\varphi))| > \epsilon\right) \le 2 \exp-\frac{N\epsilon^2}{C(T)\|\varphi\|^2}$$

for some finite constant $C(T) < \infty$. On the other hand, using Hoeffding's inequality we have that

$$\mathbb{P}\left(|\gamma_0^N(K_T(\varphi)) - \gamma_0(K_T(\varphi))| > \epsilon\right) \le 2 \exp-\frac{N\epsilon^2}{C'(T)\|\varphi\|^2}$$

for some finite constant $C'(T) < \infty$. If we combine these two bounds one concludes that

$$\mathbb{P}\left(\sup_{t \in [0,T]} |\gamma_t^N(K_{T-t}(\varphi)) - \gamma_t(K_{T-t}(\varphi))| > \epsilon\right)$$
$$\le 4 \exp-\frac{N\epsilon^2}{\max(C(T), C'(T))\|\varphi\|^2}$$

and therefore

$$\mathbb{P}\left(|\eta_t^N(\varphi) - \eta_t(\varphi)| > \epsilon\right) \le 4 \exp-\frac{N\epsilon^2}{C''(T)\|\varphi\|^2}$$

for some finite constant $C''(T) < \infty$.

Uniform convergence results are developed in [20,22]. These papers provide various stability conditions on the process X under which one can find (as in discrete-time settings, see page 459) some coefficient $\alpha \in (0, 1/2)$, such that for any $1 \le i \le N$

$$\sup_{t \ge 0} \|Law(\xi_t^i) - \eta_t\|_{\text{tv}} \le \frac{C}{N^\alpha}, \qquad C < \infty$$

and for any bounded Borel test function f, $\|f\| \leq 1$

$$\forall p \geq 1 \quad \sup_{t \geq 0} \mathbb{E}\left(\left|\eta_t^N(f) - \eta_t(f)\right|^p\right)^{1/p} \leq \frac{c(p)}{N^\alpha}$$

for some constant $c(p) < \infty$ which only depends on the parameter p.

6.2 A Branching Particle Model

We end this paper with a presentation of a novel genetic-type model based on branching selection transitions. To our knowledge this model has not been covered by the literature and its convergence analysis is still in progress. We also believe that the semi-group approach presented in [20] applies to study the convergence of this branching algorithm to the distributions (43).

In contrast to the previous Moran-type GA, the size of the population here will not be necessarily fixed but random. As a result the corresponding branching particle system will be regarded as a continuous-time process taking values in the state space

$$\mathcal{E} \stackrel{\text{def.}}{=} \cup_{p \geq 0} E^p$$

with the convention that $E^p = \{\Delta\}$ is a cemetery point if $p = 0$. The point Δ will be isolated and, by convention, all bounded Borel test functions $f \in \mathcal{B}_b(\mathcal{E} - \{\Delta\})$ will be extended to $\mathcal{E}$ by setting $f(\Delta) = 0$.

It will be also convenient to adjoin Δ to the state space E and we set $E_\Delta = E \cup \{\Delta\}$. Again, the test functions $\varphi \in \mathcal{B}_b(E)$ will be extended to E_Δ by setting $\varphi(\Delta) = 0$.

The infinitesimal generator $\mathcal{G}$ of this branching scheme is defined by

$$\mathcal{G} = \widetilde{\mathcal{G}} + \widehat{\mathcal{G}} \tag{53}$$

where, for any $f \in \mathcal{B}_b(\mathcal{E})$ and $x = (x_1, \ldots, x_p) \in E^p$, $p \geq 1$,

$$\widetilde{\mathcal{G}} f(x) = \sum_{i=1}^p L^{(i)}(f)(x)$$

and

$$\widehat{\mathcal{G}} f(x) = \lambda(x) \int_\mathcal{E} (f(y) - f(x)) \, \mathcal{Q}(x, dy)$$

with

$$\lambda(x) = \sum_{i=1}^p U(x_i)$$

the transition probability kernel Q on $\mathcal{E}$ given by

$$Qf(x) = \int f(y) \, Q(x, dy)$$

$$= \frac{1}{p} \sum_{i=1}^{p} \sum_{q \geq 0} \left\{ \int_E f(x(i,q,u)) \, \mathcal{S}(x, x_i, du) \right\} \mathcal{B}(x, x_i, q),$$

where

$$x(i, q, u) = (x_1, \ldots, x_{i-1}, \underbrace{u, \ldots, u}_{q \text{ times}}, x_{i+1}, \ldots, x_p),$$

and for any $x \in \mathcal{E}$, $\mathcal{S}(x, x_i, du)$ and $\mathcal{B}(x, x_i, q)$ are distributions on E and on $\mathbb{N}$. In our construction, the point Δ will be an absorbing point in the sense that if the process started at Δ it will stay in Δ. Therefore, for $p = 0$ we will also use the convention $\sum_\emptyset = 0$ and $Q(\Delta, \{\Delta\}) = 1$. With this convention if $p = 0$ (i.e., $x = \Delta$) we have that $\tilde{\mathcal{G}} f(\Delta) = 0$ and $\widehat{\mathcal{G}} f(\Delta) = 0$.

The distributions $\mathcal{S}(x, x_i, du)$ and $\mathcal{B}(x, x_i, q)$ will be chosen so that the following equality holds true:

$$\frac{1}{p} \sum_{i=1}^{p} \overline{\mathcal{B}}(x, x_i) \, \overline{\mathcal{S}}(\varphi)(x, x_i) = \sum_{i=1}^{p} \frac{U(x_i)}{\sum_{j=1}^{p} U(x_j)} \varphi(x_i) \tag{54}$$

for any $\varphi \in \mathcal{B}_b(E)$ where $\overline{\mathcal{B}}(x, x_i)$ and $\overline{\mathcal{S}}(\varphi)(x, x_i)$ are defined by

$$\overline{\mathcal{B}}(x, x_i) = \sum_{q \geq 0} q \, \mathcal{B}(x, x_i, q)$$

and

$$\overline{\mathcal{S}}(\varphi)(x, x_i) = \int_E \mathcal{S}(x, x_i, du) \, \varphi(u).$$

We now make this condition more precise by noting that if $f \in \mathcal{B}_b(\mathcal{E})$ is defined for any $x = (x_1, \ldots, x_p) \in E^p$, $p \geq 1$, by

$$f(x) = p.m(x)(\varphi) \quad \text{where} \quad m(x) = \frac{1}{p} \sum_{i=1}^{p} \delta_{x_i} \tag{55}$$

for some $\varphi \in \mathcal{B}_b(E)$, then

$$\mathcal{G}(f)(x) = p.m(x) \left(L_{m(x)}(\varphi) \right).$$

To see this claim we first observe that for such a bounded test function f and for any $x = (x_1, \ldots, x_p) \in E^p$, $p \geq 1$

$$\widehat{\mathcal{G}}(f)(x)$$
$$= \lambda(x) \frac{1}{p} \sum_{i=1}^{p} \sum_{q \geq 0} \left\{ \int_E (q\varphi(u) - \varphi(x_i)) \, \mathcal{S}(x, x_i, du) \right\} \mathcal{B}(x, x_i, q)$$
$$= \lambda(x) \frac{1}{p} \sum_{i=1}^{p} \sum_{q \geq 0} \left(\overline{\mathcal{B}}(x, x_i) \overline{\mathcal{S}}(\varphi)(x, x_i) - \varphi(x_i) \mathcal{B}(x, x_i, q) \right).$$

Using (54) one concludes that for any $x = (x_1, \ldots, x_p) \in E^p$, $p \geq 1$,

$$\widehat{\mathcal{G}}(f)(x) = \sum_{j=1}^{p} U(x_j) \left(\sum_{i=1}^{p} \frac{U(x_i)}{\sum_{j=1}^{p} U(x_j)} \varphi(x_i) - m(x)(\varphi) \right)$$
$$= p \cdot (m(x) (\varphi U) - m(x)(\varphi) m(x)(U)).$$

Recalling that for any bounded test function f of the form (55) and for any $x = (x_1, \ldots, x_p) \in E^p$, $p \geq 1$, we have that

$$\widetilde{\mathcal{G}}(f)(x) = \sum_{i=1}^{p} L^{(i)}(f)(x) = \sum_{i=1}^{p} L(\varphi)(x_i) = p \cdot m(x)(L(\varphi)),$$

one concludes that for any bounded test function f of the form (55) and for any $x = (x_1, \ldots, x_p) \in E^p$, $p \geq 1$

$$\mathcal{G}(f)(x) = p \cdot (m(x)(L(\varphi)) + m(x) (\varphi U) - m(x)(\varphi) m(x)(U))$$
$$= p \cdot m(x) \left(L_{m(x)}(\varphi) \right).$$

Along the same line of ideas as before, it is possible to construct inductively the branching (with interaction) particle system with generator $\mathcal{G}$. In contrast to the previous situation, the size of the population is not necessarily fixed and it will be denoted by N_t at each time t, in other words

$$\xi_t = (\xi_t^1, \ldots, \xi_t^{N_t}) \in E^{N_t}.$$

We also need to introduce a sequence $\{t_k \, ; \, k \geq 1\}$ of independent random variables with a common exponential law on $\mathbb{R}_+$ with parameter 1. If we write $\{T_k \, ; \, k \geq 0\}$ for the random times at which the competitive branching interaction occurs, the inductive description is as follows. Initially $T_0 = 0$ and the particle system $\xi_0 = (\xi_0^1, \ldots, \xi_0^{N_0})$ consists of N_0 independent random variables with common law η_0. The initial size N_0 is a non-random integer and represents the precision parameter of the scheme.

Now we assume that we have defined the branching process up to time T_{k-1} (included) for some $k \geq 1$. If $N_{T_{k-1}} = 0$ the particle system dies and we let $N_t = 0$ and $\xi_t = \Delta$ for any $t \geq T_{k-1}$. Otherwise the mutation/branching selection transition is defined as follows:

1. **Mutation**

   Between the dates T_{k-1} and T_k the particles evolve randomly and independently according to the law of the time-inhomogeneous Markov process X. That is, for any $1 \leq i \leq N_{T_{k-1}}$

   $$\xi_t^i = X_{t-T_{k-1}}^{(k,i)}\left(\xi_{T_{k-1}}^i\right) \qquad \forall t \in [T_{k-1}, T_k[$$

   where $T_k = T_{k-1} + \tau_k$ and τ_k is defined by setting

   $$t_k = \int_{T_{k-1}}^{T_{k-1}+\tau_k} \sum_{i=1}^{N_{T_{k-1}}} U\left(X_{s-T_{k-1}}^{(k,i)}\left(\xi_{T_{k-1}}^i\right)\right) ds$$

   (recall that t_k is a random variable with exponential law on $\mathbb{R}_+$ with parameter 1). During this stage the size of the system remains constant and we set

   $$N_t = N_{T_{k-1}} \qquad \forall t \in [T_{k-1}, T_k[.$$

2. **Competitive Branching Selection**

   At time $t = T_k$ a label i is chosen uniformly on $\{1, \ldots, N_{T_{k-1}}\}$, and the particle with label i dies and is replaced by a random number of offsprings q_k^i with law

   $$\mathcal{B}\left(\xi_{T_k-}, \xi_{T_k-}^i, \cdot\right)$$

   and independently, these offspring are randomly given a location u_k^i with law

   $$\mathcal{S}\left(\xi_{T_k-}, \xi_{T_k-}^i, \cdot\right).$$

   At the end of this stage the particle system ξ_{T_k} is defined as

   $$\xi_{T_k} = (\xi_{T_k-}^1, \ldots, \xi_{T_k-}^{i-1}, \underbrace{u_k^i, \ldots, u_k^i}_{q_k^i \text{ times}}, \xi_{T_k-}^{i+1}, \ldots, \xi_{T_k-}^{N_{T_{k-1}}})$$

   and the resulting population size is defined as

   $$N_{T_k} = \left(N_{T_{k-1}} - 1\right) + q_k^i.$$

Let us give some examples of branching selection laws satisfying condition (54). We assume that $x = (x_1, \ldots, x_p) \in E^p$ for some $p \geq 1$.

1. If $\mathcal{B}(x, x_i, \cdot)$ and $\mathcal{S}(x, x_i, \cdot)$ are defined by

   $$\mathcal{B}(x, x_i, \cdot) = \delta_1$$

and

$$S(x, x_i, .) = \sum_{j=1}^{p} \frac{U(x_j)}{\sum_{k=1}^{p} U(x_k)} \delta_{x_j}, \qquad (56)$$

then, since for any $\varphi \in \mathcal{B}_b(E)$ and $1 \le i \le p$

$$\overline{\mathcal{B}}(x, x_i) = 1 \quad \text{and} \quad \overline{\mathcal{S}}(\varphi)(x, x_i) = \sum_{j=1}^{p} \frac{U(x_j)}{\sum_{k=1}^{p} U(x_k)} \varphi(x_j),$$

condition (54) clearly holds. This example corresponds to the Moran-type genetic scheme presented in the previous section. Indeed, in this situation we clearly have for any $f \in \mathcal{B}_b(\mathcal{E})$ and for any $x = (x_1, \ldots, x_p) \in E^p$, $p \ge 1$

$$\widehat{\mathcal{G}}(f)(x)$$
$$= \left(\sum_{k=1}^{p} U(x_k) \right) \frac{1}{p} \sum_{i=1}^{p} \sum_{j=1}^{p} \left(f(x^{(i,j)}) - f(x) \right) \frac{U(x_j)}{\sum_{l=1}^{p} U(x_l)}$$
$$= \frac{1}{p} \sum_{i=1}^{p} \sum_{j=1}^{p} \left(f(x^{(i,j)}) - f(x) \right) U(x_j)$$

where for $1 \le i, j \le p$ and $x = (x_1, \cdots, x_p) \in E^p$, $x^{(i,j)}$ is, as usual, the element of E^p given by

$$\forall \, 1 \le k \le p, \qquad x_k^{(i,j)} = \begin{cases} x_k, & \text{if } k \ne i \\ x_j, & \text{if } k = i \end{cases}$$

2. If $\mathcal{B}(x, x_i, .)$ and $\mathcal{S}(x, x_i, .)$ are defined by

$$\mathcal{S}(x, x_i, .) = \delta_{x_i} \quad \text{and} \quad \overline{\mathcal{B}}(x, x_i) = \frac{p \, U(x_i)}{\sum_{k=1}^{p} U(x_k)} \qquad (57)$$

then for any $\varphi \in \mathcal{B}_b(E)$ we have $\overline{\mathcal{S}}(\varphi)(x, x_i) = \varphi(x_i)$ and condition (54) is again met. In this situation the size of the population may not be fixed. To highlight the connections with the discrete-time branching schemes presented in section 5.3 the reader may check that condition (57) holds for the Bernoulli and Poisson branching laws

$$\mathcal{B}(x, x_i, .) = (1 - \{\overline{\mathcal{B}}(x, x_i)\}) \, 1_{[\overline{\mathcal{B}}(x,x_i)]}(.) + \{\overline{\mathcal{B}}(x, x_i)\} \, 1_{[\overline{\mathcal{B}}(x,x_i)]+1}(.)$$

and

$$\mathcal{B}(x, x_i, .) = e^{-\overline{\mathcal{B}}(x,x_i)} \sum_{q \ge 0} \frac{(\overline{\mathcal{B}}(x, x_i))^q}{q!} 1_q(.).$$

We recall that $[a]$ (respectively $\{a\} = a - [a]$) is the integer part (respectively the fractional part) of $a \in \mathbb{R}$.

Now we return to the probabilistic analysis of this branching particle model. We have studied the asymptotic behavior of this scheme, but the corresponding publication still isn't ready. For the convenience of the reader, we only formulate a few basic results to illustrate how the methodology used for the Moran type GA can be used in this more general framework. As usual, we start by noting that for any bounded Borel test function $f \in \mathcal{B}_b(\mathcal{E})$ the stochastic process

$$M_t(f) \stackrel{\text{def.}}{=} f(\xi_t) - f(\xi_0) - \int_0^t \mathcal{G}(f)(\xi_s) \, ds$$

is a local martingale and its angle bracket is given by

$$\langle M(f) \rangle_t = \int_0^t \Gamma(f, f)(\xi_s) \, ds$$

where Γ is the "carré du champ" associated with $\mathcal{G}$

$$\Gamma(f, f) = \mathcal{G}(f^2) - 2f \, \mathcal{G}(f).$$

Now, using the decomposition (53) and the definition of $\tilde{\mathcal{G}}$ and $\widehat{\mathcal{G}}$ one can check that

$$\Gamma(f, f) = \tilde{\Gamma}(f, f) + \widehat{\Gamma}(f, f)$$

with

$$\tilde{\Gamma}(f, f) = \tilde{\mathcal{G}}(f^2) - 2f \, \tilde{\mathcal{G}}(f) \qquad \widehat{\Gamma}(f, f) = \widehat{\mathcal{G}}(f^2) - 2f \, \widehat{\mathcal{G}}(f).$$

If $f \in \mathcal{B}_b(\mathcal{E})$ is chosen so that for any $x = (x_1, \ldots, x_p) \in E^p$, $p \geq 1$, and for some $\varphi \in \mathcal{B}_b(E)$

$$f(x) = p.m(x)(\varphi), \tag{58}$$

then for any $x = (x_1, \ldots, x_p) \in E^p$, $p \geq 1$

$$\tilde{\Gamma}(f, f)(x) = p \cdot m(x) \left(\Gamma_L(\varphi, \varphi) \right)$$

with

$$\Gamma_L(\varphi, \varphi) = L(\varphi^2) - 2\varphi \left(L(\varphi) \right)$$

and

$$\widehat{\Gamma}(f, f)(x) = \widehat{\mathcal{G}} \left((f(.) - f(x))^2 \right)(x)$$
$$= \lambda(x) \, \frac{1}{p} \sum_{i=1}^{p} \sum_{q \geq 0} \int_E (q\varphi(u) - \varphi(x_i))^2 \, S(x, x_i, du) \, B(x, x_i, q).$$

Let us notice that if distributions $S(x, x_i, du)$ and $B(x, x_i, q)$ are defined by (56) then for any $x = (x_1, \ldots, x_p) \in E^p$, $p \geq 1$, we have that

$$\widehat{\Gamma}(f, f)(x) = p.m(x) \left((\varphi - m(x)(\varphi))^2 \, (U + m(x)(U)) \right)$$

and if these distributions satisfy (57) one gets that

$$\widehat{\Gamma}(f, f)(x) = p.m(x)(U).m(x) \left(\varphi^2 \sum_{q \geq 0} (q-1)^2 B(x, ., q) \right).$$

In contrast to the previous Moran-type genetic model the "carré du champ" corresponding to the selection procedure is not necessarily bounded and we need to introduce some auxiliary assumption on the mass variation of the systems, namely we will assume that

$$\sup_{x \in \mathcal{E}} m(x) \left(\sum_{q \geq 0} q^2 B(x, ., q) \right) < \infty.$$

In the special case where the test function f is given by (58) we have that the stochastic process

$$N_t \, m(\xi_t)(\varphi) - N_0 \, m(\xi_0)(\varphi) - \int_0^t N_s \, m(\xi_s) \left(L_{m(\xi_s)} \varphi \right) \, ds, \qquad t \geq 0$$

is a square integrable martingale. Of course, if $\varphi = 1$ this implies that the total mass process $\{N_t \, ; \, t \geq 0\}$ is a square integrable martingale starting at N_0.

References

1. J. Abela, D. Abramson, M. Krishnamoorthy, A. De Silval, and G. Mills. Computing Optimal Schedules for Landing Aircraft. Technical Report, Department of Computer Systems Eng. R.M.I.T., Melbourne, May 25, 1993.
2. J. Baker. Adaptive selection methods for genetic algorithms. In J. Grefenstette, editor, *Proceedings of the First International Conference on Genetic Algorithms and their Applications*, pages 101–111. Lawrence Erlbaum, Hillsdale, NJ, 1985.
3. J. Baker. Reducing bias and inefficiency in the selection algorithm. In J. Grefenstette, editor, *Proceedings of the Second International Conference on Genetic Algorithms and their Applications*, pages 14–21. Lawrence Erlbaum, Hillsdale, NJ, 1987.
4. P. Barbe and P. Bertail. The Weighted Bootstrap. *Lecture Notes in Statistics 98.* Springer-Verlag, Berlin Heidelberg New York, 1995.
5. R. Bott and J. P. Mayberry. *Matrices and Trees. Economics Activity Analysis.* Wiley, New York, 1954.
6. H. Carvalho. Filtrage Optimal Non Linéaire du Signal GPS NAVSTAR en Racalage de Centrales de Navigation. Thèse de L'Ecole Nationale Supérieure de l'Aéronautique et de l'Espace, September 1995.
7. H. Carvalho, P. Del Moral, A. Monin, and G. Salut. Optimal non-linear filtering in GPS/INS integration. *IEEE Trans. on Aerospace and Electronic Systems,* 33(3):835–850, 1997.
8. R. Cerf. Asymptotic convergence of a genetic algorithm. *C. R. Acad. Sci. Paris Sér. I Math.,* 319(3):271–276, 1994.
9. R. Cerf. The dynamics of mutation-selection algorithms with large population sizes. *Ann. Inst. H. Poincaré Probab. Statist.,* 32(4):455–508, 1996.

10. R. Cerf. Asymptotic convergence of genetic algorithms. *Adv. in Appl. Probab.*, 30(2):521–550, 1998.
11. D. Crisan, J. Gaines, and T. J. Lyons. A particle approximation of the solution of the Kushner–Stratonovitch equation. *SIAM J. Appl. Math.*, 58(5):1568–1590, 1998.
12. D. Crisan, P. Del Moral, and T. J. Lyons. Discrete filtering using branching and interacting particle systems. *Markov Processes and Related Fields*, 5(3):293–318, 1999.
13. D. Crisan, P. Del Moral, and T. J. Lyons. Interacting particle systems approximations of the Kushner–Stratonovitch equation. *Advances in Applied Probability*, 31(3):819–838, 1999.
14. D. Crisan and M. Grunwald. Large deviation comparison of branching algorithms versus re-sampling algorithms. Preprint, Imperial College, London, 1998.
15. D. Dawson. Measure-valued Markov processes. In P. L. Hennequin, editor, *Ecole d'Eté de Probabilités de Saint-Flour XXI-1991*, Lecture Notes in Mathematics 1541. Springer-Verlag, Berlin Heidelberg New York, 1993.
16. D. Delahaye, J.-M. Alliot, M. Schoenauer, and J.-L. Farges. Genetic algorithms for automatic regrouping of air traffic control sectors. In J. R. McDonnell, R. G. Reynolds, and D. B. Fogel, editors, *Proceedings of the 4th Annual Conference on Evolutionary Programming*, pages 657–672. MIT Press, Cambridge, MA, March 1995.
17. P. Del Moral, J. Jacod, and Ph. Protter. The Monte-Carlo method for filtering with discrete-time observations. *Publications du Laboratoire de Probabilités*, 453, Paris VI, France, June 1998.
18. P. Del Moral and J. Jacod. The Monte-Carlo method for filtering with discrete-time observations. Central Limit Theorems. *Publications du Laboratoire de Statistiques et Probabilités*, 7, Toulouse III, France, 1999.
19. P. Del Moral and J. Jacod. Interacting particle filtering with discrete observations. *Publications du Laboratoire de Statistiques et Probabilités*, 8, Toulouse III, France, 1999.
20. P. Del Moral and L. Miclo. Branching and interacting particle systems approximations of Feynman–Kac formulae with applications to non-linear filtering. *Publications du Laboratoire de Statistiques et Probabilités*, 5, Toulouse III, France, 1999.
21. P. Del Moral and L. Miclo. On the convergence and the applications of the generalized simulated annealing. *SIAM Control and Optimization*, 37(4):1222–1250.
22. P. Del Moral and L. Miclo. Asymptotic stability of nonlinear semigroups of Feynman–Kac type. *Publications du Laboratoire de Statistiques et Probabilités*, 4, Toulouse III, France, 1999.
23. P. Del Moral and L. Miclo. A Moran particle system approximation of Feynman–Kac formulae. *Publications du Laboratoire de Statistiques et Probabilités*, 11, Toulouse III, France, 1998.
24. P. Del Moral and A. Guionnet. Large deviations for interacting particle systems. Applications to nonlinear filtering problems. *Stochastic Processes and their Applications*, 78:69–95, 1998.
25. P. Del Moral and A. Guionnet. On the stability of measure-valued processes. Applications to nonlinear filtering and interacting particle systems. *Publications du Laboratoire de Statistiques et Probabilités*, 3, Toulouse III, France, 1998.
26. P. Del Moral. Nonlinear filtering: interacting particle solution. *Markov Processes and Related Fields*, 2(4):555–580, 1996.
27. P. Del Moral. Measure valued processes and interacting particle systems. Application to nonlinear filtering problems. *Ann. Appl. Probab.*, 8(2):438–495, 1998.

28. P. Del Moral, J. C. Noyer, and G. Salut. Résolution particulaire et traitement non-linéaire du signal: application Radar/Sonar. *Revue du Traitement du Signal*, Septembre 1995.
29. P. Del Moral, G. Rigal, J. C. Noyer, and G. Salut. Traitement non-linéaire du signal par reseau particulaire: application radar. *Quatorzième colloque GRETSI*, Juan les Pins, 13–16 Septembre 1993.
30. P. Del Moral, G. Rigal, and G. Salut. *Estimation et commande optimale non lineaire*. Contract D.R.E.T.-DIGILOG-LAAS/CNRS, SM.MCY/685.92/A, 89.34.553.00.470.75.01, Report No. 2, 18 Mars 1992.
31. S. Ethier and T. Kurtz. *Markov Processes, Characterization and Convergence*. Wiley series in probability and mathematical statistics. John Wiley, New York, 1986.
32. M. I. Freidlin and A. D. Wentzell. *Random Perturbations of Dynamical Systems*. Grundlehren der math. Wissenschaften, vol. 260. Springer-Verlag, Berlin Heidelberg New York, 1984.
33. D. E. Goldberg. Genetic algorithms and rule learning in dynamic control systems. In J. Grefenstette, editor, *Proceedings of the First International Conference on Genetic Algorithms*, pages 8–15. Lawrence Erlbaum, Hillsdale, NJ, 1985.
34. D. E. Goldberg. *Genetic Algorithms in Search, Optimization and Machine Learning*. Addison–Wesley, Reading, MA, 1989.
35. N. J. Gordon, D. J. Salmon, and A. F. M. Smith. *Novel Approach to Non-Linear/Non-Gaussian Bayesian State Estimation*. IEEE, 1993.
36. C. Graham and S. Méléard. Stochastic particle approximations for generalized Boltzmann models and convergence estimates. *The Annals of Probability*, 25(1):115–132, 1997.
37. J. H. Holland. *Adaptation in Natural and Artificial Systems*. University of Michigan Press, Ann Arbor, 1975.
38. J. Jacod and A. N. Shiryaev. *Limit Theorems for Stochastic Processes*. Grundlehren der math. Wissenschaften, vol. 288. Springer-Verlag, Berlin Heidelberg New York, 1987.
39. J. M. Johnson and Y. Rahmat-Samii. Genetic algorithms in engineering electromagnetics. *AP-S Magazine*. 39:7–25, 1997.
40. C. Kane and M. Schoenauer. Topological optimum design using genetic algorithms. *Control and Cybernetics*, 25(5), 1996.
41. G. Kitagawa. Monte-Carlo filter and smoother for non-Gaussian nonlinear state space models. *Comput. and Graphical Stat.*, 5(1):1–25, 1996.
42. J. W. Kwiatkowski. *Algorithms for Index Tracking*. Department of Business Studies, University of Edinburgh, UK, 1991.
43. R. S. Liptser and A. N. Shiryayev. *Theory of Martingales*. Kluwer Academic, Dordrecht, 1989.
44. S. Méléard. Asymptotic behaviour of some interacting particle systems; McKean-Vlasov and Boltzmann models. In D. Talay and L. Tubaro, editors, *Probabilistic Models for Nonlinear Partial Differential Equations, Montecatini Terme, 1995*, Lecture Notes in Mathematics 1627. Springer-Verlag, Berlin Heidelberg New York, 1996.
45. C. Musso and N. Oudjane. Regularized Particle Schemes applied to the Tracking Problem. Preprint, ONERA Chatillon, 1998.
46. C. Musso and N. Oudjane. Regularization schemes for branching particle systems as a numerical solving method of the nonlinear filtering problem. Preprint, ONERA Chatillon, 1998.
47. Y. Nishiyama. Some central limit theorems for l^∞-valued semimartingales and their applications. *Probability Theory and Related Fields*, 108:459–494, 1997.

48. D. Revuz and M. Yor. *Continuous Martingales and Brownian Motion*. Grundlehren der math. Wissenschaften, vol. 293. Springer-Verlag, Berlin Heidelberg New York, 1991.
49. J. Shapcott. Index Tracking: Genetic Algorithms for Investment Portfolio Selection. EPCC–SS92–24, September 1992.
50. T. Shiga and H. Tanaka. Central limit theorem for a system of Markovian particles with mean field interaction. *Z.f. Wahrscheinlichkeitstheorie verw. Geb.*, 69:439–459, 1985.
51. M. Schoenauer, L. Kallel, and F. Jouve. Mechanical inclusions identification by evolutionary computation. *Finite Elem.*, 5(5–6):619–648, 1996.
52. M. Schoenauer, F. Jouve, and L. Kallel. Identification of mechanical inclusions. In D. Dasgupta and Z. Michalewicz, editors, *Evolutionary Algorithms in Engineering Applications*, pages 479–496. Springer-Verlag, Berlin Heidelberg New York, 1997.
53. D. Treyer, D. S. Weile, and E. Michielsen. The application of novel genetic algorithms to electromagnetic problems. *Applied Computational Electromagnetics Symposium Digest*, 2:1382-1386. Monterey, CA, 1997.
54. A. Trouvé. *Parallélisation massive du recuit simulé*. Thèse de Doctorat, Université Paris XI, January 1993.

Index

$(\mu/\rho \overset{+}{,} \lambda)$-ES, 110, 240
(1+1)-ES, 189

ADDM problem, 223
ant colony optimization, 150
autocorrelation, 190

Baluja problem, 363
basin of attraction, 183, 208, 343
basin with a barrier, 60, 208
Bayes–Dirichlet score, 168
Bayesian information criterion, 169
Bayesian network, 164
breeder GA, 225
breeding, 148
building block hypothesis, 154

camel function, 375
co-evolving populations, 21, 338
competing conventions problem, 419
computing resources, 3
connectivity, 197
constrained optimization problem, 15
constraint
– handling, 13
– satisfaction problem, 15, 196
convergence
– for large population, 444, 456
– long-time, 446, 460
– time, 213, 223
correlation length, 190
crossover, *see* recombination
– choice of operator, 332
– complementary, 184
– cyclic permutation, 395
– effects of, 75, 94, 211
– for graph coloring, 388
– linear assignment, 392
– multi-parent, 307, *see* scanning
– uniform, 424
cumulants, 63, 94, 209

design search, 2
disorder line, 269

distributions
– approximation of, 209
– factorization of, 164, 193
– Gibbs, 91, 289
– multinomial, 83
domain walls, 201

elitism, 7, 392
engineering design, 1
ensemble, 63, 96
epistasis, 332, 424
– bit-wise, 373
– variance, 179, 372
error threshold, 251, 261
Euclidean dimension, 198
experimental design, 178

factorization complexity, 199
FDA, 136, 163, 193
Feynman–Kac formula, 455, 456, 474, 475
finite population effects, 40, 59, 71, 87, 165, 212, 219, 471
first improvement algorithm, 183
first passage time, 213
fitness distance
– correlation, 376
– relation, 344
fitness landscape, *see* landscape, 175, 343
– time-dependent, 261
fixed-point, 36, 38, 265
founder effects, 93
four peaks problem, 298
Fourier transform, 82, 178, 335
free energy, 290
free optimization problem, 15
free search space, 15
frustration, 201

GA, *see* genetic algorithm
genetic algorithm
– continuous-time, 450, 474
– focused, 39
– steady state, 307

genetic drift, 36, 167, 209, 282, 307
- self-amplifying, 319
- strength of, 328
genetic repair, 247
- principle, 126
gradient search, 145, 290
Gram–Charlier expansion, 83
graph
- coloring, 19, 196, 389, 409, 425
- critical, 202
- isomorphism, 411
- Laplacian, 185
- partitioning, 425
- representation, 409
- theory, 409
Gray code, 182, 332, 363
greedy search, 105, 299, 392

hypersphere, 113

interaction
- benign, 180
- between variables, 178
- coefficients, 195
- effect, 180
- graph, 195
- malign, 180
- order, 179
- second-order, 424
Ising model, 424
isolation, 187
isomorphy, 409

Jackson's algorithm, 336

KM algorithm, 338
Kullback–Leibler divergence, 289

landscape
- decomposition, 184
- definition of, 182
- isomorphism of, 184
leading ones problem, 378
LFDA, 136, 168
linkage equilibrium, 101, 138
local optimum, 182
local search, 26, 208, 298, 345, 381, 391
longpath problem, 189

macroscopic
- description, 114, 149
- dynamics, 59, 87, 209
Markov chain, 115, 167, 289
- time-inhomogeneous, 441
master sequence, 251
maximum entropy, 63, 89
microgenetic algorithm, 21
MIMIC, 197
minimal description length, 168
mixing matrix, 48
multimodality, 187
mushy region, 197
mutation
- calculating effect of, 66
- Cauchy distributed, 113
- fitness dependent transitions, 470
- free operator, 14
- Gaussian distributed, 112
- greedy flip, 392
- optimal rate, 225, 259, 261
- rate, 218
- rescaled, 246
- strength, 241

nearest neighbor interaction functions
- one-dimensional, 200, 424
needle-in-a-haystack, 187, 195, 374
- moving, 255, 262
neighborhood structure, 183
neural network, 409
- structure optimisation, 409
- topology, 409
niching, 7
NK-model, 106, 190, 373
no-free-lunch theorem, 372
noise, 242
nonlinear filtering, 454
NP-hard problem class, 195

one-max, 59, 188, 223, 333, 364, 436
- permuted, 364
ones counting, *see* one-max

PAC model, 334
PBIL, 150, 197, 333
Perron vector, 265
Perron–Frobenius theorem, 47, 265
phase diagram, 268

phase transition, 25, 92, 196
pinning, 201
population dynamics
- fixed-points, 255
- infinite, 252, 442
population size
- critical, 463
- optimal, 225
prisoner's dilemma, 301
problem solving environment, 2
progress rate, 115, 241

quadratic assignment problem, 151
quality gain, 115, 241
quasispecies model, 251, 261
- generalized, 265

random fitness function, 188
random number partitioning problem, 426
recombination, *see* crossover
- dominant, 111
- free operator, 15
- harmonic, 336
- intermediate, 111
regular change, 265
reinforcement learning, 291
relaxation methods, 288
replicator equation, 151
representation, 3, 332, 372, 388, 409, 423
- order-based, 18
ridge function, 114
Robbins' proportions, 138
royal road problem, 80, 96, 154
royal staircase, 223
ruggedness, 190
running intersection property, 194

sawtooth landscape, 158
scanning
- fitness-based, 308
- occurrence-based, 307

- uniform, 307
selection
- Baker, 93, 209
- Boltzmann, 93, 163, 194
- branching, 473, 484
- comparison, 217
- fitness proportional, 35, 93, 251
- greedy, 392
- ranking, 66, 93, 209, 365
- remainder stochastic, 471
- stochastic universal sampling, 74, 209, 257
- tournament, 65, 93, 147, 225
- truncation, 147
- weak, 88
shortcut, 189
sigmoid, 294
simulated annealing, 105, 201, 207, 424, 446
SK-model, 195
statistical mechanics, 59, 88, 288
steepest ascent, 346
stepwise adaptation of weights, 22
subset sum problem, 88
success
- probability, 118, 243
- rule, 243
symmetric problem, 387
symmetry, 201, 387, 409, 423
- spin-flip, 423

tightness, 197
time-dependent search parameters, 104
traveling salesman problem, 14, 151, 191, 426

UMDA, 136, 193
unimodality, 188

Walsh transform, 54, 177, 335

Natural Computing Series

W.M. Spears: Evolutionary Algorithms. The Role of Mutation and Recombination. XIV, 222 pages, 55 figs., 23 tables. 2000

H.-G. Beyer: The Theory of Evolution Strategies. XIX, 380 pages, 52 figs., 9 tables. 2001

L. Kallel, B. Naudts, A. Rogers (Eds.): Theoretical Aspects of Evolutionary Computing. X, 497 pages. 2001

M. Hirvensalo: Quantum Computing. XI, 190 pages. 2001

M. Amos: Theoretical and Experimental DNA Computation. Approx. 200 pages. 2001

L.F. Landweber, E. Winfree (Eds.): Evolution as Computation. DIMACS Workshop, Princeton, January 1999. Approx. 300 pages. 2001